NELSON BLACKIE

MATHS IN ACTION

Mathematics in Action Group

Members of the Mathematics in Action Group associated with this book:
D. Brown, R. D. Howat, E. C. K. Mullan, K. Nisbet, A. G. Robertson

STUDENTS' BOOK

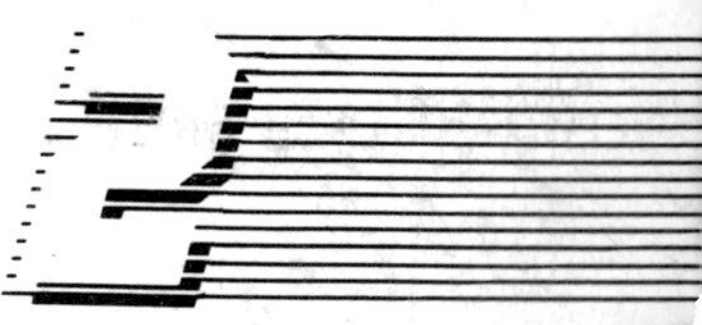

Thomas Nelson and Sons Ltd
Nelson House Mayfield Road
Walton-on-Thames Surrey
KT12 5PL UK

Thomas Nelson Australia
102 Dodds Street
South Melbourne
Victoria 3205 Australia

Nelson Canada
1120 Birchmount Road
Scarborough Ontario
M1K 5G4 Canada

Cover photograph by Darryl Williams/Steelcase Strafor plc

First published by Blackie and Son Ltd 1986
New edition published by Thomas Nelson and Sons Ltd 1994

I(T)P® Thomas Nelson is an International
Thomson Publishing Company

I(T)P® is used under licence

ISBN 0-17-431420-5
NPN 9 8 7 6 5

Printed in China

CONTENTS

INTRODUCTION

Maths in Action—New Edition provides a course in mathematics that covers the Mathematics 5-14 National Guidelines in Scotland, the Northern Ireland Curriculum and the National Curriculum in England and Wales.

The new edition builds on experience gained in the classroom with the original series, and particular attention has been paid to providing a differentiated course with exercises graded at three distinct levels—A, B and C. Every chapter starts with a Looking Back exercise, which can be used for revision and to assess readiness for the topic, and ends with a Check-up exercise giving a further element of revision and assessment. Investigative work features prominently in each chapter in the many puzzles, projects, challenges, brainstormers and investigations. Answers to every question (except puzzles, challenges, brainstormers and investigations) are to be found at the back of this book.

Each *Students' Book* is supported by a *Teacher's Resource Book* and, in the first two years, by revised books of *Extra Questions* and *Further Questions*.

The *Teacher's Resource Book* contains 5-14, Northern Ireland Curriculum and National Curriculum references for every exercise, photocopiable worksheets, notes and suggestions for further activities, and the answers to the puzzles, challenges, brainstormers and investigations in the *Students' Book*. In addition, there are grids which may be photocopied and used to record and assess students' progress.

Extra Questions 2 and *Further Questions 2* consist of exercises which are directly related to those in *Students' Book 2*. *Extra Questions* contains easier questions than those found in the *Students' Book*, and *Further Questions* consists of harder questions to extend the more able.

1 NUMBERS IN ACTION

LOOKING BACK

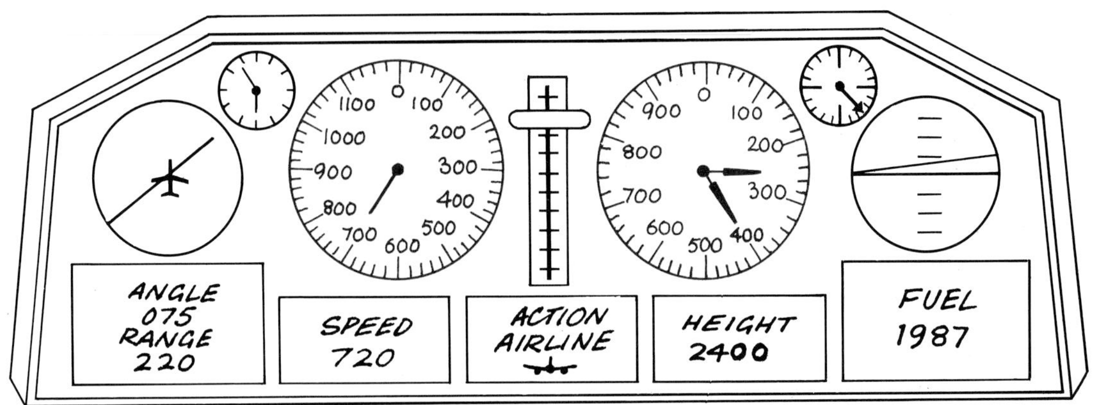

1 a Find the new readings on the aircraft's flight panel if:
(i) speed increases by 200 km/h
(ii) height decreases by 400 m
(iii) angle increases by 25°
(iv) range is halved.

b At take-off the plane had 4500 litres of fuel. How much has been used?

c How far will the plane fly in two hours at the speed shown?

d The total fare paid by all the passengers on board is £11 730. Each of the 85 passengers paid the same amount of money. How much?

2 How many?

a 
in £200

b 
in £400

c 
in £175

d 
in £2.50

3 Without using a calculator, find the value of:
a $15+15-10$ **b** $6\times4\times2$ **c** $(5\times8)+10$
d $(7+9)\div8$

4 The populations of Great Glen and Small Glen are 3527 and 3492. Round each population to the nearest: **a** 10 **b** 100.

5 This table shows the cost (per person) of short holidays in Paris.

	Nov 1–Mar 31		Apr 1–Oct 31	
	3 days	7 days	3 days	7 days
Aberdeen	£395	£444	£410	£462
Birmingham	292	310	301	337
Southampton	245	357	266	381

What is the cost for one person, flying from:
a Aberdeen for 3 days in November
b Birmingham for 7 days in May
c Southampton for 7 days in January?

6 Find the values of these, without using a calculator.
a £8.50 + £7.50 **b** £6.40 − £4.60 **c** £12.25 × 2
d £13.74 ÷ 3

7 Write these sums of money in £s and pence, to the nearest penny.

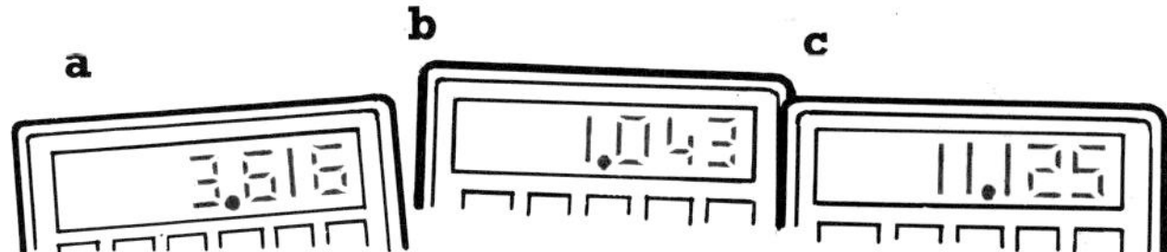

8 Phil thinks that he needs 100 tiles to cover his wall. The tiles are sold in boxes of eight. How many boxes should he buy?

EVERYDAY NUMBERS

EXERCISE 1A

This extract from a brochure shows the cost of a package holiday at the Costa Holiday Camp in Spain.

COME TO THE COSTA HOLIDAY CAMP — SUN&SEA

	May	June	July	Aug
Adult-7 Nights	£319	£331	£372	£365
Child-7 Nights	£195	£208	£234	£228
Adult-14 Nights	£369	£405	£443	£438
Child-14 Nights	£224	£266	£288	£274

Adult: 15 and over *Child:* Under 15

1 Write down the cost of a holiday for:
- **a** one adult for 7 nights in June
- **b** one child for 7 nights in August
- **c** one adult for 14 nights in May
- **d** one child for 14 nights in July.

2 a What is the cheapest 7-night charge for an adult? In which month?
- **b** What is the dearest 14-night charge for a child? In which month?

3 Mr and Mrs Owen book a 7-night holiday in August. How much do they pay?

4 The Harvey family go to the camp for 14 nights in June. Mr and Mrs Harvey have two children, aged 5 and 8. How much do they pay?

5 Mr and Mrs Kumar have three children, aged 10, 15 and 17. How much would they pay for a holiday at the camp if they stay from 3rd to 10th of July?

6

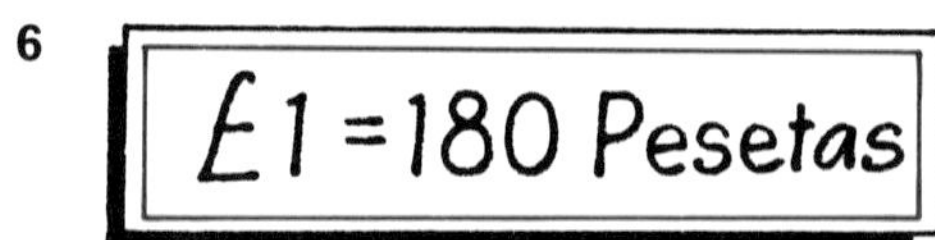

- **a** The Kumars change £500 to pesetas. How many pesetas do they receive?
- **b** A family meal costs 3600 pesetas. How much is this in £s?

7 a The flight from London takes $2\frac{1}{2}$ hours. The plane flies at an average speed of 700 km/h. How long is the journey, in kilometres?
- **b** (i) The plane leaves at 10 30 hours. When does it arrive (in 24 hour time)?
  (ii) Give both these times in am/pm time.

8 How much would it cost your family to spend 7 nights at the camp in July?

EXERCISE 1B

1 The Taylors' house was worth £87 200 when they bought it, but only £85 800 a year later. By how much had its value fallen?

2 A Boeing 747 is flying at 27 500 feet, and an Airbus is 18 300 feet above the Boeing. Calculate the height of the Airbus.

3 Mr Seth earns £1980 per month. Estimate, then calculate, how much he earns in a year.

4

- **a** What is the total playing time of five videos: (i) in minutes (ii) in hours?
- **b** How many videos could Dave buy with £25, and how much change would he be given?

5 Susan takes a job with a salary of £12 600 a year, and is promised an increase of £650 at the end of each year. Calculate her salary at the end of one, two, three, four and five years.

6 Charlotte and her friend Kate fly to Greece for a holiday.

FLY TO GREECE
£1 = 320 Drachmas

a Estimate, then calculate, the number of drachmas they would receive if they change £240.

b When they come home they have 4800 drachmas left. How many £s would they get for these?

7 Eight people share a national lottery prize of £$1\frac{1}{2}$ million. Calculate the sum of money each receives.

8 Mr and Mrs Wilson's net monthly salaries are £1350 and £375.50 (part-time). Their average monthly spending is:

Mortgage	£345
Life insurance	£64.60
Gas	£45.70
Telephone	£22.05
House insurance	£37.25
Council tax	£65.33
Electricity	£38
Motor car	£125

How much do they have left each month for food, etc?

EXERCISE 1C

1 The first nine holes of Lochview Golf Course average 355 yards in length. The second nine average 384 yards. Calculate:

a the total length of the course

b the average length of a hole.

2 The table shows the number of £15 tickets sold each night for a concert.

Night	Mon	Tue	Wed	Thu	Fri
Number sold	170	260	330	290	450

Can you describe *two* ways of calculating the total value of the sales? What is this value?

3 The United Kingdom has a population of 55.5 million, and an area of 244 000 square kilometres. Calculate the average number of people per square kilometre, to the nearest 10.

4 *Instructions* **Fahrenheit to Celsius**

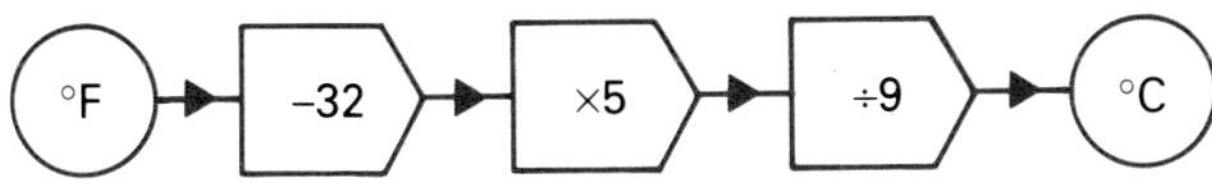

a Convert to Celsius: (i) 140°F (ii) 68°F.

b Draw a similar diagram for °C to °F. Check it, using your answers to part **a**.

If your calculator has a memory key it would be useful in questions **5** and **6**.

5 1 inch ≑ 2.54 cm.
Calculate the number of:

a centimetres, to the nearest cm, in 12 inches, 30 inches and 50 inches

b inches, to the nearest inch, in 10 cm, 84 cm and 288 cm.

6 1 dollar = £1.76.
Estimate, then calculate, the number of:

a £s, to the nearest £, you would get for \$50, \$500 and \$1200

b \$s, to the nearest \$, you would get for £100, £1250 and £1 million.

7 Money spent, in millions of £s, by Britons on holiday in a recent year:
abroad £8525m
in England £6275m
in Scotland and Wales £1575m.

a Calculate:
(i) how much was spent, in total
(ii) how much more money was spent abroad than in England, Scotland and Wales

b Taking a population of 55 million, calculate the average amount spent per person. Round your answer sensibly.

APPROXIMATION

(i) **Whole numbers**
Measurements are rarely exact. Often it is useful to round them to the nearest 1, 10 or 100 units.

Example
To the nearest centimetre:
1.2 cm → 1 cm, 2.5 cm → 3 cm, 4.7 cm → 5 cm

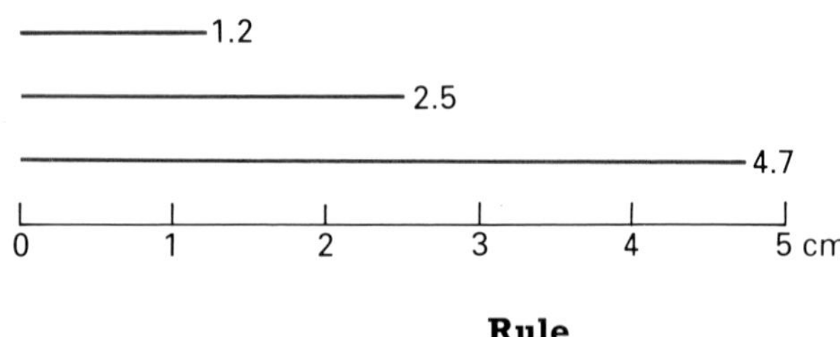

Rule
If the next figure is 5 or over, round **up**.

EXERCISE 2A

1 Write these lengths to the nearest cm.

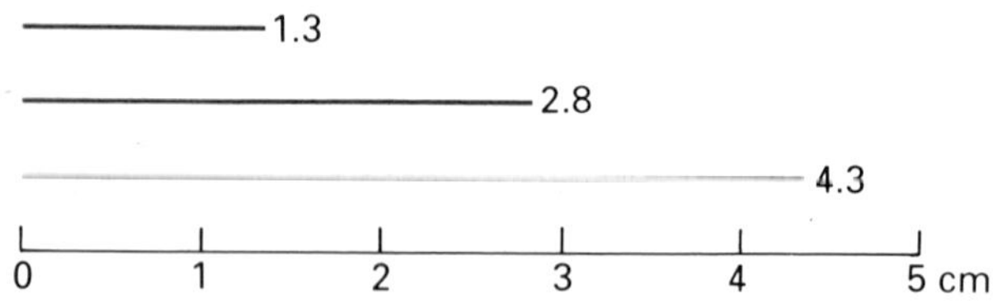

2 Write these times to the nearest second.
a 6.8 s **b** 1.1 s **c** 9.4 s **d** 3.6 s **e** 2.5 s
f 10.3 s

3 Write these volumes to the nearest ml.
a 1.7 ml **b** 2.9 ml **c** 3.0 ml **d** 4.4 ml
e 7.7 ml **f** 9.5 ml

4 Write these measurements to the nearest whole unit.
a 7.8 cm **b** 12.4 km **c** 36.9 g **d** 20.2 m
e 1.5 kg **f** 81.2 s

5 a Measure the length and breadth of this page to the nearest tenth of a centimetre (1 decimal place).
b Write down the length and breadth to the nearest centimetre.

6 Repeat question **5** for a page of your notebook.

Example
To the nearest 10 cm:
21 cm → 20 cm, 48 cm → 50 cm
Also: 123 cm → 120 cm, 275 cm → 280 cm.

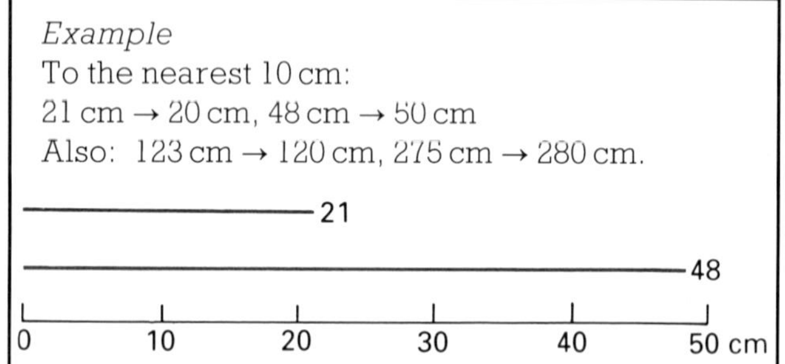

7 Write these lengths to the nearest 10 cm.
a 28 cm **b** 41 cm **c** 75 cm **d** 82 cm
e 36 cm **f** 99 cm

8 Give these distances to Edinburgh to the nearest 10 km.
a Dundee 51 km **b** Peebles 37 km
c Dumfries 72 km **d** Stirling 58 km

9 Write these weights to the nearest 10 g.
a 132 g **b** 246 g **c** 309 g **d** 471 g
e 815 g **f** 101 g

10 Write these measurements to the nearest 10 units.
a 36 litres **b** 45 kg **c** 342 m **d** 107 m
e 11 g **f** 987 cm

Example
To the nearest 100 cm:
140 cm → 100 cm, 353 cm → 400 cm.

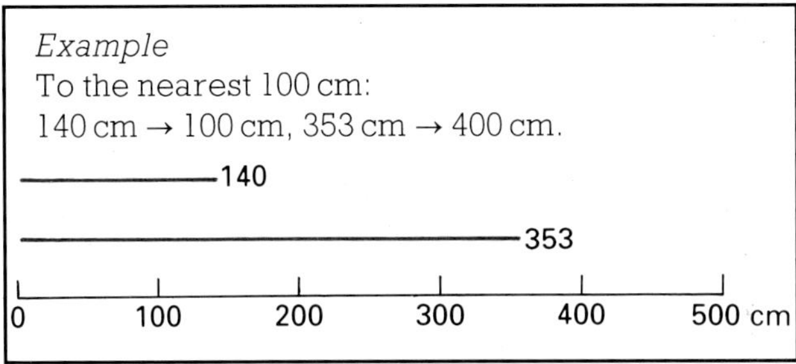

11 Write these lengths to the nearest 100 cm.
a 120 cm **b** 580 cm **c** 360 cm **d** 850 cm
e 161 cm **f** 749 cm

12 Write these measurements to the nearest 100 units.
a 567 g **b** 121 hours **c** 306 ml **d** 881 m
e 250 g **f** 474 s

13 Give these distances to London to the nearest 100 km.
a Glasgow 634 km **b** Carlisle 483 km
c Liverpool 317 km **d** Cardiff 258 km

EXERCISE 2B/C

1 Write the distances on the signpost to the nearest:
a 10 miles **b** 100 miles.

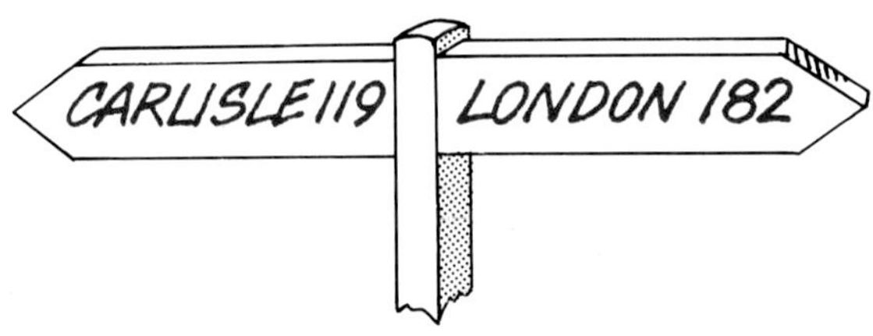

2 Write each sum of money to the nearest £10, and to the nearest £100.
a £666 **b** £123 **c** £98 **d** £919 **e** £2345

3 'At a height of 125 km the temperature of the atmosphere is 157°C.' Give the height and temperature to the nearest:
a 10 km and 10°C **b** 100 km and 100°C.

4 a Calculate the number of pupils in each school in the table, and the total number in all three schools.

School	Number of	
	girls	**boys**
Mount High	505	469
St John	781	825
Riverside	218	209

b Write the totals to the nearest 10 and the nearest 100.

5 Write the heights of these mountains to the nearest 10, 100 and 1000 feet.

Scotland	Ben Nevis	4406 feet
Wales	Snowdon	3561 feet
England	Scafell Pike	3210 feet

6 The population of Scotland is given as 4 597 861.
a Why is this figure unlikely to be correct?
b What is the population of Scotland to the nearest:
(i) 10 (ii) 100 (iii) 1000 (iv) 10 000
(v) 100 000 (vi) million?
c What approximate figure would you give for the population?

INVESTIGATION

a *Round the cost of each item on this supermarket bill to the nearest 10p and then total it up.*

SWISS CEREAL	1.27	POTATO BAKING	0.32
PLN CHOC RICH T	0.59	PRUNES IN JUICE	0.56
DAIR. I/CRM	2.25	ROUND HAM	0.59
OATMEAL BREAD	0.54	CHICKEN THIGH X2	0.52
ORIG CRCHY BREK	0.75	BAKED BEANS	0.20
PINEAPPLE PIECES	0.28	CHICKEN SOUP	0.42
MUSHROOM SOUP	0.30	CLEMENTINE	0.91
FRUIT COCKTAIL	0.63	GRAPES WHITE	0.77
		TOTAL	10.90

b *Compare your total cost with that on the bill, and comment on the result.*

(ii) **Decimal places**

The distance on the car's mileometer is 125.68 km.
Rounded to the nearest tenth of a kilometre, this is 125.7 km.
Same rule again: If the next figure is 5 or over, round up.
This rule is used for approximations to any number of decimal places.

Examples
a Rounded to 1 decimal place: 5.87 → 5.9, 36.84 → 36.8, 1.05 → 1.1
b Rounded to 2 decimal places: 9.829 m → 9.83 m, 0.345 kg → 0.35 kg

Note 'Rounding a number to 1 decimal place' is the same as 'giving the number correct to 1 decimal place.'

EXERCISE 3A

1 Round these mileometer readings to $\frac{1}{10}$ km (1 decimal place).

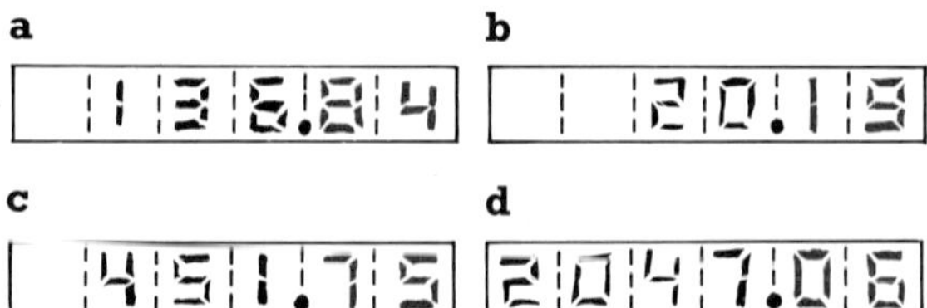

2 Round to 1 decimal place:
a 3.19 **b** 5.42 **c** 6.14 **d** 4.77 **e** 8.15
f 5.09 **g** 7.55 **h** 11.36 **i** 9.02 **j** 1.25
k 1.05 **l** 0.94

3 Round to one tenth of a unit (1 decimal place):
a 2.54 cm **b** 8.27 g **c** 2.35 km **d** 16.32°

4 **a** For a year with 365 days, divide by 12 to find the average number of days in a month. Give your answer correct to:
(i) one decimal place
(ii) the nearest number of days.
b Repeat **a** for a leap year.

5 First write down estimates for the answers to parts **a**–**d** below. Then calculate them, correct to 1 decimal place.
For example, 30 ÷ 8.
Estimate 4.
Calculation 30 ÷ 8 = 3.75, or 3.8 correct to 1 decimal place.
a 28 ÷ 3 **b** 44 ÷ 7 **c** 100 ÷ 9 **d** 11 ÷ 4

6 Give each sum of money below to the nearest penny (2 decimal places).
a £1.751 **b** £8.327 **c** £5.049 **d** £6.115

7 Round these to 2 decimal places.
a 8.127 **b** 1.513 **c** 2.965 **d** 0.294 **e** 10.126

8 A school football team sends for 15 jerseys and pays a total of £88.75 for them. The first day they are worn one team member loses his and is told that he must pay to replace it. How much must he pay?

9 Calculate $\frac{1}{6}$, $\frac{1}{3}$ and $\frac{2}{3}$ as decimal fractions, rounded to 2 decimal places. (*Hint* $\frac{1}{6} = 1 \div 6$)

10 **a** Two towns are 31 km apart. A single rail fare costs £2.80. Find the cost in pence per kilometre, correct to 2 decimal places.
b For two other towns, 43 km apart, the fare is £3.90. What is the cost in pence per kilometre this time?
c Which journey is better value, and by how much per kilometre?

11 Calum, Saeed and Jamie order different meals in a cafe, but agree to share the cost of £16 equally. How much should each pay?

12 A racing car completed laps in 61.3, 58.7, 59.2, 57.9 and 56.8 seconds. Add these times and divide by 5 to find the average lap time, to the nearest tenth of a second.

EXERCISE 3B

1 Absolute zero temperature is −273.15°C, or −459.67°F. Write these correct to:
a 1 decimal place **b** the nearest degree.

2 One mile ≑ 1609.341 metres. Calculate the length of five miles in metres, correct to 2 decimal places.

3 The length of a year, based on the seasons, is 365.2422 days, correct to 4 decimal places. Compare the length of 100 of these years with 100 years of 365 days and then comment.

4 Round to the nearest tenth of a second:
a 11.17 s **b** 23.84 s **c** 56.35 s **d** 9.89 s

5 Round to the nearest hundredth of a second:
a 10.147 s **b** 56.606 s **c** 28.015 s **d** 0.123 s

6 Estimate, then calculate, correct to 2 decimal places:
a 41.95 ÷ 20 **b** 246.3 ÷ 8 **c** 90.01 ÷ 11
d 34.5 ÷ 16 **e** 94.26 ÷ 32

7 The hockey team manager orders eleven hockey skirts for £178.50. How much must Jane pay if she wants to buy one of the skirts?

8 Here are the competitors' times in a 100 m race, correct to $\frac{1}{1000}$ second.

Competitor	A	B	C	D	E
Time (s)	10.002	9.857	10.006	9.951	9.866

a Arrange the competitors in order, winner first.
b What would the order be if the times were given correct to $\frac{1}{10}$ second?

EXERCISE 3C

The width of the page is 8 cm, to the nearest cm.

So the width is between 7.5 cm and 8.5 cm to 1 decimal place.

1 Between what lengths, to 1 decimal place, do these measurements lie?
a 6 cm **b** 10 cm **c** 8 m **d** 13 mm **e** 123 km

2 What are the greatest and least values, to 1 decimal place, of these measurements?
a 9 seconds **b** 32 g **c** 5 ml
d 48 hours **e** 17 cm²

3 a A leaf is 4 cm long, to the nearest cm. What are its greatest and least lengths:
(i) to 1 decimal place (ii) in millimetres?
b A corridor is 22 m long, to the nearest m. What are its greatest and least lengths:
(i) to 1 decimal place
(ii) in metres and centimetres?

4 The side of a square is 6 cm long, to the nearest cm. Compare its area with those of squares with sides 5.5 cm and 6.5 cm long, to 1 decimal place.

5 Calculate the largest and smallest possible areas of a rectangle whose length 12 cm and breadth 8 cm are given to the nearest cm.

6 The area of a square is 28 cm².
a Estimate the length of its side, to 1 decimal place (x cm).
b Multiply x by x to find how close your estimate was.
c Using only the [×] key on your calculator, find better approximations for the length, until you can define it correct to 1 decimal place.

7 Repeat question **6** for:
a squares of areas:
(i) 20 cm² (ii) 40 cm² (iii) 80 cm²
b rectangles of areas and breadths:
(i) 50 cm², 6.2 cm (ii) 95 cm², 8.5 cm

PRACTICAL PROJECTS

1 *Measure the lengths and breadths of the following:*
a *your desk or table* ***b*** *the floor of a room*
c *a sports pitch or court.*
Decide on the best units to use, and explain the approximations you use.

2 *Measure the thickness of the pages of this book, not counting the covers. Use this measurement, and the number of sheets of paper making the pages, to calculate the thickness of one sheet in mm, correct to 2 decimal places.*

3 *Repeat* ***2*** *for a telephone directory, or Yellow Pages.*

4 *Describe the accuracy of these measuring instruments, for example, to the nearest unit, 0.1 mm, 0.5 seconds, etc:*
a *your ruler* ***b*** *a height chart*
c *a 10 m tape measure* ***d*** *a trundle wheel*
e *your watch* ***f*** *a thermometer*
g *the volume of petrol from a pump*
h *a tyre pressure gauge.*

(iii) **Significant figures**

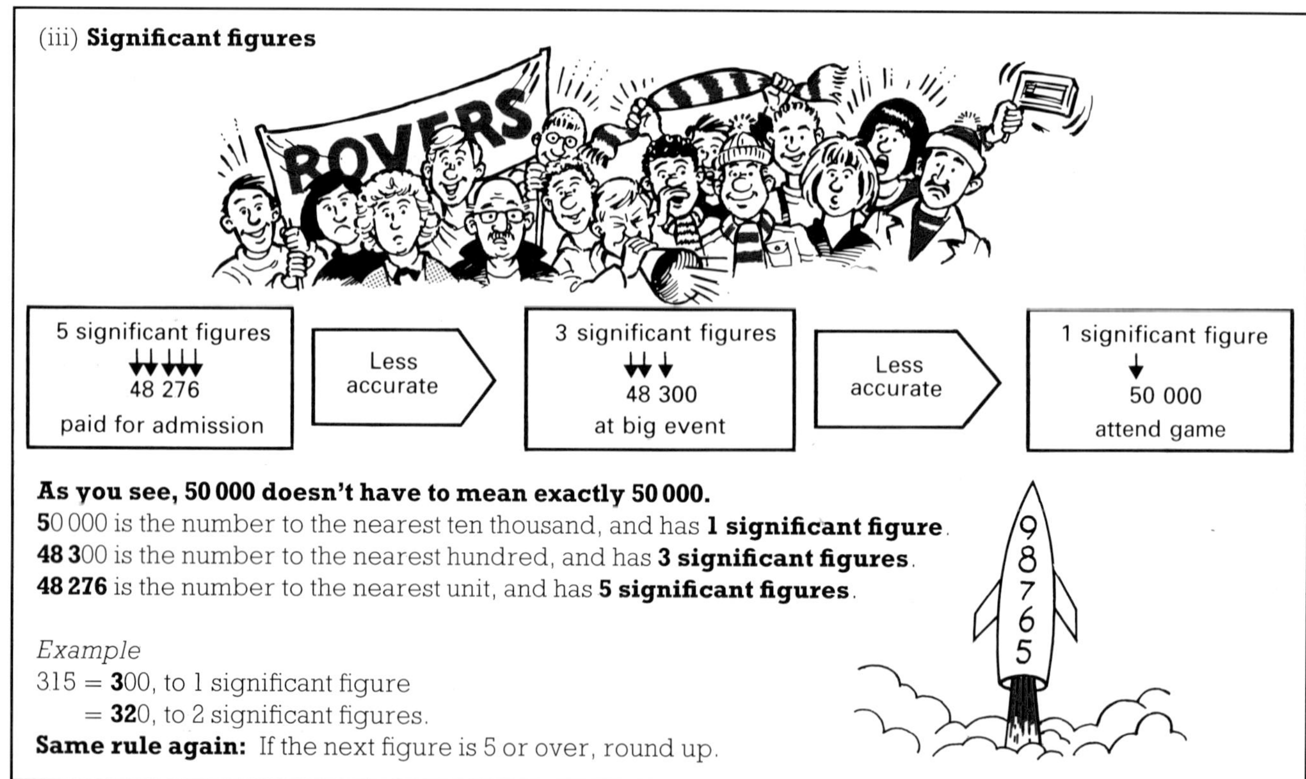

As you see, 50 000 doesn't have to mean exactly 50 000.
50 000 is the number to the nearest ten thousand, and has **1 significant figure**.
48 300 is the number to the nearest hundred, and has **3 significant figures**.
48 276 is the number to the nearest unit, and has **5 significant figures**.

Example
315 = **3**00, to 1 significant figure
= **32**0, to 2 significant figures.
Same rule again: If the next figure is 5 or over, round up.

EXERCISE 4A

1 Round to 1 significant figure:
a 19 **b** 43 **c** 75 **d** 91

2 Round to 2 significant figures:
a 234 **b** 128 **c** 509 **d** 315

3 27 359 people watched an athletics meeting. One paper reported this as 'about 20 000', another as 'about 30 000'. Which was the better approximation?

4 Write these attendances as approximations with 1 significant figure.
a 4900 **b** 18 000 **c** 9240 **d** 826 **e** 54 000

5 Give the engine size of each car to 1 and 2 significant figures.

	Car	Engine size
a	Astra 1.4L	1389 cc
b	Escort 1.8 Estate	1753 cc
c	Rover 825	2498 cc
d	Metro 1.1	1120 cc
e	Rolls Royce	6750 cc
f	Nissan Micra 1.0	993 cc

6 How many significant figures are in each of these?
a 13 years **b** 8 kg **c** 132 litres **d** 1047 cm
e 0 pupils

7 If the school roll is exactly 930, how many significant figures is this?

Quick calculations
Using 1 significant figure for each number, estimate the answers to **a** and **b** below, then check each one with a calculator.
a 79×9
Using 1 significant figure: $80 \times 9 = 720$
By calculator: $79 \times 9 = 711$
b $58 \div 9$
Using 1 significant figure: $60 \div 9 \doteqdot 7$
By calculator: $58 \div 9 = 6.4$, to 2 s.f.

8 Use 1 significant figure to estimate each answer, then check by calculator, to 2 significant figures if necessary.
a 68×7 **b** 83×8 **c** 9×29 **d** 21×21
e $43 \div 8$ **f** $58 \div 6$ **g** $33 \div 7$ **h** $121 \div 5$

9 Use question **8**'s method to estimate, then calculate, the areas of the rectangles with the given lengths and breadths. Again, give 2 significant figures in answers.
a 38 cm, 6 cm **b** 62 mm, 9 mm
c 126 m, 7 m **d** 289 cm, 12 cm

EXERCISE 4B/C

8 mm
= 0.8 cm
= 0.008 m

8 is the significant figure in all three lengths.
The zeros are **not** significant; they merely show the position of the decimal point.

Examples
a 104.2 cm, 4 s.f. **b** 0.097 kg, 2 s.f. **c** 0.351 m, 3 s.f.
d 0.1005 m, 4 s.f. **e** 5.40 cm, 3 s.f. (this zero **is** significant; it shows 40 mm)

1 How many significant figures are in each measurement?
a 5.8 m **b** 0.38 km **c** 107 litres
d 2048 cm **e** 0.06 kg **f** 4578 g
g 5.60 hours **h** 0.000 75 m **i** 1006 seconds
j 2.05 ml

2 Round each number to 2 significant figures.
a 21.3 **b** 5.09 **c** 986 **d** 0.715 **e** 0.0123

3 Round these conversion factors to the given number of significant figures.
a 1 yard = 0.91440 m (2)
b 1 ton = 1016.05 kg (4)
c 1 mile = 1609.341 m (5)
d 1 pound = 0.45359 kg (4)
e 1 gallon = 4.5459 litres (3)
f 1 inch = 25.39998 mm (5)

4 An engineer is asked to make a blade 0.70 mm thick. Why is it described as 0.70 mm, and not 0.7 mm?

5 a Use 1 significant figure to estimate the area of each parking bay in the diagram.

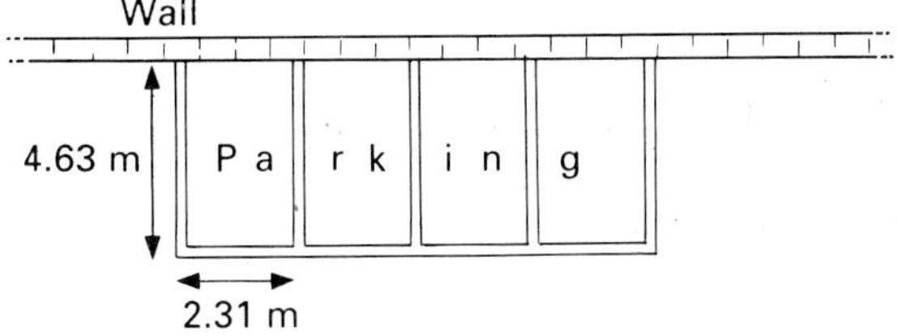

b Calculate the area of each bay, to 3 significant figures. Why not more than 3?
c White tape is used to mark the edges of the bays. Estimate, then calculate, the total length required for the four bays in the diagram.

6 Calculate the volumes of these objects.

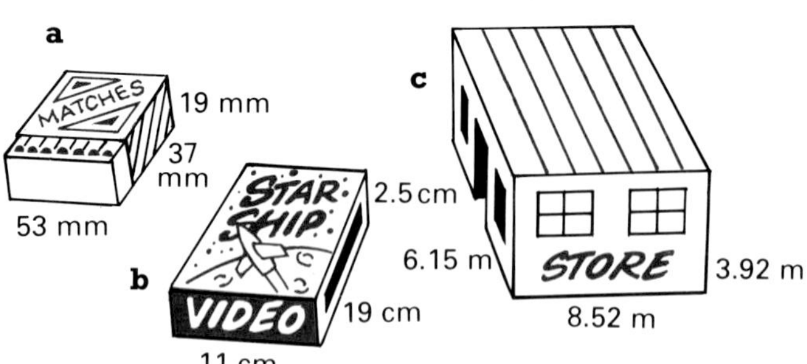

7 Write down the upper and lower limits of the measurements given below. All of them have been given correct to 1 significant figure.

a Car's speed, 50 km/h
b River, 700 km long
c Race time, 9 seconds
d Population, 60 000 people
e Juice, 0.9 litre
f 2000 passengers

KEEPING NUMBERS IN ORDER

What is the value of $2+3\times4$? 20, or 14?
The rule is shown in the red triangle, so

$$2+3\times4=2+12=14$$

Can your calculator do it? Scientific calculators can!

EXERCISE 5

1 Without using a calculator, find the values of these. Remember the warning in the red triangle!

a $2\times5+1$ **b** $3\times4-5$ **c** $10+2\times7$
d $12-3\times3$ **e** $20+3\times5$ **f** $18-4\times4$
g $7\times9-3$ **h** $10-5\times2$ **i** $6+9\div3$
j $20\div10+5$ **k** $12-10\div2$ **l** $18\div2-9$

2 If you have a scientific calculator, check your answers to question **1**. (*Check* It should show $10+2\times7=24$)

3 Calculate without, then with, a scientific calculator:

a $1+2\times2$ **b** $13-3\times3$ **c** $6\times5+2$
d $18\div6-3$ **e** $10+10\div10$ **f** $4\times4-4$
g $10-3\times3$ **h** $1+1\times1$

4 a Using the numbers 1, 2 and 3 once each, in any order, and the + and × signs, try to get the answer 9.
b If you can't find a solution, try $(1+2)\times3$.

$(1+2)\times3=3\times3=9$
The rule is in the red triangle.

WARNING
( )
before
× ÷
before
+ −

Examples
a $(3\times4)\div2=12\div2=6$
b $5\times(4-3)=5\times1=5$

5 Without using a calculator, find the value of:

a $(5+4)\times3$ **b** $5+(4\times3)$ **c** $(12-6)\div2$
d $12-(6\div2)$ **e** $(16\div4)\div2$ **f** $16\div(4\div2)$
g $(2\times3)+(4\times5)$ **h** $(9+6)\div(8-3)$

Remember the order! **Brackets; multiplication and division; addition and subtraction.**

6 Calculate, with or without the aid of a scientific calculator:

a $4+1\times5$ **b** $(4+1)\times5$ **c** $16-8\div4$
d $(16-8)\div4$ **e** $(10+8)\div9$ **f** $6-(10-9)$
g $5\times9-5$ **h** $7+9\div3$

7 Copy these calculations, and then put brackets in the correct places to give the answers shown.

a $2+3\times4=20$ **b** $5-3\times6=12$
c $7\times5-3=14$ **d** $12\div3+1=3$

8 In a science experiment, Amy had to calculate $\dfrac{72\times56}{37+27}$.
She used her calculator to try two different methods — one used brackets, $(72\times56)\div(37+27)$, and the other used the memory. Try both methods for yourself.

9 Tariq finds that the length of this block is given by $\frac{39\,375}{35 \times 25}$.

Calculate the length:

a using brackets
b using the memory
c another way.

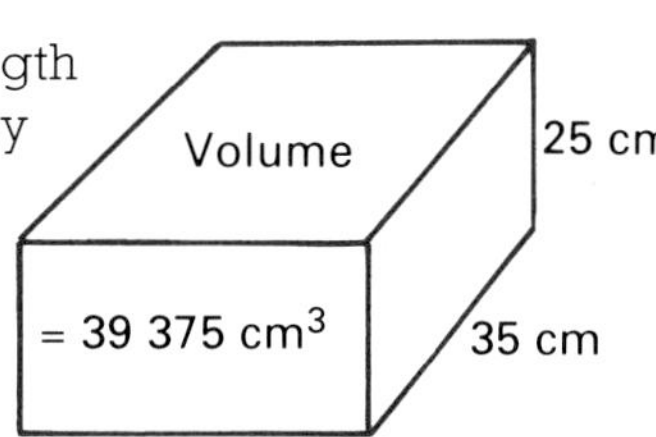

10 Use a calculator for these, then check by calculation.

a $\frac{75\,000}{125 \times 75}$ **b** $\frac{75\,000}{125+75}$ **c** $\frac{645+387}{645-387}$

d $\frac{24\,000}{(125+75) \times 8}$

BRUSH UP YOUR NUMBER WORK

EXERCISE 6

Calculator out of action! Do these calculations mentally, or using pencil and paper.

1 a $2+4+6$ **b** $8+7+6$ **c** $9+9+9$
d $15-6-2$ **e** $17-9-1$ **f** $20-10-7$

2 a $4 \times 3 \times 2$ **b** $5 \times 2 \times 7$ **c** $8 \times 5 \times 2$
d $2 \times 4 \times 6$ **e** $5 \times 5 \times 5$ **f** $4 \times 5 \times 9$

3 a $12+47$ **b** $58+22$ **c** $39+19$
d $87-63$ **e** $63-35$ **f** $100-75$

4 a 8×10 **b** 80×10 **c** 35×100
d 9×1000 **e** 230×10 **f** 1000×0

5 a 25×2 **b** 40×3 **c** 25×4 **d** 50×5
e 62×6 **f** 22×7

6 a $120 \div 10$ **b** $200 \div 100$ **c** $40 \div 10$
d $300 \div 20$ **e** $5000 \div 1000$ **f** $6000 \div 30$

7 a $32 \div 8$ **b** $56 \div 7$ **c** $55 \div 11$ **d** $60 \div 5$
e $80 \div 4$ **f** $90 \div 6$

8 a $265+125$ **b** $265-125$ **c** $407+23$
d $407-23$ **e** $562+534$ **f** $562-534$

9 a $5 \times 19 \times 2$ **b** $4 \times 6 \times 25$ **c** $50 \times 17 \times 2$
d 99×5 **e** 999×8

10 a $2 \times 5+3 \times 5$ **b** $6 \times 4-2 \times 4$
c $19 \times 6+19 \times 4$ **d** $25 \times 15-25 \times 5$

11 a 235×5 **b** 307×8 **c** 35×15
d 24×22 **e** $235 \div 5$ **f** $456 \div 8$

12 a 235×12 **b** 206×23 **c** 567×35
d $625 \div 25$ **e** $352 \div 44$ **f** $364 \div 52$

CHALLENGES

1 **a** *Using the numbers 1, 2, 3 once each, in any order, along with any of the symbols +, −, ×, ÷, make as many numbers as you can.*
Example $3+1 \times 2 = 5$
b *Try part* **a** *again for the numbers 1, 2, 3 and 4.*

2 **a** *Arrange 1, 2, 3, . . . , 8 in two groups with equal sums.*
b *Repeat* **a** *for three groups, then four groups.*

CHECK-UP ON NUMBERS IN ACTION

1 A golfer won £10 000 in a competition. His expenses were: hotel £275.50; travel £95.25. If he also bought a new set of golf clubs for £487, how much of his prize money was left?

2 This is the price table of spring bulbs at the Green Gates Garden Centre.

Bulb \ Number	10	25	50	100
Daffodil	£1.75	£4.00	£7.75	£14.00
Tulip	2.75	6.50	11.75	22.00
Crocus	1.50	3.50	6.00	10.50
Snowdrop	0.90	1.75	3.80	6.50

Calculate the cost of:
a 50 daffodils
b 10 daffodils and 10 tulips
c 50 snowdrops and 50 crocus
d 100 daffodils and 150 crocus.

3 Write each distance given below to the nearest 10 km, and to the nearest 100 km.
a 127 km **b** 841 km **c** 569 km **d** 605 km
e 984 km

4 Round each measurement to 1 decimal place.
a 2.16 m **b** 8.84 cm **c** 10.05 seconds
d 1.01 kg

5 Round to 2 decimal places:
a 5.163 km **b** 1.019 litres **c** 0.305 m
d 10.856 kg

6 Express correct to 2 significant figures:
a 246 **b** 503 **c** 215 **d** 1090 **e** 3644

7 Using 1 significant figure for each number, *estimate* the area and perimeter of a square with side 8.85 cm long. Check by calculator.

8 Write 26.36 correct to:
a 1 decimal place **b** 1 significant figure.

9 Calculate the cost of 1 g or 1 ml of these to decide which is the most economical purchase in each case.

a

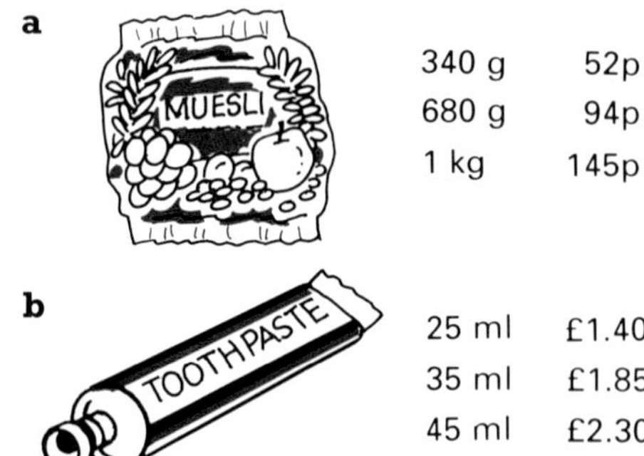

b

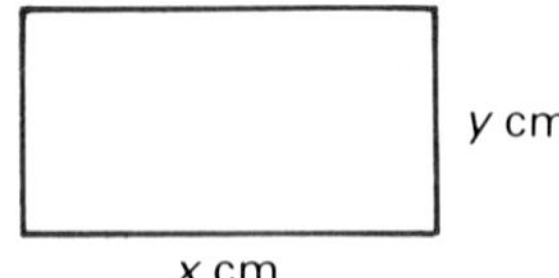

25 ml £1.40
35 ml £1.85
45 ml £2.30

10 Without using a calculator, find the value of:
a $34-4\times8$ **b** $2\times4+3\times5$ **c** $9\times9-7\times7$
d $48\div(6\div2)$

11 Copy these, and put in brackets to give the answers shown.
a $7-4\times9=27$ **b** $6\times4+5=54$
c $20\div4+1=4$ **d** $55\div5\times5=55$

12 The perimeter P cm of the rectangle can be found from either formula, $P=2x+2y$ or $P=2(x+y)$.

y cm

x cm

Calculate P, using both formulae, when $x=169$ and $y=88$.

2 ALL ABOUT ANGLES

LOOKING BACK

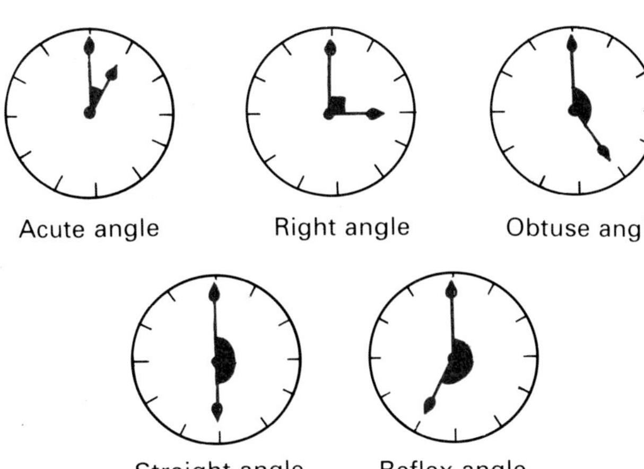

1 What kind of angle is marked in each diagram below?

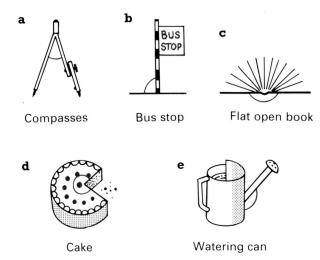

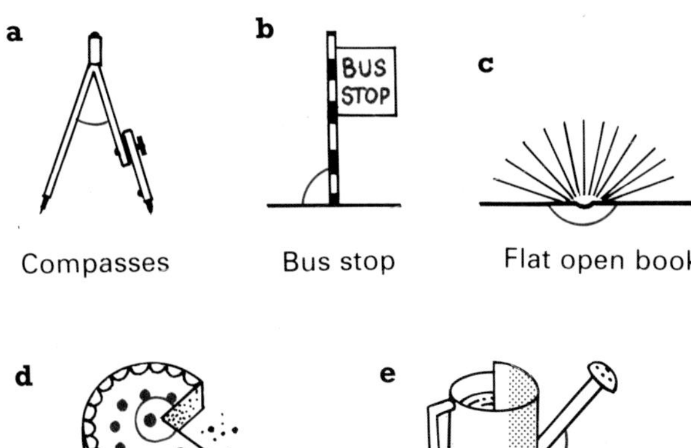

Cake

Watering can

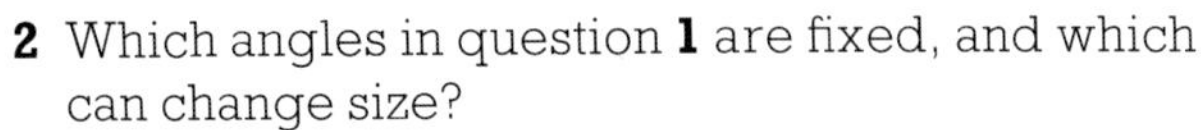

2 Which angles in question **1** are fixed, and which can change size?

3 Which kind of angle is each of these?
a 123° **b** 74° **c** 45° **d** 195° **e** 96° **f** 90° **g** 180°

4 a How many degrees are there in a complete turn?
b What fraction of a complete turn is:
(i) 180° (ii) 90° (iii) 45° (iv) 60° (v) 120°?

5 Name and calculate the size of each angle marked with an arc.

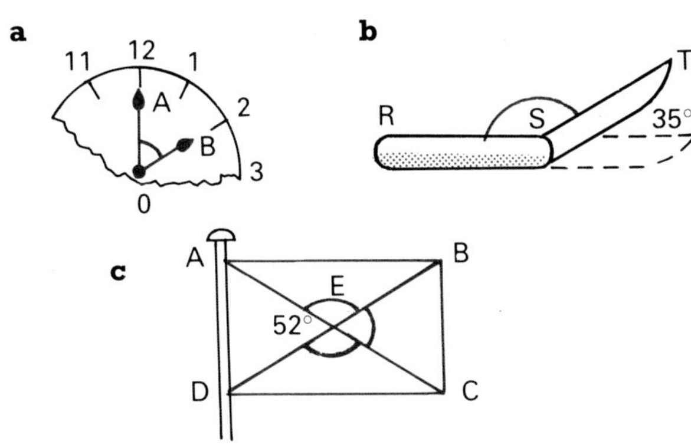

6 A circular cake is cut into three slices. The angles at the centre of two of them are 70° and 140°. Calculate the angle at the centre of the third slice.

7 Estimate the size of each angle below, then measure it with a protractor.

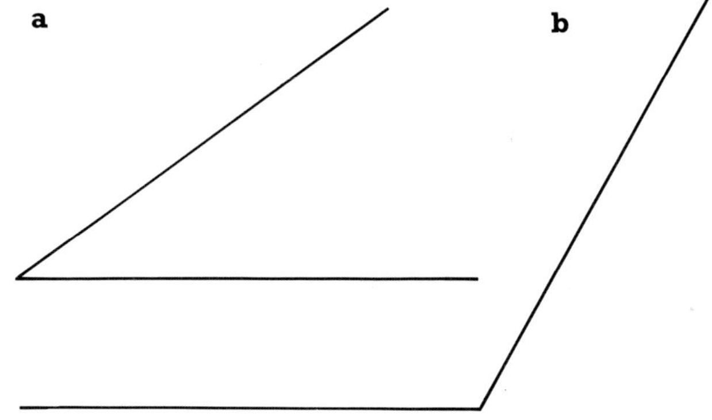

8 The shop window is an octagon with all its sides the same length.

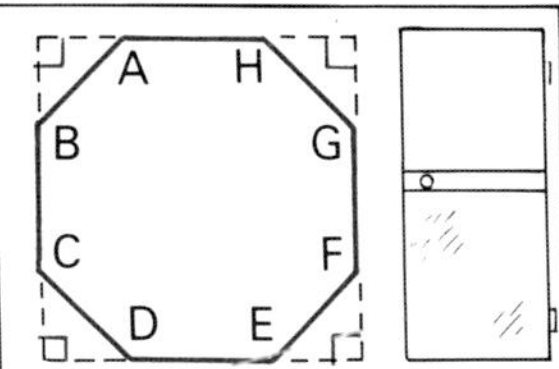

Name pairs of:
a vertical lines **b** horizontal lines
c parallel lines **d** perpendicular lines.

9 Draw angles of: **a** 25° **b** 77° **c** 90° **d** 132°

COMPLEMENTARY ANGLES

Angles often come in pairs. There are hundreds of pairs in these pictures.

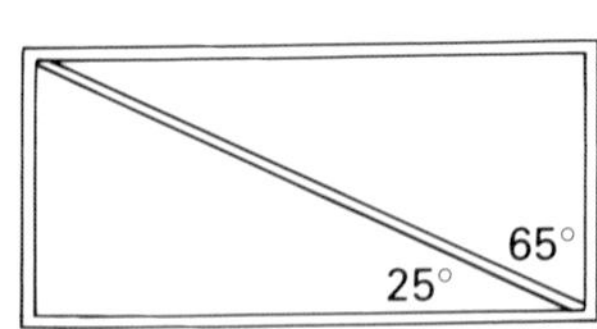

Each corner of the gate is a right angle.
25° + 65° = 90°.
25° and 65° are complementary angles.

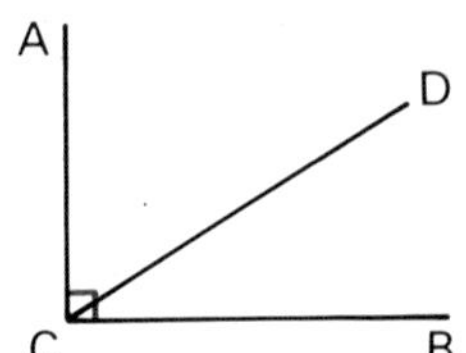

Two angles which make a right angle are **complementary**.
∠ACD is the **complement** of ∠BCD.

EXERCISE 1A

1 Write down the name and size of each angle marked with an arc. For example, ∠XYZ = 25°.

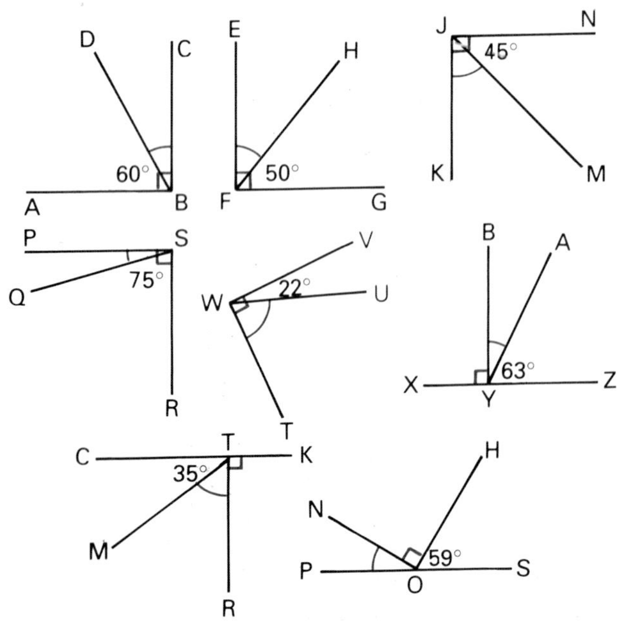

2 What is the complement of:
a 80° **b** 55° **c** 89° **d** 17° **e** 39° **f** 5°?

3 The scale is marked from 0° to 90°.

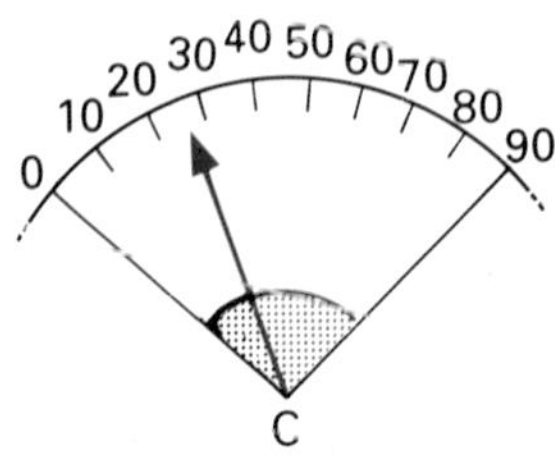

a What is the sum of the black angle and the red angle as the needle turns about C?
b Calculate the size of the red angle when the black angle is:
(i) 10° (ii) 80° (iii) 20° (iv) 90°

4 Through what angle must the aerial turn to be vertical?

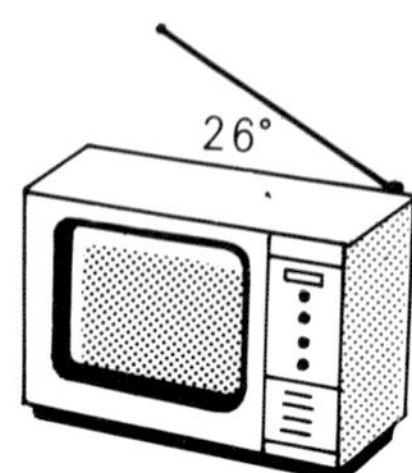

5 Through what angle must the bridge turn to be horizontal?

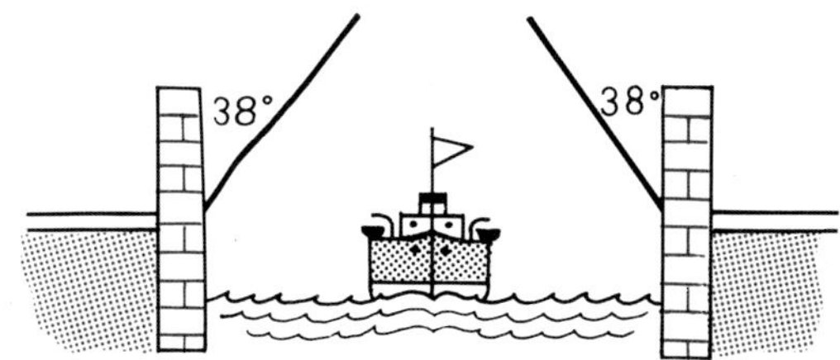

6 These two 'complementary dominoes' fit together because the angles 37° and 53° are complementary.

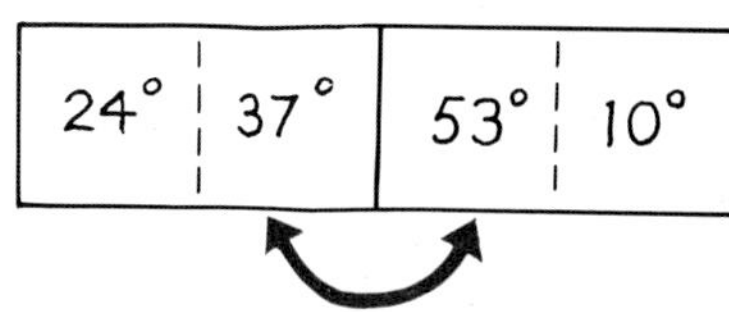

Fit these six 'dominoes' together in a straight line, pairing complementary angles.

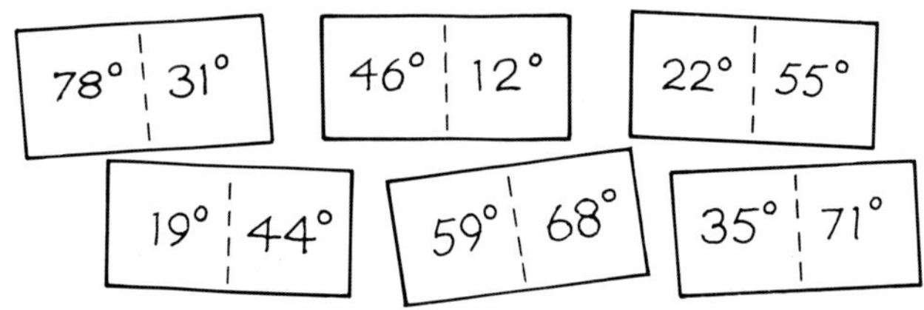

7 Which angle is the complement of:
a ∠ABF **b** ∠EBD?

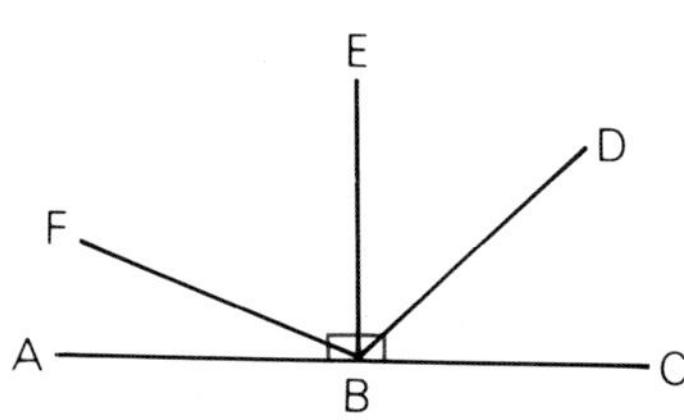

8 Name four pairs of complementary angles in this flag.

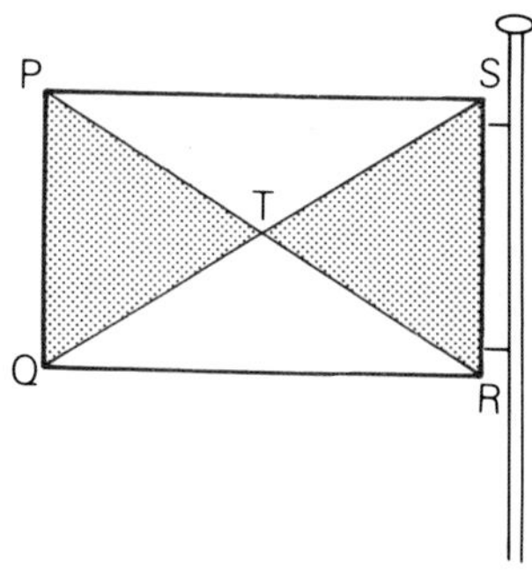

EXERCISE 1B/C

1 Name the complement of each of these angles, and calculate its size:
a ∠AOC **b** ∠BOD **c** ∠EOF

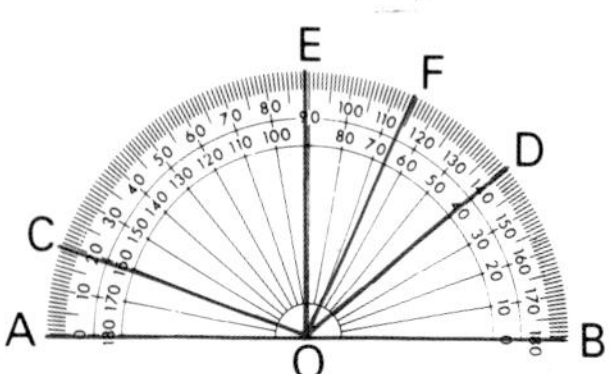

2 The leaves of the table can turn from the vertical to the horizontal. What angle has a leaf still to turn through after turning:
a 10° **b** 25° **c** 48° **d** 77° **e** 90°?

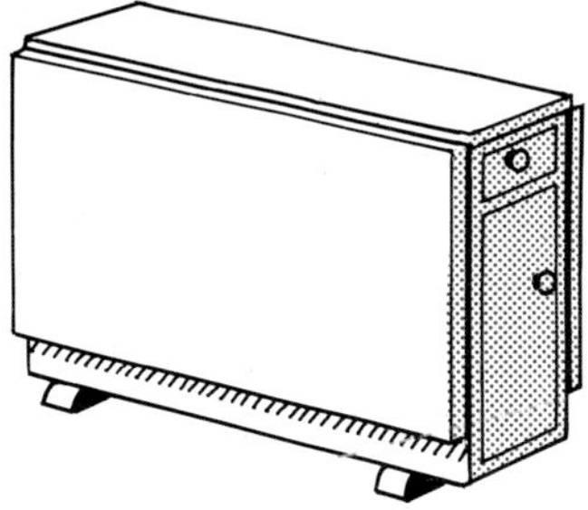

3 Seven pipes meet at a junction O. Which angle is complementary to:
a ∠KOM **b** ∠ROS?

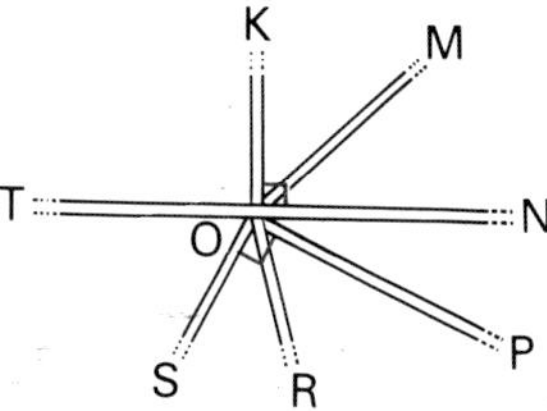

4 a Explain why ∠ABC is the complement of ∠EBD.

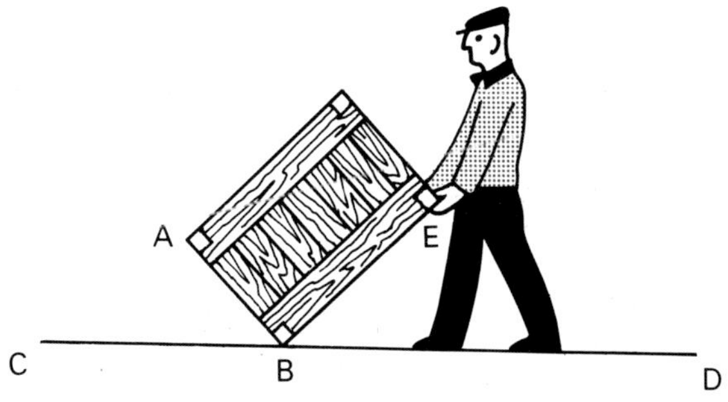

b ∠EBD = 43°. What is the size of ∠ABC?

5 a A waiter pushes the IN door open by 51°. What angle has it still to turn through to reach the door-stop?

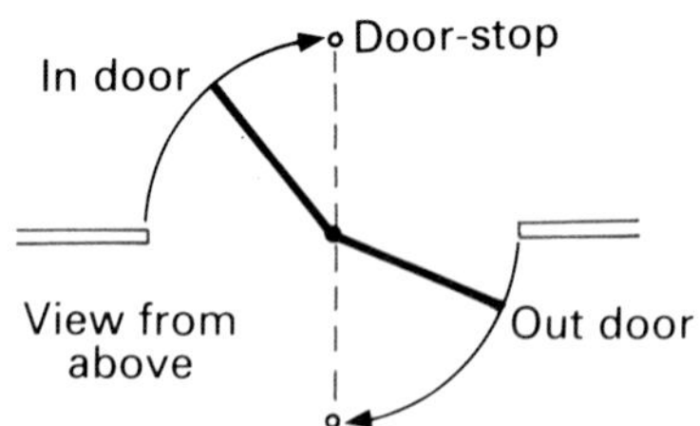

b Repeat **a** for the OUT door, which has been opened 27°.

c As one waiter goes into the kitchen through the IN door, another leaves by the OUT door. Calculate the size of the obtuse angle between the doors in each case below.

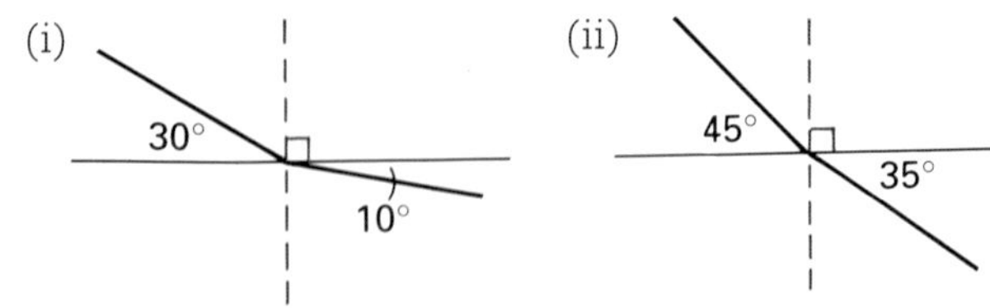

6 a Copy and complete the tables.

$x°$	$(90-x)°$
20°	
40°	
60°	
75°	

$x°$	$(90-x)°$	Complement of $(90-x)°$
10°		
30°		
45°		
60°		

b What is the complement of: (i) $x°$ (ii) $(90-x)°$?

SUPPLEMENTARY ANGLES

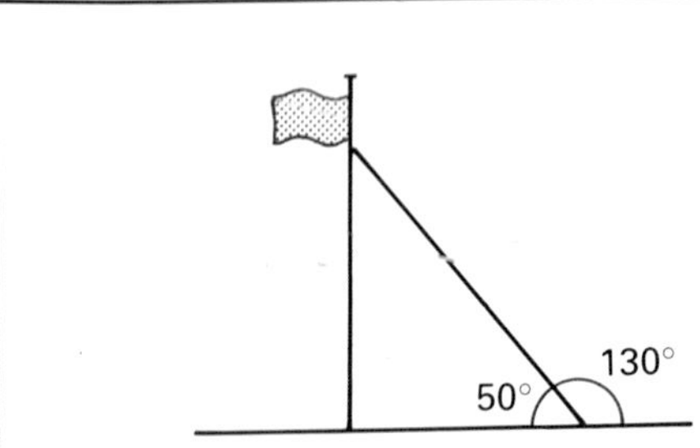

The angle at the foot of the wire is a straight angle.

$$50° + 130° = 180°.$$

50° and 130° are supplementary angles.

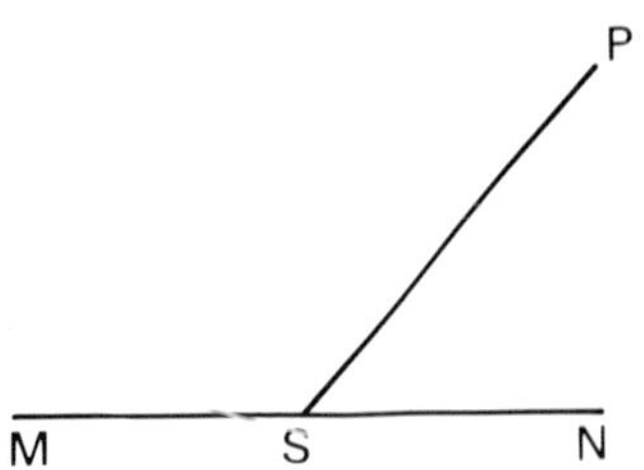

Two angles which make a straight angle are **supplementary**.
∠MSP is the **supplement of** ∠NSP.

EXERCISE 2A

1 Write down the name and size of each angle marked with an arc.

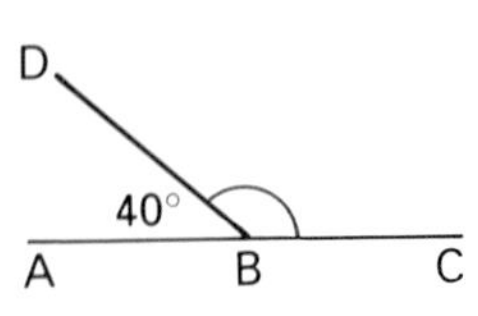

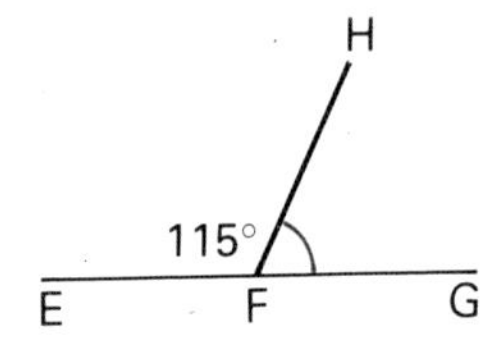

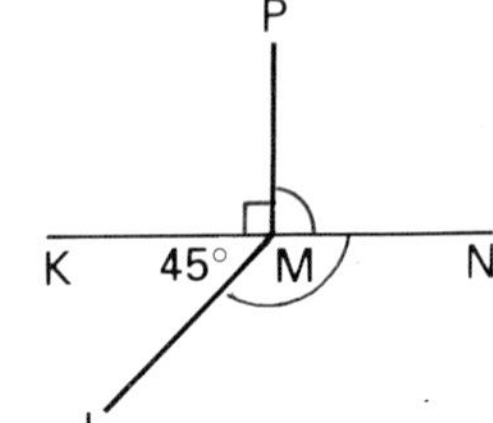

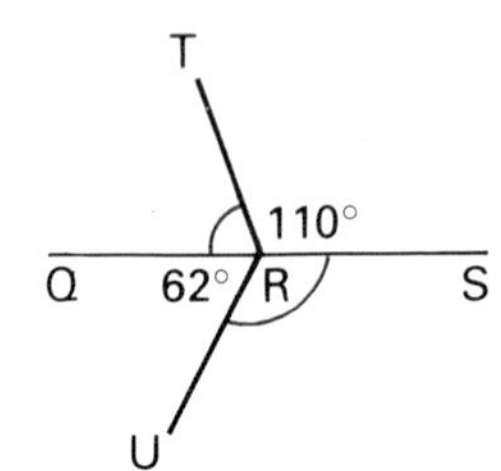

2 Calculate the supplement of:
a 150° **b** 105° **c** 80° **d** 95° **e** 1° **f** 175°

3 Measure the angles in each pair of supplementary angles below.

a

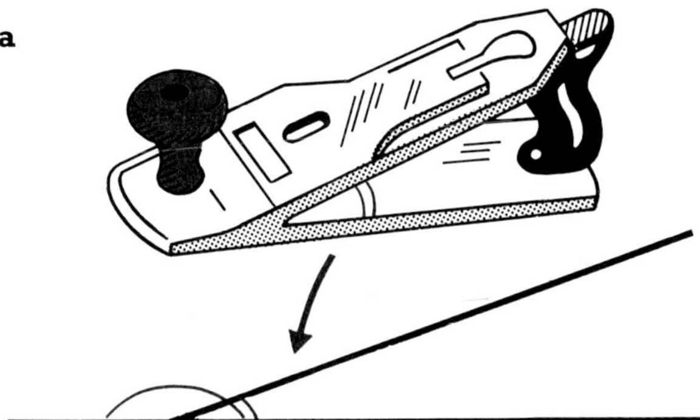

b

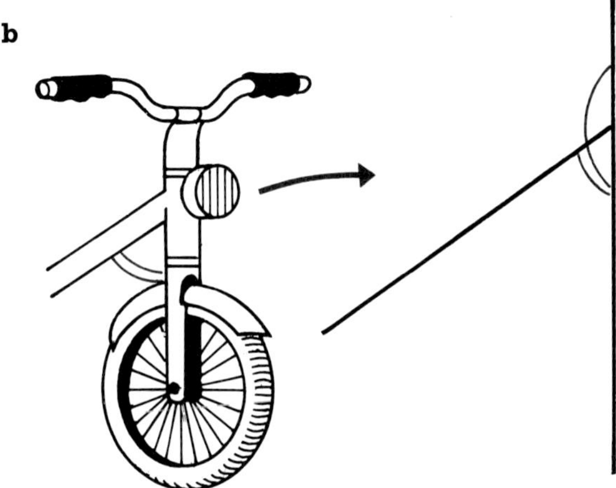

4 Make larger copies of these letters. Mark the pairs of supplementary angles formed by the parts of the letters.

5 Name, and calculate the supplement of:
a ∠SQR **b** ∠TQR **c** ∠PQS.

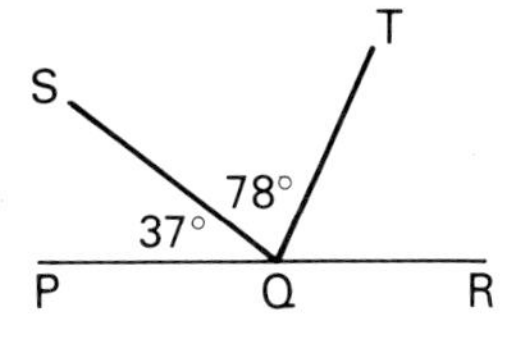

6 Calculate the acute angle between the ladder and:
a the ground **b** the wall.

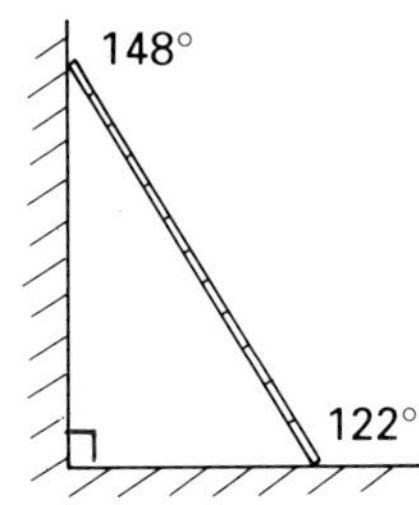

7 As the hand-brake in the car is pulled up, it gives a click every 10°.
a Copy and complete the table up to 60°.

Position	a°	**Supplement**
Start	0°	180°
First click	10°	170°
Second click	20°	

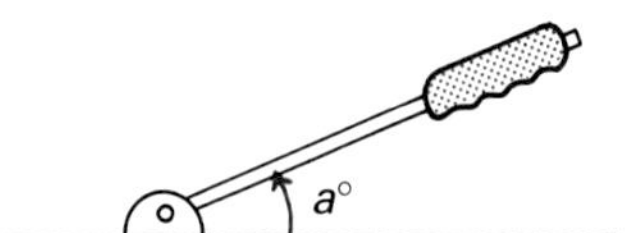

b Copy and complete: As angle a° increases by 10°, its supplement . . . by . . .°.

8 These two 'supplementary dominoes' fit together because the angles 68° and 112° are supplementary.

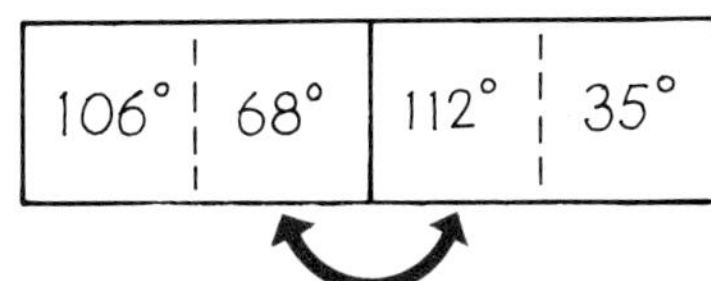

Fit these six 'dominoes' together in a straight line, pairing supplementary angles.

EXERCISE 2B/C

1 Name the supplement of each of these angles, and calculate its size:
a ∠XYV **b** ∠XYU **c** ∠WYZ **d** ∠VYZ

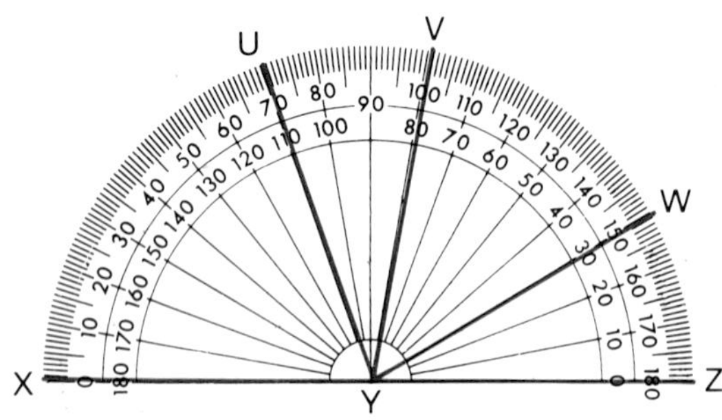

2 a Calculate the acute angle between each cable and the horizontal.

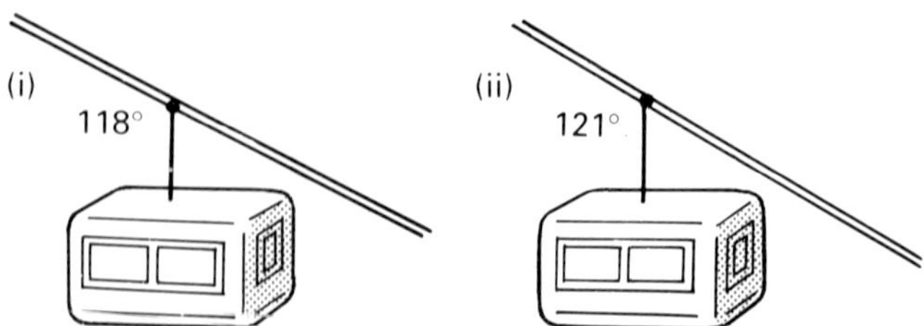

b Which cable has the greater slope?

3 Calculate the sizes of angles x, y and z in this roof structure.

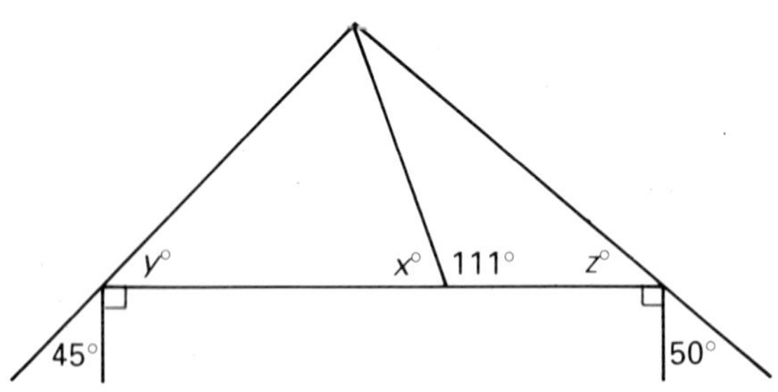

4 Calculate a, b and c. What is the sum of the angles of the triangle?

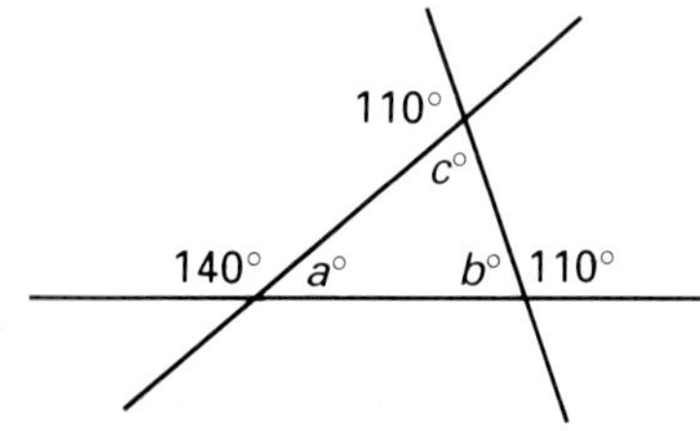

5 a Copy and complete the tables.

$x°$	$(180-x)°$
30°	
60°	
90°	
120°	

$x°$	$(180-x)°$	Supplement of $(180-x)°$
20°		
40°		
60°		
80°		

b What is the supplement of:
(i) $x°$ (ii) $(180-x)°$?

6 Complementary or supplementary pairs of dominoes can be fitted together at A or B.

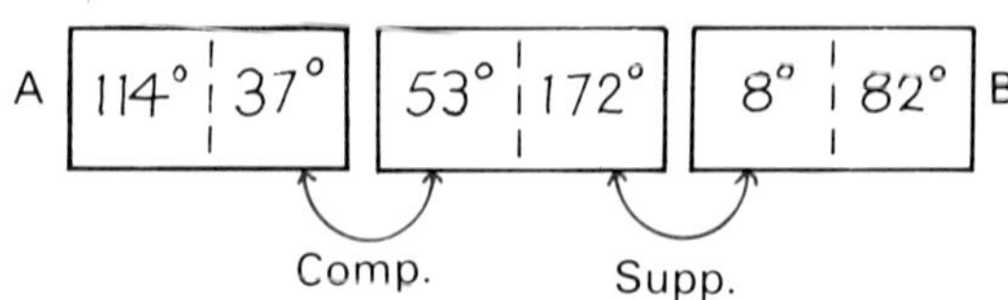

Ian has:

98° 114°	23° 49°	84° 1°	66° 73°

Waheed has:

8° 8°	10° 15°	1° 89°	117° 79°

Shona has:

8° 98°	43° 77°	11° 8°	88° 120°

a Explain each person's possible moves at A or B if it is his or her turn to play.
b If Ian plays first, he can prevent Waheed from being able to play. Explain how.

VERTICALLY OPPOSITE ANGLES

When two straight lines cross, the angles on opposite sides of the vertex are called **vertically opposite** angles. ∠s AOC and BOD are vertically opposite. So are ∠s AOB and COD.
Can you find all the pairs of vertically opposite angles in these pictures?

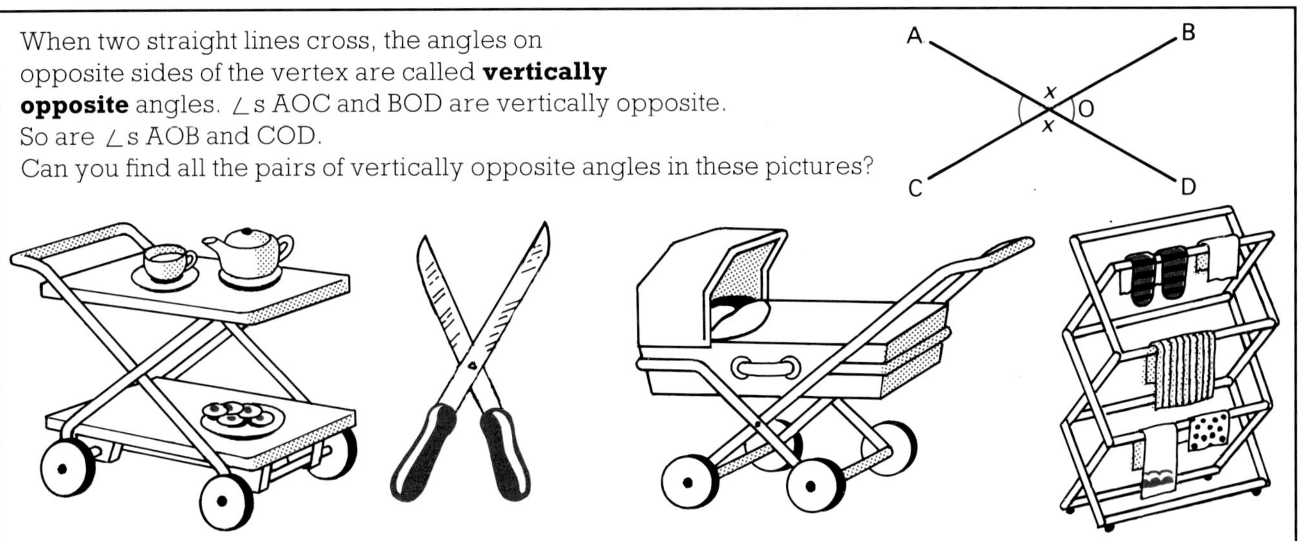

EXERCISE 3A

1 Copy these diagrams and mark pairs of vertically opposite angles with dots, crosses, arcs, etc.

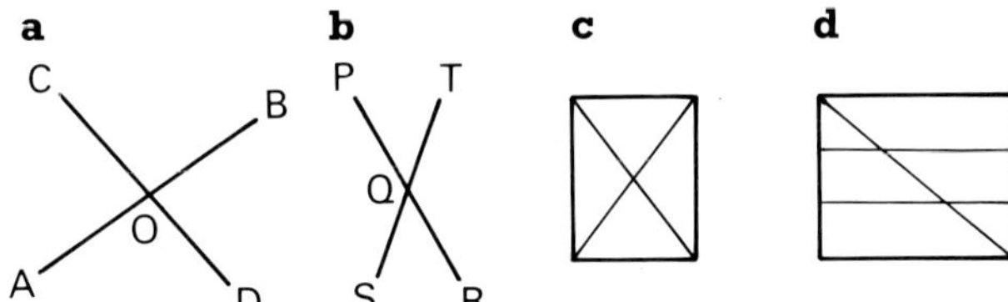

2 Name two pairs of vertically opposite angles in question **1a**, and two pairs in **1b**.

3 a Draw two straight lines which cross each other near their centres.
b Measure the vertically opposite angles. What do you find?

4 Repeat question **3** for a different pair of straight lines.

5 a Copy these diagrams and fill in all the angles, in the order x, y, z:

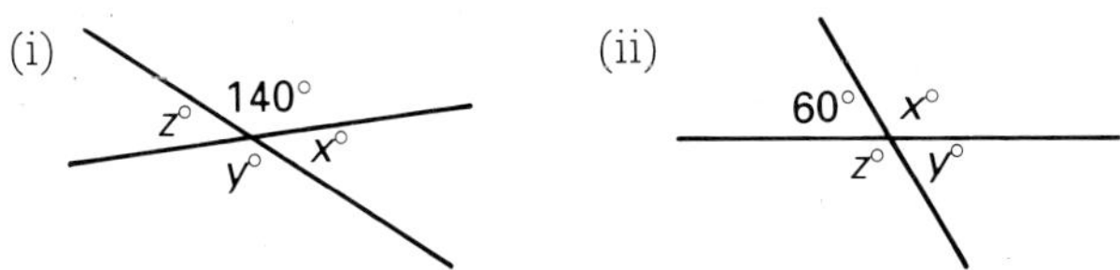

b Do you find that the angles in each pair of vertically opposite angles are equal?

Vertically opposite angles are equal.

6 Copy these diagrams, and fill in all the angles:

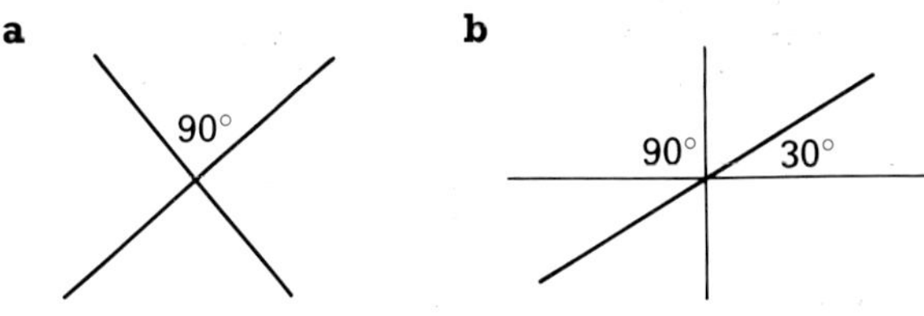

7 Name and calculate the sizes of all the acute angles in the second diagram.

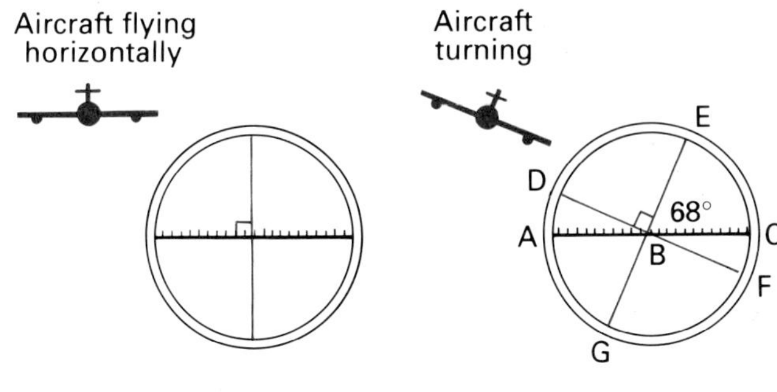

8 Copy the diagram, and fill in the sizes of all the angles in the triangle.

EXERCISE 3B/C

1 a Write down the sizes of:
(i) ∠AED and ∠AEB (ii) ∠FJI and ∠GJH
(iii) ∠QST and ∠PST

b What happens to the vertically opposite angles as the shears are closed?

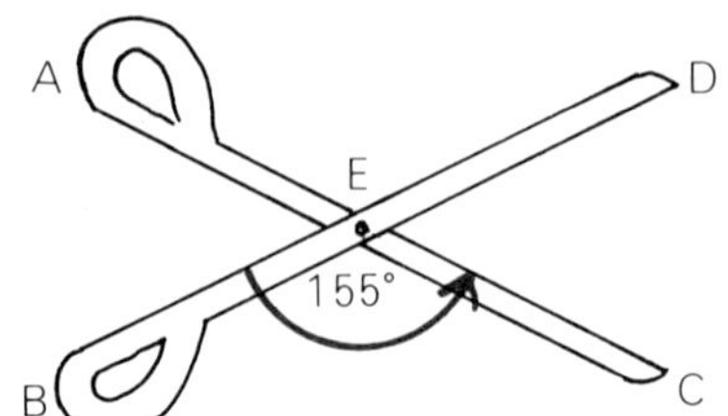

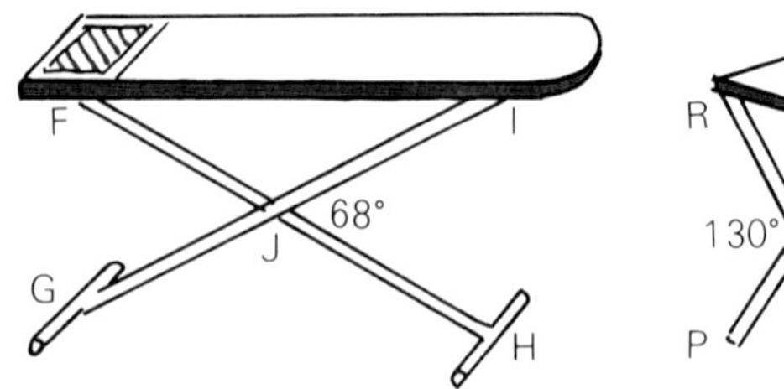

2

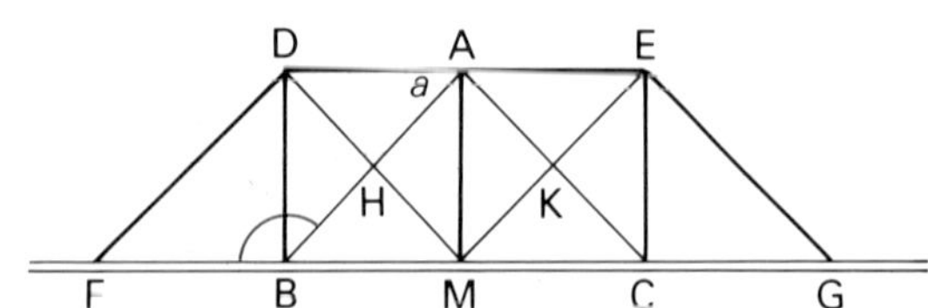

Copy this girder bridge, and mark each:

a acute angle at A, with letter *a*

b obtuse angle at B and C, with an arc

c pair of complementary angles at D and E, with *c*

d pair of supplementary angles at F and G with *s*

e pair of vertically opposite angles at H and K, with *v*.

3 Copy and complete this table for angles $a°$, $b°$, $c°$ and $d°$ which are made by the two crossing lines.

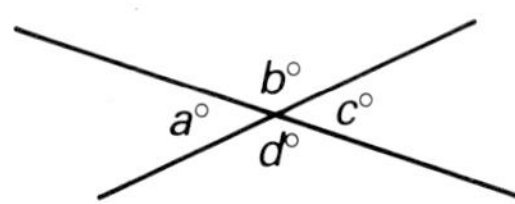

$a°$	$b°$	$c°$	$d°$
40°			
25°			
$x°$			
Increase by $y°$			

4 a *Prove* that vertically opposite angles are always equal, like this:
∠AOD + ∠AOB = 180° (why?)
∠BOC + ∠AOB = . . .° (why?)
So ∠AOD = ∠. . .

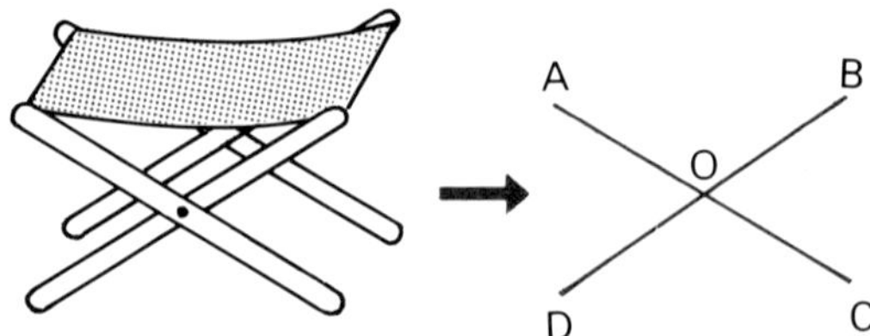

b In the same way, prove that ∠AOB = ∠DOC.

PRACTICAL PROJECT

Draw a 3 by 3 tiling of rectangles. Mark pairs of vertically opposite angles 1, 1; 2, 2; . . . , or colour code them.

CHALLENGE

a *Calculate a, b, c, . . .*

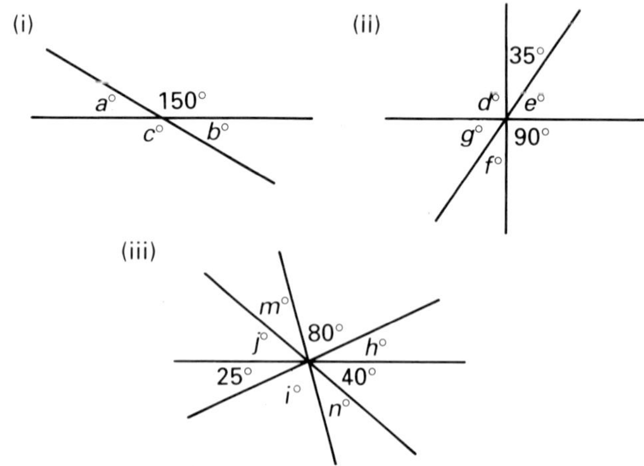

b *Copy and complete this table:*

Number of lines	2	3	4	5	. . .	n
Number of angles marked with numbers or letters	4				. . .	
Number of angles you had to be given in order to calculate all the others	1				. . .	

CORRESPONDING ANGLES

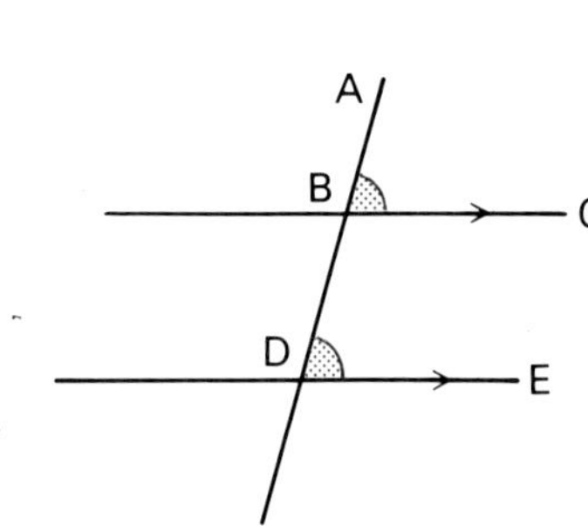

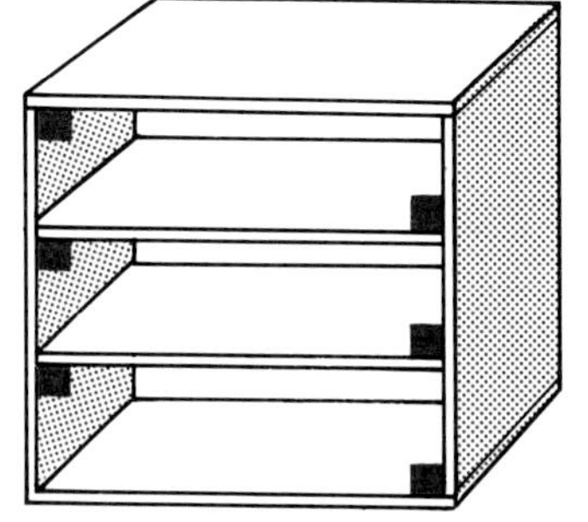

Angles ABC and ADE are **corresponding** angles. They correspond in position.

There are two sets of corresponding angles in the bookcase.

Where can you see sets of corresponding angles in these pictures?

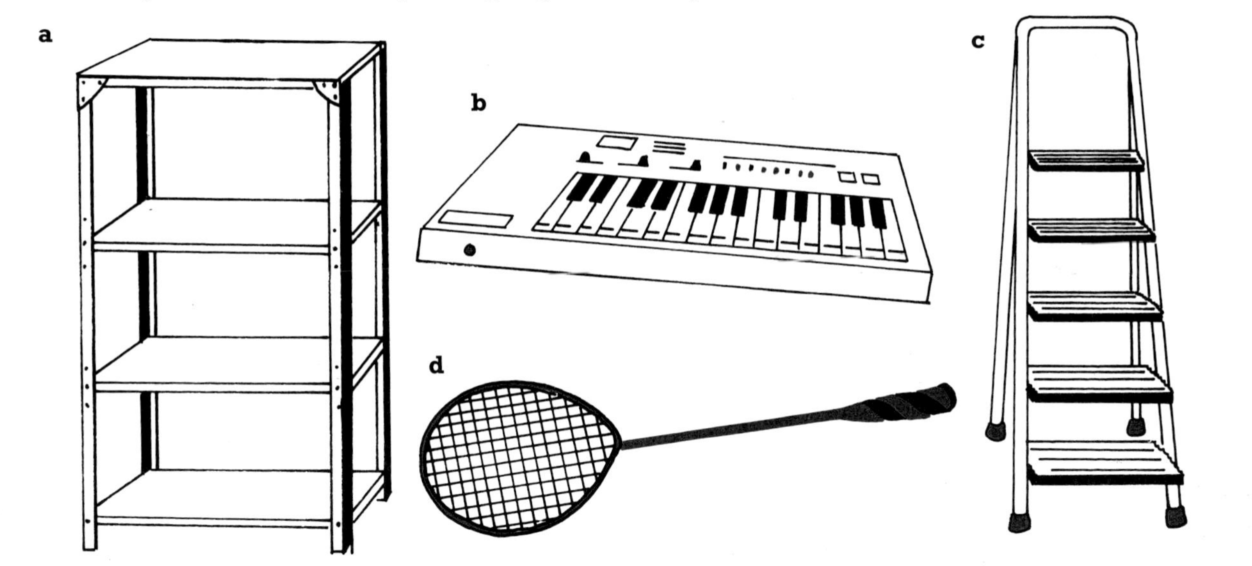

EXERCISE 4A

1 Copy the diagrams, and mark all the pairs of corresponding angles with dots, crosses, arcs, etc, like the examples shown.
(Pairs of corresponding angles will both be 'above, to the left,' or 'below, to the right,' etc.)

a

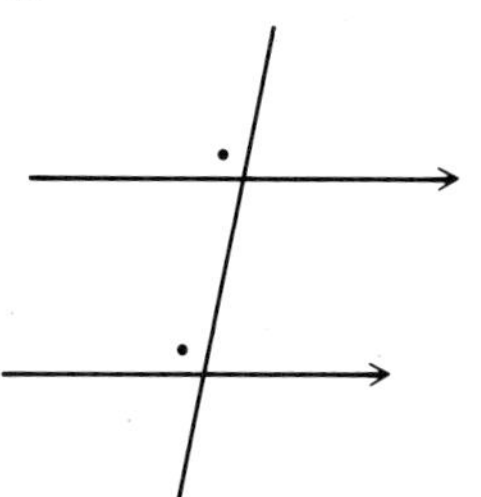

b

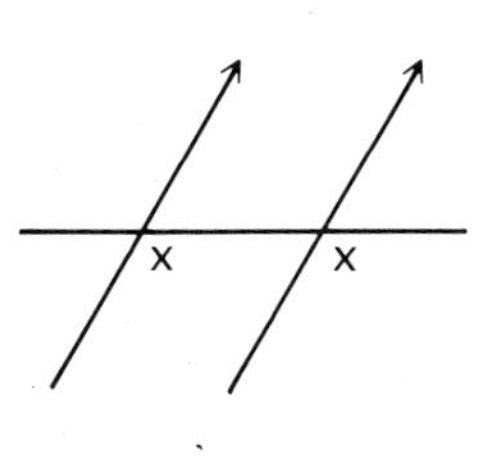

2 Name the pairs of corresponding angles here:

a

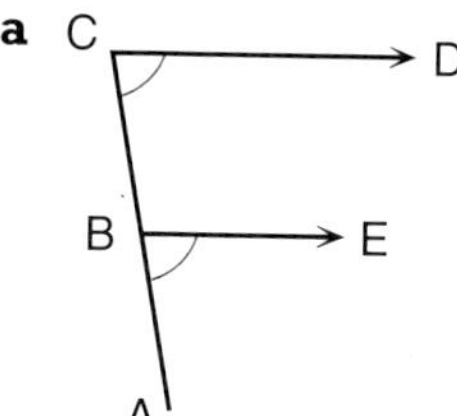

b

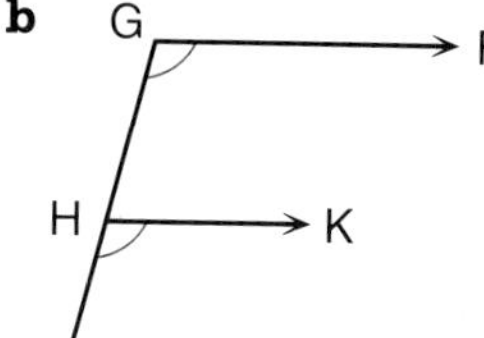

c

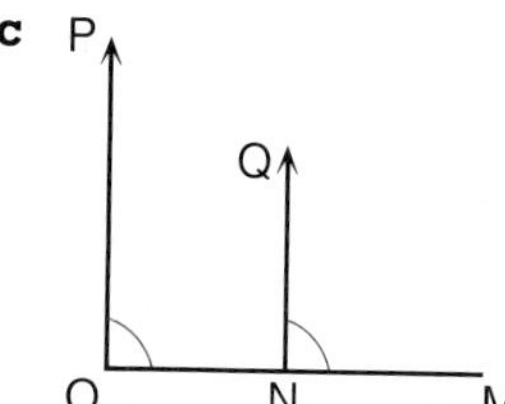

d

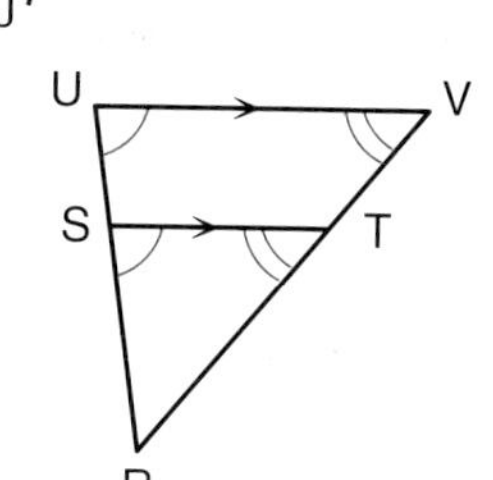

3 Write down the number of the angle in the diagram which corresponds to angle:
a 1 **b** 2 **c** 5 **d** 6

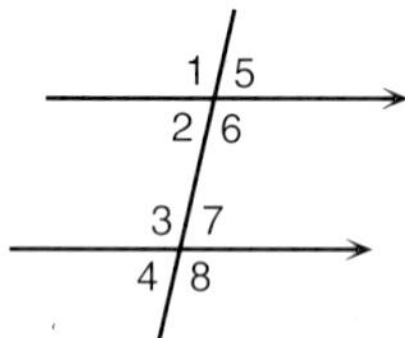

4 Copy the diagrams, and mark the angles which correspond to the given ones.

a 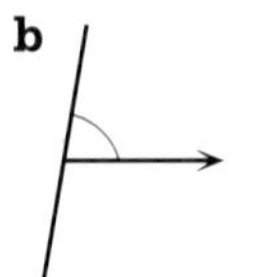 **b**  **c**

d  **e** 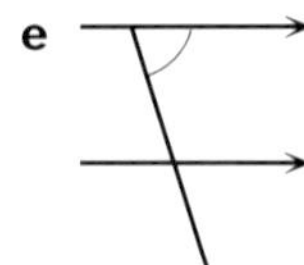 **f** 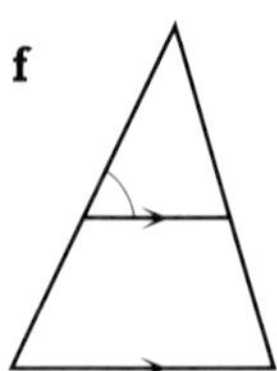

5 a Measure the two corresponding angles marked with dots. What do you find?

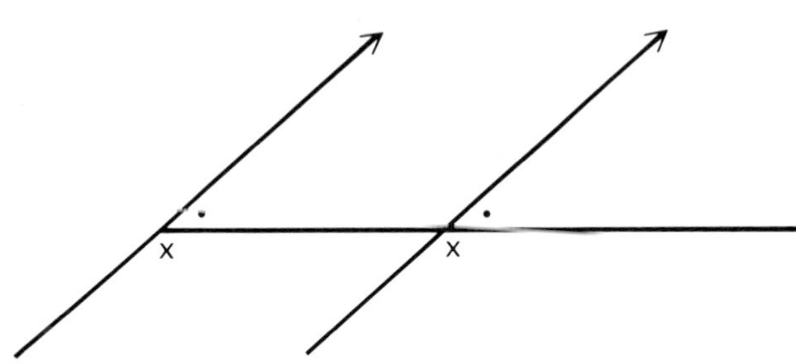

b Repeat part **a** for the corresponding angles marked with crosses.

6 (i) Make a tracing of each diagram.
(ii) Slide the tracing along the crossing line AB. Do the marked angles fit onto each other? Are they equal?

a 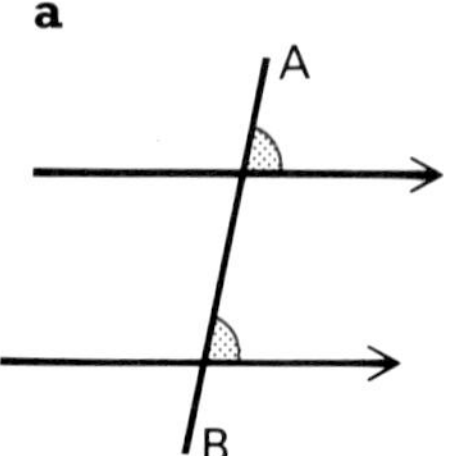

b 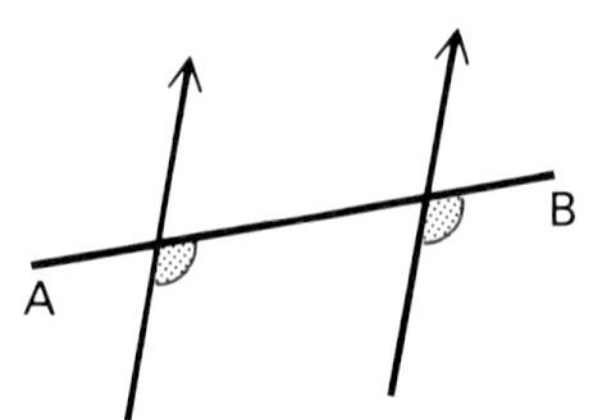

7 The lines are parallel, so the shaded tile can slide along the row, or down the column.

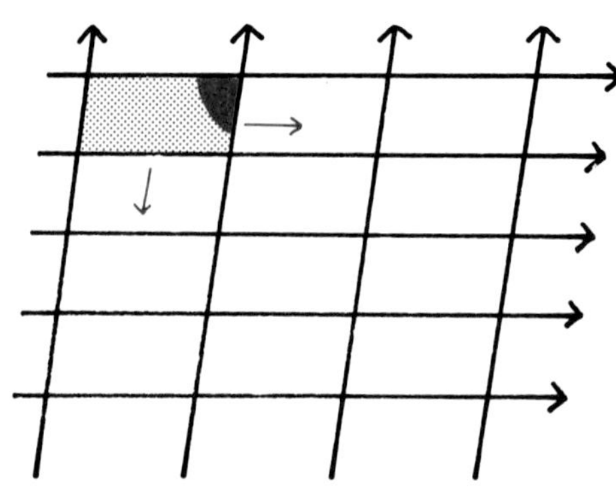

Trace the diagram, and mark in the corresponding angles as the tile moves:
a along **b** down.

8 The diagram in question **7** is a tiling of congruent shapes. What can you say about the sets of corresponding angles in it?

Corresponding angles are equal on parallel lines.

9 Find the sizes of the angles marked $a°$, $b°$, $c°$, etc.

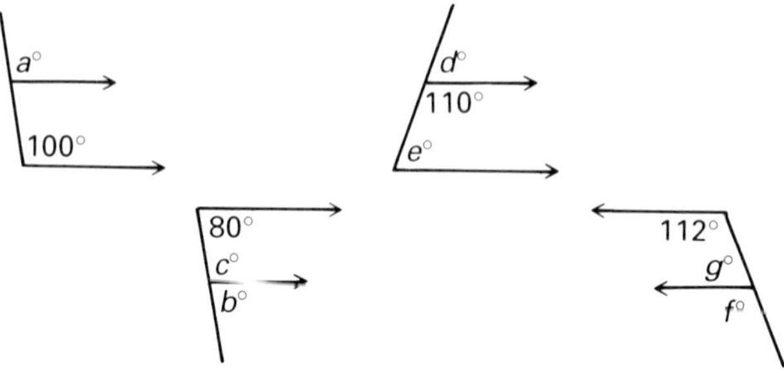

10 Copy this parallelogram tiling, and fill in the sizes of all the angles in it.

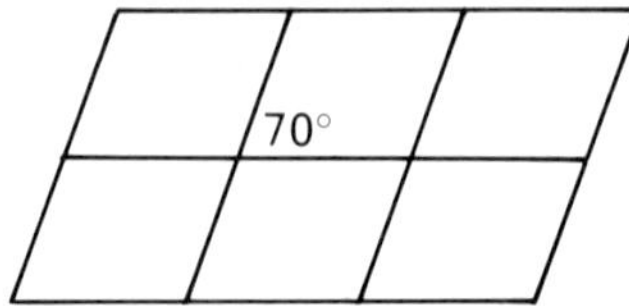

11 Copy these diagrams, and fill in the sizes of all the angles in them.

a 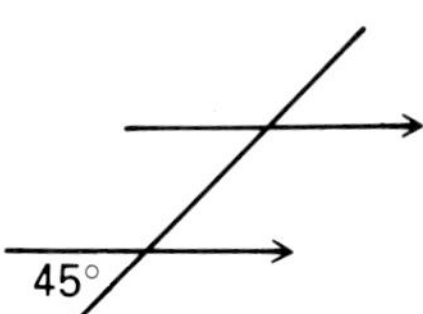

b 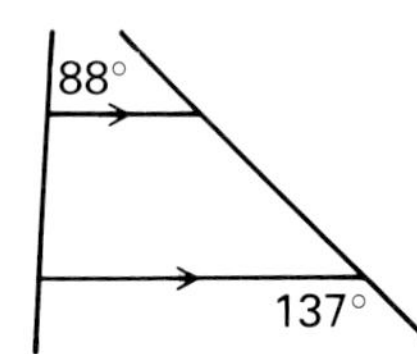

EXERCISE 4B/C

1 Calculate a, b, c, etc.

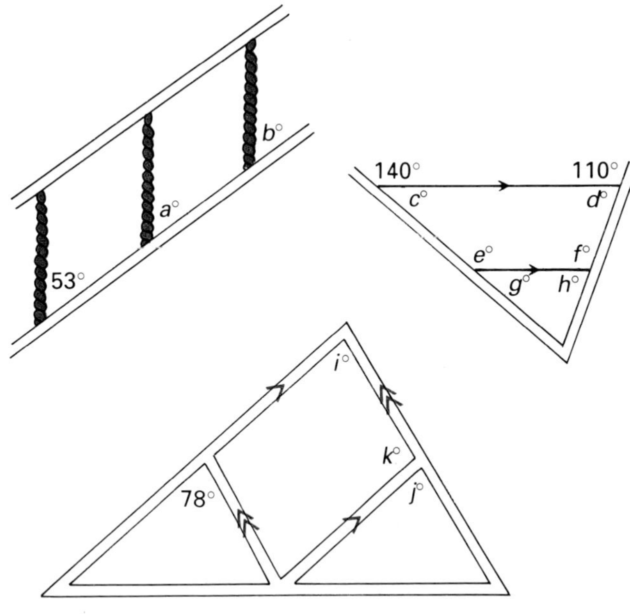

2

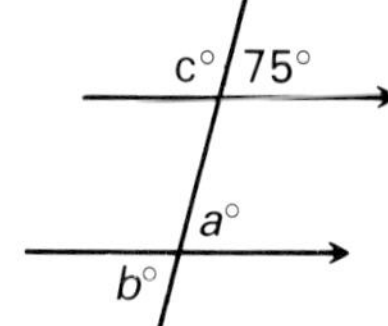

$a = 75$ (corresponding)
$b = 75$ (vertically opposite)
$c = 105$ (supplementary)

Calculate a, b, c in each diagram, giving a reason for your answer to each one.

(i)

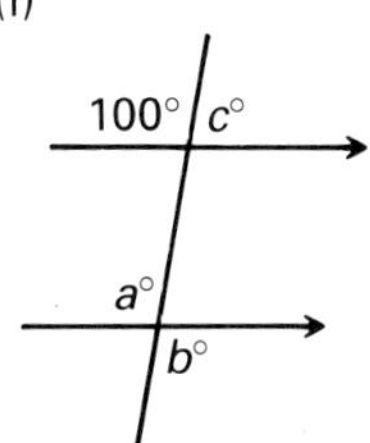

(ii)

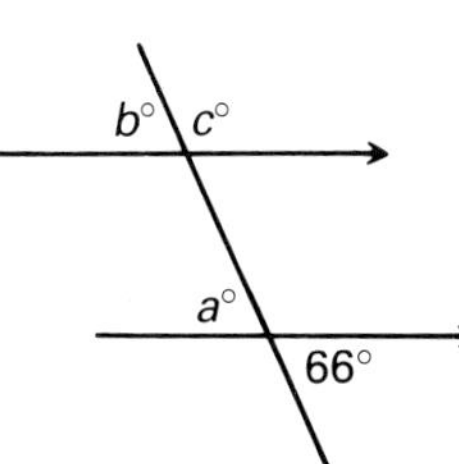

3 Calculate a, b, c, etc., in order, and give a reason for each, as in question **2**.

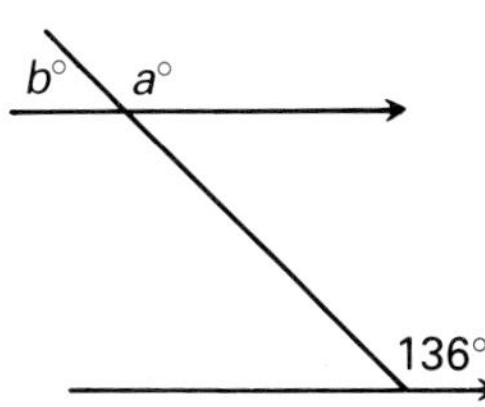

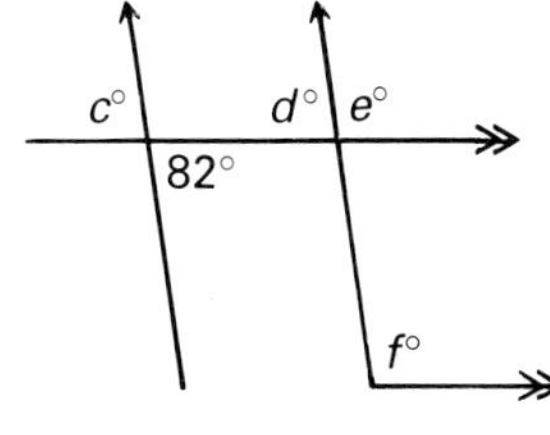

4 a Calculate p, q, r, s, and give a reason for each.

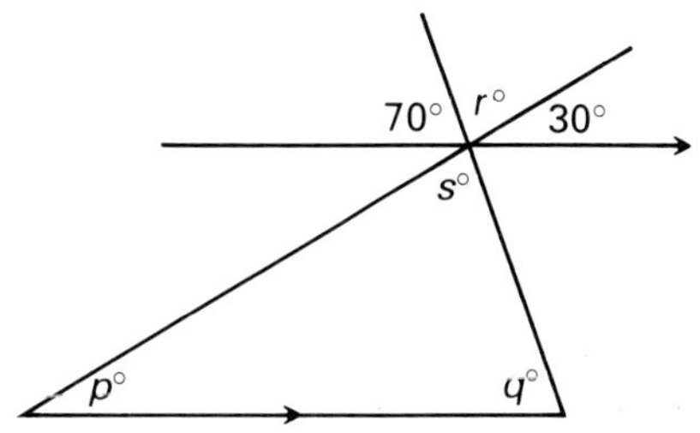

b What is the value of $p° + q° + s°$?

5 Copy each diagram, and calculate u, v, w, etc. You will have to add a line or extend a line in each diagram.

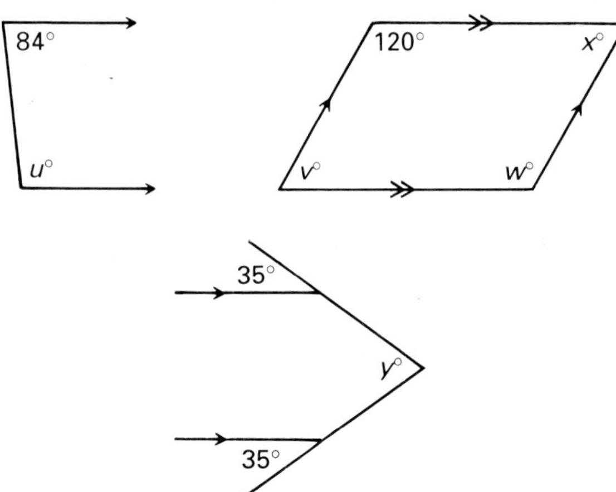

PRACTICAL PROJECT

Draw a 3 by 3 tiling of parallelograms. Mark sets of corresponding angles 1, 1, 1, 1; 2, 2, 2, 2; . . . ; or colour code them.

ALTERNATE ANGLES

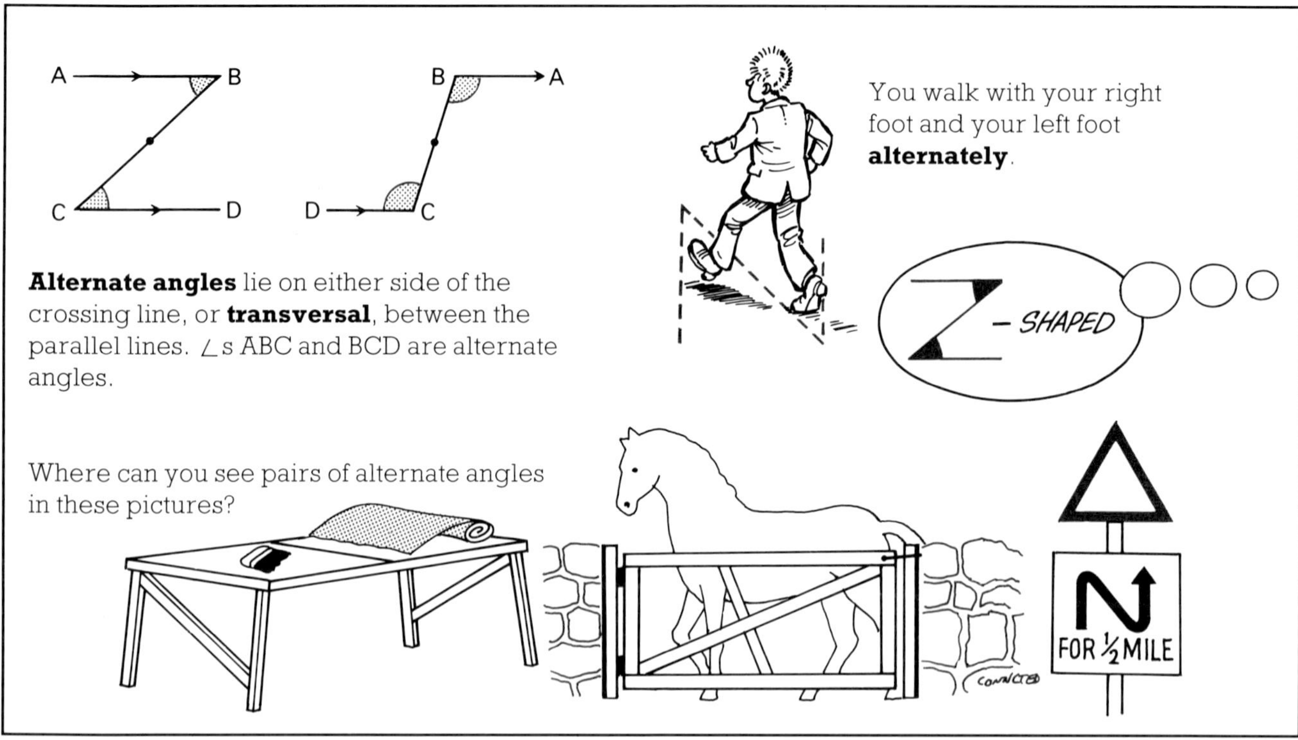

Alternate angles lie on either side of the crossing line, or **transversal**, between the parallel lines. ∠s ABC and BCD are alternate angles.

Where can you see pairs of alternate angles in these pictures?

EXERCISE 5A

1 Copy the diagrams, and mark the angles which are *alternate* to the given ones.

2 a Measure the two alternate angles marked with dots. What do you find?

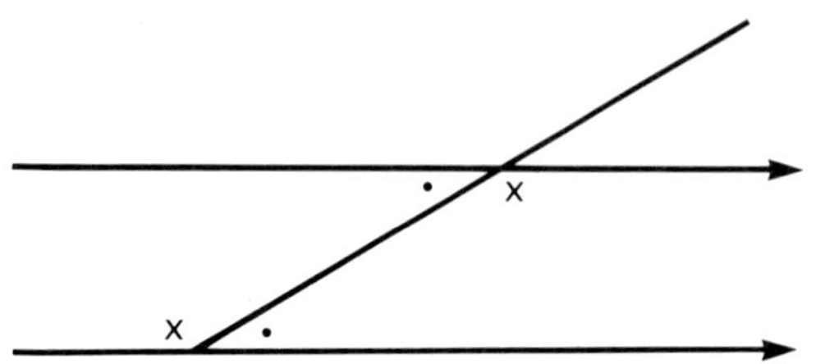

b Repeat part **a** for the alternate angles marked with crosses.

3 Trace each of the two Z-shapes in the box at the top of this page. Give them a half turn about the dot in the centre.
The Z-angles change places, so what can you say about their size?

4 Draw a large Z-shape, with its opposite sides parallel.
Measure each angle with your protractor. What do you find?

5 a Copy the diagram, and fill in the size of the angle which is:
(i) vertically opposite 70°
(ii) corresponding to 70°.

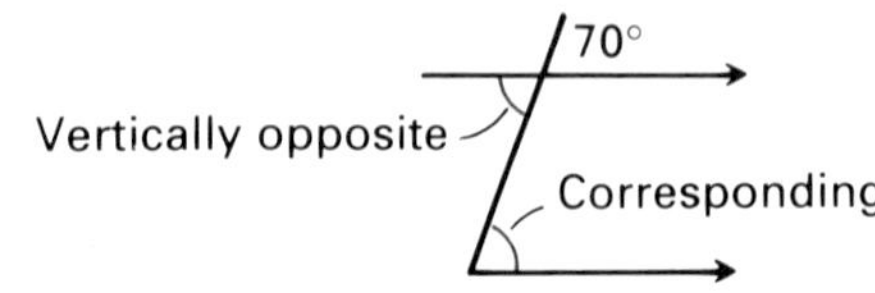

b Do you find that the two alternate angles are equal?

6 Repeat question **5** for these diagrams:

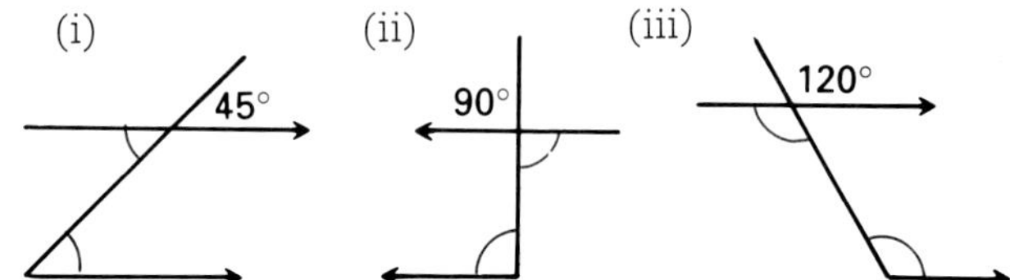

Alternate angles are equal between parallel lines.

7 Look for alternate angles and calculate a, b, c, . . .

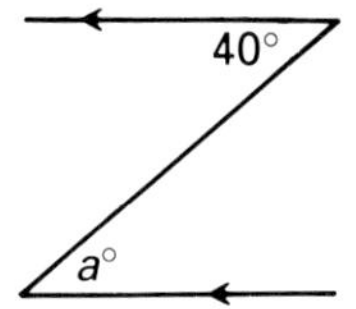

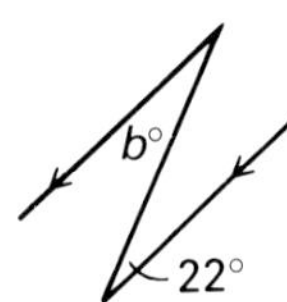

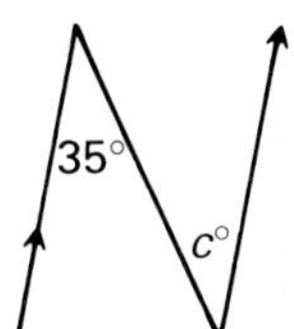

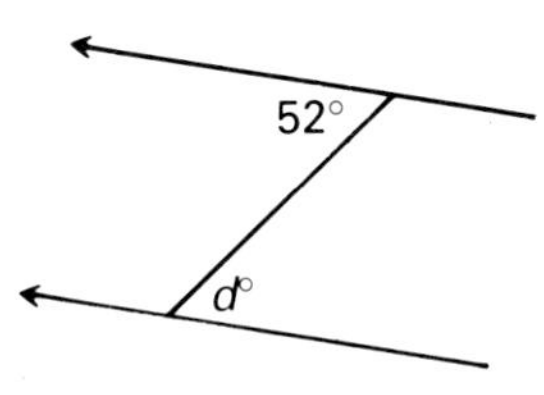

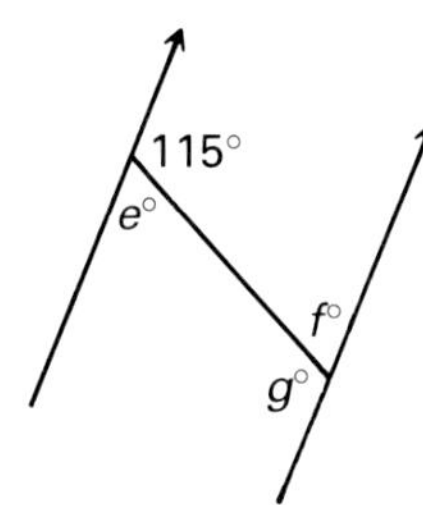

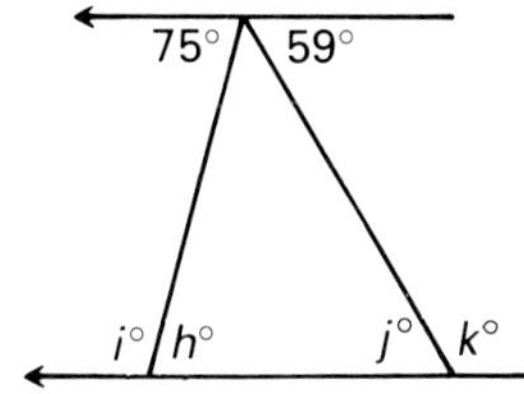

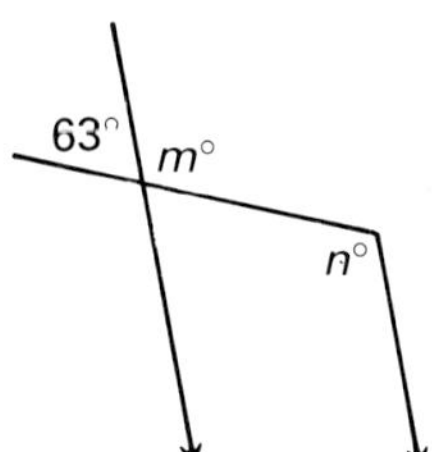

8 Nicola measured $\angle ABC = 23°$.

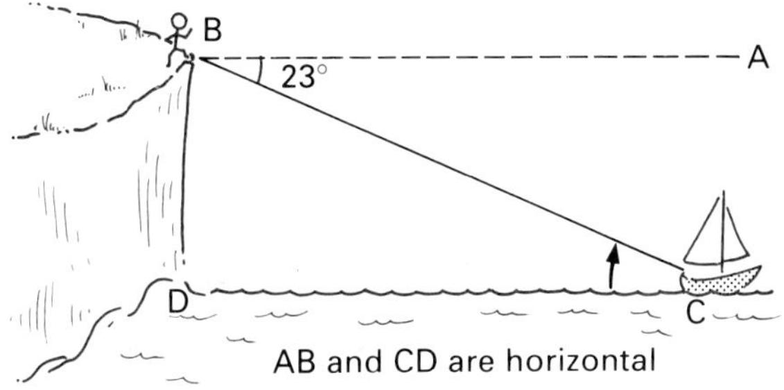

a What size is $\angle BCD$?
b Which property of angles did you use to find the answer?

9 Copy the front of the fire escape at these flats, and mark all the equal alternate angles.

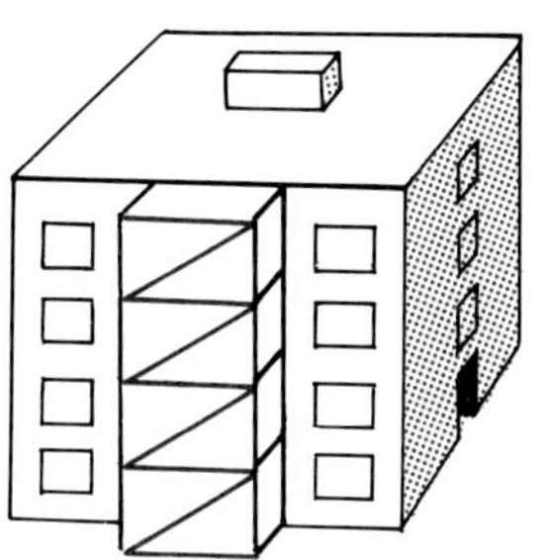

10 Make larger copies of these letters, and mark the pairs of alternate angles.

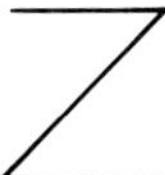

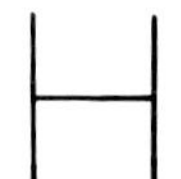

11 Copy this railway crossing diagram, and fill in the sizes of all the angles.

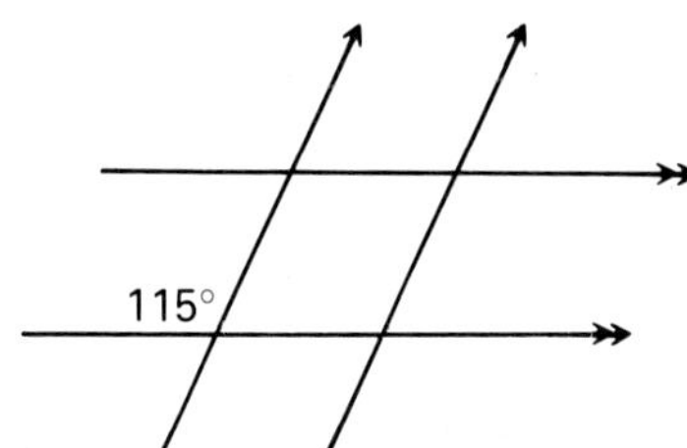

12 Name all the pairs of equal alternate angles in these diagrams.

a

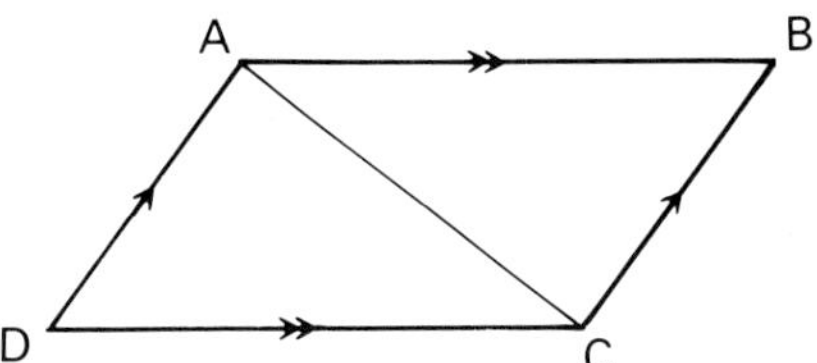

b

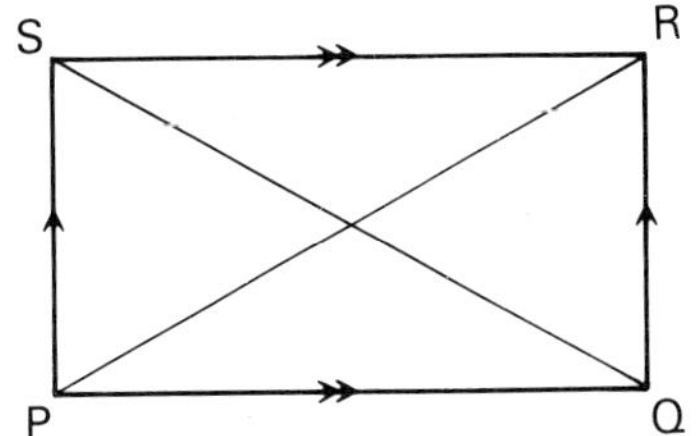

EXERCISE 5B/C

1 Copy these diagrams. Use pairs of equal alternate angles to help you fill in all the angles. (Dots signify equal angles.)

a

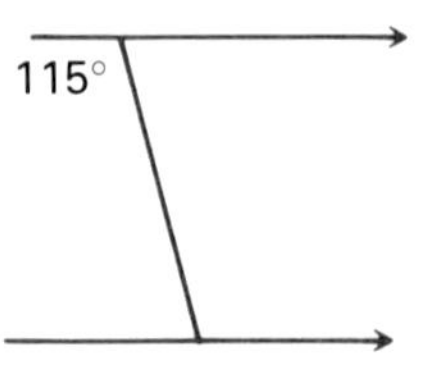

b

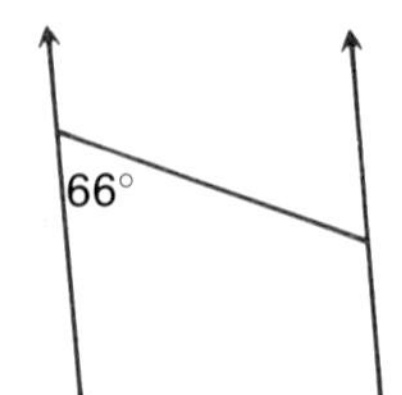

c

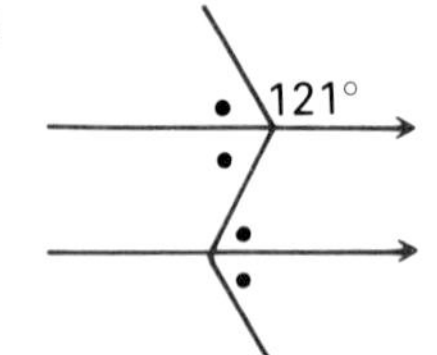

d

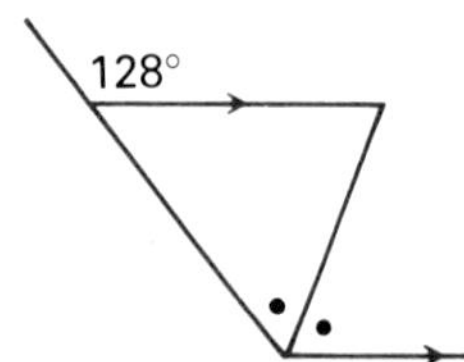

2 Calculate *a*, *b*, *c* and *d* for this symmetrical roof structure.

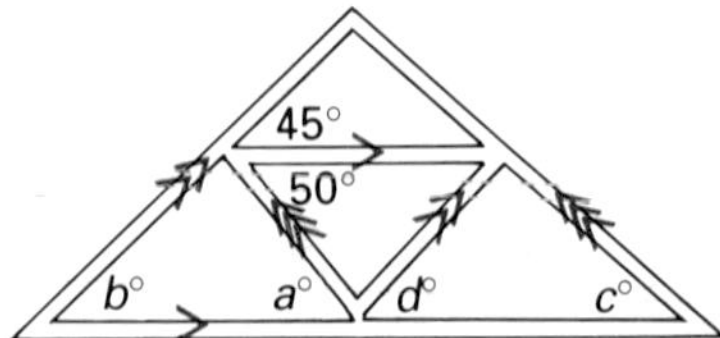

3 Calculate the values of *a*, *b*, *c*, . . . , in order, giving a reason for each one.

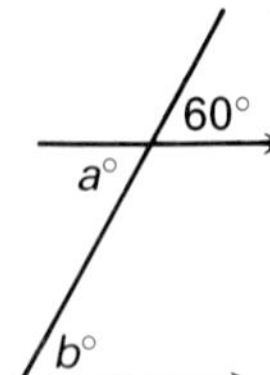

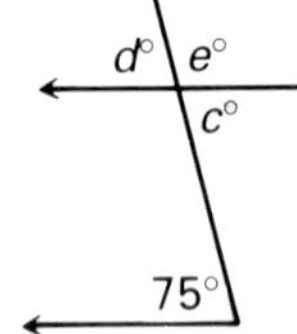

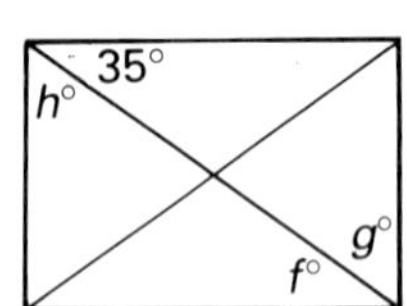

4 Copy the ray diagram. By drawing another line you should be able to calculate *x*°.

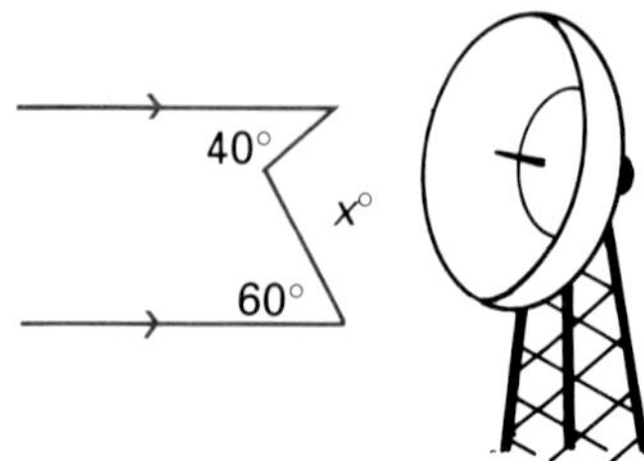

5 Repeat question **4** for given angles of 35° and 45°.

6 Find the sizes of the angles listed, giving a reason for each answer. For example, ∠CBD = 35° (alternate to ∠BDE).

a

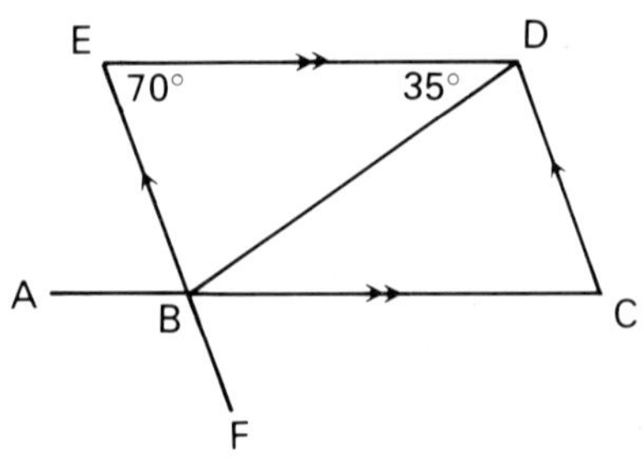

Angles ABE, EBD, BCD

b

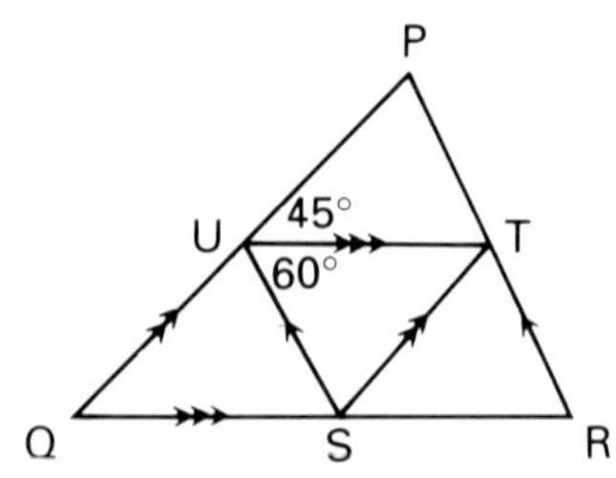

Angles PQR, USQ, PRQ, PTU

PRACTICAL PROJECT

Draw a tiling of kites, and mark the sets of parallel lines.
Now mark sets of alternate angles 1, 1, . . . ; 2, 2, . . . ; or colour code them.

CHALLENGE

How many pairs of alternate angles are there in diagrams like these, with:

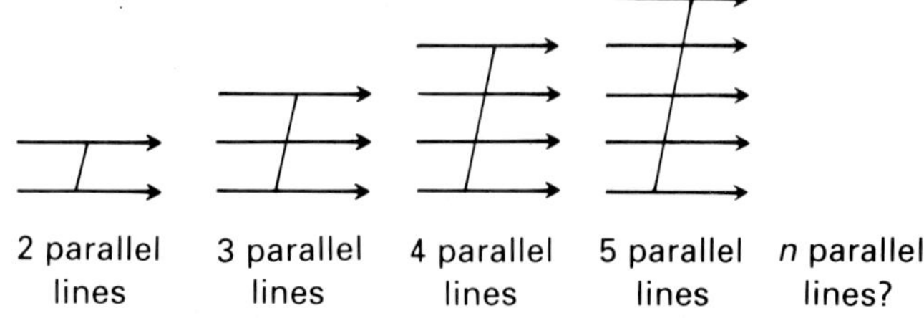

2 parallel lines | 3 parallel lines | 4 parallel lines | 5 parallel lines | *n* parallel lines?

PAIRS OF ANGLES SUMMARY

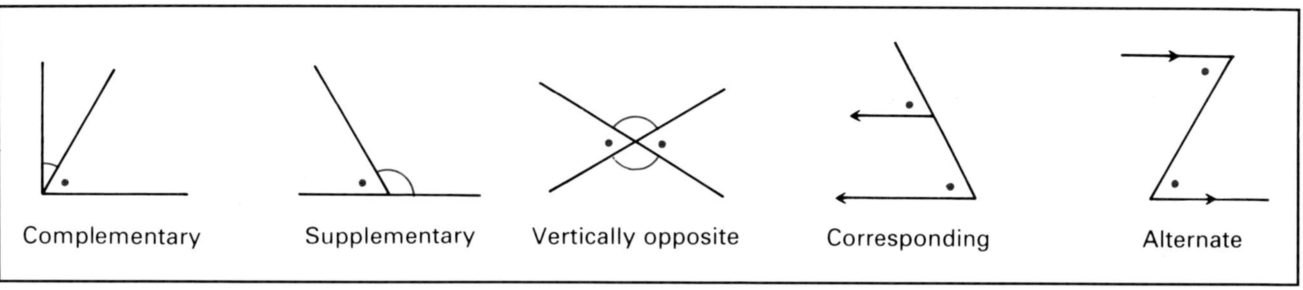

EXERCISE 6

1 In the diagrams in the box above, each angle marked with a dot is 60°. Copy the diagrams, and fill in the sizes of all the angles.

2 Write down the numbers of as many *pairs* of supplementary, vertically opposite, corresponding and alternate angles as you can find in the diagrams below.

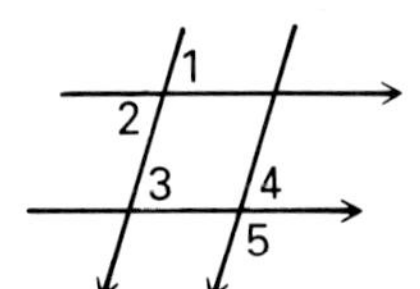

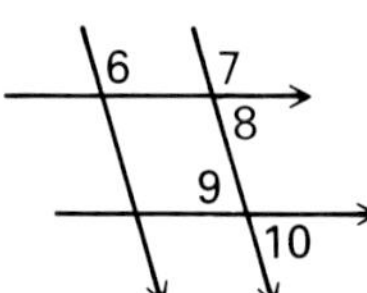

3

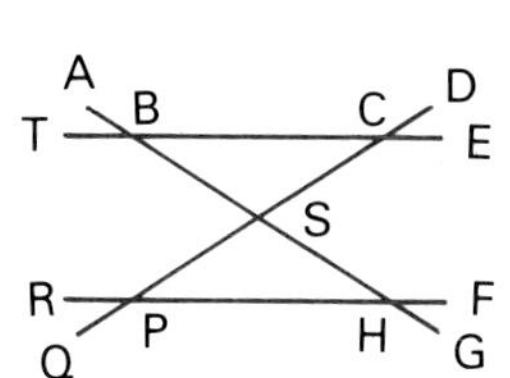

Name an angle which is:

a corresponding to ∠ABT
b alternate to ∠BCP
c vertically opposite ∠FHG
d supplementary to ∠BCS.

4 Calculate a, b, c, . . .

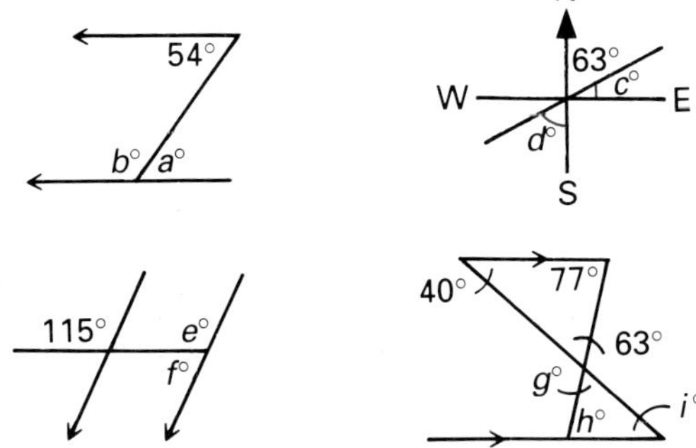

5 Copy these 'side view' diagrams, and fill in the sizes of all the angles.

a

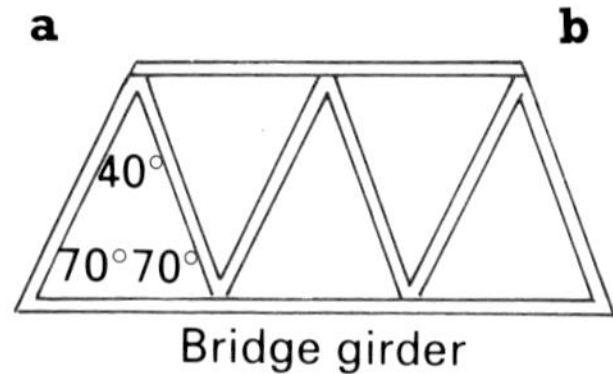

Bridge girder

b

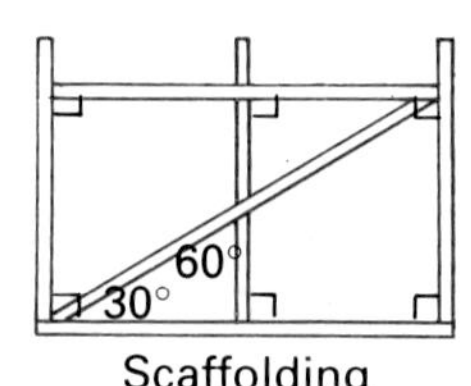

Scaffolding

6 Copy these diagrams, and fill in the sizes of all the angles.

a **b**

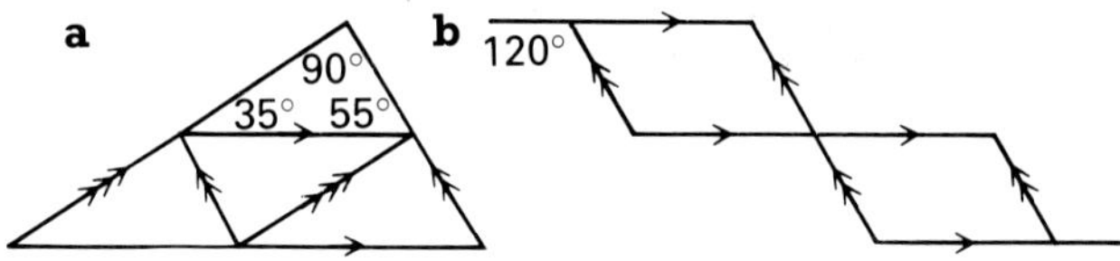

7 Prove that the sum of the angles of the triangle is 180°. Give a reason for each statement you make about angles.

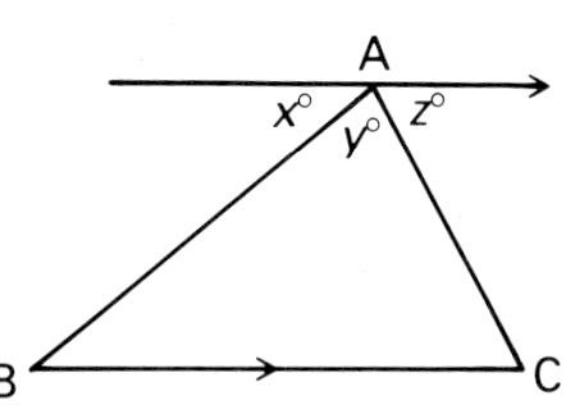

INVESTIGATIONS

1 *The mirrors or prisms in a periscope, which is used for seeing over crowds of people, or in a submarine, make use of parallel line angles. Investigate how the system works.*

2 **a** *Make a trellis, like those in question 2 of Exercise 6, with strips of card and paper fasteners, or wood and nails.*

b *Change the shape of the trellis, and watch all the supplementary, vertically opposite, corresponding and alternate angles in action.*

CHECK-UP ON ALL ABOUT ANGLES

1 Write down the size of:
a the supplement of ∠RST
b the complement of ∠RST.

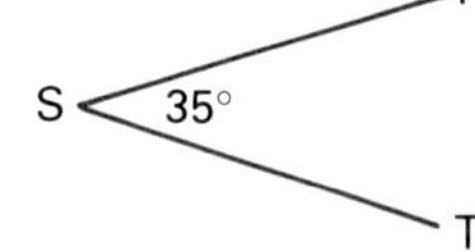

2 a How many pairs of supplementary angles can you see on this envelope?

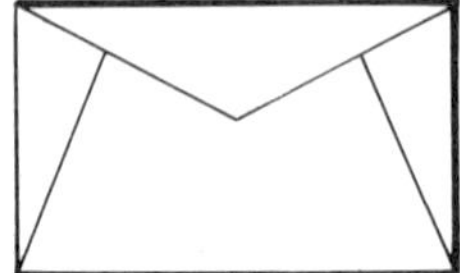

b How many pairs of complementary angles are there?

3 a Name two pairs of vertically opposite angles in this butterfly brooch.

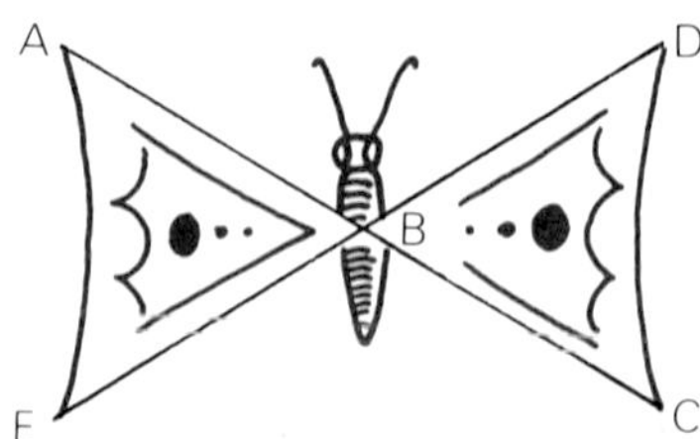

b ∠ABD = 115°. Write down the sizes of the other angles with vertex B.

4 Calculate x in each diagram.

a

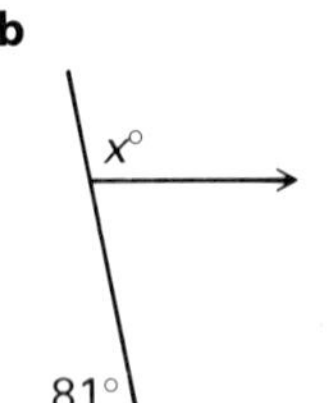

b

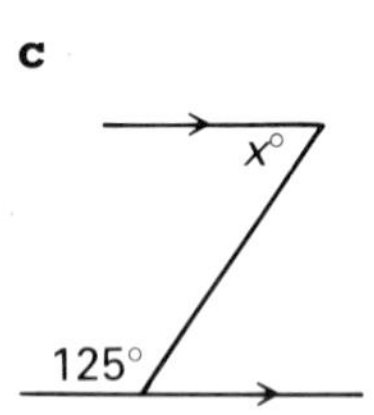

c

125° x°

d

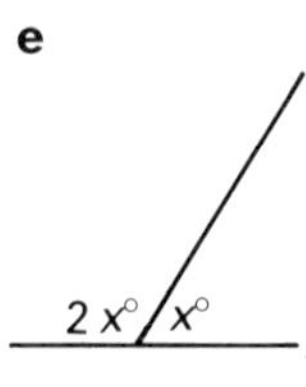

e

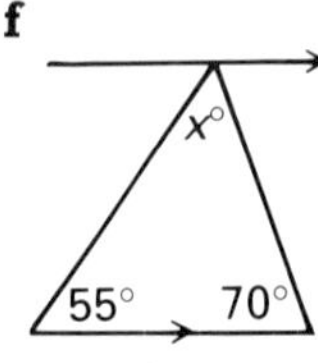

f

x° 55° 70°

5 a Calculate a, b and c in this rectangle.

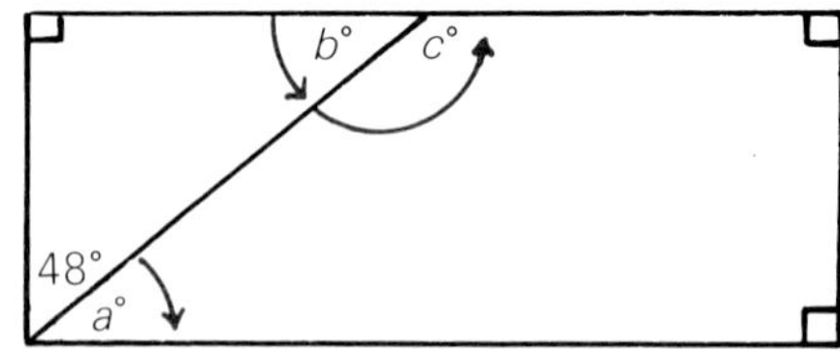

b Which pair of angles is complementary, and which pair is supplementary?

6 Copy these diagrams, and fill in the sizes of all the angles.

a 40°

b 110° 45°

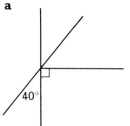

7 The parts of this handrail are parallel to each other.
a Name four pairs of corresponding angles.
b ∠PSK = 101°. Copy the diagram and fill in the sizes of all the angles.

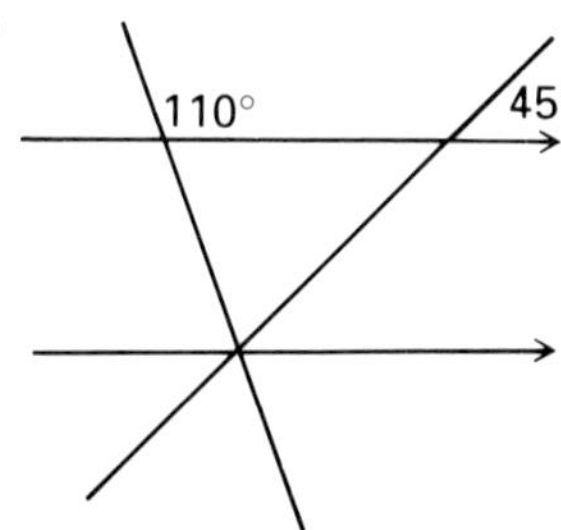

8 a How many pairs of alternate angles are there in this sketch of a gate?

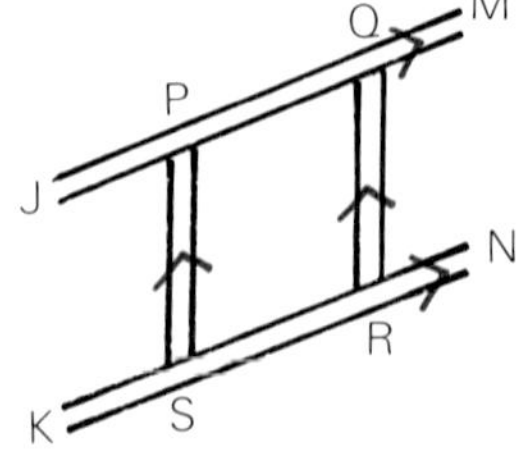

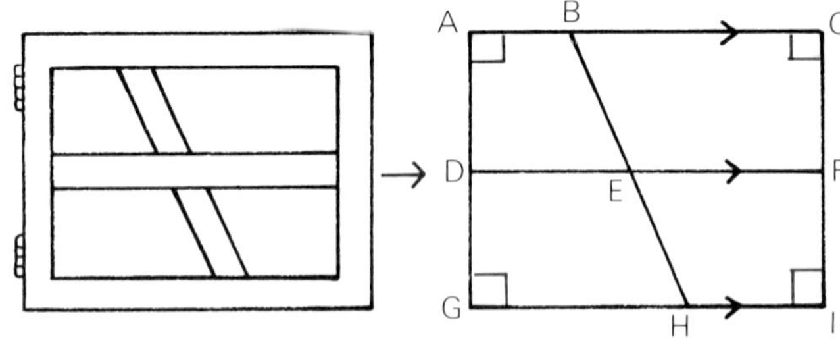

b Calculate all the angles, given ∠ABE = 125°.

9 A corner is cut off this rectangular plate as shown. Copy the diagram and, by drawing a suitable line, calculate the size of ∠BCD.

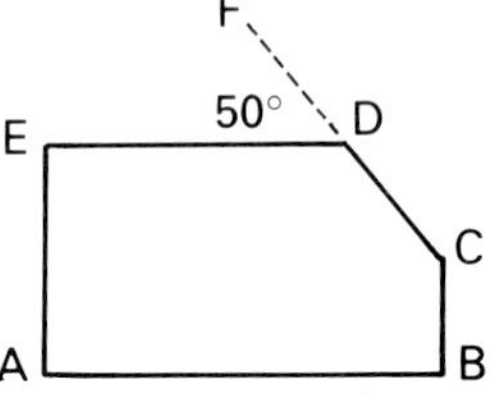

LETTERS AND NUMBERS

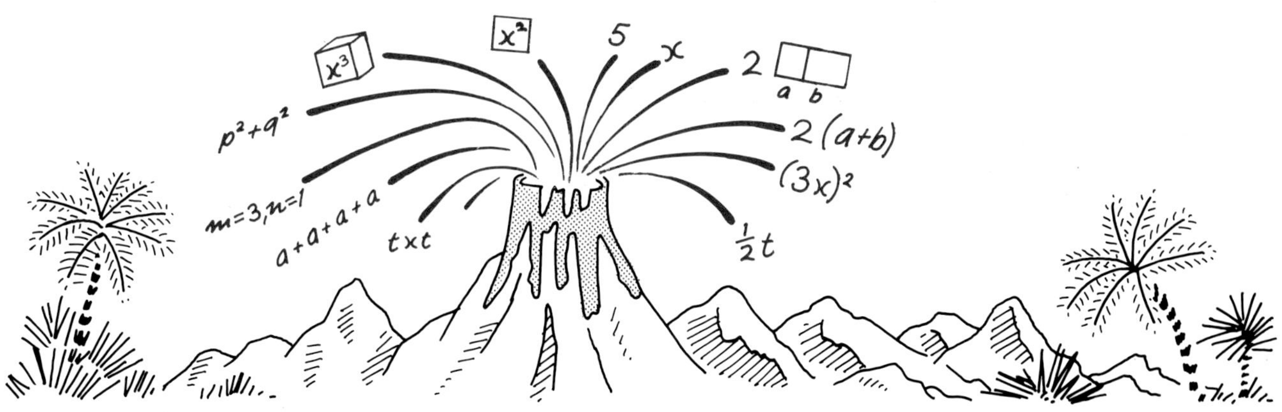

LOOKING BACK

1 $x = 10$. Calculate the value of:
a $x+4$ **b** $2x$ **c** $x-10$ **d** $3x+1$ **e** $12-x$

2 $a = 5$ and $b = 3$. Calculate:
a $a+b$ **b** $a-b$ **c** ab **d** $2ab$ **e** $2a+3b$

3 Calculate: **a** the perimeter of the rectangle
b the area of the rectangle.

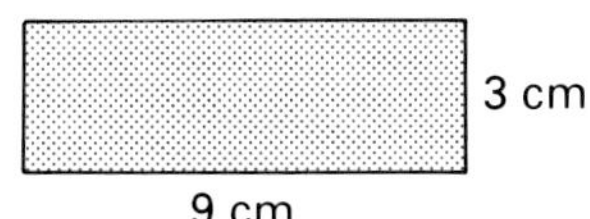

4 Write down the length of each straw. The lengths are in centimetres.

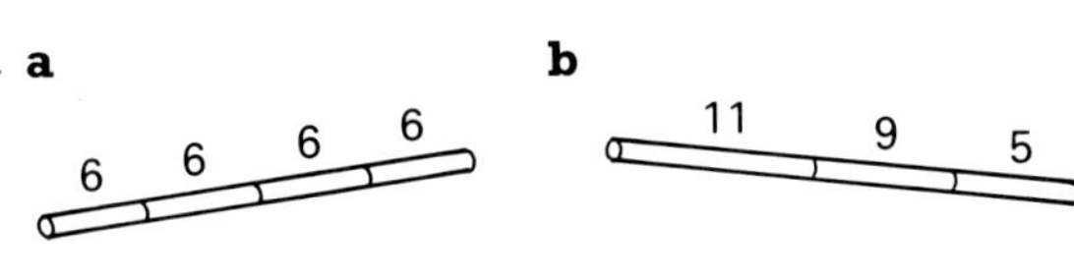

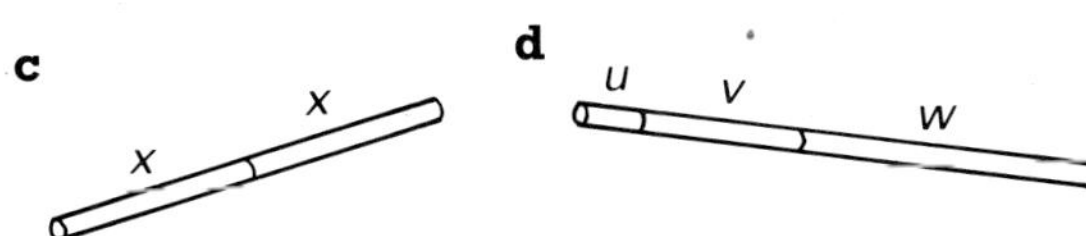

5 Copy and complete these tables.

a

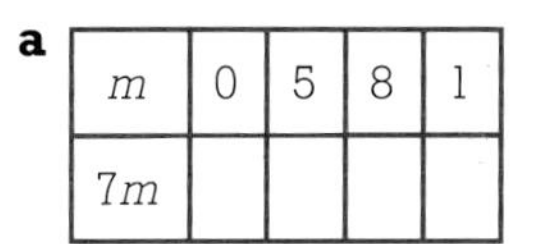

m	0	5	8	1
$7m$				

b

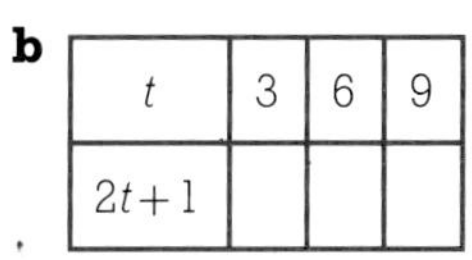

t	3	6	9
$2t+1$			

6 Write down the OUT numbers or letters.

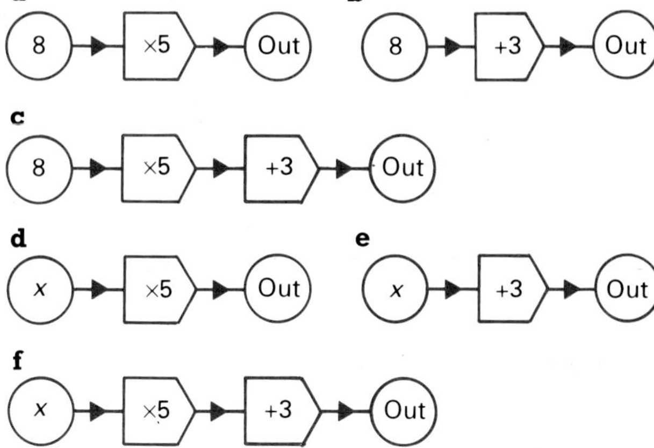

7 Remember the rule? 'Brackets, *then* multiplication and division, *then* addition and subtraction.' Calculate:
a $(5+4)\times 3$ **b** $5+4\times 3$ **c** $6\times 3-2$
d $6\times(3-2)$ **e** $8\div 4-2$

8 The side of a square is 6 cm long. Calculate the perimeter and area of the square.

9 Calculate:
a $\frac{1}{2}\times 10$ **b** $\frac{1}{3}\times 12$ **c** $\frac{1}{4}\times 20$ **d** $\frac{2}{3}\times 9$ **e** $\frac{3}{4}\times 12$

10 Write each of these in a shorter form.
a $x+x$ **b** $2x+x$
c $3y-y$ **d** $5y-4y$
e $5t-3t+2$ **f** $3+y+y$
g $4x+1+2x$ **h** $t+4-t$
i $x+y+x-y$ **j** $3m+1-m+1$
k $n+2-n-2$ **l** $9+4p-7+p$

ADDING AND MULTIPLYING

Examples

a longer form $2+2+2+2+2$

shorter form 5×2 ('5 lots of 2')

b longer form $x+x+x+x$

shorter form $4x$ ('4 lots of x')

EXERCISE 1A

1 Write in shorter form; for example, $2+2+2+2 = 4 \times 2$.

a $2+2+2$ **b** $3+3+3+3$
c $7+7$ **d** $8+8+8+8+8$
e $4+4$ **f** $9+9+9$
g $11+11+11+11$ **h** $1+1+1+1+1+1$

2 Write in shorter form; for example, $a+a+a+a = 4a$.

a $d+d$ **b** $n+n+n$
c $k+k+k+k$ **d** $x+x+x+x+x$
e $c+c+c$ **f** $d+d+d+d$
g $v+v$ **h** $t+t+t+t+t+t$

3 Write in shorter form:

a m^2+m^2 **b** $x^2+x^2+x^2$ **c** $a^2+a^2+a^2+a^2$
d $t^2+t^2+t^2+t^2+t^2$

4 Write these in shorter form, then calculate their values if $u = 15$.

a $u+u$ **b** $u+u+u+u$ **c** $u+u+u+u+u+u$

Examples

a longer form $2 \times 2 \times 2$

shorter form 2^3

b longer form (i) $m \times m$ (ii) $t \times t \times t \times t$

shorter form m^2 t^4

5 Write in shorter form; for example $2 \times 2 \times 2 \times 2 \times 2 = 2^5$.

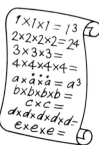

a 6×6 **b** $3 \times 3 \times 3 \times 3$
c $4 \times 4 \times 4$ **d** 7×7
e $m \times m$ **f** $y \times y$
g $x \times x$ **h** $a \times a$
i $n \times n \times n$ **j** $t \times t \times t$
k $p \times p$ **l** $v \times v \times v \times v \times v$
m $t \times t$ **n** $d \times d \times d$
o $c \times c \times c \times c$ **p** $a \times a \times a \times a \times a \times a$

6 Copy and complete these lists.

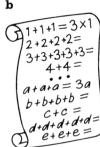

a

1+1 = 2×1
2+2 = 2×2
3+3 =
4+4 =
...
x+x = 2x
y+y =
a+a =
b+b =
v+v =

b

c

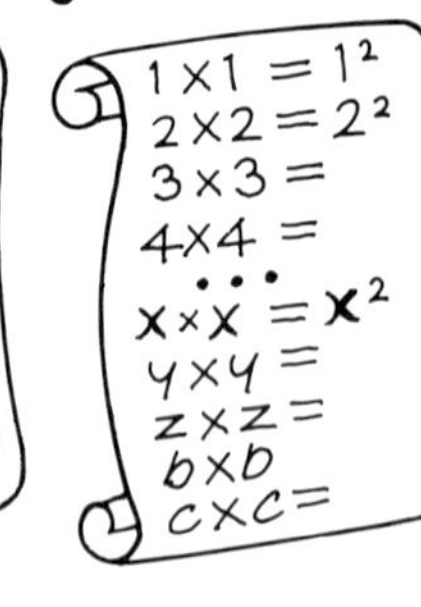

d

a+a+a=3a
x+x =
y+y+y =
x×x = x²
y×y×y =
d×d×d×d =
n×n =
m×m×m =

e

1×1×1 = 1³
2×2×2×2=2⁴
3×3×3=
4×4×4×4=
a×a×a = a³
b×b×b×b =
c×c =
d×d×d×d×d=
e×e×e =

f

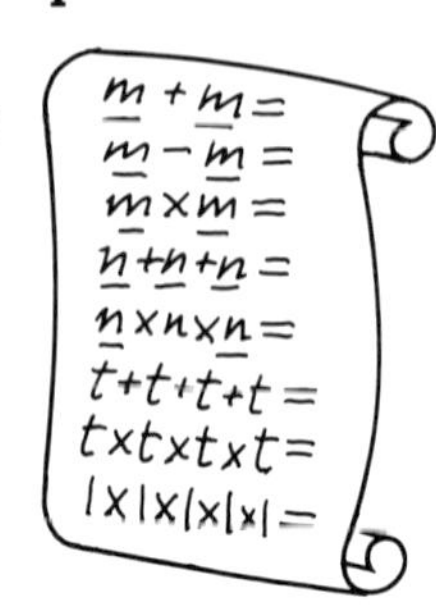

7 Look at these (+/×) targets:

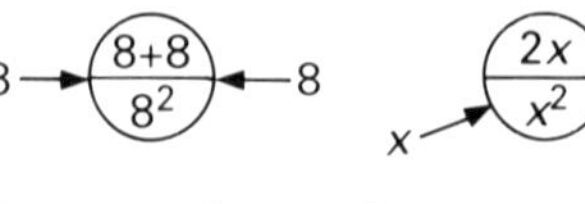

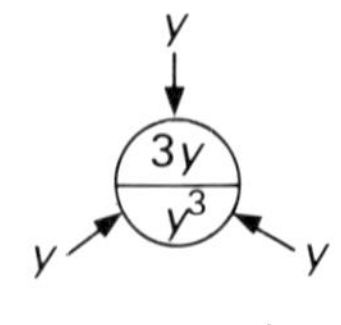

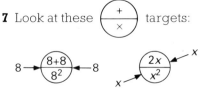

Copy and complete:

a **b**

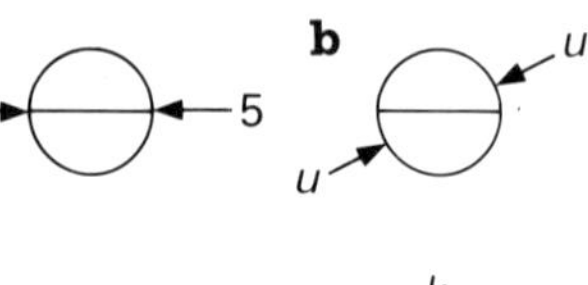

c **d**

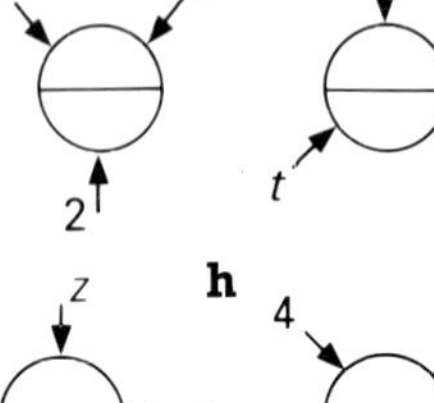

e **f**

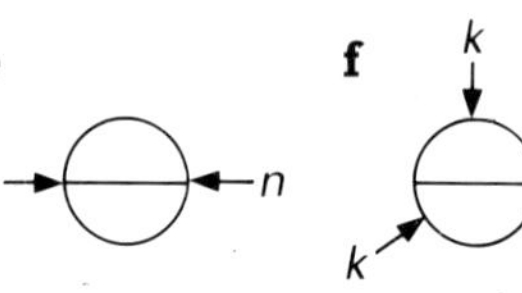

g

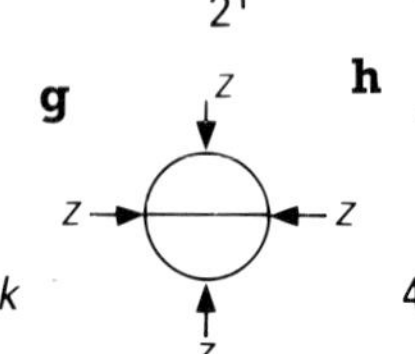

h

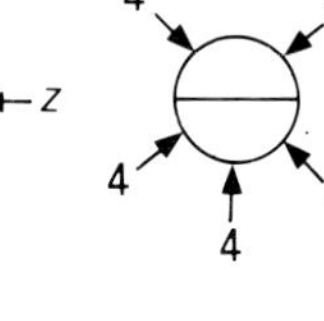

8 Make up some 'add and multiply' targets of your own.

EXERCISE 1B

Examples

a $3a+2a=5a$ **b** $6x-x=5x$
c $a+b+a-b=2a$

1 Simplify:

a $x+x$ **b** $2x+x$ **c** $2x-x$ **d** $3x+2x$
e $3a+a$ **f** $b+4b$ **g** $5c+5c$ **h** $4a-a$
i $6c-c$ **j** $2n-2n$ **k** $5a-4a$ **l** $6y-3y$
m x^2+x^2 **n** $3y^2+y^2$ **o** $7z^2-z^2$ **p** $5t^2-4t^2$

2 Simplify:

a $3x+2x+4$ **b** $2y-y+1$ **c** $4x+2+x$
d $5+2m+m$ **e** $a+b+a$ **f** $m+n-m$
g $x+y+2x$ **h** $2x+y-x$ **i** $2a+1+a+1$
j $3c+5+c-2$ **k** $p+q+p-q$ **l** $2s+3t-s-t$

3 Simplify these, *then* calculate their values when $u=9$ and $v=3$.

a $u+2u+v+2v$ **b** $5u-4u+2v-v$
c $6u+2v-6u+3v$

Examples

a $2a\times a=2\times a\times a=2a^2$

b $5c\times 2c=5\times c\times 2\times c=10c^2$

4 Simplify:

a $2c\times c$ **b** $3a\times a$ **c** $2\times 3y$
d $4\times 2y$ **e** $5z\times 3$ **f** $4s\times s$
g $2t\times 2t$ **h** $3m\times 3m$ **i** $5c\times 2c$
j $4r\times r$ **k** $5\times y$ **l** $e\times 4e$
m $y\times 2y\times 3y$ **n** $5x\times 2x\times x$ **o** $2m\times 2m\times 2m$
p $a\times a\times 2a\times 3a$

5 Simplify these, and list them in order: $a, 2a, \ldots a^4$.
$a+a$, $a\times a$, $a+a+a$, $a\times a\times a$, $2\times a\times a$, $1\times a$, $a+3a$, $a^3\times 2$, $2a+3a$, $a\times a\times a\times a$, $a\times 3\times a$, $2a\times 3$, $2a\times 3a$, $2a\times 2a$, $a+2a+4a$, $a\times 5a$

EXERCISE 1C

Examples

a $2\times y+3$
$=2y+3$

b $2x+3\times x$
$=2x+3x$
$=5x$

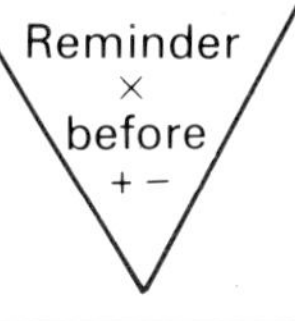

1 Simplify:

a $2\times y+3$ **b** $4\times k-1$ **c** $n\times n+7$
d $m\times m-2$ **e** $8+2\times d$ **f** $5-3\times c$
g $1+t\times t$ **h** $3-p\times p$ **i** $3+a\times 2$
j $b\times 5-1$ **k** $4-2\times n$ **l** $3\times z+3$
m $2x+x+3$ **n** $3y-y+1$ **o** $p+5+p$
p $2s+1-s$ **q** $2y\times y+4$ **r** $3z\times z-2$
s $3u+2\times u$ **t** $5v-4\times v$ **u** $9n-3\times 3n$
v $3a+5+2a$ **w** $3a\times 5+2a$ **x** $3a+5\times 2a$
y $3a\times 5\times 2a$

Examples

a $\frac{x}{2}=\frac{1}{2}x=\frac{1}{2}\times x$ **b** $\frac{2}{3}\times y=\frac{2}{3}y=\frac{2y}{3}$

2 Write each of these in two more ways, as in the examples above.

a $\frac{a}{2}$ **b** $\frac{1}{3}y$ **c** $\frac{3}{4}\times t$ **d** $\frac{1}{6}x$ **e** $\frac{4m}{5}$ **f** $\frac{d}{12}$ **g** $\frac{5}{7}\times b$

3 a Starting at x, do both routes lead to the same expression?
b Start with $x=5$, and list the values, stage by stage.

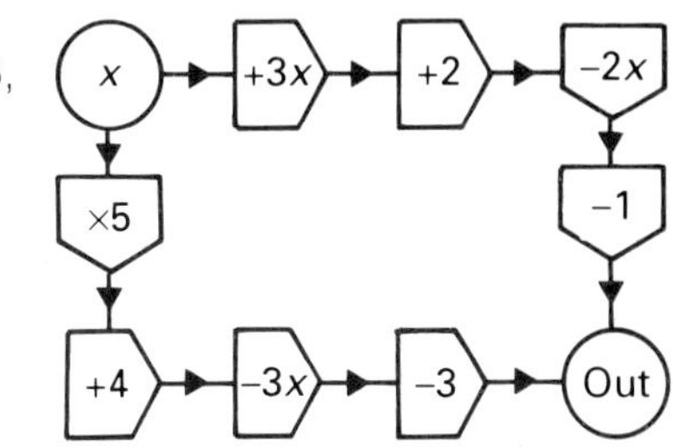

INVESTIGATION

In how many different ways can you make 2x, 3x, 4x, 5x and 6x by adding x terms?
For example:
$2x=x+x$ . . . 1 way
$3x=2x+x$, or $x+x+x$ . . . 2 ways, and so on.

CHALLENGE

Arrange these in groups of equal expressions.

USING ADDITION AND MULTIPLICATION

EXERCISE 2A

Examples (All lengths are in cm)

a

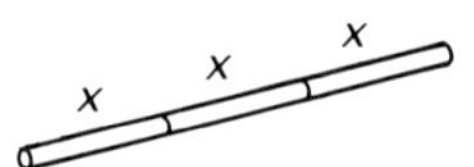

Total length $= 3x$ cm

b

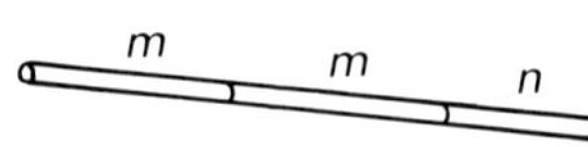

Total length $= 2m+n$ cm

Examples (All lengths are in cm)

a

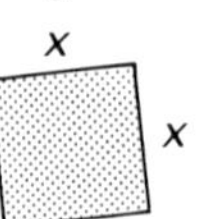

Area of square $= x \times x = x^2$ cm^2

b

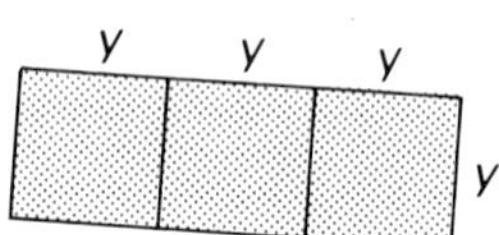

Area $= y^2 + y^2 + y^2 = 3y^2$ cm^2

1 Write down the total length of each cane, in cm.

a

b

c

d

e

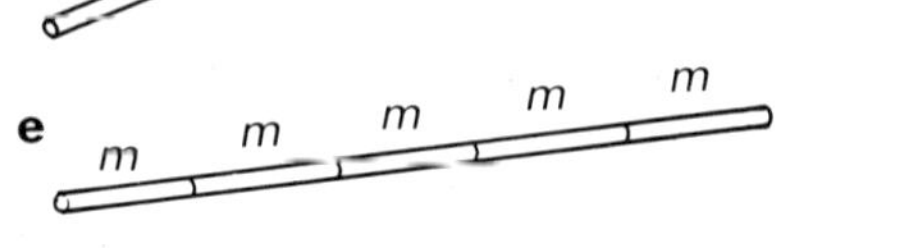

f

g

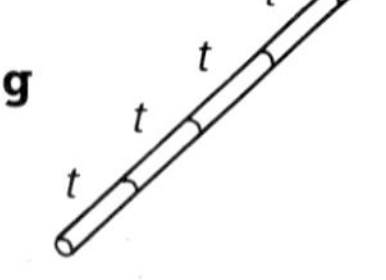

h

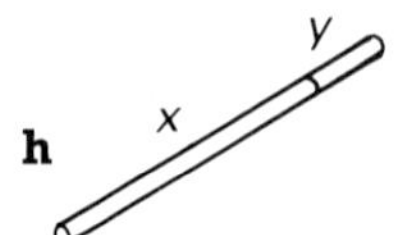

i

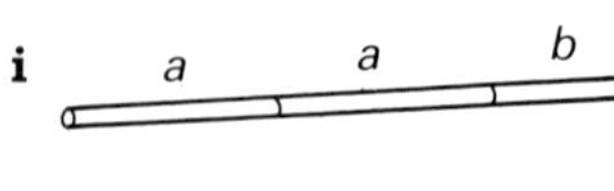

j

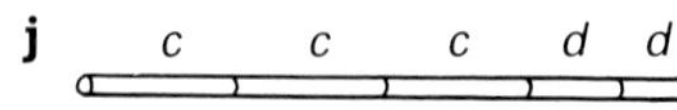

k

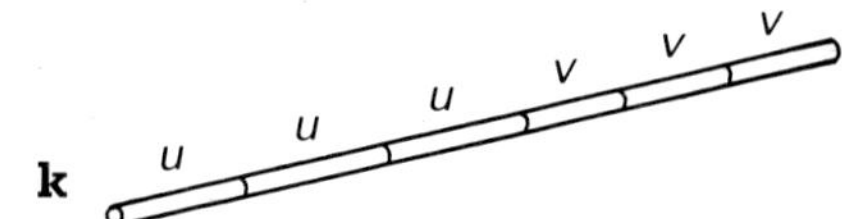

2 Write down expressions for the areas of these shapes which are made of squares. Lengths are in centimetres.

a

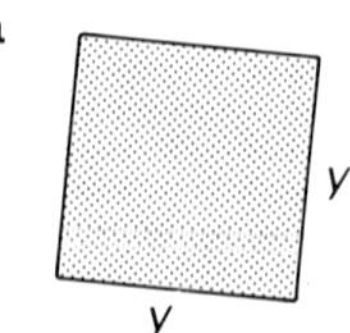

b

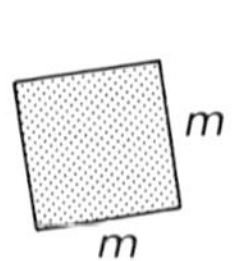

c

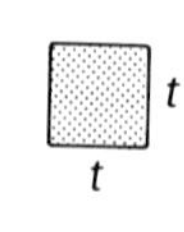

d

e

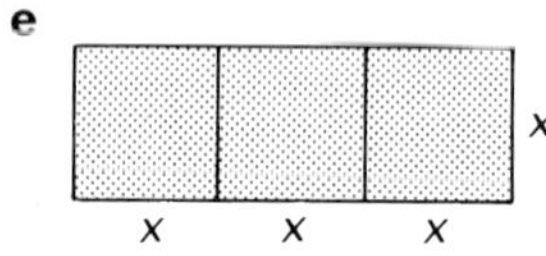

3 $x = 3$ and $y = 1$. Find the values of these. For example, $5x = 5 \times x = 5 \times 3 = 15$.

a $2x$ **b** $3y$ **c** $x+y$ **d** $x-y$
e $x \times y$ **f** $3x+1$ **g** $2y-1$ **h** $4x+y$
i $x+2y$ **j** $5x+5y$

4 $a = 5$, $b = 4$ and $c = 2$. Calculate these. For example, $a^2 = a \times a = 5 \times 5 = 25$.

a ab **b** bc **c** ac **d** abc
e $a+b+c$ **f** $a+b-c$ **g** b^2 **h** c^2
i a^2+c^2 **j** b^2-c^2

Example

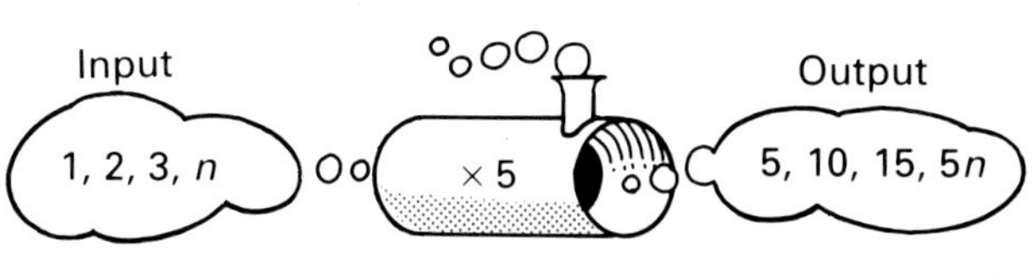
Input
1, 2, 3, n
× 5
Output
5, 10, 15, 5n

a **b**

c **d**

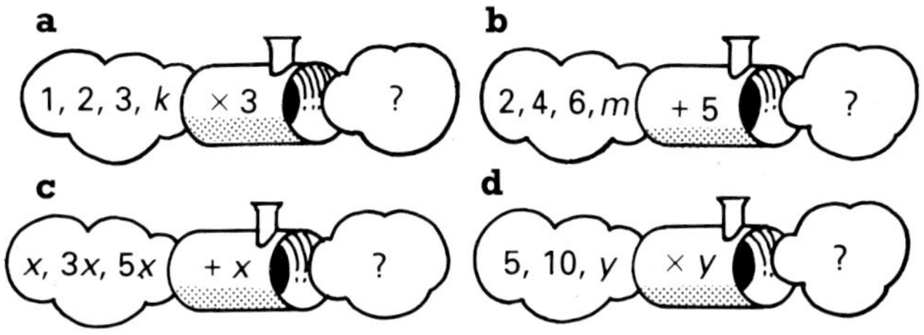
1, 2, 3, k
× 3
?
2, 4, 6, m
+ 5
?
x, 3x, 5x
+ x
?
5, 10, y
× y
?

5 Find the output from each machine opposite.

6 Make a machine of your own, and show some inputs and outputs.

EXERCISE 2B/C

Examples

a If $a = 9$ and $b = 6$, then:

(i) $3(a+b)$
$= 3(9+6)$
$= 3 \times 15$
$= 45$

(ii) $(a-b)^2$
$= (9-6)^2$
$= 3^2$
$= 9$

b

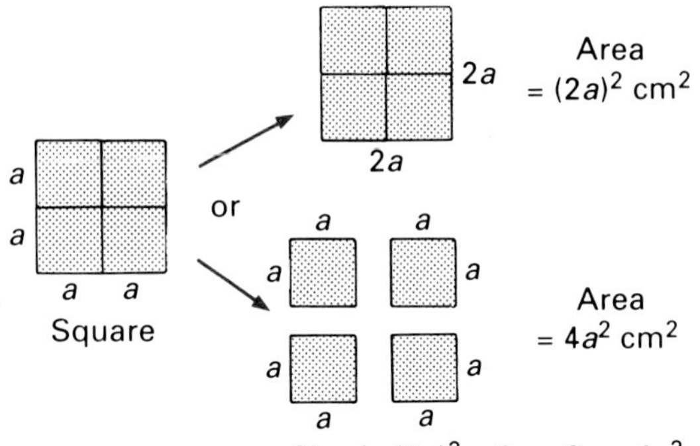
2a
Area
= (2a)² cm²
2a
a
a
or
a a
a a
Square
a a
a a
Area
= 4a² cm²
a a
a a

Check: $(2a)^2 = 2a \times 2a = 4a^2$

1 $x = 7$ and $y = 3$. Find the values of:
a $2(x+y)$ **b** $3(x-y)$ **c** $x(x+1)$ **d** $y(y-1)$
e $y(x+y)$ **f** $(x+y)^2$ **g** $(x-y)^2$ **h** $(x+5)^2$
i $(y-3)^2$ **j** $(x-6)^2$

2 Find two different expressions for the area of each large square below, and check that they are equal. Lengths are in centimetres.

a **b** **c**

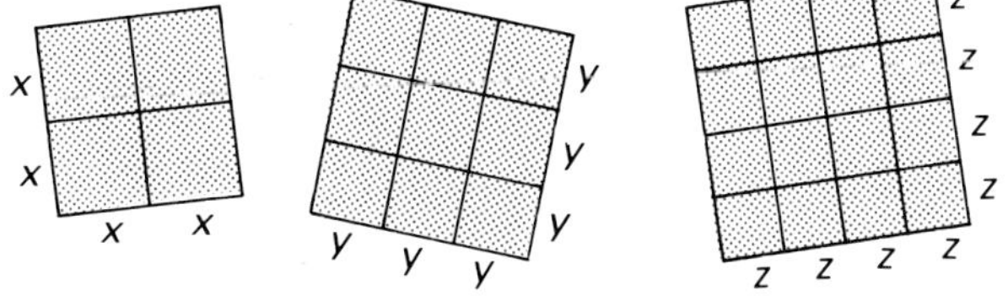
x
x
x x
y
y
y
y y y
z
z
z
z
z z z z

3 $p = 3$, $q = 2$ and $r = 1$. Calculate:
a p^2 **b** $2p^2$ **c** $(2p)^2$ **d** $4p^2$
e q^2 **f** $2q^2$ **g** $(2q)^2$ **h** $4q^2$
i $2r^2$ **j** $(2r)^2$ **k** $(pq)^2$ **l** p^2q^2
m pqr **n** $(pqr)^2$ **o** $(p+q+r)^2$

4 Find the outputs from these machines.

a

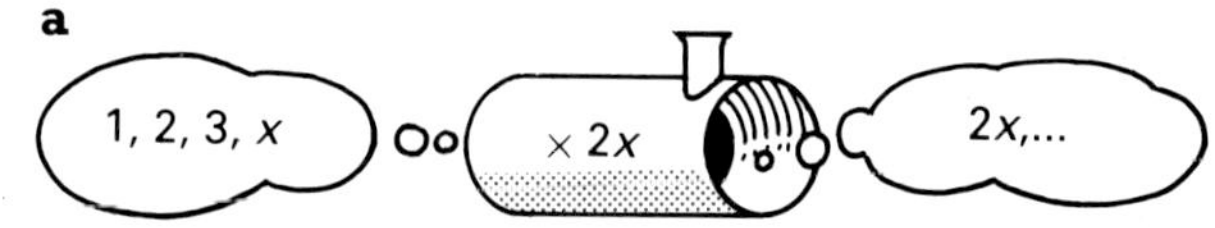
1, 2, 3, x
× 2x
2x,...

b

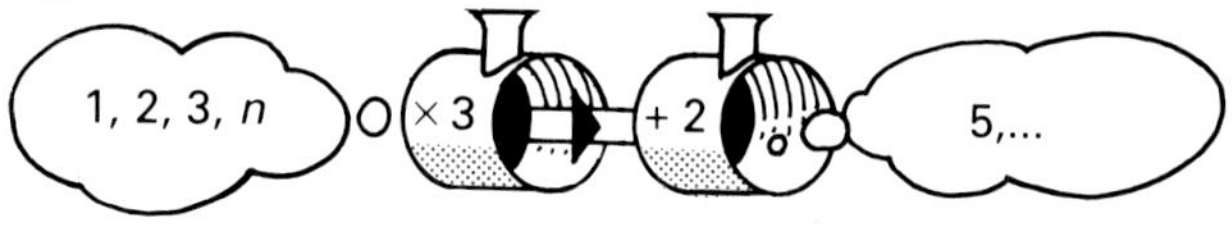
1, 2, 3, n
× 3
+ 2
5,...

c

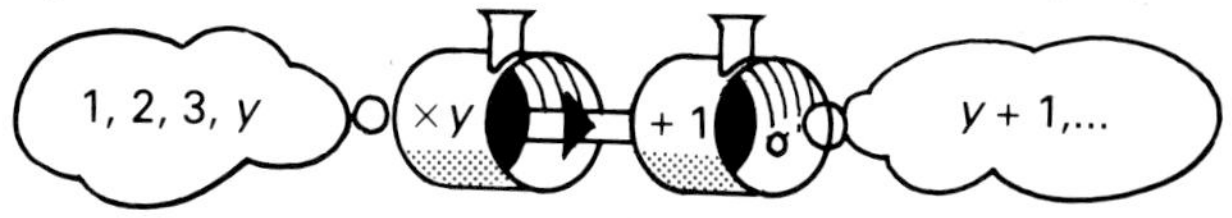
1, 2, 3, y
× y
+ 1
y + 1,...

d

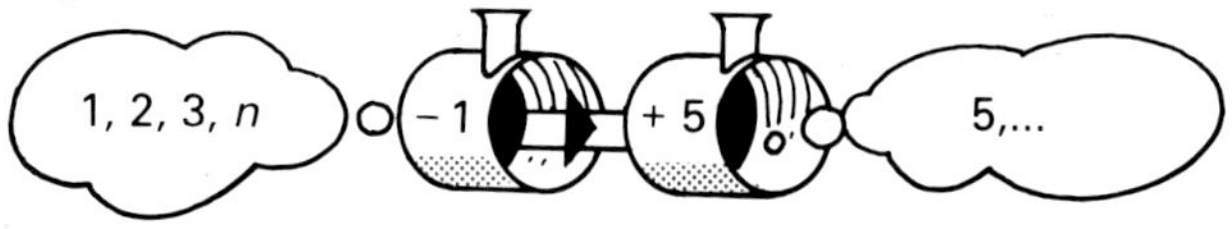
1, 2, 3, n
- 1
+ 5
5,...

5 If $n = 5$, find the values of:
a $(n+1)^2$ **b** n^2+n **c** $(n-2)^2$
d $n(n+1)$ **e** n^2+2n+1 **f** n^2-2n-3
g $(n+1)(n-3)$ **h** n^2-4n+4

6 a Which pairs of expressions in question **5** appear to be equal?
b Check, by using different values of n.

Examples

If $a = 10$ and $b = 12$, then:

a $\frac{1}{2}a = \frac{1}{2} \times 10 = 5$, and also $\frac{a}{2} = \frac{10}{2} = 5$

b $\frac{1}{2}(a+b) = \frac{1}{2}(10+12) = \frac{1}{2} \times 22 = 11$, and also $\frac{a+b}{2} = \frac{10+12}{2} = \frac{22}{2} = 11$

7 $x = 2$, $y = 3$ and $z = 6$. Calculate:

a $\frac{1}{2}x$	**b** $\frac{1}{2}z$	**c** $\frac{1}{3}z$	**d** $\frac{1}{3}y$
e $\frac{2}{3}y$	**f** $\frac{1}{2}(x+z)$	**g** $\frac{1}{3}(y+z)$	**h** $\frac{1}{4}(z-x)$
i $\frac{1}{3}(z-y)$	**j** $\frac{3}{4}(x+z)$	**k** $\frac{x+y}{5}$	**l** $\frac{y+z}{3}$
m $\frac{x+z}{4}$	**n** $\frac{x+y+z}{11}$	**o** $\frac{x+2y+z}{2}$	

BRAINSTORMER

Which IN number would give the same OUT number?

CHALLENGE

Copy and complete this cross-number puzzle.
$x = 23$, $y = 7$, $z = 11$, $a = 120$, $b = 3$.

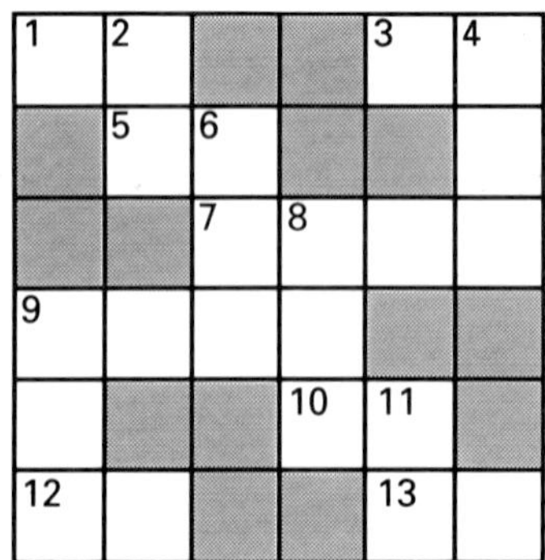

Across	**Down**
1 $3x$	***2*** $a-x$
3 $3y$	***4*** $a+2b$
5 yz	***6*** $6a$
7 $4x^2$	***8*** z^2+x
9 $8a+4z$	***9*** $3y^2$
10 y^2	***11*** $3x+4y$
12 $a-2x$	
13 y^2+3b^2	

INVESTIGATION

The height of a cricket ball after t seconds is $t(8-t)$ metres.

a *Copy and complete the table.*

Time (s)	0	1	2	3	4	5	6	7	8
Height (m)	0	7							

b *Use the axes and scales shown to plot the points and draw a graph.*

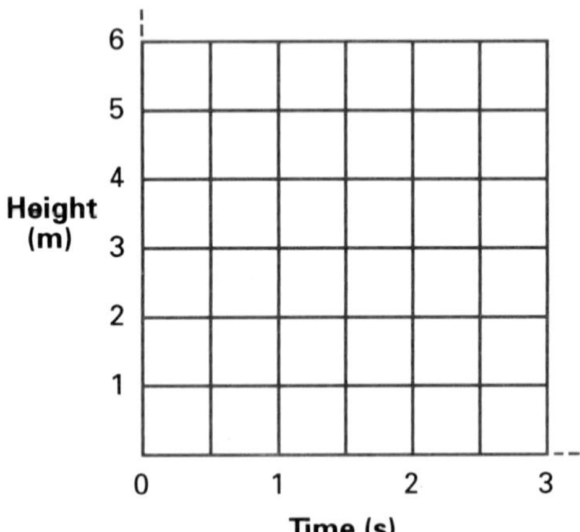

c *Investigate the height of the ball at different times, and write a few sentences about it.*

LENGTHS, AREAS AND VOLUMES

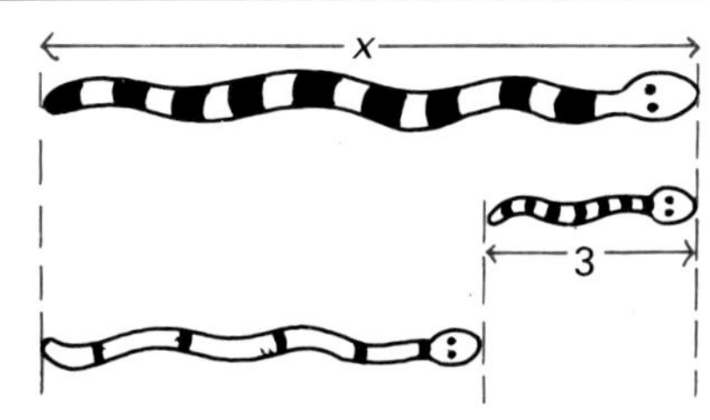

This snake has length x metres,

and this snake has length 3 metres.

So this snake must have length $x-3$ metres.

EXERCISE 3A

Find expressions for the lengths of the unmarked ladders, crocodiles and snakes below. The lengths are all in metres.

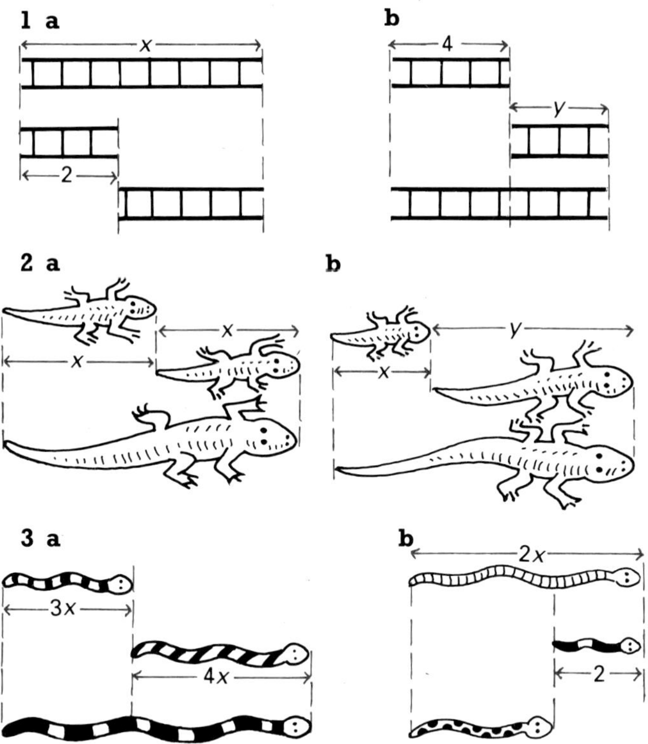

4 Calculate the perimeters of these shapes.

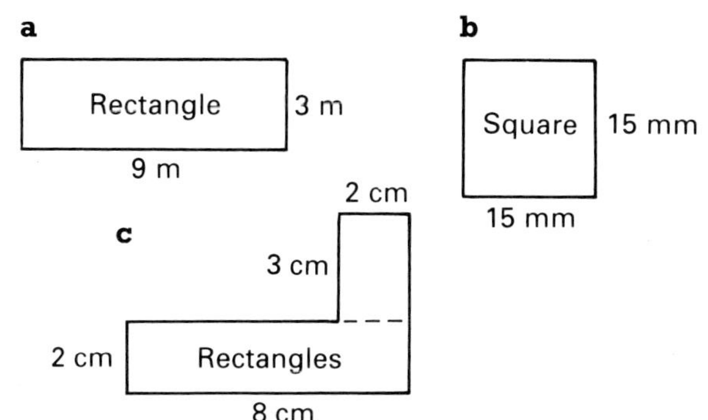

5 Find a formula for the perimeter, P cm, of each rectangle and square. For example, $P = 2m+2n$.

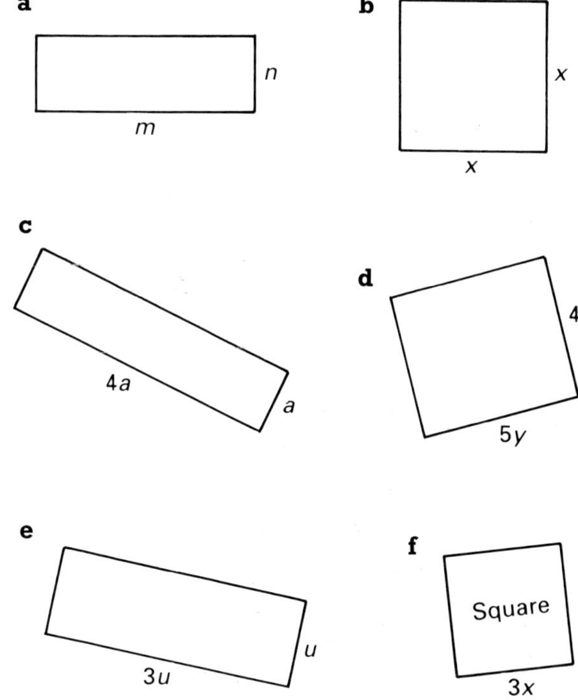

6 Calculate the area of each shape in question **4**.

7 Find a formula for the area, A cm^2, of each rectangle and square in question **5**. For example, $A = xy$.

EXERCISE 3B

1 Find expressions for the lengths, in centimetres, of the canes (i), (ii) and (iii).

a

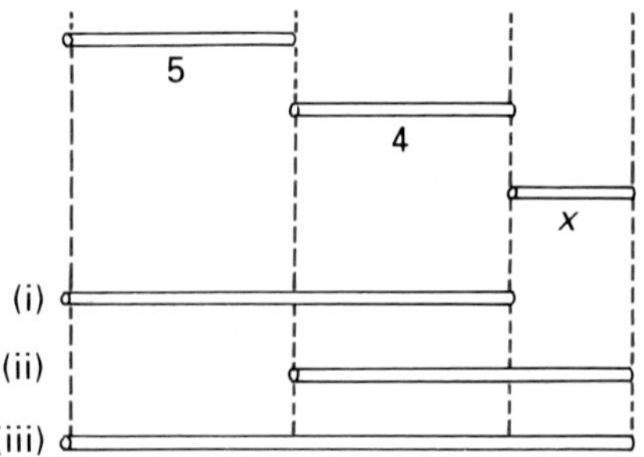

b

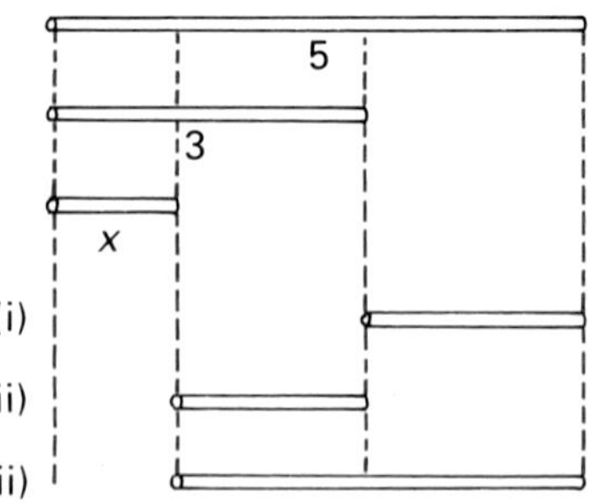

c

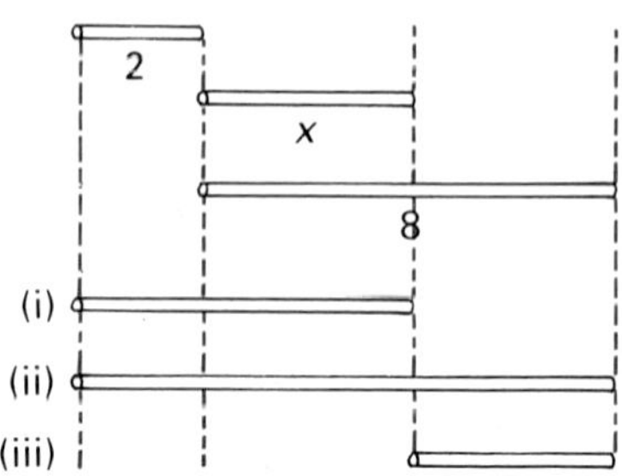

2 Find a formula for the perimeter, P cm, of each shape.

a

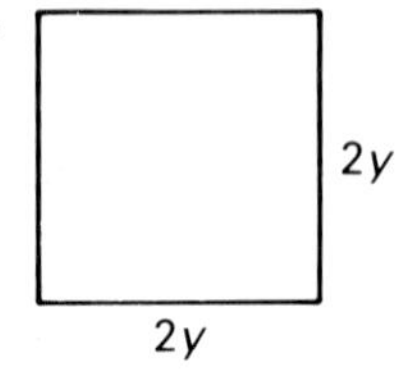

b

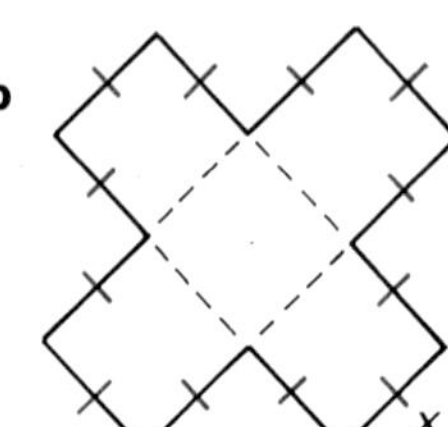

c

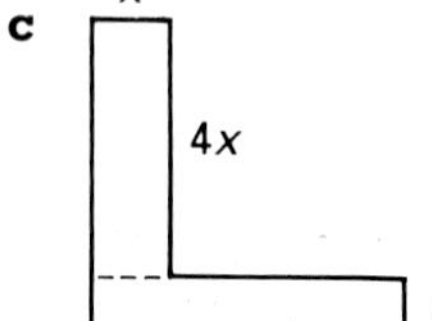

3 Find a formula for the area, A cm², of each shape in question **2**.

4 Calculate the volumes of these solids.

a **b**

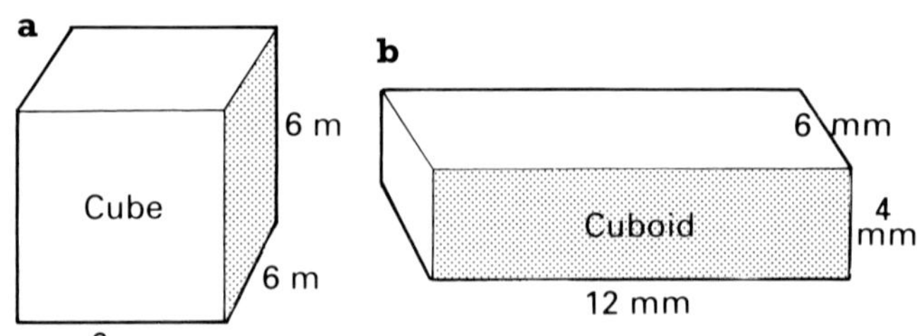

c

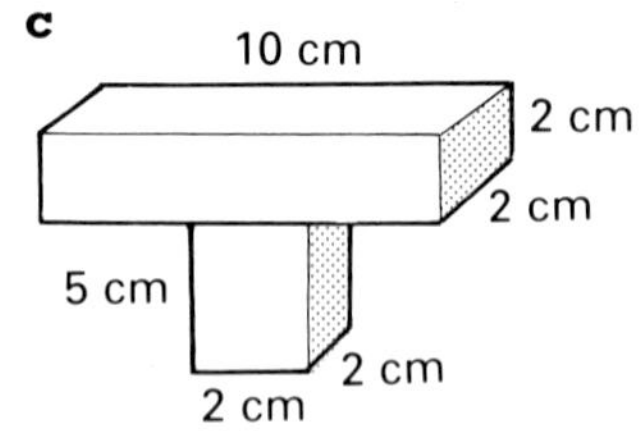

5 **a** Find a formula for the volume of:

(i) the cube in question **4a**, if each edge is $3a$ metres long.

(ii) the cuboid in question **4b**, if its edges are $2a$ mm, $3a$ mm and $6a$ mm long.

b Calculate the volumes of the cube and cuboid if $a = 4$.

6 For this skeleton cuboid, find formulae for:

a the perimeter, P cm, of all its edges

b its volume, V cm³

c the area, A cm², of plastic needed to cover all six sides.

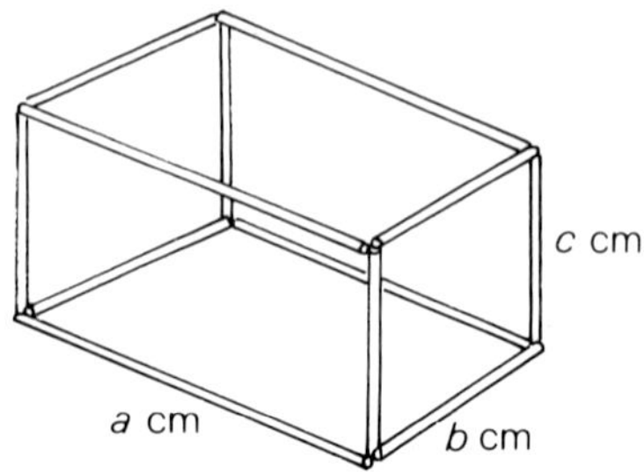

EXERCISE 3C

1 Find expressions for the lengths, in centimetres, of the canes (i), (ii) and (iii).

a

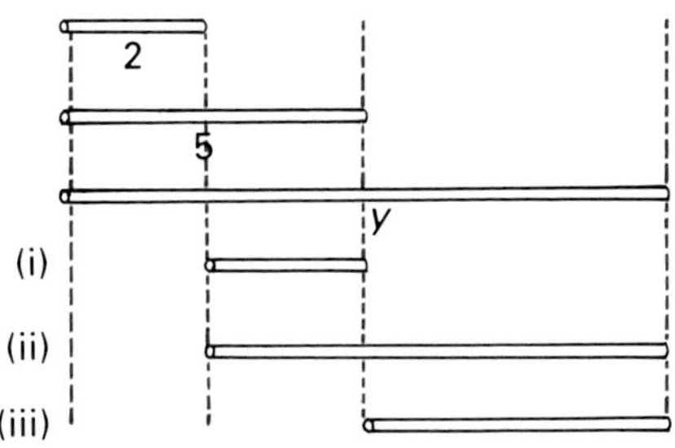

b

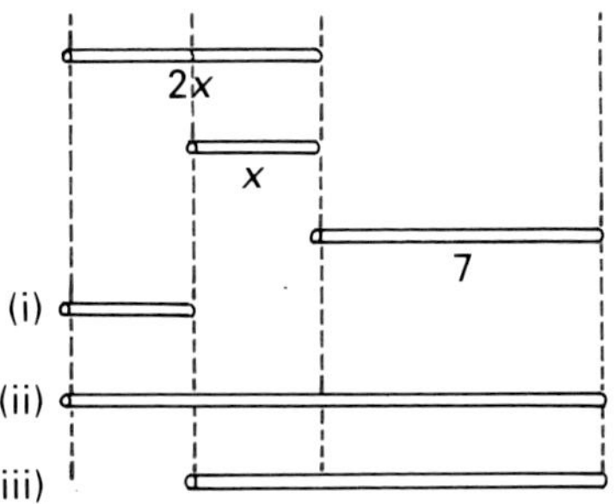

c

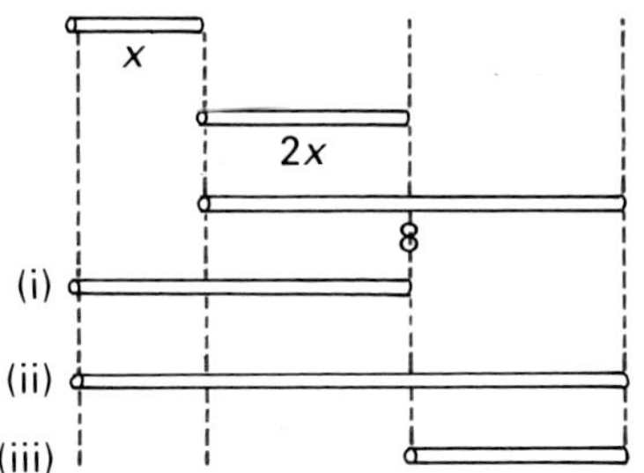

2 Find a formula for the perimeter, P cm, of each shape below.

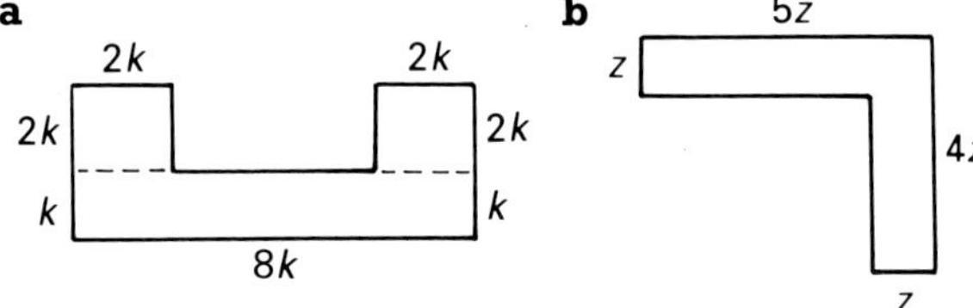

c

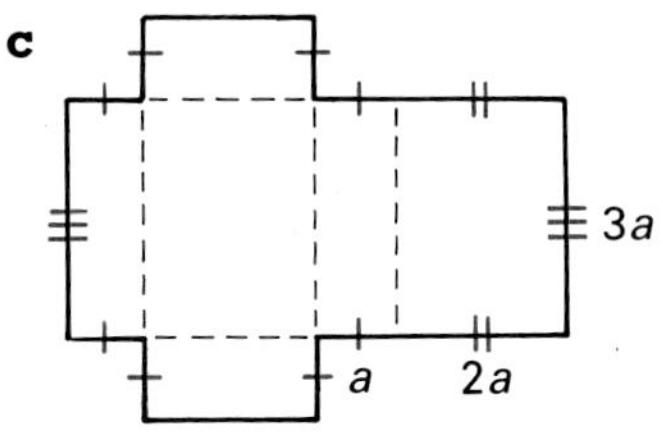

3 Find a formula for the area, A cm^2, of each shape in question **2**.

4 Find a formula for the volume, V cm^3, of each solid.

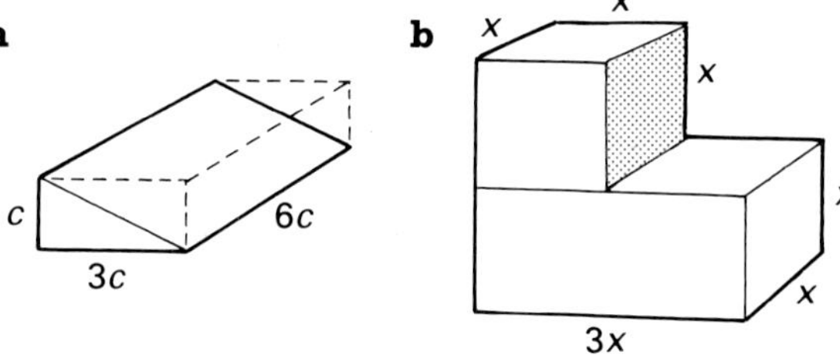

c

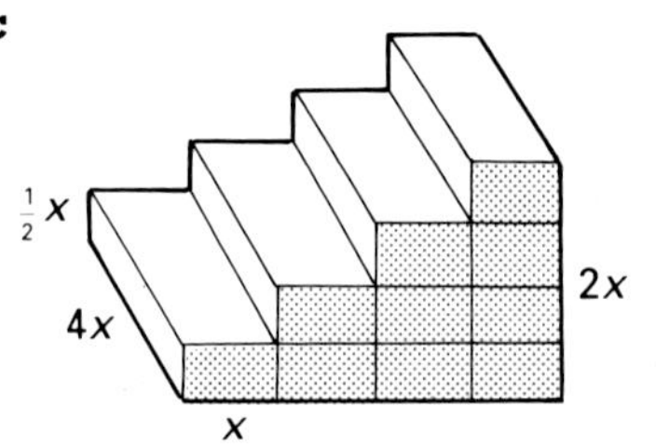

BRAINSTORMERS

1 *The dimensions of the crate are in metres.*

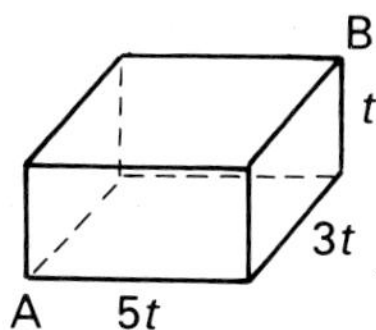

a *Find an expression for:*

(i) the shortest distance along the edges from A to B

(ii) the total length of all the edges.

b *Calculate the values of the expressions when $t = 3$.*

2 **a** *A wire frame is made for a model truck. Find an expression for the length of wire needed. The lengths are in centimetres.*

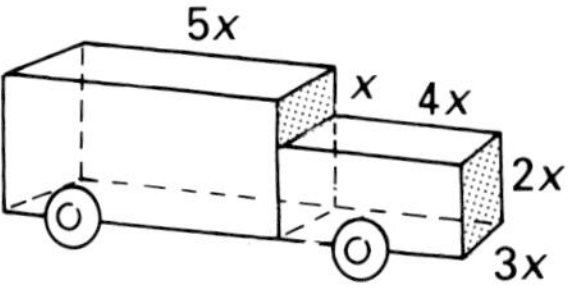

b *The truck is then made in solid wood. Find expressions for the total area and volume of wood in the model truck.*

BRACKETS

EXERCISE 4A

Alf and Nadeen are trying to calculate the area of their rectangular garden. The lengths are in metres.

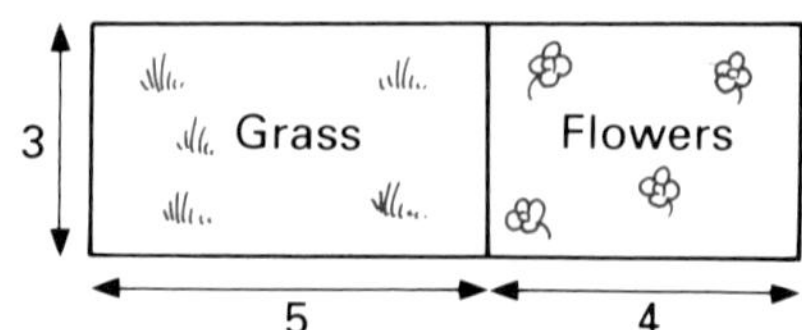

Alf's method

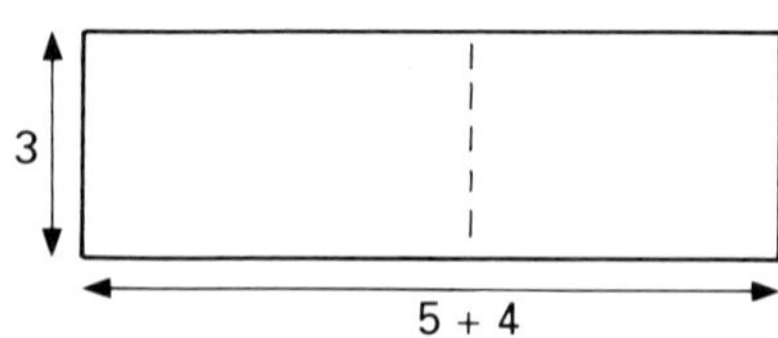

Area $= 3 \times (5+4) = 3 \times 9 = 27\,\text{m}^2$

Nadeen's method

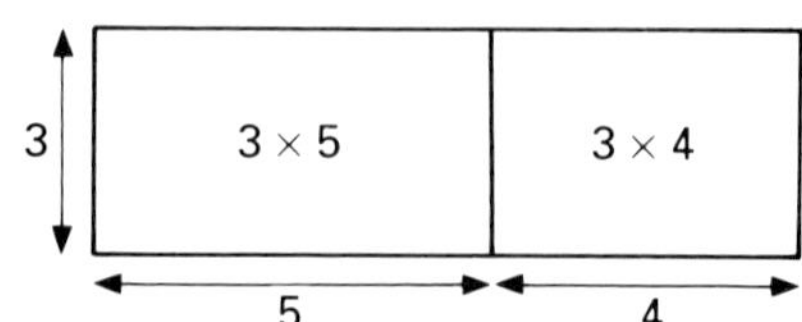

Area $= 3 \times 5 + 3 \times 4 = 15 + 12 = 27\,\text{m}^2$

1 (i) Sketch these gardens, showing the areas of grass and flowers.
(ii) Calculate the area of each garden by using both methods above, ie by adding the two areas, and also by multiplying the length of the whole garden by its breadth.

a

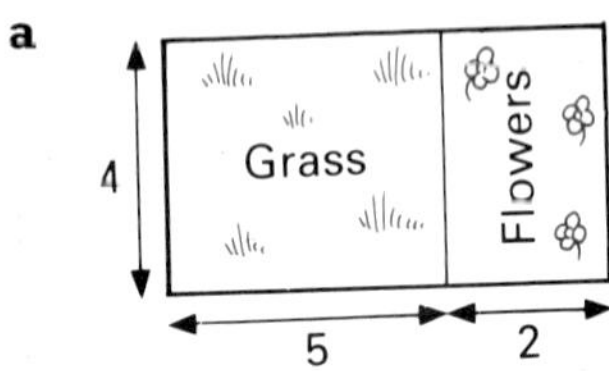

b

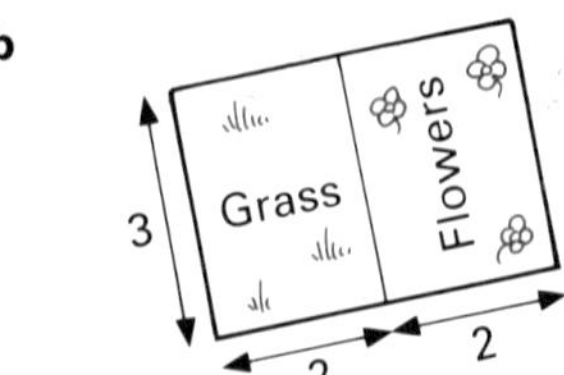

c

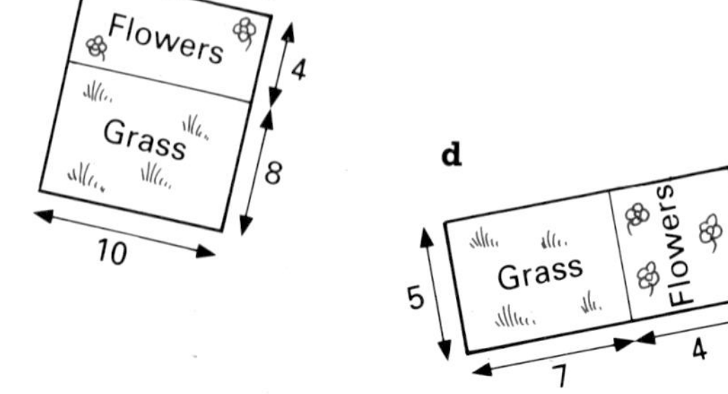

d

2 Calculate the areas of these gardens using the same two methods that Alf and Nadeen used.

a

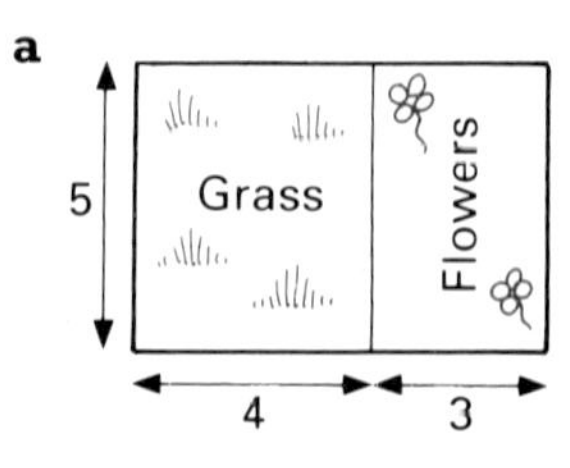

b

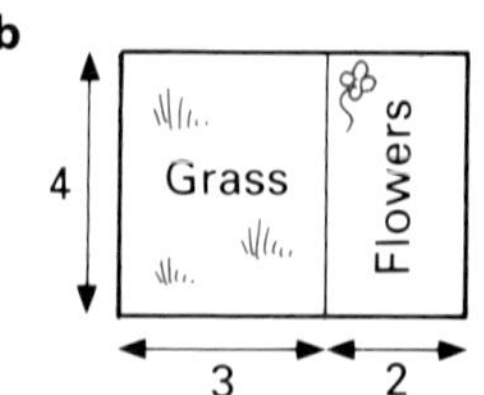

c

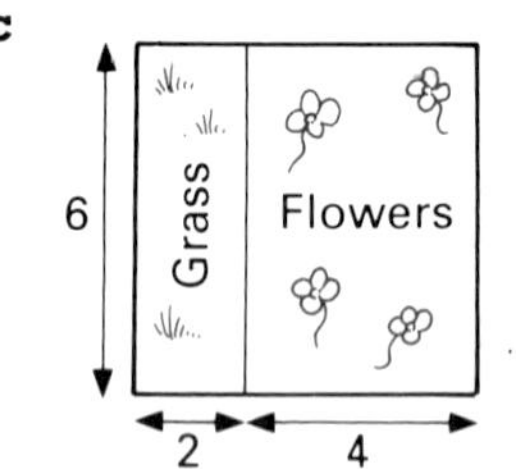

d

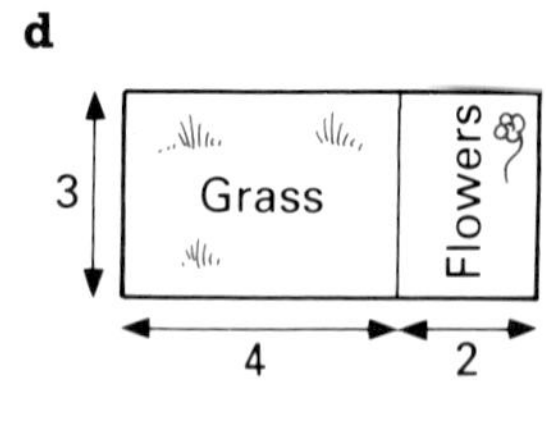

e

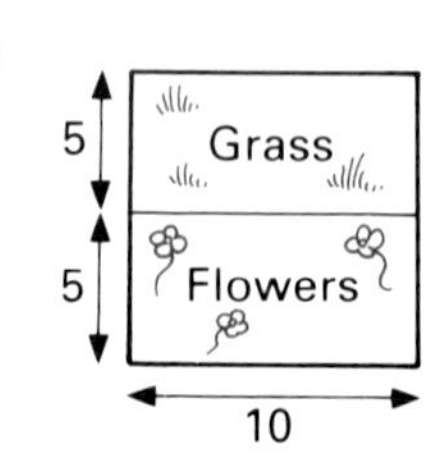

f

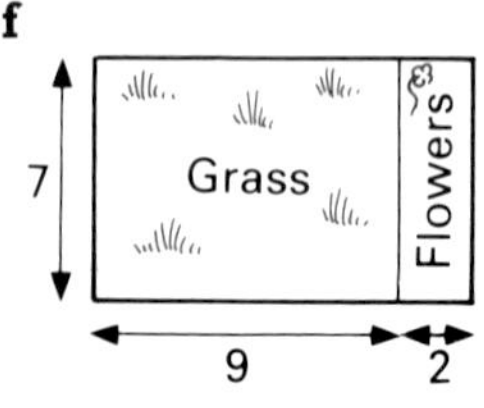

3 (i) Draw gardens which illustrate these calculations of equal areas.

a $5 \times (3+1) = 5 \times 3 + 5 \times 1$
b $4 \times (2+2) = 4 \times 2 + 4 \times 2$
c $8 \times (5+3) = 8 \times 5 + 8 \times 3$

(ii) Mark in the areas.

EXERCISE 4B/C

Alf and Nadeen want to extend the length of their lawn, but are not sure how far. They plan it like this:

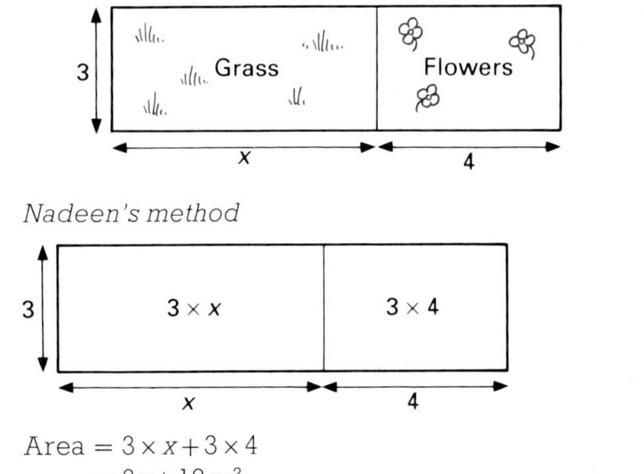

Alf's method

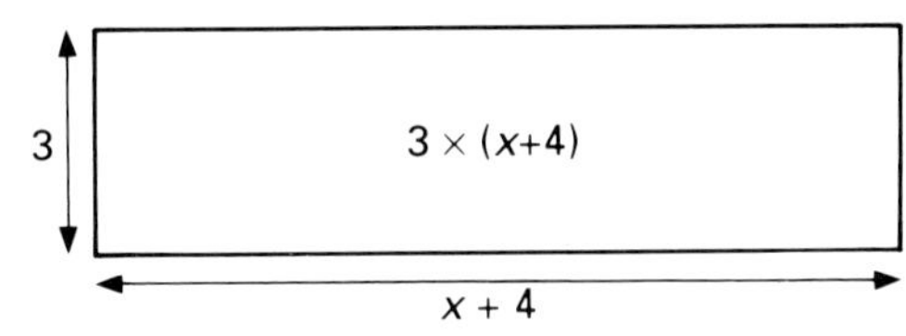

Nadeen's method

Area $= 3 \times (x+4)$
$= 3(x+4)\text{m}^2$

Area $= 3 \times x + 3 \times 4$
$= 3x + 12\,\text{m}^2$

They both agree that $3(x+4) = 3x+12$.

1 Find two expressions for the area of each garden, using Alf's and then Nadeen's method.

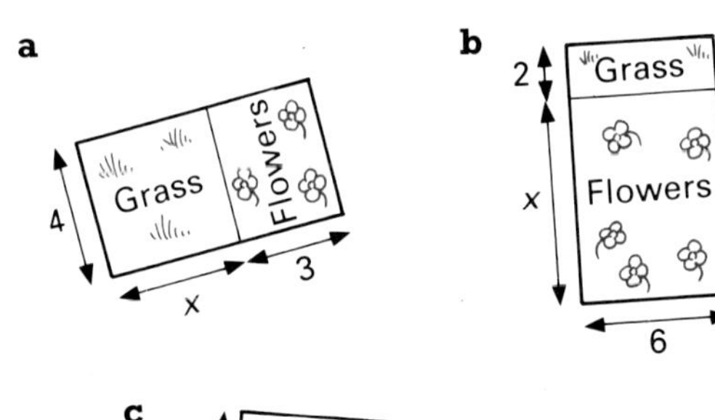

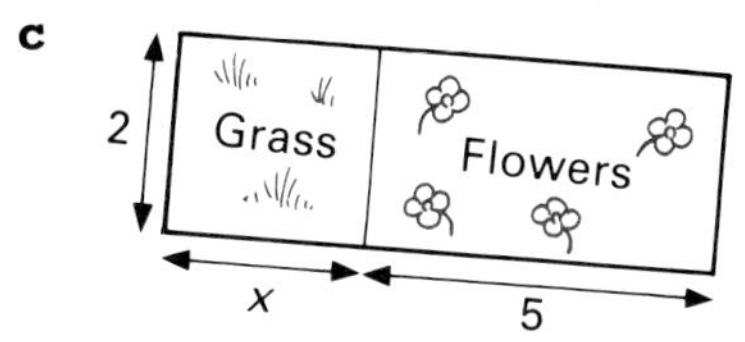

2 Draw gardens which illustrate these pairs of equal areas.

a $3(x+2) = 3x+6$ **b** $5(x+1) = 5x+5$
c $8(x+3) = 8x+24$

3 Find two different expressions for the area of each garden shown in **a–l**.

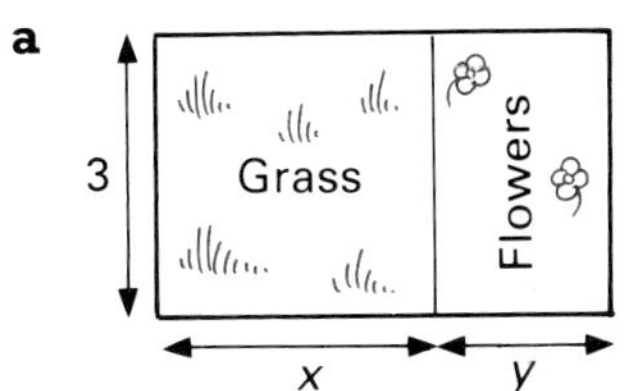

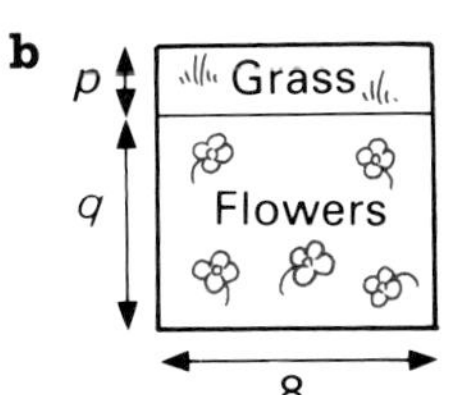

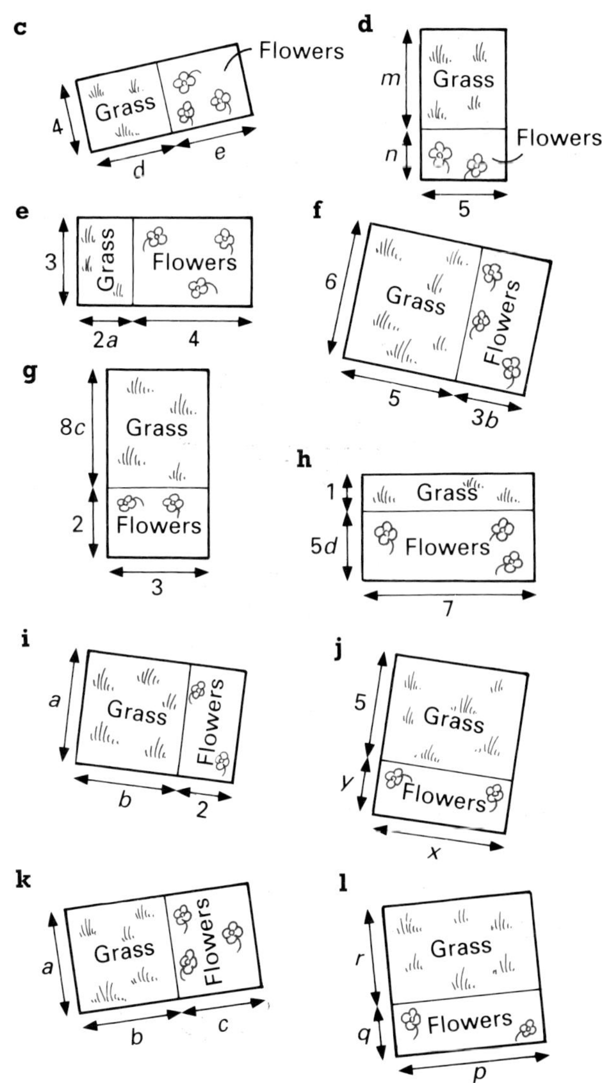

REMOVING BRACKETS

EXERCISE 5A

$a(b+c) = ab+ac$ $\quad \overset{\frown}{a(b+c)}$

Also, $a(b-c) = ab-ac$ $\quad \overset{\frown}{a(b-c)}$

Examples
a $5(x+3) = 5x+15$ **b** $3(2y-1) = 6y-3$

Write these without brackets:

1 a $2(x+3)$ **b** $3(x+4)$ **c** $4(x+2)$ **d** $5(x+1)$
e $8(x+2)$ **f** $6(y+5)$ **g** $3(y+2)$ **h** $7(y+1)$
i $2(y+4)$ **j** $4(y+6)$ **k** $2(a+9)$ **l** $3(b+7)$
m $8(c+5)$ **n** $5(k+2)$ **o** $4(m+9)$

2 a $3(x-1)$ **b** $2(x-3)$ **c** $4(y-2)$ **d** $6(m-1)$
e $8(n-4)$ **f** $7(k-3)$ **g** $5(m-6)$ **h** $9(n-7)$
i $10(x-1)$ **j** $4(y-11)$

3 a $2(5+x)$ **b** $3(4+y)$ **c** $5(1+t)$ **d** $4(7+u)$
e $6(8+v)$ **f** $3(1-x)$ **g** $2(5-y)$ **h** $4(2-m)$
i $7(1-n)$ **j** $8(4-p)$

4 Copy and complete:

a
$2(x+5)=2x+10$
$3(x+2)=3x+$
$4(y+1)=$
$5(y+3)=$
$6(b+2)=$

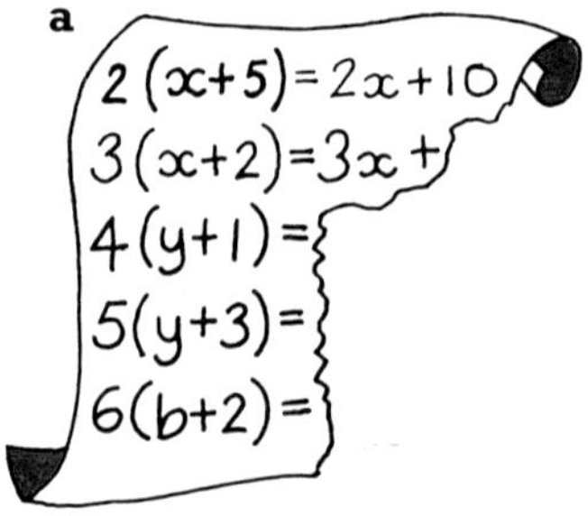

b
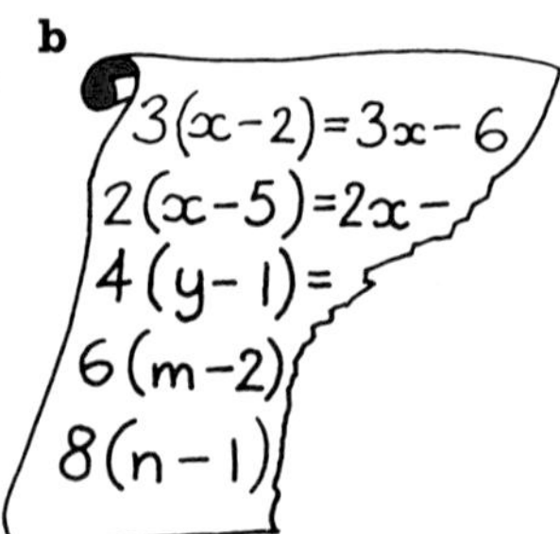

Remove the brackets in questions **5** and **6**.

5 a $3(2x+1)$ **b** $5(2y+1)$ **c** $4(3a+1)$
d $6(4b+1)$ **e** $7(5c+1)$ **f** $2(3x+4)$
g $5(2y+3)$ **h** $4(5k+2)$ **i** $3(4m+5)$
j $6(2t+3)$ **k** $5(1-2x)$ **l** $3(5-2y)$
m $4(3-4k)$ **n** $6(2-3u)$ **o** $9(6-5v)$

6 a $2(x+y)$ **b** $7(c+d)$ **c** $10(p-q)$
d $5(s-t)$ **e** $7(m+n)$ **f** $3(2a+b)$
g $4(2c+d)$ **h** $5(x+2y)$ **i** $2(u+3v)$

EXERCISE 5B

More examples

a $a(x+3) = ax+3a$ **b** $t(t-2) = t^2-2t$

c $u(u+v) = u^2+uv$

Write without brackets:

1 a $a(x+4)$ **b** $b(y+2)$ **c** $c(k-1)$
d $d(m-5)$ **e** $e(n+2)$ **f** $n(2+x)$
g $m(3+y)$ **h** $t(1-x)$ **i** $u(4-x)$
j $v(1-2y)$

2 a $x(x+1)$ **b** $y(y-1)$ **c** $t(t+2)$
d $u(u-3)$ **e** $v(v-8)$ **f** $t(1-t)$
g $s(3-s)$ **h** $k(1+k)$ **i** $m(8-m)$
j $n(n-7)$

3 a $x(x+y)$ **b** $a(a-b)$ **c** $m(m+n)$
d $u(u-v)$ **e** $t(s+t)$ **f** $a(b-a)$
g $y(y-x)$ **h** $x(1-3x)$ **i** $y(2y+5)$
j $z(4-7z)$

4 Write down expressions for the areas of these rectangles:
(i) with brackets (ii) without brackets
Lengths are in centimetres.

a
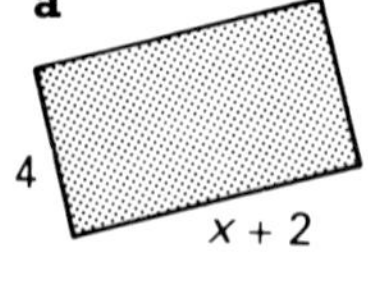

b
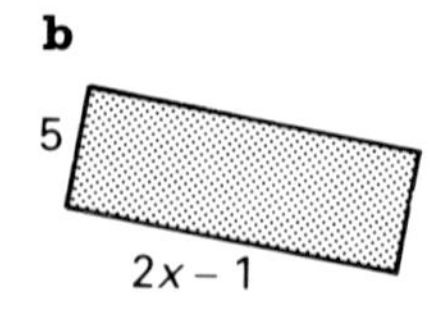

c
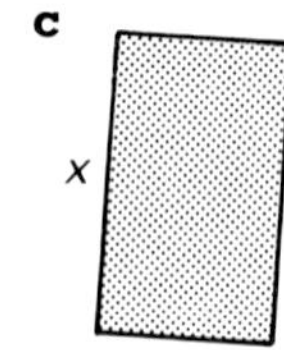

d
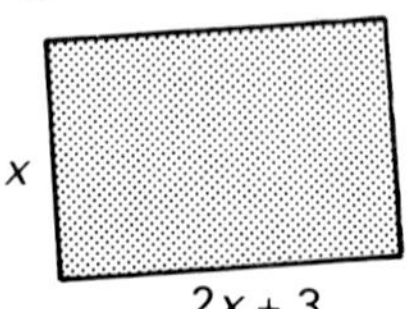

e
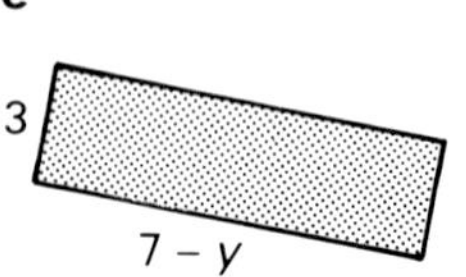

f
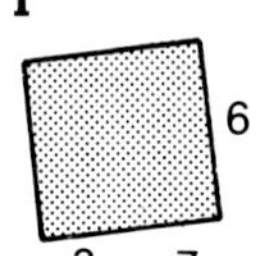

g

h
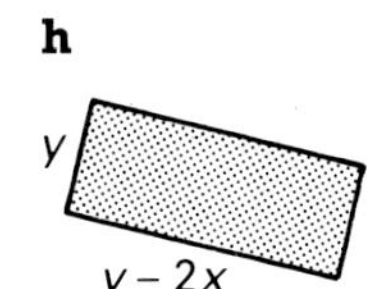

EXERCISE 5C

1 Remove the brackets, and then simplify, where possible.

a $2(x+2)+4$ **b** $3(x-1)+12$ **c** $13+4(x-2)$
d $2(y+3)+7$ **e** $3(p+4)-5$ **f** $5(y-2)-2y$
g $x(x+2)+x$ **h** $y(1+y)+y$ **i** $4x+x(1+x)$
j $2x+x(x-1)$ **k** $a(a+b)-ab$ **l** $y+y(y-1)$

2 **a** $2(x+y+3)$ **b** $5(m+n-2)$ **c** $3(2x-y-3)$
d $4(2a+3b-4c)$ **e** $x(x^2-1)$ **f** $y(y^2-3)$
g $a(a^2+b)$ **h** $c(d-c^2)$

3 **a** $2(x+4)+3(x+1)$ **b** $3(y+1)+2(y-1)$
c $4(y+3)+3(y-4)$ **d** $5(a+2)+2(a+5)$
e $3(2c+1)+2(3c-1)$ **f** $2(2x+3)+3(4x-1)$

4 A path one metre wide goes right round the outside of Derek's rectangular garden. He has to order gravel for it.

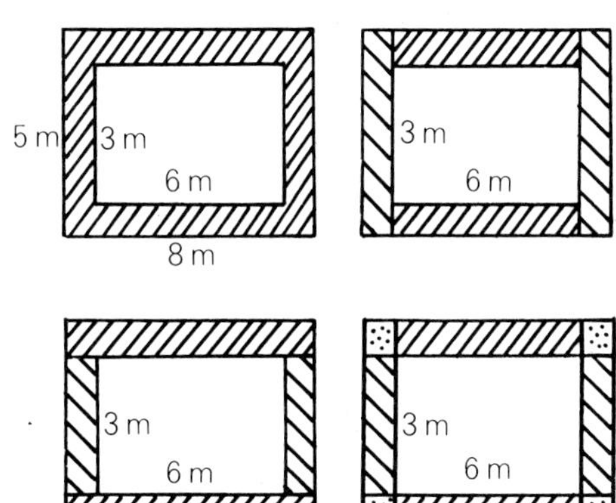

a The pictures suggest four different ways that he could calculate the area of the path. Investigate these.

b Draw the pictures, and change the inside length to 7 m. The path is one metre wide. Investigate the area now.

c Try it again for an inside length of x m and path width of one metre.

CHALLENGES

1 *Find the area of this rectangle. Your final expression should not have brackets in it. Check your answer for $x = 10$. All lengths are in metres.*

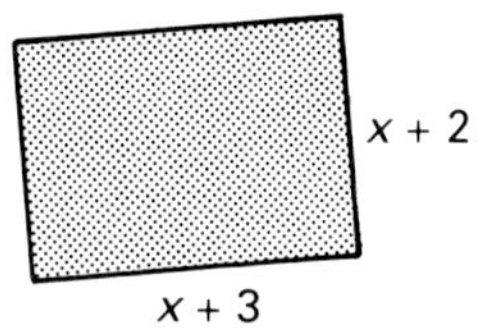

2 *This large area of potatoes is $2x$ km long and x km broad. Find an expression for the area that is not affected by flood or virus damage. Check for $x = 8$. The lengths are in kilometres.*

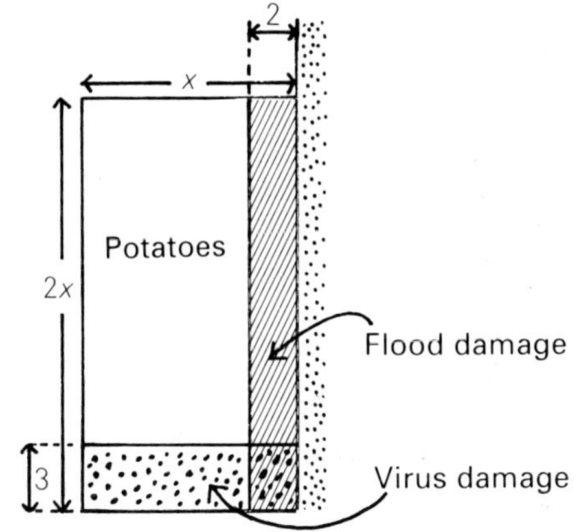

3 ***a*** *Can you remove these brackets?*
(i) $a(b+c+d)$
(ii) $(a+b)(c+d)$

b *Check your answers by putting $a = 2$, $b = 3$, $c = 4$ and $d = 5$.*

CHECK-UP ON LETTERS AND NUMBERS

1 Write in shorter form:

a $m+m$ **b** $y+y+y$ **c** $t+t+t+t$ **d** $n\times n$
e $k\times k\times k$

2 Write down the length of each cane below. Lengths are in centimetres.

a

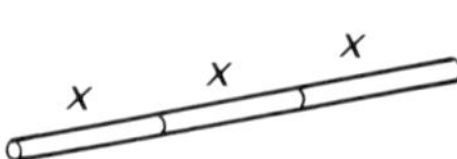

b

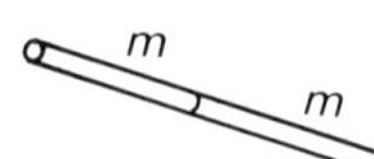

c

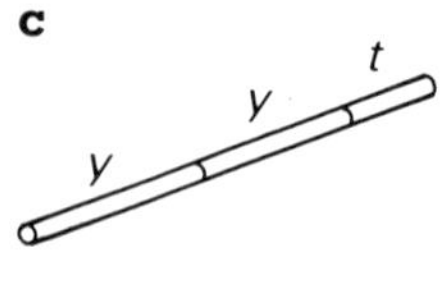

d

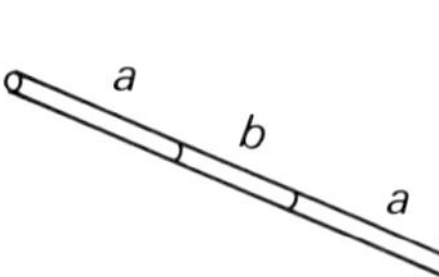

3 $p = 2$, $q = 3$ and $r = 1$. Find the values of:

a $5p$ **b** $2q+1$ **c** $3q+2r$ **d** p^2+q^2
e $pq-r^2$

4 These shapes are made of squares and rectangles. Find a formula for each perimeter P cm, and area A cm^2.

a

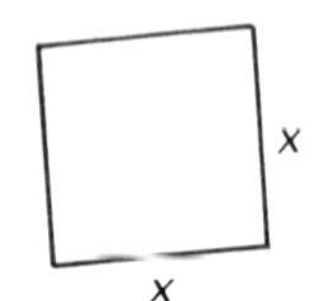

b

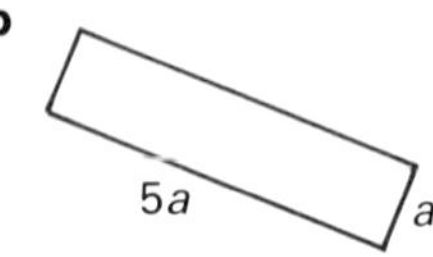

c

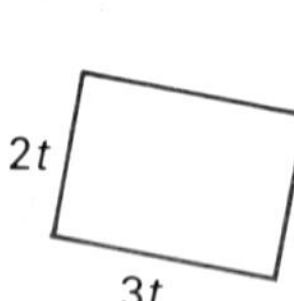

d

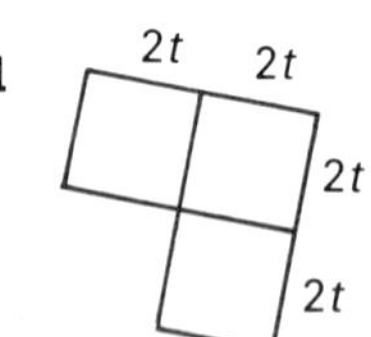

5 Write without brackets:

a $5(x+4)$ **b** $2(y-3)$ **c** $3(t+1)$ **d** $4(8-x)$
e $5(2-m)$

6 Copy this rectangular garden, and fill in all the lengths and areas that illustrate $6(x+12) = 6x+72$.

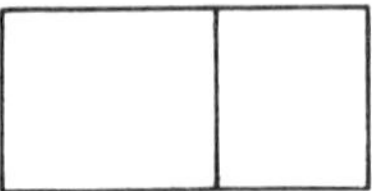

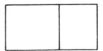

7 Write without brackets:

a $2(2t+1)$ **b** $3(5n-1)$ **c** $4(1+2k)$ **d** $5(a+b)$
e $6(2c-3d)$

8

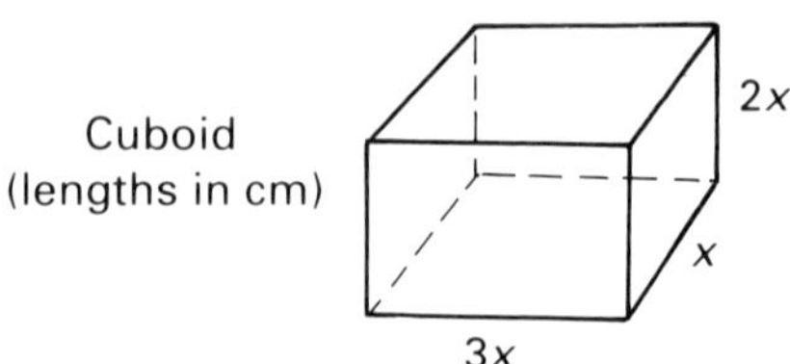

Find a formula for:

a the length, L cm, of all the edges
b the area, A cm^2, of all the faces
c the volume, V cm^3, of the cuboid.

9 Simplify:

a $5x+3-x+2$ **b** $2a+5b-2a-b$
c $4\times 2t$ **d** $2m\times 3n$
e $a^2-a\times a$ **f** $4+t\times 2t$
g $\frac{1}{2}\times p\times p$ **h** $\frac{1}{3}\times a\times a\times a$
i $2d+3d+4d$ **j** $2d\times 3d\times 4d$

10 $x = 5$, $y = 4$ and $z = 2$. Find the values of:

a z^2 **b** $2z^2$ **c** $(2z)^2$
d $4x^2$ **e** $(3y)^2$ **f** $y(y+4)$
g $(x-3)(x+2)$ **h** $(z+3)^2$ **i** $\frac{1}{3}(x+y)$
j $\frac{1}{2}(y-z)$

11 Remove brackets, then simplify:

a $2(x+3)-2x$ **b** $3(y+1)+2$
c $n(n+5)+2n$ **d** $k(k-1)+k$
e $a(a+b)-ab$ **f** $t(10+t^2)-10t$
g $a(a-b)+b(a-b)$ **h** $x^2(x+1)+x(1-x^2)$

4 MAKING SENSE OF STATISTICS 1

LOOKING BACK

1 The pictogram shows sales of Christmas trees on 20th December.

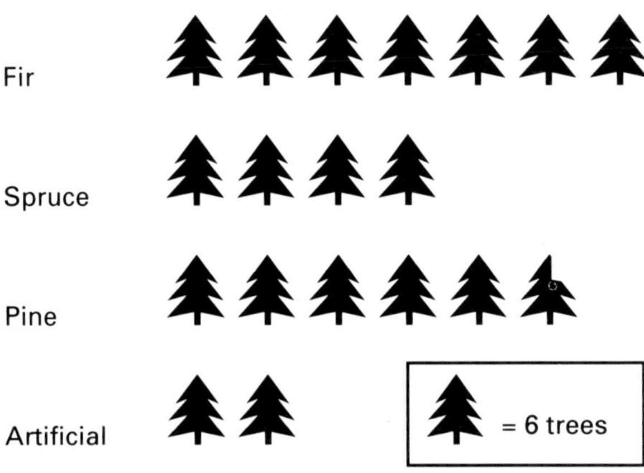

a Which type of tree was the best seller?
b How many fir trees were sold?
c How many more spruce than artificials were sold?

2 The table shows the sales on 22nd December. Illustrate the data in a pictogram.

Fir	Spruce	Pine	Artificial
30	23	20	5

3 The bar graph shows the rainfall one year.

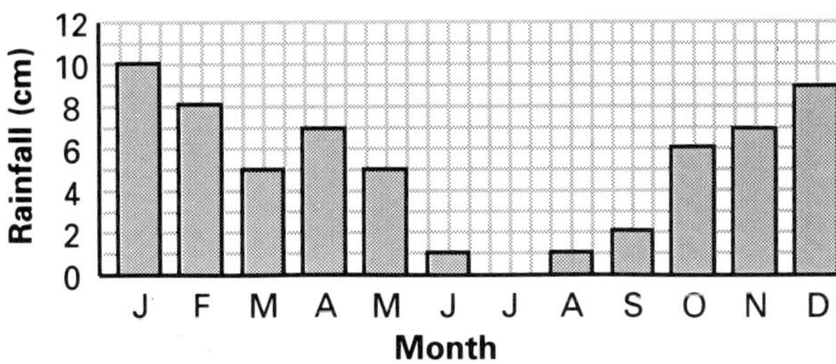

a Which was the wettest month?
b How much rain fell altogether in February, March and April?
c In how many months did more than 4 cm of rain fall?

4 This graph shows the flow of traffic through a village from 8am until 9.40am. The main cause of early morning traffic is people travelling to work or to school. The school day starts after the work day.

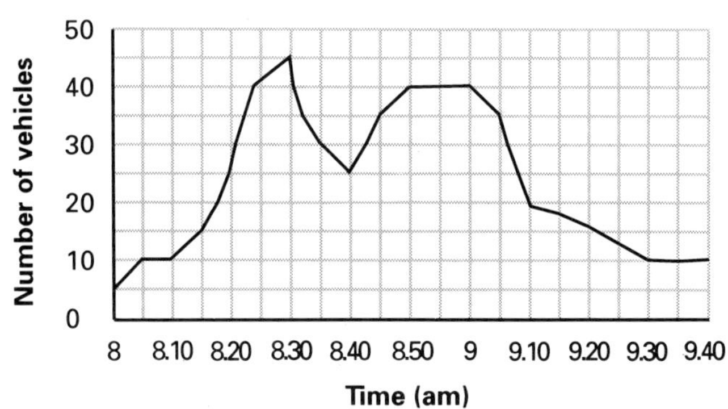

a How many vehicles passed at 8.40?
b How many passed at the 'peak' time?
c Suggest a starting time for the local:
(i) factory (ii) school.

5 Twelve people were asked what they liked to eat for breakfast. The pie chart shows their preferences.

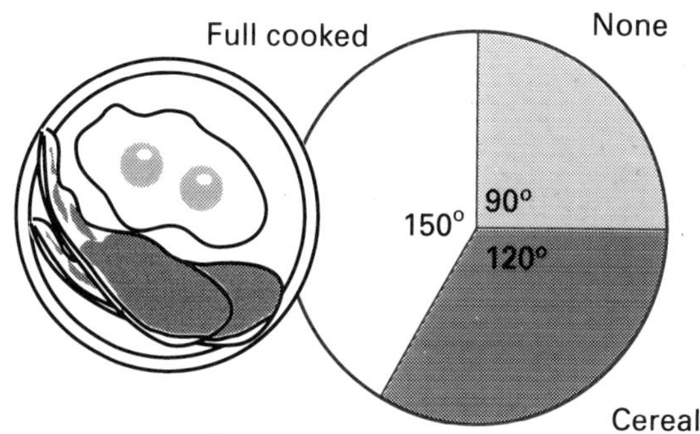

a What fraction of the group took no breakfast?
b How many people chose:
(i) cereal (ii) a full cooked breakfast?

6 If, from another group of twelve, $\frac{1}{2}$ took a full cooked breakfast, $\frac{1}{3}$ cereal and the rest nothing, what would the sizes of the angles in the pie chart be?

GRAPHS AGAIN

EXERCISE 1A

1 A paper plane was thrown from an upstairs window. This graph shows its flight.

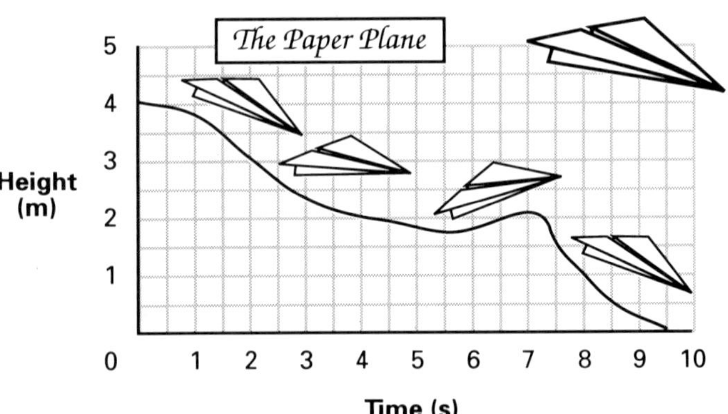

a How high was the window?
b How high was the plane after:
(i) 2 seconds (ii) 7.5 seconds?
c Between what times did the plane climb?

2 Neil carried out a survey of the day of the week on which each member of his class was born.

Here are his results:
Mon, Tue, Wed, Fri, Fri, Sat, Tue, Sun, Sat, Mon, Tue, Thu, Wed, Tue, Fri, Fri, Mon, Tue
a Copy and complete the table.

Day	Mon	Tue	Wed	Thu	Fri	Sat	Sun
Number							

b Draw a bar graph.
c Which was the most frequent 'birth day' for the class?

3 This graph shows how a spring stretches as weights are put on it.

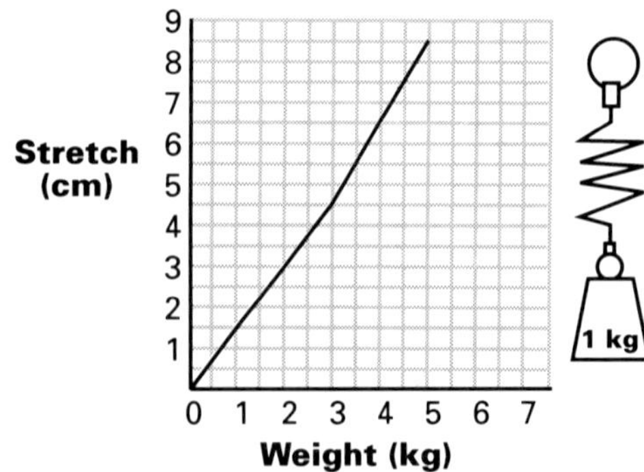

a A 2 kg weight stretches it 3 cm. How far will it stretch for weights of: (i) 1 kg (ii) 4 kg?
b What weight would stretch it by $4\frac{1}{2}$ cm?

4 The exam results are out. Simon's class marks have been given in a table. Mary's class marks have been organised in a line graph.

Simon's Class

Score	1–10	11–20	21–30	31–40	41–50
Number	5	8	7	4	3

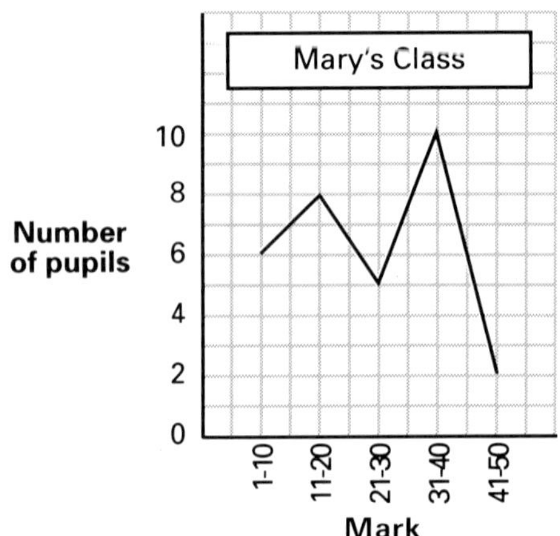

a How many pupils are there in:
(i) Simon's class (ii) Mary's class?
b What fraction of Simon's class scored over 40?
c Which band of marks in Mary's class contains most pupils?
d How many pupils in her class scored less than 21?
e How many more pupils scored more than 30 in Mary's class than in Simon's class?

EXERCISE 1B

> On their own, numbers may not mean very much. But in a graph, among other numbers, they often tell a story.

1 For each graph:

a What is the temperature at 3 pm?

b What are the temperatures one hour before 3 pm and one hour after 3 pm?

c Describe the trend of the temperature from noon until 7 pm.

(i)

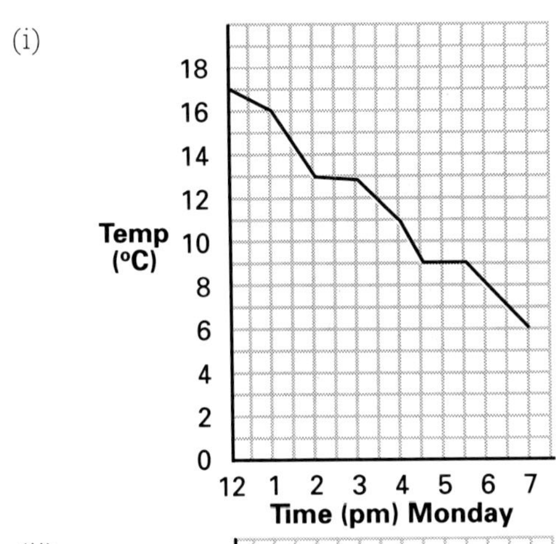

(ii)

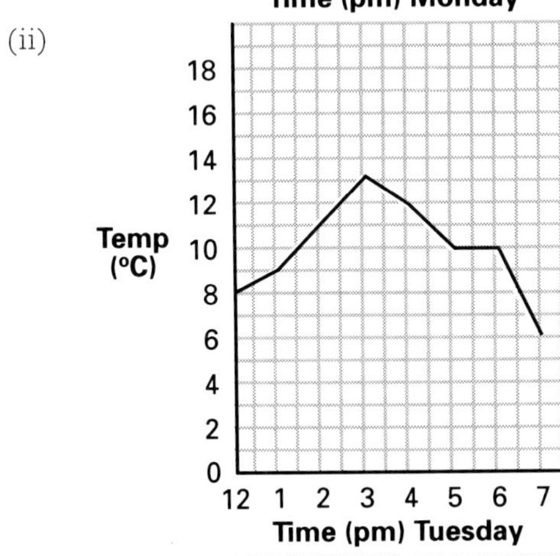

(iii)

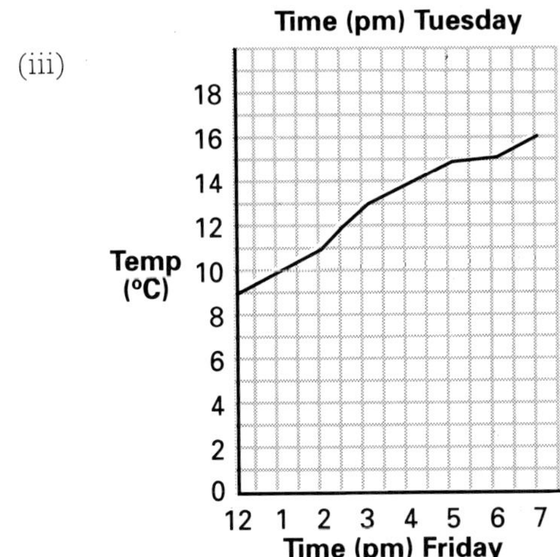

2 This is the graph of an earthquake which began at 1201.

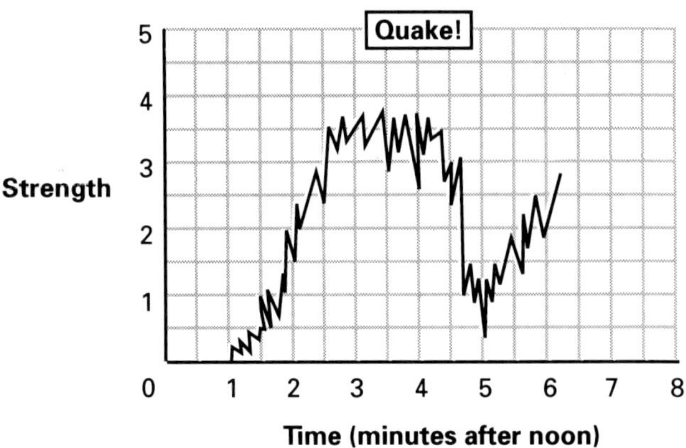

a Between what times was the tremor:
(i) getting worse (ii) staying about the same
(iii) calming down?

b Estimate the strength of the tremor at:
(i) 1201 (ii) 1203 (iii) 1205 (iv) 1206.

c Can you estimate the probable strength at:
(i) 1206$\frac{1}{2}$ (ii) 1208?

3 On March 3, the moon's phase was $\frac{7}{8}$ and it was waxing (growing).

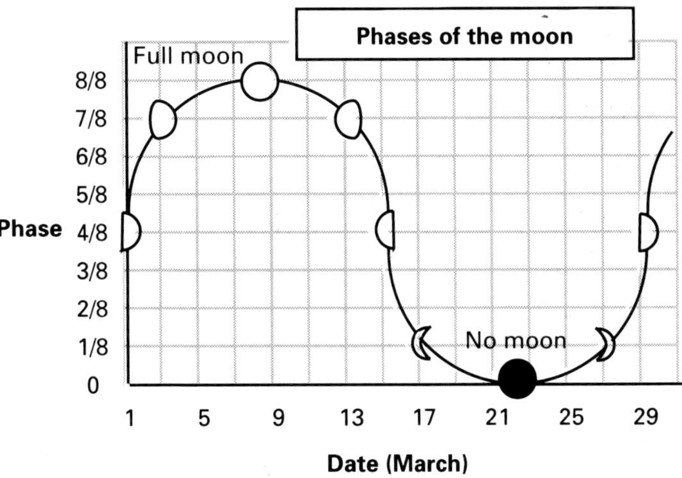

a What is the date of:
(i) full moon (ii) no moon?

b How many days are there between full moon and no moon?

c On what date was the moon's phase $\frac{1}{2}$ and waning (shrinking)?

d Describe its phase, and whether waxing or waning on: (i) March 1 (ii) March 29.

EXERCISE 1C

1 Tariq's class were given a test which was marked out of 50. Here are their marks:
7, 26, 39, 42, 31, 47, 12, 24, 36, 8, 36, 17, 24, 14, 28, 16, 23, 34, 23, 27, 43, 35, 26, 45, 26, 25, 19, 9, 32, 35

a Copy and complete the table.

Mark	1–10	11–20	21–30	31–40	41–50
Tally	III				
Number	3				

b Draw a bar graph to illustrate the data.

2 Graeme's class thought they'd done better than Tariq's class.

a Organise their marks in the same kind of table.
8, 18, 26, 36, 29, 32, 35, 9, 24, 31, 35, 41, 17, 37, 42, 16, 42, 46, 18, 19, 36, 27, 33, 21, 34, 21, 28, 33, 25, 14

b Draw a bar graph, and comment on the claim of Graeme's class.

3 a Copy this graph of the lifetime of the Standard Battery.

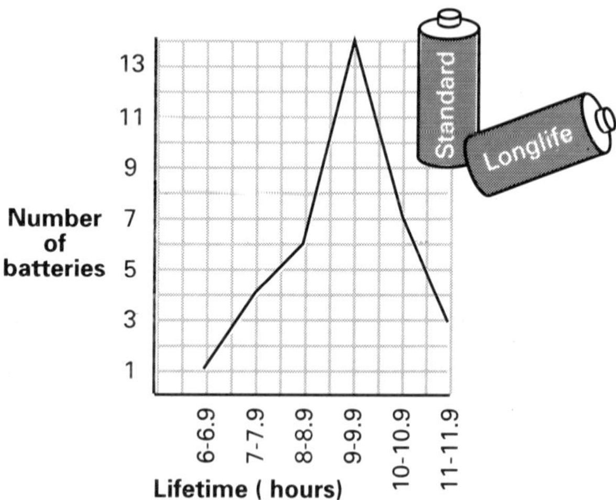

b Using the table, draw the graph of the Longlife battery on the same diagram.

Lifetime (h)	6–6.9	7–7.9	8–8.9	9–9.9	10–10.9	11–11.9
Number of batteries	1	1	3	10	15	5

c Is the name 'Longlife' justified? Comment.

d A man bought a Longlife and a Standard Battery. He complained that the Standard one lasted longer. Is that possible?

4 Here is Leah's petrol usage over a period of 30 weeks, in kilometres per litre.
10.8, 11.2, 12.6, 12.4, 12.9, 13.2, 12.5, 10.9, 11.8, 14.2, 13.1, 11.3, 13.4, 12.5, 13.7, 11.9, 12.2, 14.1, 12.6, 10.7, 12.8, 13.8, 12.4, 11.4, 13.7, 11.2, 10.6, 12.4, 13.1, 12.4

a Organise the data into suitable groups, and then draw a line graph.

b In town, Leah gets less than 12 km per litre. How many weeks of town driving does she do?

PRACTICAL PROJECT

a *Collect your own data to find out. Make a survey of the distances travelled by a sample group of pupils, using a map if necessary. Draw up a table, and then a bar graph.*

b *Now carry out a survey of the ways in which the pupils in your sample travel to school. Again make a table and graph.*

c *Use your graphs to find any connection between the results of your two surveys.*

MAKING NONSENSE OF STATISTICS

EXERCISE 2

1 'We have halved unemployment since coming to power.'

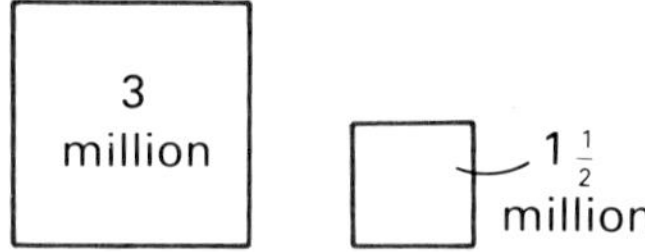

One square is half the length of the other, but why is the picture misleading?

2 'Our Galaxy Buster video game is the most popular one in the universe, as the diagram shows.'

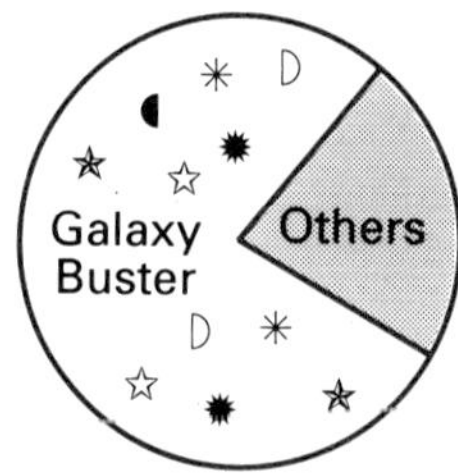

Does the diagram show this?

3 'Since advertising in our newspaper, Sky-Hi's profits have doubled.'

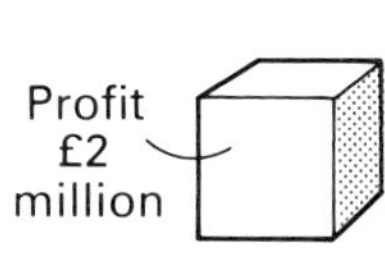

One cube is twice the length of the other, but . . . what's wrong here?

4 'Since we took over the company in 1988, profits have soared.'

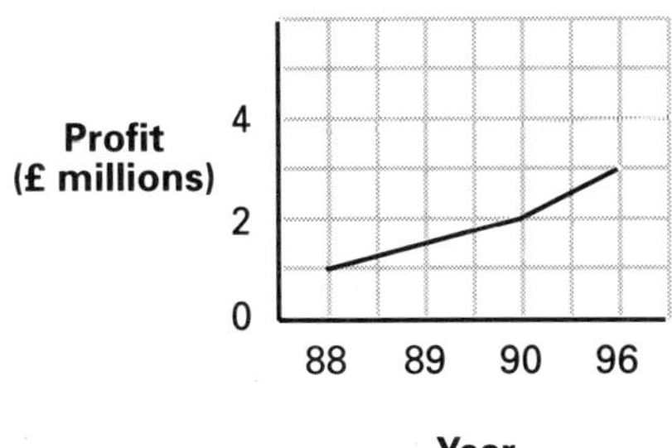

Have they? What is wrong with the graph?

5 'More people pass their driving test with us than with any other company.'

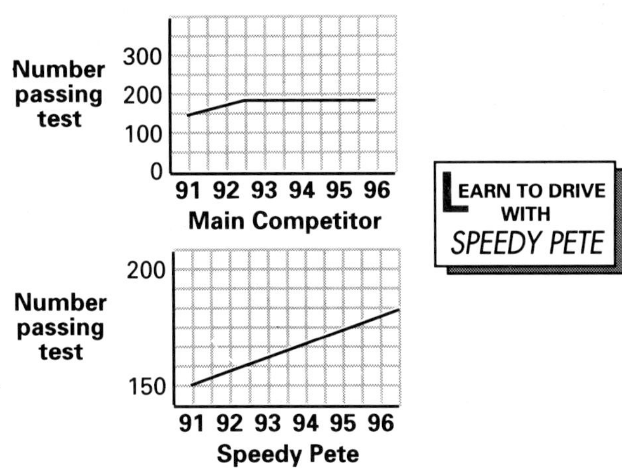

Why are these graphs misleading?

6 'Prices slashed!'

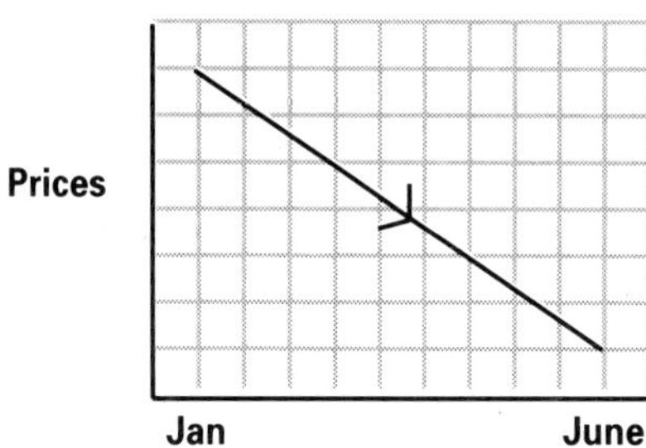

What have you to say about this claim?

7 At first sight, which company is doing better? Look carefully! Have you changed your mind?

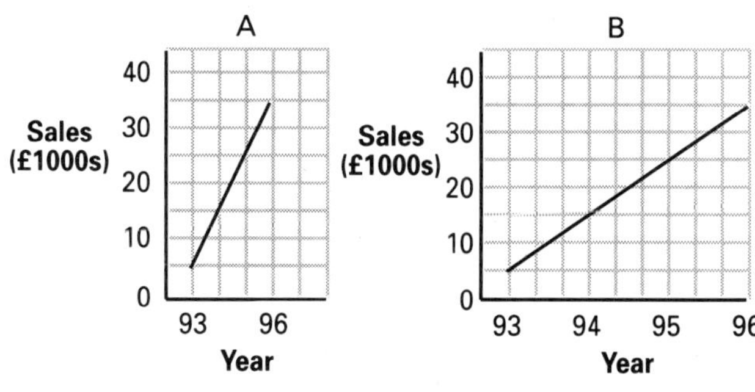

> Study graphs carefully before you believe what they tell you!

CALCULATING AVERAGES: (i) THE MEAN

CLASS DISCUSSION

In her maths exam:
Tania scored 59. Was this good?
She scored 59%. What would you say now?
She scored 59%, *and the class average was* 48%. And now?
She scored 59%, *the class average was* 48%, *the lowest mark was* 40% *and the highest mark was* 60%. How did she do?

Would you agree that the last statement, giving Tania's mark, the class average, and the top and bottom marks, gives the best indication of her performance?

This kind of average is often called the **mean**.
The difference between the highest and lowest scores is the **range**.

Example
Find the mean and range of the lengths 20 cm, 25 cm, 27 cm, 30 cm and 33 cm.

$$\text{Mean} = \frac{20+25+27+30+33}{5}\text{ cm} = \frac{135}{5}\text{ cm} = 27\text{ cm}$$

Range = 33 − 20 cm = 13 cm

EXERCISE 3A

1 Calculate the mean of each set below.
a 1, 3, 5, 7 **b** 36, 63, 51 **c** 2, 5, 3, 7, 9, 4
d 5 kg, 6 kg, 4 kg
e 1.2 cm, 4.6 cm, 7.8 cm, 9.2 cm
f 9p, 19p, 29p, 39p
g 25%, 33%, 47%, 47%, 54%, 64%

2 Calculate the mean value of the coins in each collection.

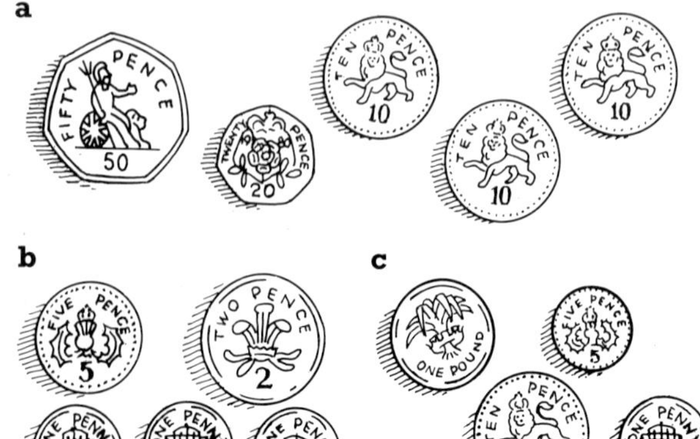

3 Calculate the mean length of these three centipedes.

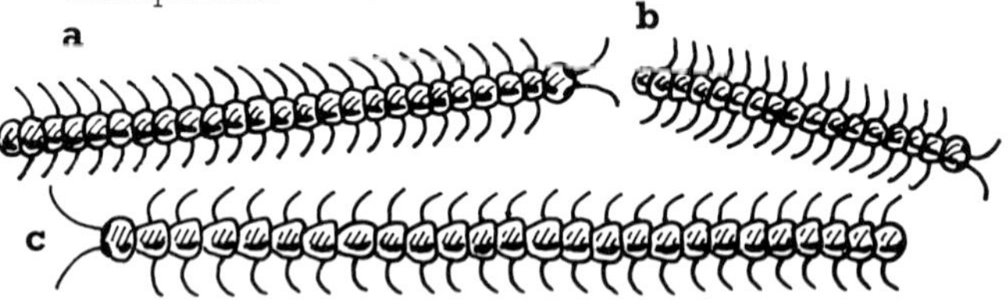

4 Calculate the mean score on each dartboard.

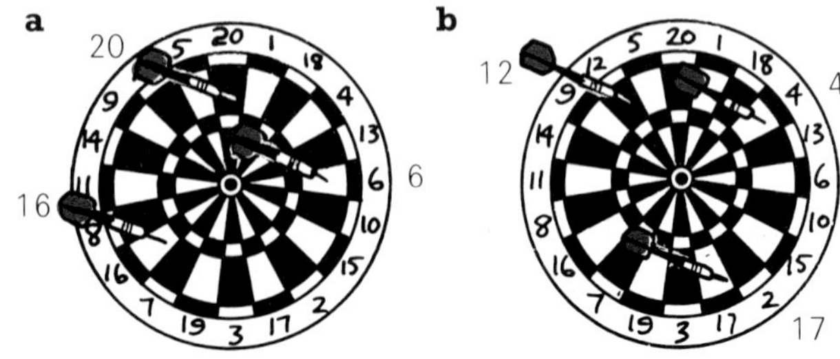

5 a Calculate the mean value, to the nearest £, of these amounts:
£23, £59, £64, £43, £94, £174, £181, £87, £57, £44, £23
b How many are below the mean, and how many are above it?

EXERCISE 3B

1 Calculate the mean and range of each of the following:

a Darts scores: 5, 6, 22, 3, 9, 18, 1, 40

b Plant heights: 34 cm, 32 cm, 32 cm, 33 cm, 31 cm, 35 cm, 32 cm, 35 cm

c Cost of VHS tapes: £3.50, £4.75, £4.25, £4.10

d Throws of a dice: 1, 3, 2, 3, 6, 4, 5, 6, 4, 4, 6

2 The Everbrite Battery Company tested a sample of its batteries in a pocket Space Invaders game. They hoped to be able to claim an average battery life of 18 hours.

The batteries lasted, in hours:
12, 19, 16, 11, 18, 17, 19, 14, 16, 14, 19, 16, 18, 13, 15, 17.

a Calculate the mean life of a battery, correct to 1 decimal place.

b How many hours life will the company claim?

3 May was listening to the craft teacher. She decided to test 12 strips of wood. They broke under loads, in kilograms, of: 5.2, 5.1, 4.9, 5.5, 4.8, 4.7, 4.9, 5.0, 5.3, 5.6, 5.5, 5.2

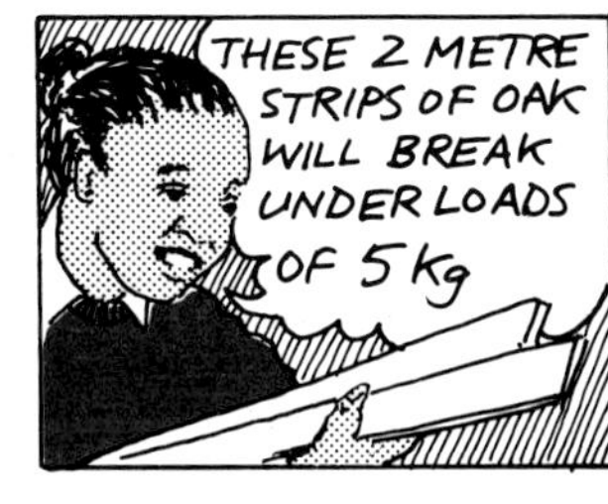

a Calculate the mean and range of the breaking loads, correct to 1 decimal place.

b What is the maximum safe load that might be recommended for these wooden strips?

4

The weather centre checked its records for Silver Sands. Last June the noon temperatures (°C) were:
16, 16, 19, 18, 20, 21, 16, 12, 11, 12, 15, 14, 16, 16, 18, 17, 18, 15, 14, 12, 15, 18, 18, 17, 18, 14, 18, 18, 17, 15.

a Calculate the mean (to the nearest °C) and range of these temperatures.

b If you wrote to the Tourist Office at Silver Sands, what could you tell them about their advertisement?

EXERCISE 3C

1 a Calculate the mean mark and the range of marks for John's class in English and French.

b What might appear in the comments column for these subjects?

Name John Stevens	**Section** 2A	**Date** October 15
Subject	**Score** (%)	**Comment**
Mathematics	69	above average
English	50	
French	70	

2 A machine fills matchboxes. Samples are taken, and if the average falls below 48 the machine is speeded up; above 52, the machine is slowed down. Calculate the mean for each sample shown, and say what action should follow.

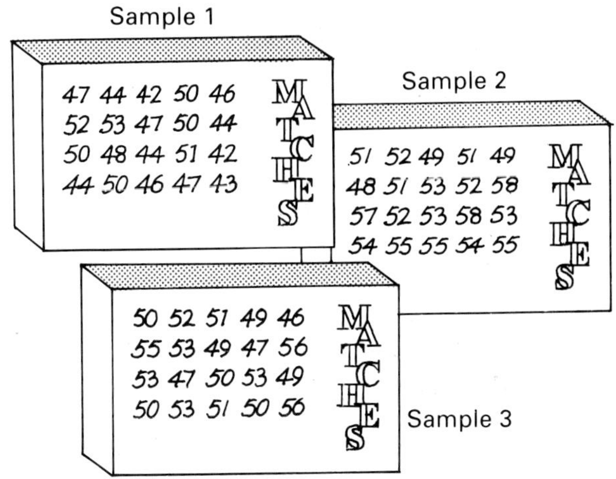

3 A scientist redesigns a lightbulb and claims that it now has a longer life. Sets of old and new bulbs are tested to destruction. The results of the experiments are shown, in numbers of hours.

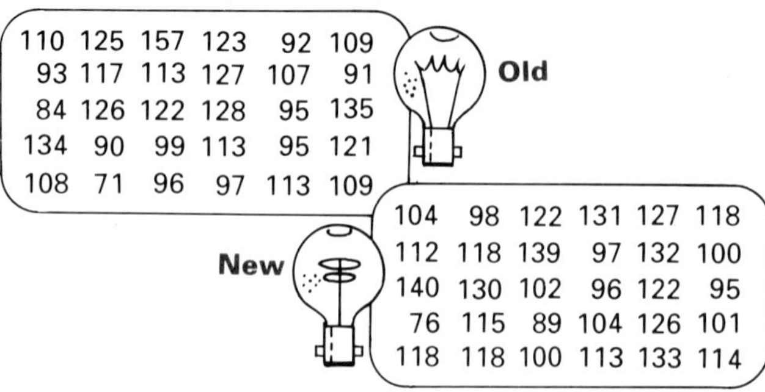

Old

110	125	157	123	92	109
93	117	113	127	107	91
84	126	122	128	95	135
134	90	99	113	95	121
108	71	96	97	113	109

New

104	98	122	131	127	118
112	118	139	97	132	100
140	130	102	96	122	95
76	115	89	104	126	101
118	118	100	113	133	114

Calculate the mean lifetime of each type of bulb, and decide whether the new bulb does last longer.

CLASS STATISTICS

Choose one of the topics below and make a table for your class. Don't lose it—you'll need it for later parts of the investigation.

Heights *to the nearest centimetre*	***Weights*** *to the nearest kilogram*	***Ages*** *to the nearest month*

1 *Calculate:*

a *the mean height, weight or age of the pupils in your class*

b *the range of heights, weights or ages.*

2 *Compare your own height, weight or age with the mean.*

PRACTICAL PROJECT

You have been asked to design a new railway carriage.

- *The carriage should be high enough to give the average pupil 30 cm headroom.*
- *The handrail should be at head level.*
- *The seats should be designed so that the average pupil can sit comfortably with his or her feet on the floor and his or her back in contact with the back of the seat.*
- *The carriage should be wide enough to let two pupils sit face to face and one to stand between them without a crush.*

Collect suitable statistics in your class to help you calculate the requirements.
Write a report for the 'management' giving tables, charts and recommended sizes related to the mean of your sample.

CALCULATING AVERAGES: (ii) THE MODE AND MEDIAN

Lorraine noted the number of children in each of the families of the pupils in her class:
2, 1, 3, 1, 2, 4, 1, 1, 1, 4, 2, 1, 1, 2, 2, 5, 3, 2, 1, 1, 3, 4, 1, 3, 1, 1, 1, 1, 1, 1, 1, 1, 1, 1, 2, 2, 2, 2, 2, 2, 3, 3, 3, 3, 4, 4, 4, 5

The mean is 2.1, Mrs Jenkins.

Arrange them in order, Lorraine, and you'll be surprised to see the most common number.

The **entry that occurs most often**, 1, is called the **mode**, or the modal number of children.

Lorraine's uncle was foreman in a local workshop.
He told her the salaries of the people who were employed there:
Managing director—£65 000
Secretary—£18 000
Foreman—£25 000
Assistants—£20 000, £16 000, £16 000
Manager—£45 000

Lorraine arranged them in order:
£16 000 £16 000 £18 000 **£20 000** £25 000 £45 000 £65 000

The mean salary = £203 000 ÷ 7 = £29 000
The modal salary = £16 000

She did not think that either of these represented the true picture, and decided that the middle one, £20 000, would be a better one to take.

The **median** is the middle entry in an odd number of entries arranged in order, *or* the average of the middle two in an even number of entries.

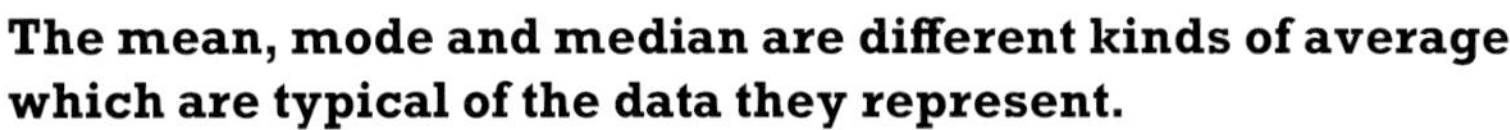

The mean, mode and median are different kinds of average which are typical of the data they represent.

Example
Calculate the mean, mode and median of these marks: 2, 3, 5, 7, 8, 8, 9, 10.

The **mean** $= \frac{52}{8} = 6.5$

The **mode** $= 8$

Since there is an even number of marks, the median is the average of the middle two.

The **median** $= \frac{7+8}{2} = 7.5$

EXERCISE 4A

1 Arrange the data in each of the following in order, from smallest to largest. Then write down the mode and the median.

a Test marks: 7, 5, 4, 7, 3, 6, 8, 5, 7, 6, 1

b Bus fares (pence): 60, 40, 50, 45, 40, 50, 55, 45, 40

c Football goals one Saturday: 0, 3, 0, 1, 2, 0, 3, 0, 2, 0, 1, 0, 0, 2, 1, 0, 0, 2

2 Calculate the mean, mode and median of these bags of money.

3 Calculate the mean, mode and median for the heights, in millimetres, of these toy soldiers.

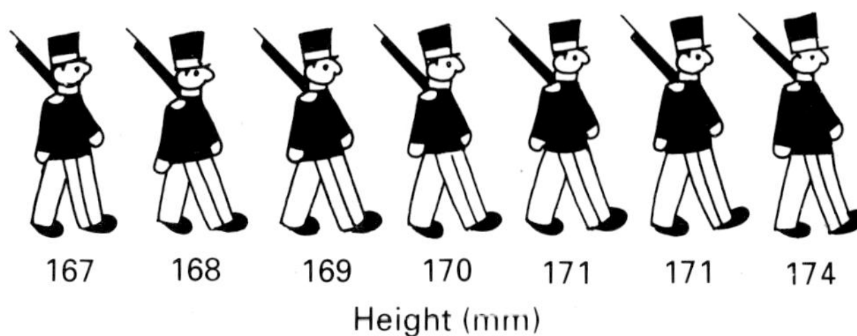

4 Calculate the mean and median scores for these ten throws of the dice.

1 4 2
4 1 4
6 1 5 2

5 Compare these two sets of scores in a game by finding their totals, means, medians and ranges.

a 10, 12, 28, 45, 55 **b** 25, 27, 28, 34, 36

EXERCISE 4B/C

1 Find the mean, median and mode of each list, correct to 1 decimal place where necessary.

a 3, 4, 5, 5, 6, 6, 6, 7, 8
b 3, 4, 4, 5, 5, 6, 6, 6, 7, 8
c 6, 1, 3, 4, 2, 5, 1, 4, 6, 2, 3, 5 (No mode)
d Grades D, C, E, B, A, C, C, B (No mean)

2 In the book *Treasure Island* eight men and a boy meet Long John Silver, who has only one leg.

a Calculate the mean, median and mode of the number of legs.
b Does the boy really have more than the 'average' number of legs?

3 A Tyre and Exhaust Centre has thirteen employees, including the manager and assistant manager. Their weekly wages are:
£150, £150, £150, £150, £150, £160, £160, £160, £170, £170, £180, £300, £400

a Calculate the mean, median and modal wage.
b What fraction of the employees had 'above average' earnings?

4 Coffee is sold by weight (grams) in different sizes of jars. A shopkeeper makes a note of twenty sales, for a survey.
150, 250, 150, 150, 500, 300, 150, 150, 250, 150, 250, 100, 150, 500, 300, 150, 250, 150, 300, 150

a Calculate the mean, median and modal sizes of jar.
b Which 'average' is most likely to interest the shopkeeper?

CLASS STATISTICS

1 *Take the class data in the table you made for* Class Statistics *on page 50, and arrange them in order of size, smallest first.*

2 *Write down the mode and median. Compare these with the mean which you calculated before.*

BRAINSTORMER

For bus 38 to town Alison had to wait 8, 9, 7, 5 and 6 minutes one week. Next week she took the 38A, and had to wait 2, 6, 11, 1 and 5 minutes. Find the mean and range for each set of times, and explain which route you would choose, and why.

INVESTIGATION

'This paragraph is not typical of what you would find in a British book. Do a count. Look for a most common symbol, a modal symbol. What is it that turns this paragraph into an oddity?'

Take a paragraph from any text and do a survey to find how often each letter appears. What is the modal letter?

Repeat the process with the paragraph quoted above. Comment.

CHECK-UP ON MAKING SENSE OF STATISTICS 1

1 This graph shows how the weight of a plant changes as it grows.

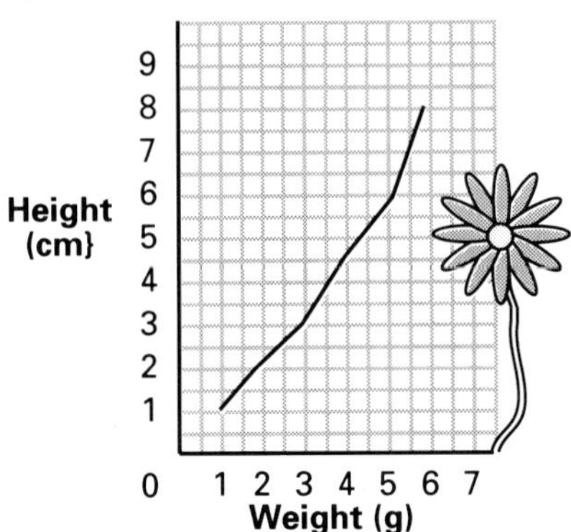

a What weight was the plant at 2 cm?
b Estimate the weight of the plant when it was: (i) 1 cm (ii) 6.5 cm high.
c What was its height when it weighed 3 g?
d Estimate its height when it weighed: (i) 0.5 g (ii) 6.5 g.

2 The History papers have been marked but the marks are unsorted.
10, 20, 28, 40, 31, 33, 36, 43, 15, 39, 45, 18, 26, 13, 39, 19, 21, 28, 33, 35, 44, 48, 19, 24, 35, 28, 34, 21, 35, 22

a Copy and complete the table.

Mark	1–10	11–20	21–30	31–40	41–50
Number					

b Make a bar graph to illustrate the figures.

3 The mean height of Alan's sunflowers was 1 metre. He tried a new fertilizer which claims to increase the average height of a plant by at least 10%.
This table shows the heights his flowers reached then.

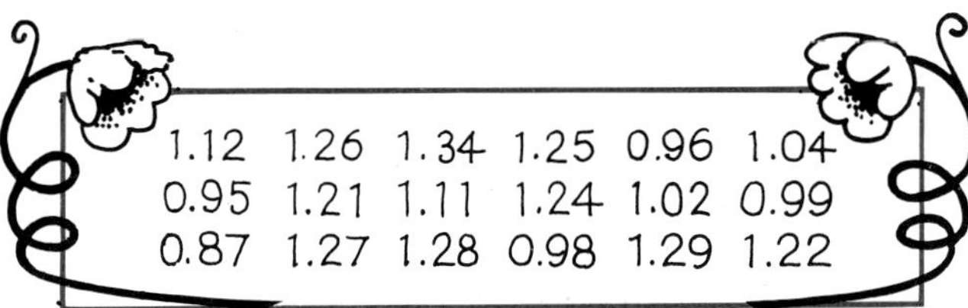

1.12	1.26	1.34	1.25	0.96	1.04
0.95	1.21	1.11	1.24	1.02	0.99
0.87	1.27	1.28	0.98	1.29	1.22

Calculate the mean height of the 18 sunflowers, and say whether it supports the maker's claim or not.

4

Calculate the mean and range of the weights of the five passengers' baggage.

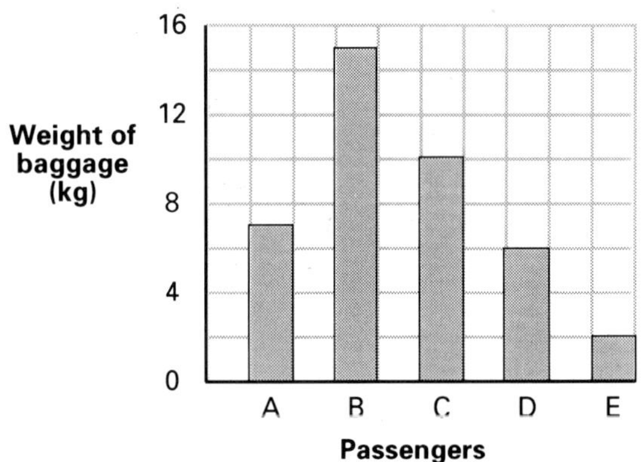

5 Find the mean, median, mode and range for each list below. In each case choose the average which best represents the list.
a 9, 9, 10, 11, 11, 11, 12, 16, 145
b 9, 9, 9, 9, 11, 11, 12, 32, 60
c 9, 9, 10, 11, 12, 13, 15, 18, 20

6 Organise this set of marks into:
a 5 groups **b** 10 groups.

8, 21, 37, 42, 56, 62, 61,
76, 18, 84, 68, 34, 47, 42, 86,
71, 30, 78, 97, 28, 56, 25, 58,
91, 53, 77, 31, 41, 51, 69

5 FRACTIONS, DECIMALS AND PERCENTAGES

LOOKING BACK

1 In the number 352.1, the 2 stands for 2 units. What does 2 stand for in:
a 207 **b** 23.4 **c** 81.24 **d** 903.72?

2 Copy and complete:

a $9\% = \frac{9}{\ldots}$ **b** $5\% = \frac{5}{\ldots} = \frac{1}{\ldots}$

3 Calculate: **a** 50% of £6 **b** 25% of £4
c 75% of £20

Questions **4–8** refer to the pictures at the top of the page.

4 How much is saved on a train fare of £100?

5 What is the deposit on a suite priced £1200?

6 Air consists of the gases shown in the balloon.
a What is the sum of the three percentages?
b What fraction is made up of rare gases?

7 What percentage of the toothpaste is *not* sodium fluorophosphate?

8 What fraction of the material in the shirt is cotton?

9 Which percentages are the arrows pointing to?

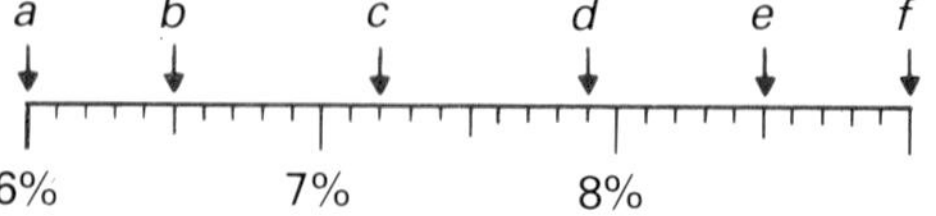

10 One eighth of a class of 32 pupils are left-handed. How many is this? How many are not left-handed?

11 Which is greater?
a 0.01 or 0.10 **b** 9.1 or 8.9 **c** $5\frac{1}{8}$ or $5\frac{1}{4}$

12 Write each fraction below in its simplest form.
a $\frac{3}{6}$ **b** $\frac{2}{8}$ **c** $\frac{8}{12}$ **d** $\frac{4}{10}$ **e** $\frac{15}{20}$ **f** $\frac{15}{18}$

REMEMBERING DECIMAL FRACTIONS

EXERCISE 1 (A CHECK-UP)

Examples

a 123.45 = 1 hundred, 2 tens, 3 units, 4 tenths and 5 hundredths

b 0.4 = 4 tenths = $\frac{4}{10}$

c (i) 123.45 × 10 = 1234.5
(ii) 123.45 ÷ 10 = 12.345

d £1.385 = £1.39, to the nearest penny ('5 or over, round up')

1 Write these numbers in figures:
a sixteen point two
b one hundred and five point eight
c nought point one four.

2 What numbers are these arrows pointing to?

a

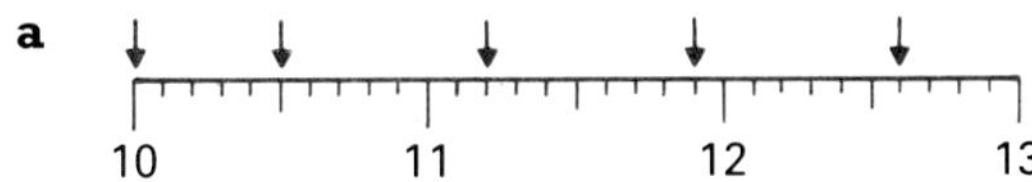

b

3 Arrange these numbers in order, smallest first:
a 15.6, 16.5, 17.0, 16.1, 16
b 9.96, 10.81, 9.68, 11.12

4 Calculate:
a 3.5 + 2.5, and 3.5 − 2.5
b 5.67 + 2.34, and 5.67 − 2.34

5 Calculate:
a 2.25 + 1.76 − 0.56
b 0.68 + 0.79 − 0.54
c 123.4 − 90.2 + 77.6

6 In a diving competition the gold medallist scored 49.68, 47.52 and 60.03 points, and the silver medallist scored 46.86, 51.84 and 55.38 points in the final. Find the total score of each. By how many points was the gold medallist the winner?

7 Steven drives 236.5 km, 168.7 km, 97.6 km and 302.8 km. How much farther does he have to travel to complete a total distance of 1000 km?

8 Calculate the perimeter of this L-shape. Would you have been able to do this if the 5.5 cm and 8.9 cm lengths had not been given?

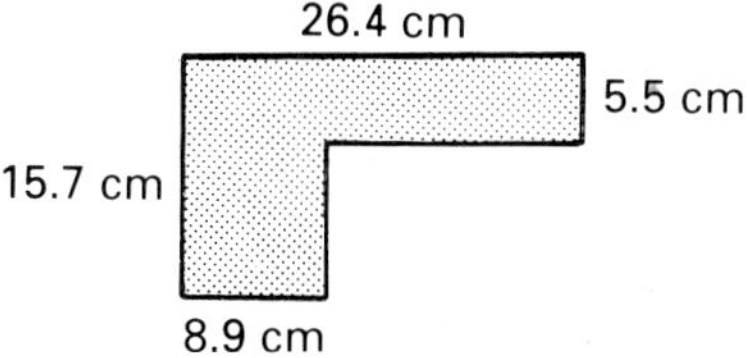

9 Multiply each number by 10, and by 100.
a 3.4 **b** 12.5 **c** 1.09 **d** 234.5 **e** 0.1

10 Estimate, then calculate:
a 7.8 × 5 **b** 3.2 × 9 **c** 12.4 × 4 **d** 25.3 × 6
e 1.89 × 3

11 Divide each number in question **9** by 10, and by 100.

12 Estimate, then calculate:
a 14.7 ÷ 3 **b** 26.5 ÷ 5 **c** 49.5 ÷ 9 **d** 10.4 ÷ 4
e 6.4 ÷ 8

13 Round to the nearest penny:
a £5.632 **b** £1.246 **c** £0.092 **d** £7.125
e £3.199

14 Colour films cost £2.95, or £9.74 for four. How much would you save by buying four?

15 Calculate the cost of this telephone bill.

METER READING		CHARGES	AMOUNT
PRESENT	PREVIOUS		
15499	14994	RENTAL	£19.00
		..... units at 5.65p each →	£
		TOTAL	£

16 The distance by rail between two towns is 31 km. A day return ticket costs £5.60. Calculate the cost in pence per km.

REMEMBERING FRACTIONS AND PERCENTAGES

EXERCISE 2 (A CHECK-UP)

Examples

a (i) $\frac{\cancel{4}^{2}}{\cancel{10}_{5}} = \frac{2}{5}$

(ii) $\frac{2}{5}$ of £12 $= £\frac{2}{5} \times \frac{12}{1} = £\frac{24}{5} = £4.80$

b **Per cent means per hundred**; 1 per cent, or 1%, means 1 per 100; $1\% = \frac{1}{100} = 0.01$

c (i) $17\% = \frac{17}{100} = 0.17$ (ii) $6\% = \frac{6}{100} = 0.06$

d $10\% = \frac{1}{10}$, $25\% = \frac{1}{4}$, $50\% = \frac{1}{2}$, $75\% = \frac{3}{4}$, $100\% = 1$, $150\% = 1\frac{1}{2}$

e 18% of £20 $= 0.18 \times £20 = £3.60$

1 What fraction of each shape below is shaded?

a 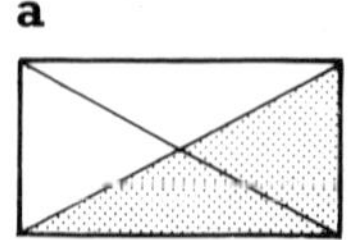 **b** 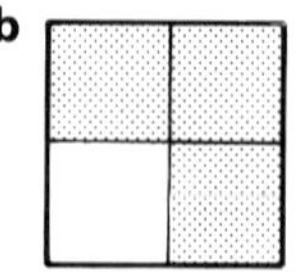 **c** 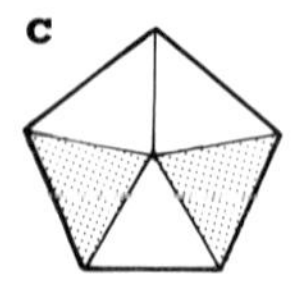

d 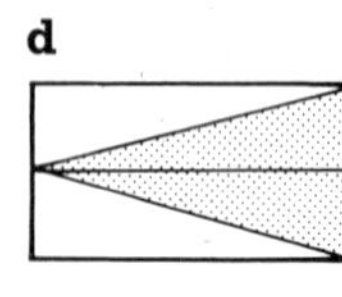 **e** 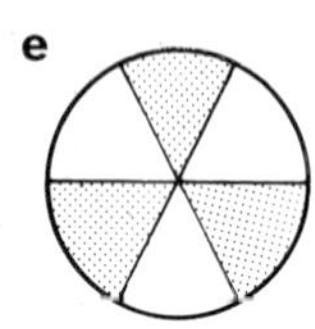 **f** 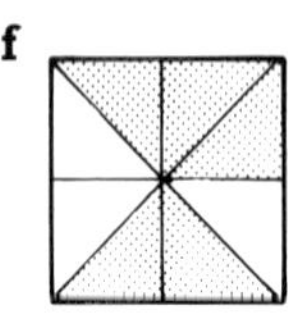

2 What do we call:
a $\frac{1}{12}$ of a year **b** $\frac{1}{60}$ of a minute **c** $\frac{1}{24}$ of a day?

3 Simplify these fractions:
a $\frac{6}{8}$ **b** $\frac{10}{20}$ **c** $\frac{10}{15}$ **d** $\frac{12}{16}$ **e** $\frac{30}{100}$ **f** $\frac{40}{100}$ **g** $\frac{64}{72}$

4 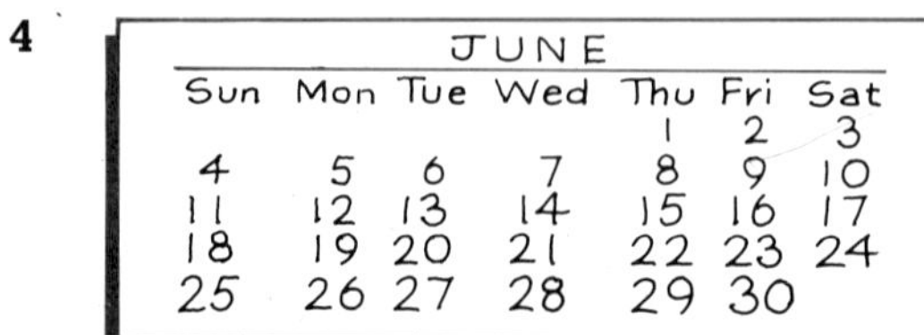

Look at this calendar and find, in simplest form, the fraction of the days in June that are:
a Sundays **b** Fridays
c weekends **d** weekdays

5 Calculate:
a $\frac{1}{10}$ of £1 **b** $\frac{3}{10}$ of £2 **c** $\frac{4}{5}$ of £1 **d** $\frac{1}{2}$ of £5
e $\frac{7}{10}$ of £80 **f** $\frac{2}{3}$ of £12

6 A car park can hold 180 cars.
a How many cars are in it when it is one quarter full?
b There are 120 cars in it. What fraction of the spaces are: (i) used (ii) free?

7 In each shape below what fraction, and what percentage, is shaded?

a 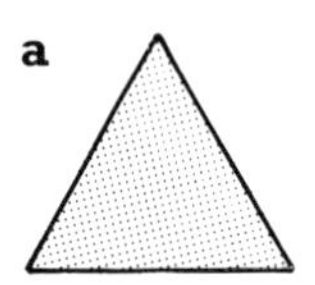 **b** 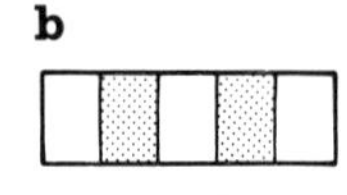 **c** 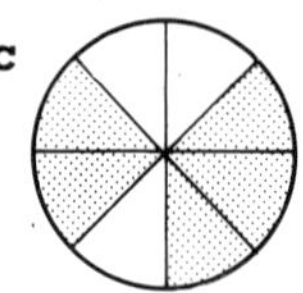

d 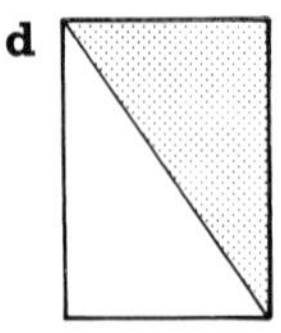 **e** 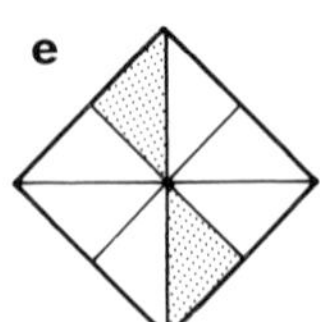 **f** 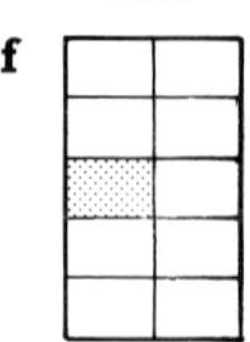

8 Calculate:
a 5% of £200 **b** 10% of £120 **c** 25% of £60
d 60% of £25 **e** 1% of £1000 **f** 150% of £10

9 600 patients were admitted to a hospital one month. 90% were treated successfully. How many patients were cured?

10 a How much reduction will there be on a carpet marked £160?
b What is its sale price?

11 Fit these 'dominoes' together in a straight line.

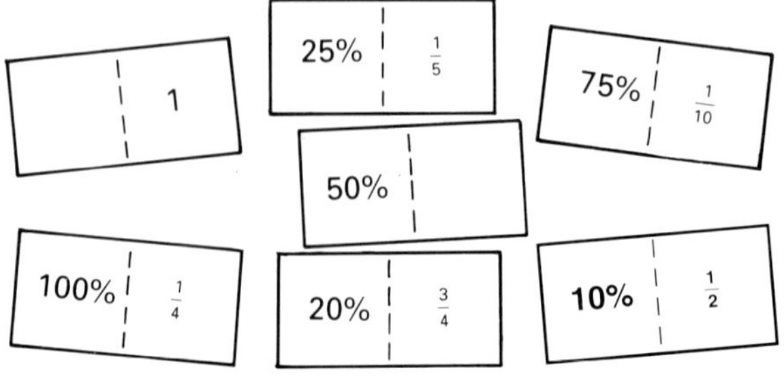

12 60 000 tickets are available for a football match. Each club is allocated 15 000 tickets. What percentage of tickets:
a does each club get
b can be sold to others?

SPENDING AND SAVING

EXERCISE 3A

Sales and discounts

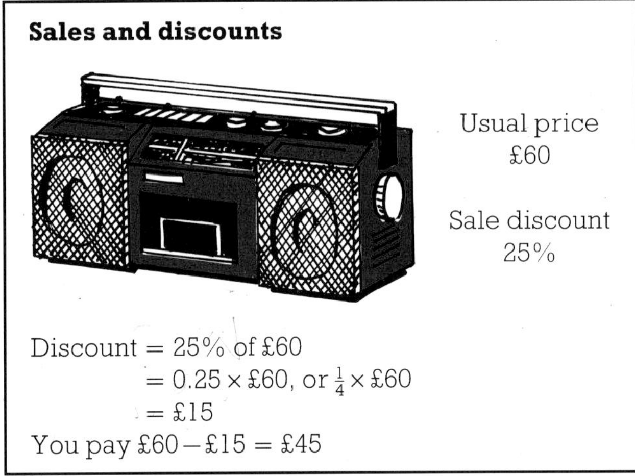

Usual price £60

Sale discount 25%

Discount = 25% of £60
= 0.25 × £60, or $\frac{1}{4}$ × £60
= £15
You pay £60 − £15 = £45

1

The sale discount is still 25%. What discount does Margo get on:

a a notebook, priced £1.80
b a pencil, marked 60p
c books, priced: (i) £10 (ii) £2.60?

2 Calculate these discounts:

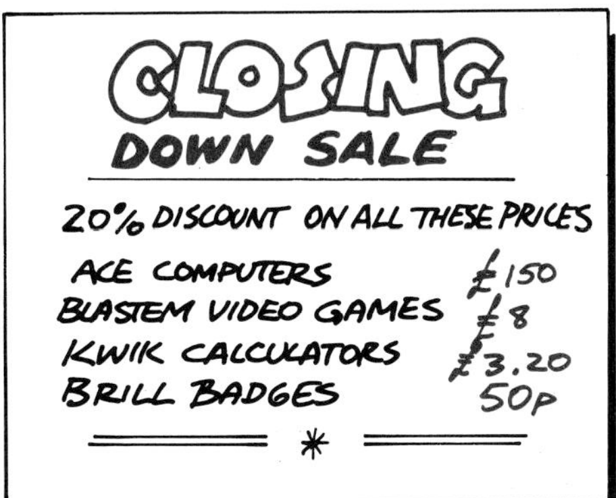

3 Calculate: (i) the discount
(ii) the price you pay, for each item below.

a

b **c**

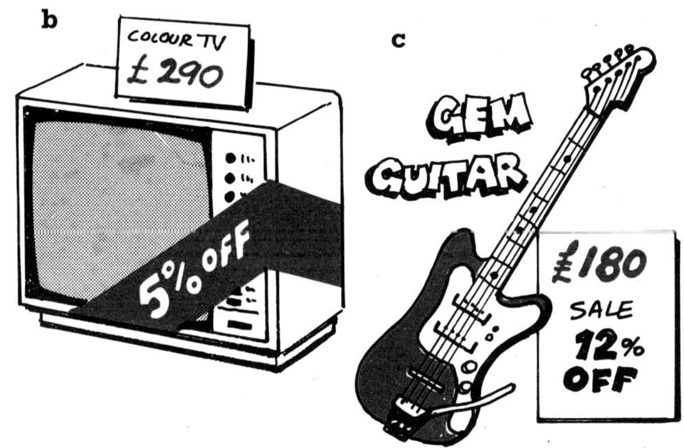

4 **a** If a discount of 20% is given, what percentage of the price must be paid?
b How much will be paid for each of these?
(i) Trainers, usually £45; discount 20%.
(ii) Set of golf clubs, usually £240; discount 15%.

5 Lynda wonders which computer is cheaper. Which one *is* cheaper, and by how much?

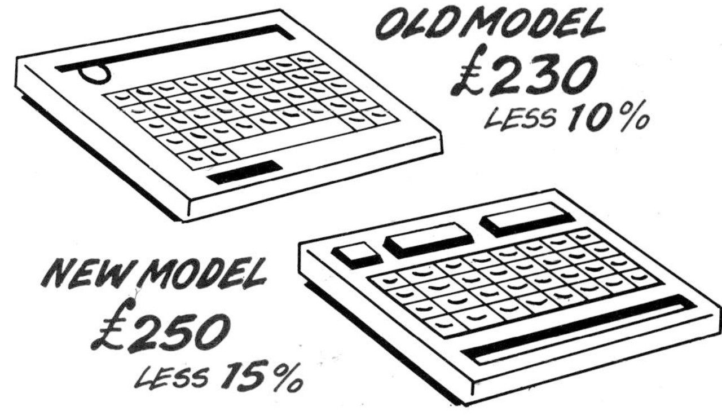

Interest from banks and building societies
Do you ever save any money?
If you do, put it in an account in a bank or a building society and get **interest** on it.

The Bank of Brit pays a **rate of interest** of 8% per year.
8% means 8 per hundred, so you would get £8 interest on £100 after one year.

6 Write down the interest you would receive on £100 for one year at these rates of interest:
a 5% **b** 7% **c** 9% **d** 12% **e** $2\frac{1}{2}$%

7 How much interest would you get after one year at 8% on:
a £100 **b** £200 **c** £500 **d** £1000 **e** £50?

8 Calculate the interest on £75 for one year at these rates:
a 8% **b** 10% **c** 12% **d** 4% **e** 15%

9 The Safety First Bank offers 4% interest. Maxine has £250 in her account.
a How much interest would she get after one year?
b How much money would she have then?

10 Victor is going to put £450 into the Union Jack Building Society.

a How much interest will he receive after one year?
b How much money will he have altogether?

11 The Union Jack Building Society reduces its rate of interest to 7.5% before Victor puts in his money. Answer question **10**, parts **a** and **b** again for the new rate of interest.

12 Compare the interest on £1000 after one year in the National Savings Bank at 9.5%, and in a High Street Bank at 7% per annum.

EXERCISE 3B

Profit and loss

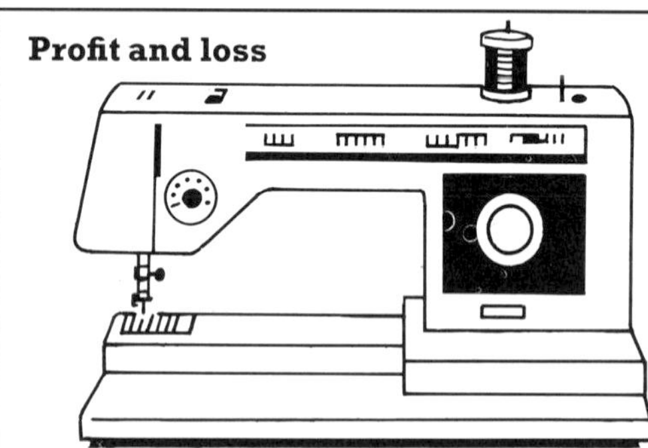

A store buys sewing machines for £175 each, and sells them for £205.
Their profit = £205 − £175 = £30.

$$\textbf{Percentage profit} = \frac{\textbf{profit}}{\textbf{cost price}} \times \textbf{100}\%$$
$$= \frac{30}{175} \times 100\%$$
$$= 17.1\%, \text{ correct to 1 decimal place}$$

1 Calculate the profit and the percentage profit based on the cost price of:
a pencils costing 30p each, selling at 36p
b books costing £5, selling at £7
c ice cream costing 40p, selling at 70p
d anoraks costing £64, selling at £80
e records costing £4.80, selling at £5.40
f sweets costing 80p, selling at £1

2 Bottles of lemonade are bought by a supermarket for 60p each, and sold at 75p. Calculate the percentage profit.

3

This pen was bought for £1, and sold for £1.50. Calculate the profit, and the percentage profit.

4 This camera was bought for £100, and sold for £101. Calculate the profit, and the percentage profit.

5 A store buys boxes of crisps for £18, and sells them at £24. Calculate their percentage profit, correct to 1 decimal place.

6 This car cost the dealer £8000.

a He sold it to Mr Watson for £8800. Calculate his percentage profit.
b Three years later, Mr Watson sold the car for £6600. Calculate his percentage loss.

7 This table shows the stock sold in the school shop.

Item	Cost price	Selling price
300 packets of Chewy	£50	25p a packet
280 chocbars	£63	30p a bar
10 dozen apples	£15	15p each
420 tubs of Fruito	£100	30p a tub

What was the percentage profit:
a on each item
b overall, correct to the nearest 1%?

EXERCISE 3C

Percentage errors, increases and decreases

Julie's weight is 40 kg, but she estimates it at 35 kg.
Her error = 40 kg − 35 kg = 5 kg.

Her **percentage error** = $\dfrac{\textbf{error}}{\textbf{actual weight}} \times \mathbf{100}\% = \dfrac{5}{40} \times 100\% = 12.5\%$

1 Julie's height is 150 cm, but she estimates it at 156 cm. Calculate her percentage error, based on her actual height.

2 The width of Dan's desk is 60 cm. He reckoned it was 66 cm. Calculate his percentage error, based on the actual width.

3 When a metal bar was heated, its length increased from 80 cm to 81 cm. Calculate the percentage increase in length, based on the bar's original length.

4 **a** Look at the table. Who won the local election each year?

	1991	1994
Alison Roberts	3100	3420
Mel Thomson	2047	1866

b What was the percentage increase or decrease in each person's votes between 1991 and 1994, correct to the nearest 1%?

5 In a heatwave, a rail 10 m long increased in length by 1 cm. Calculate its percentage increase in length.

6 A square of new fabric has sides 10 cm long. When stretched, its length and breadth increase by 10%. Calculate:
a the original area, and the new area, of the fabric
b its percentage increase in area.

7 Find the percentage errors in these approximations, correct to 2 decimal places. The actual values are given to 3 decimal places. Which are the best and which are the worst approximations?

Measure	Approximation	Actual value
a 1 inch	2.5 cm	2.540 cm
b 1 mile	1.5 km	1.609 km
c 1 pound	0.5 kg	0.454 kg
d 1 litre	2 pints	1.780 pints
e 1 gallon	4.5 litres	4.546 litres

8 Estimate the length, breadth, area of cover and weight of this book. Measure them, and calculate your percentage errors.

LINKING DECIMALS, FRACTIONS AND PERCENTAGES

EXERCISE 4

Percentages to fractions

1 Change these percentages to fractions in their simplest form.

For example, $55\% = \frac{\cancel{55}^{11}}{\cancel{100}_{20}} = \frac{11}{20}$

a 35% **b** 15% **c** 70% **d** 60% **e** 9%

2 Change the percentages in question **1** to decimal fractions. For example,
$55\% = \frac{55}{100} = 0.55$, and $3\% = \frac{3}{100} = 0.03$

3 Calculate:

a

10% of the money

b

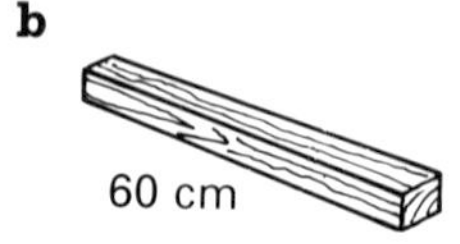

25% of the length

c

30% of the juice

d

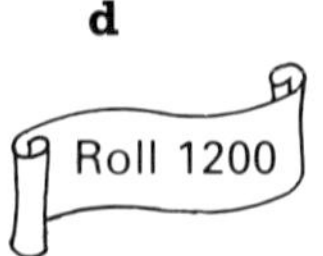

75% of the pupils

e

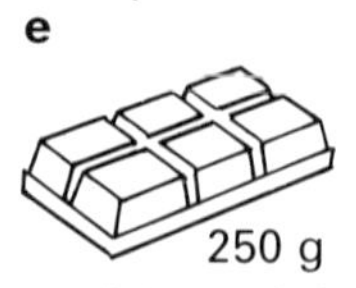

5% of the weight

f

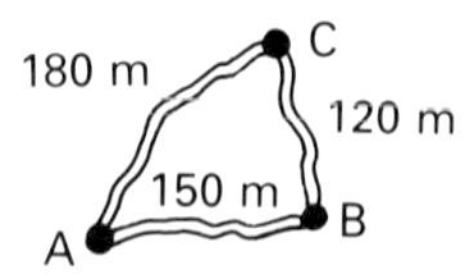

40% of each distance

Fractions to percentages

4 Change these fractions to percentages. For example, $\frac{3}{5} = \frac{3}{5} \times 1 = \frac{3}{5} \times 100\% = 60\%$, or $\frac{3}{5} = \frac{3}{5} \times 100\% = 60\%$

a $\frac{1}{2}$ **b** $\frac{3}{4}$ **c** $\frac{1}{10}$ **d** $\frac{2}{5}$ **e** $\frac{9}{10}$ **f** $\frac{3}{20}$

5 Change these decimal fractions to percentages. For example, $0.87 = 0.87 \times 100\% = 87\%$

a 0.47 **b** 0.22 **c** 0.93 **d** 0.8 **e** 0.08 **f** 1.23

6 Change these to percentages.

a $\frac{1}{5}$ **b** $\frac{1}{50}$ **c** $\frac{7}{10}$ **d** 0.35 **e** 0.5 **f** 0.01

Fractions to decimal fractions

7 Change these to decimal fractions. For example, $\frac{3}{5} = 3 \div 5 = 0.6$

a $\frac{3}{4}$ **b** $\frac{2}{5}$ **c** $\frac{4}{5}$ **d** $\frac{3}{10}$ **e** $\frac{1}{2}$ **f** $\frac{1}{4}$ **g** $\frac{7}{10}$ **h** $\frac{1}{5}$

8 Change these to decimal fractions, correct to 2 decimal places. For example, $\frac{2}{3} = 2 \div 3 = 0.666\ldots = 0.67$, correct to 2 decimal places

a $\frac{2}{7}$ **b** $\frac{5}{8}$ **c** $\frac{1}{3}$ **d** $\frac{5}{6}$ **e** $\frac{1}{11}$ **f** $\frac{8}{9}$

9 Write each shaded part below as a fraction, decimal fraction and percentage of the whole shape.

a **b** **c**

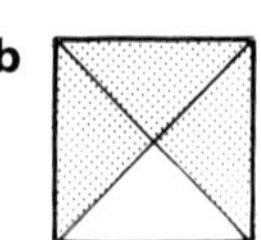

10 Match up pairs of equal numbers—a fraction and a percentage in **a**, a decimal fraction and a percentage in **b**.

a

$\frac{1}{2}$ $\frac{1}{4}$ $\frac{1}{3}$ $\frac{1}{10}$ $\frac{3}{10}$ $\frac{3}{4}$	10% 25% 30% 75% 50% $33\frac{1}{3}\%$

b

0.2 0.25 0.5 0.01 0.1 0.05	10% 1% 25% 5% 50% 20%

11 Copy and complete this table, putting fractions in simplest form.

Fraction	$\frac{1}{2}$			$\frac{2}{5}$		
Decimal		0.75			0.3	
Percentage			25%			7%

A FRACTION-DECIMAL-PERCENTAGE GAME

Any number can play. Take turns at rolling a dice twice. Using the first number as the numerator and the second as the denominator, write down your score as a fraction. Convert this to a decimal fraction, then to a percentage, correct to 1 or 2 decimal places if necessary. The winner of each round is the person with the highest percentage score.

INVESTIGATIONS

1 a *Write down the next five terms of the sequence* $\frac{1}{2}, \frac{1}{4}, \frac{1}{8}, \frac{1}{16}, \ldots$

b *Use a calculator to investigate the sum* $\frac{1}{2}+\frac{1}{4}+\frac{1}{8}+\frac{1}{16}+\ldots$ $(1 \div 2 + 1 \div 4 + \ldots)$

c *What happens as you add more fractions? Will the sum ever be 1?*

2 *A tree is 4 m tall. In successive years it grows by* $\frac{1}{5}, \frac{1}{6}, \frac{1}{7}, \ldots$ *of its height at the beginning of that year. In how many years will its height be doubled?*

FRACTIONS, DECIMALS AND PERCENTAGES IN ACTION

EXERCISE 5A

1 A school team won 50% of their games, and drew 20% of them.
What percentage of the games did the team lose?

2 In a class of 25 pupils, 4 are absent.
a What fraction, and what percentage, are absent?
b What percentage of the pupils are present?

3 What fraction, decimal fraction and percentage of the windows in this block of flats are dark?

4 The jar holds one litre of liquid. What decimal fraction, fraction and percentage of the jar:
a contains liquid
b is empty?

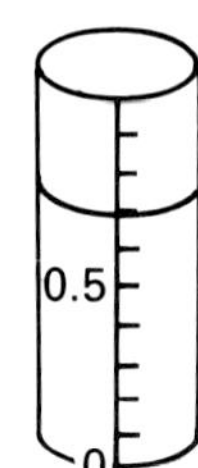

5 This picture and pie chart show how heat is lost from a typical house.

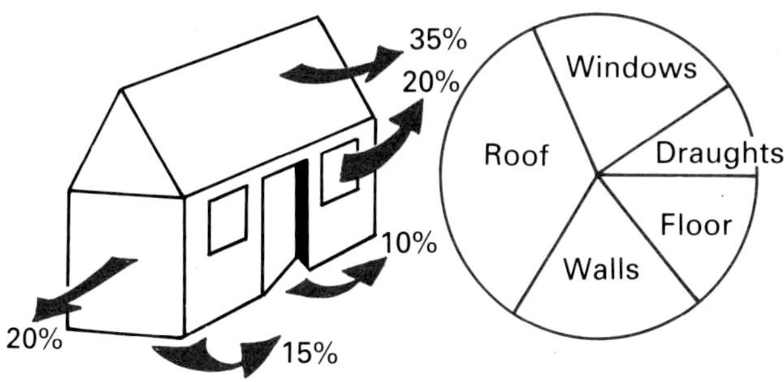

a What percentages should be written in beside walls, roof and windows in the pie chart?
b Where is most heat lost? How could this be reduced?
c This house loses heat to the value of about £500 each year. How much of this is:
(i) in draughts (ii) through the walls
(iii) through the roof?

6 Mr Guha spent four nights in New York. He paid £100 a night for his hotel, plus 10% service charge. Calculate:
a the service charge for four nights
b his total bill.

7 The Bluebird Bus Company claim that only 6% of their buses are late.
a (i) What fraction is late?
(ii) What decimal fraction is late?
b If 150 buses are on the road, how many would the Company expect to be late?

8 Calculate the perimeter and area of the video tape and the carpet.

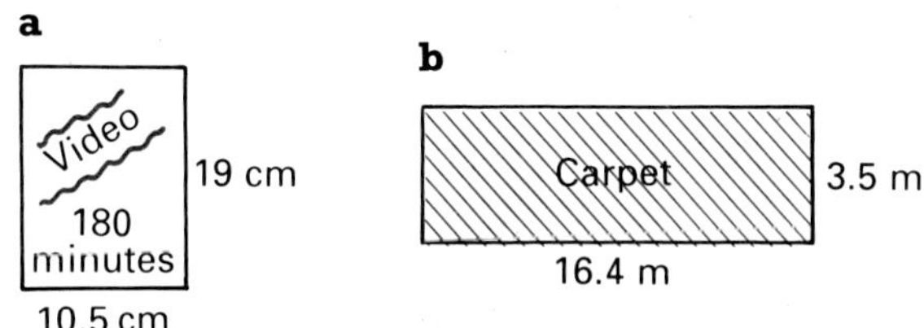

9 80 pupils in Allford High School took part in the Duke of Edinburgh's Award Scheme. Their results are shown in this pie chart.

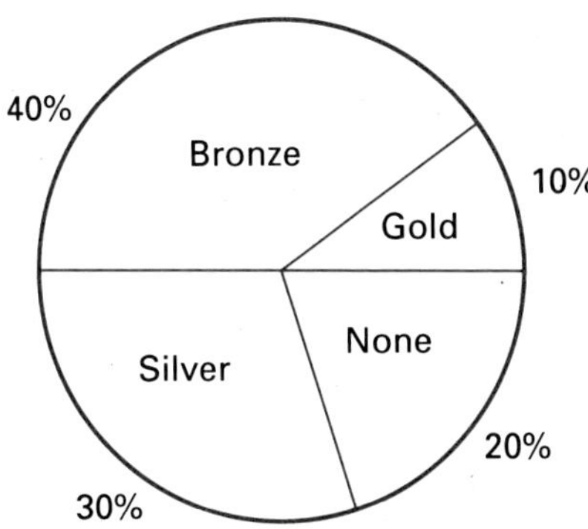

a 10% of the pupils achieved the Gold award. How many pupils was this?
b How many received Silver and Bronze awards?

10 Here are Harry's marks in his class tests.
Test 1: $\frac{18}{25}$ Test 2: $\frac{18}{24}$ Test 3: $\frac{12}{15}$ Test 4: $\frac{21}{30}$
a Can you tell which is his best mark?
b Change them to percentages, and list them in order, best first.

EXERCISE 5B/C

1 What fraction, decimal fraction and percentage of the target:
a has been reached
b is still left?

2 Some Middle Eastern countries' oil reserves, as percentages of the world's resources, are:
Iran 9.2%, Iraq 9.9%, Kuwait 9.3%,
Saudi Arabia 25.2%, United Arab Emirates 9.1%.
a Are their total reserves more or less than half of the world's? By how much?
b What percentage of these countries' reserves are in Saudi Arabia?

3

What fraction, decimal fraction and percentage of these coins are:
a 50ps **b** 10ps **c** 2ps **d** 1ps?

4 The three top selling items in a school shop are divided like this: crisps 50%, soft drinks 30%, chocolate bars 20%.
The angle needed on a pie chart for the crisps is 50% of 360° = 180°.

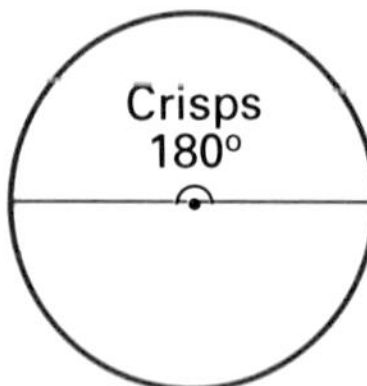

Find the angle for the soft drinks and chocolate bars, and draw the pie chart.

5 In a school of 800 pupils, 60% are girls and 40% are boys.
a Draw a pie chart which illustrates this information.
b How many girls and how many boys are in the school?

6 Calculate the percentage increase in the fare from winter to summer.

7 A survey showed that, to the nearest 5%, the time that people spent watching television was divided up like this:
BBC1 30%, BBC2 10%, ITV 45%, Channel 4 10%, Sky 5%.
Illustrate these data in a pie chart.

8 The number of goals scored by 80 teams in the English leagues one Saturday were shown in Monday's paper in a pie chart like this.

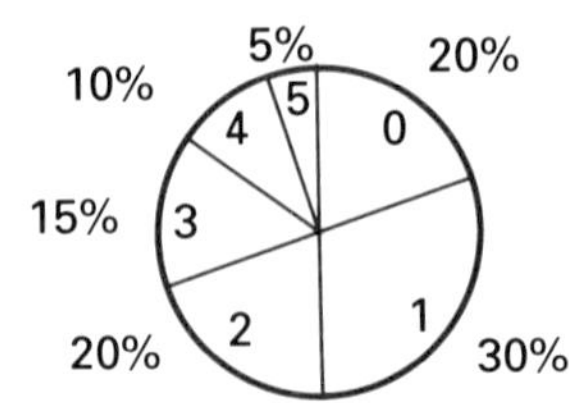

a What was the most common number of goals scored?
b How many teams scored five goals?
c Calculate the total number of goals scored.
d What was the average number of goals scored?

BRAINSTORMERS

1 *£1000 is invested at 10% per year. At the end of each year the interest is added on, and the following year's interest is calculated on this total. In which year would the £1000 invested double itself?*

2 *Each edge of a metal plate expands by 1% of its length when the plate is heated. What is the percentage increase in its area?*

INVESTIGATION

The games master at Firth Academy has a problem. He has to award a trophy to the team which has had the most successful season. The teams' results were:

	Games played	***Games won***	***Games lost***	***Games drawn***
Rugby	*25*	*17*	*7*	*1*
Hockey	*20*	*11*	*6*	*3*
Golf	*12*	*8*	*3*	*1*
Basketball	*10*	*7*	*3*	*0*
Netball	*15*	*11*	*4*	*0*
Football	*18*	*11*	*3*	*4*

Investigate ways in which he could choose the winner. Which team would you choose? Explain why.

CHECK-UP ON FRACTIONS, DECIMALS AND PERCENTAGES

1 Calculate:
a 40% of 650 g **b** 25% of £14 **c** 5% of £1.80

2 Arrange in order, smallest first: 41%, 0.39, $\frac{2}{5}$.

3 Personal stereos costing £45 are reduced by 10% in a sale.
Calculate:
a the discount
b the price you pay.

4 Calculate the interest on £250 for one year at 7%.

5 Which earns more in a year:
£150 at 11% or £180 at 9%?

6 A car's price of £9850 is increased by 6%.
Calculate:
a the increase in price **b** the new price.

7 What fraction of an hour is 48 minutes:
a in simplest form **b** as a decimal fraction?

8 Calculate this gas bill.
Gas used: 1523 kW at 1.507p per kW
Standing charge: 82 days at 10.3p per day

9 Calculate these marks to the nearest 1%:
25 out of 40, 44 out of 60, 18 out of 25, 26 out of 35.

10 Last year Caremore High School pupils saved £2500 for charity. The money was divided between several charities, like this:

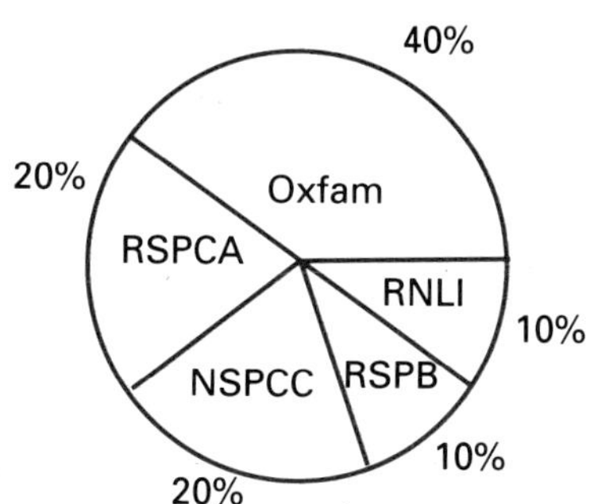

a How much did each charity receive?
b Do you know the full names of any of these charities?

11 Copy and complete this table:

Fraction	$\frac{4}{5}$			$\frac{3}{8}$		
Decimal		0.7			0.04	
Percentage			7%			64%

12 Cartons of ice-cream are bought for 56p and sold for 70p. Calculate:
a the profit on each carton
b the percentage profit.

13 A building plot is 60 m long and 50 m wide. New plans increase the length by 20% and decrease the width by 20%. Calculate:
a its original area **b** its new area
c the percentage change in area.

14 Mr Hope spent £1000 buying Rocket shares. Their value rose by 20%, then fell by 20% of the new value. Were the shares now worth more or less than at the beginning?

15

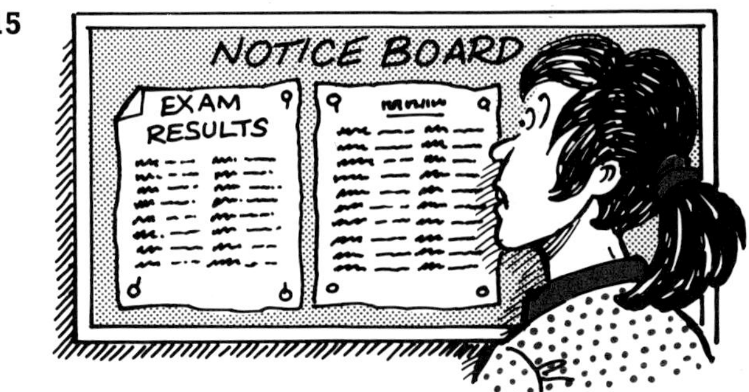

If Tanya chooses to study her five best subjects next year, which ones will these be, based on this record?

English	**Maths**	**History**	**Geography**
56/80	56/70	33/60	35/50
French	**Science**	**Art**	**Technology**
48/75	51/75	22/40	30/40

6 DISTANCES AND DIRECTIONS

LOOKING BACK

1 Use your ruler and protractor to measure:
a PQ, PR, QR (correct to 1 decimal place)
b angles PQR, QPR and PRQ.

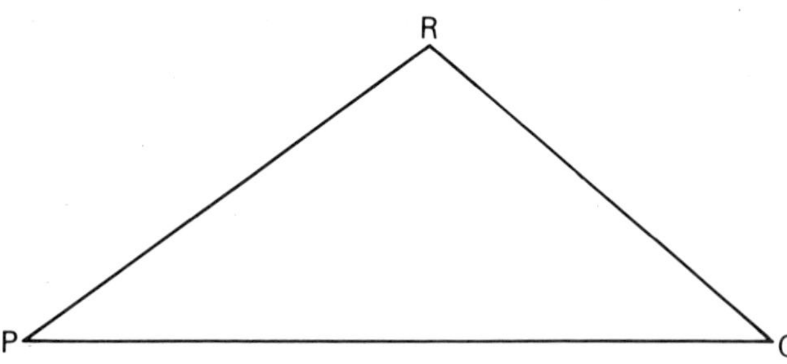

2 a Draw △ABC full size.

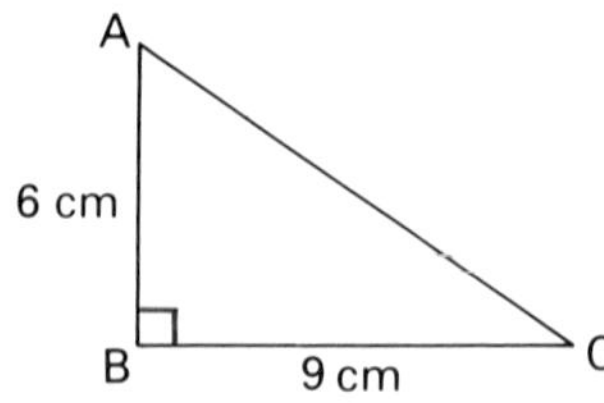

b Measure ∠BCA, ∠CAB and the length of AC.

3 Through how many degrees does the minute hand of the clock turn in:
a quarter of an hour
b half an hour
c an hour?

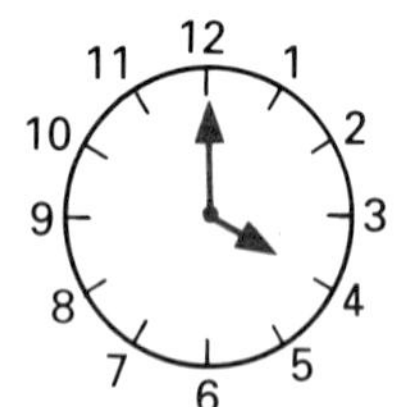

4 Karina, the discus thrower, is facing north. In which directions are arrows 1, 2 and 3 pointing?

5 On the island, the symbols represent:
PO, post office; ⧍, campsite; [castle symbol], castle; [i], information; [airport symbol], airport; [flag], golf course; †, church; [telephone symbol], telephone.

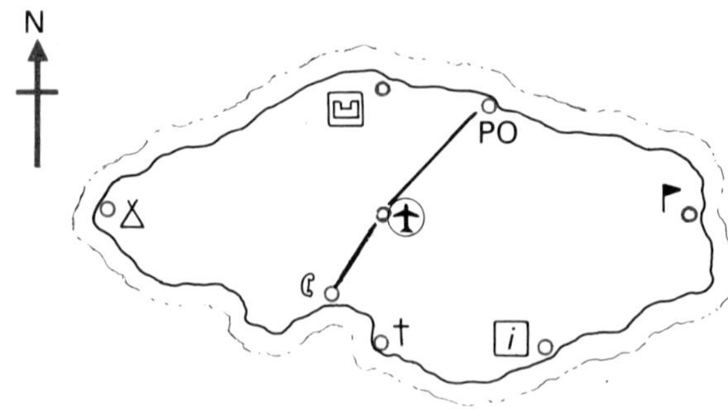

Which of the features listed above are due:
a north of the airport
b east of the campsite
c south of the castle
d west of the information office?

6 The computer-controlled 'turtle' follows these instructions:
FORWARD 10 (cm); RIGHT 90 (degrees); FORWARD 6; RIGHT 90; FORWARD 10.

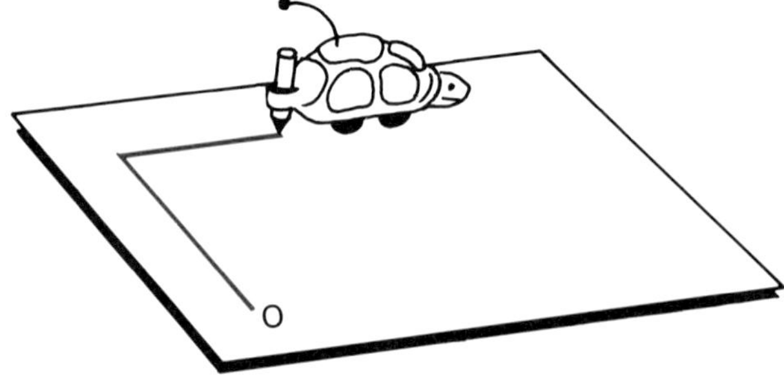

a Sketch the turtle's journey.
b What two instructions will take it back to point O?
c Write out instructions for:
(i) a square of side 5 cm
(ii) an equilateral triangle of side 8 cm, starting from O each time. Check, by sketching the shapes.

SCALE DRAWINGS

Scale drawings can show lengths or distances which are too great to draw full-size.

Examples

a A distance of 3 km
Scale: 1 cm to 1 km
Scale drawing:
0 1 2 3 km

b A length of 40 m
Scale: 1 cm to 10 m
Scale drawing:
0 10 20 30 40 m

EXERCISE 1A

1 Measure each line below in cm. Then, using the scale 1 cm to 1 m, write down the length in metres that each line represents.

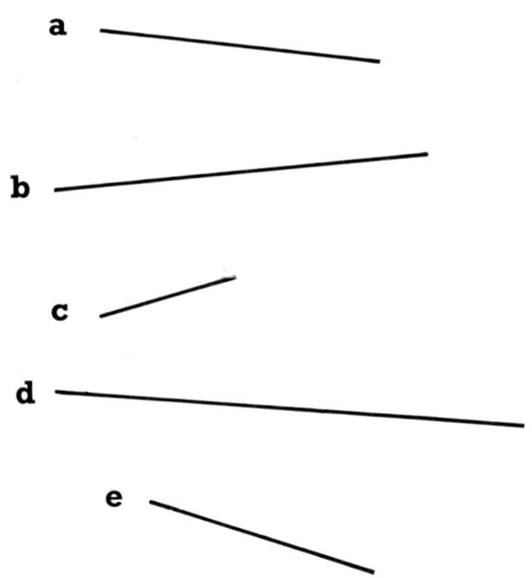

2 *Scale:* 1 cm to 2 m. What length does each line represent?

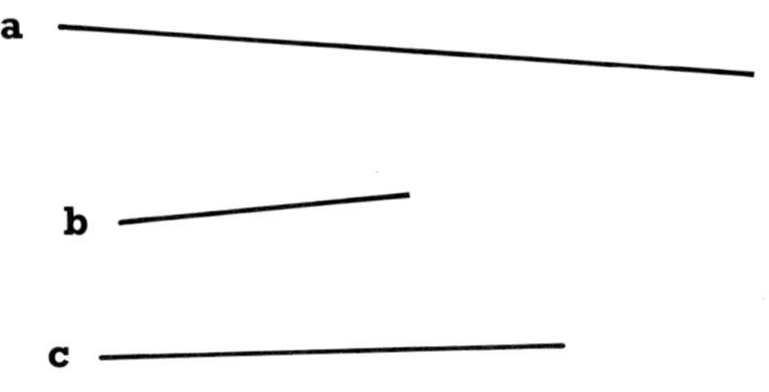

3 Draw lines to represent these distances, using the given scales.

a *Scale:* 1 cm to 1 km.
Distance: (i) 2 km (ii) 6 km

b *Scale:* 1 cm to 5 m.
Distance: (i) 10 m (ii) 25 m

c *Scale:* 1 cm to 2 km.
Distance: (i) 2 km (ii) 7 km

4 Describe these four journeys, giving the distance in km and the direction of each part.

a *Scale:* 1 cm to 1 km

b *Scale:* 1 cm to 1 km

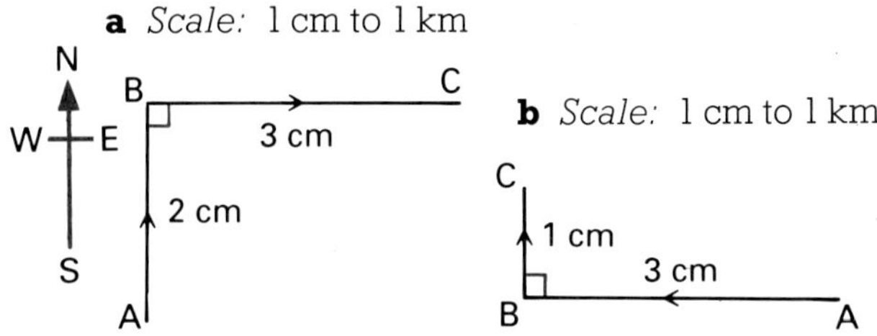

c *Scale:* 1 cm to 2 km

d 1 cm to 2 km

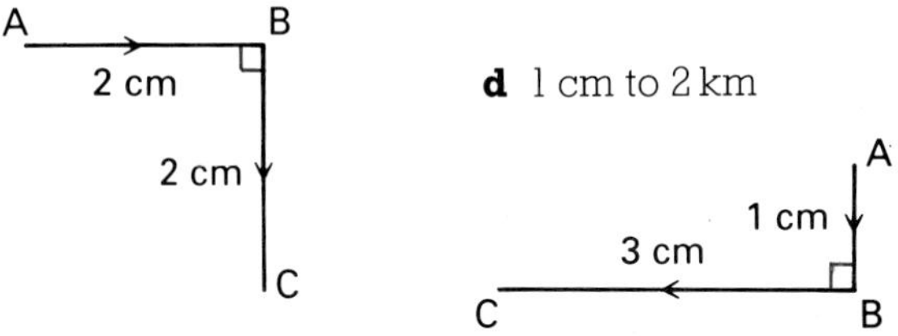

5 (i) Draw the scale drawings in question **4** accurately.
(ii) Measure AC in cm, correct to 1 decimal place, in each drawing, and write down the distance in km that it represents.

6 A yacht sails 8 km north, then 6 km east from its marina M.

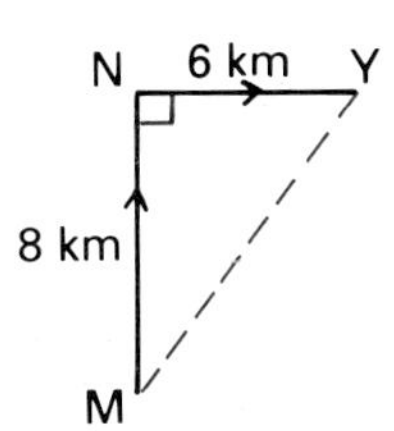

a Using a scale of 1 cm to 1 km, make a scale drawing of its voyage.

b Measure MY. How far is the yacht from M?

7 Repeat question **6** for a voyage 7 km south, then 7 km east, from M.

EXERCISE 1B

An angle of elevation or depression measures the angle between the horizontal and a line to an object which is above or below the horizontal.

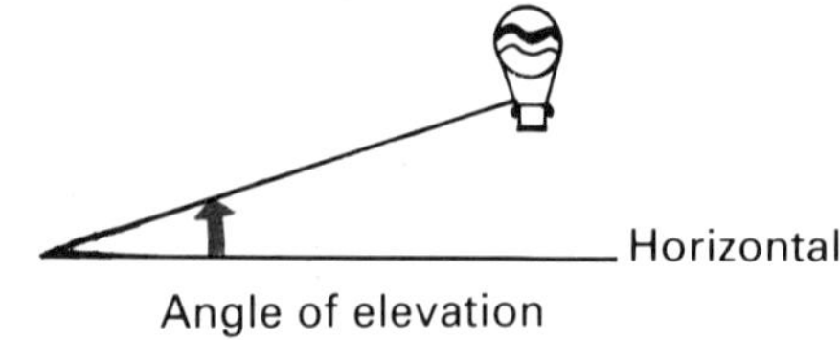

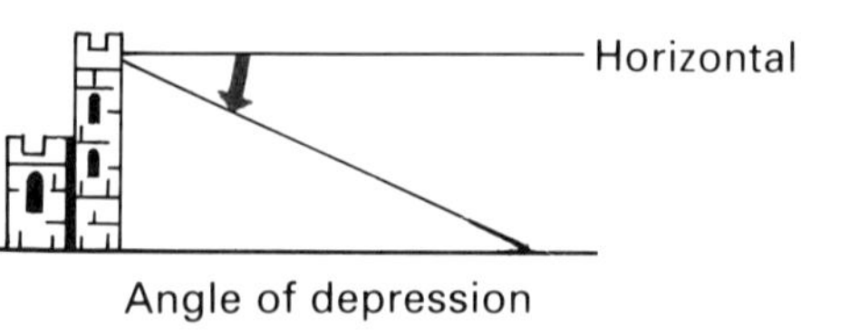

1 The angle of elevation of the top of the school is 40°.

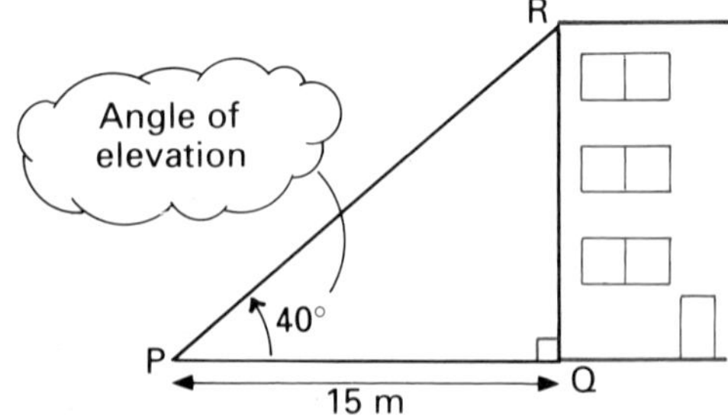

a Use a scale of 1 cm to 3 m to make a scale drawing.
b Measure QR. What is the height of the school?

2 Use scale drawings to find the heights of the cliff and the flagpole, shown below, to 2 significant figures.

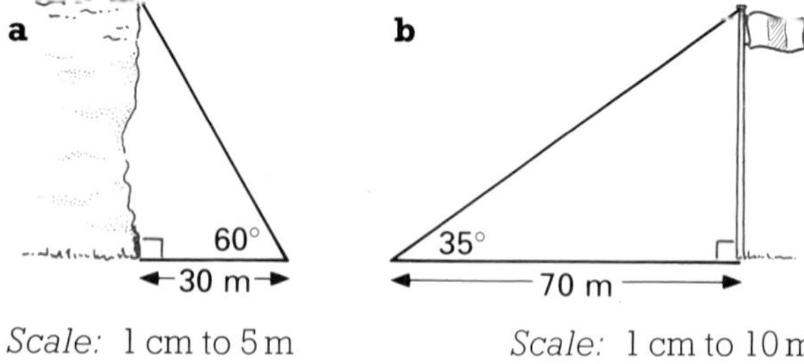

Scale: 1 cm to 5 m   *Scale:* 1 cm to 10 m

3 The angle of depression of the rock is 28°.

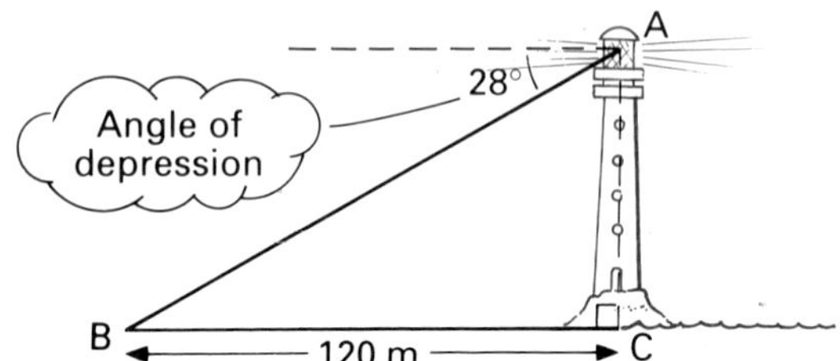

a What size is ∠ABC?
b Using a scale of 1 cm to 20 m, prepare a scale drawing.
c Calculate the height of the lighthouse.

4 From the top of the Eiffel Tower, Mark sees his friend Jeff on the ground at an angle of depression of 75°. Jeff is 80 m from the foot of the tower.

Use a scale of 1 cm to 20 m to make a scale drawing. Find the height of the Eiffel Tower, to the nearest metre.

5 Rasheen measures the height of the spire, using an anglemeter: 41°.

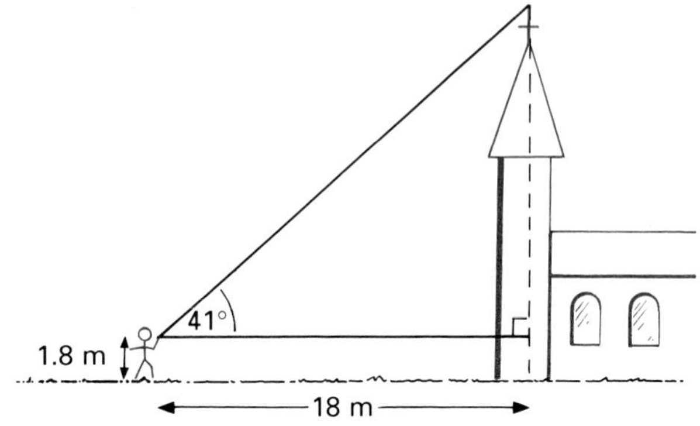

a Use a scale of 1 cm to 3 m for a scale drawing.
b Find the height of the spire, but remember to add Rasheen's height, 1.8 m.

EXERCISE 1C

Make a scale drawing in each question in this exercise.

1 Find the height of the balloon.

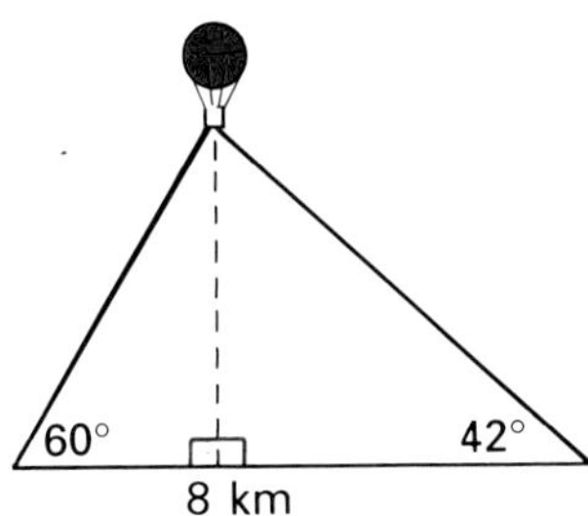

2 Calculate the distance of the fishing boat, B, from the harbour, H.

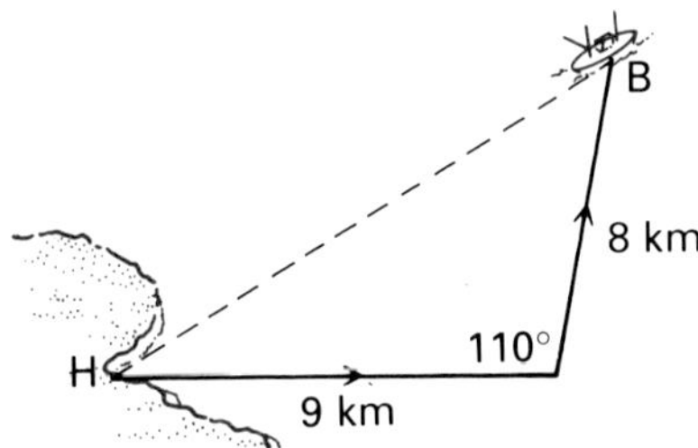

3 Find the distance from the pier, P, to the island, R.

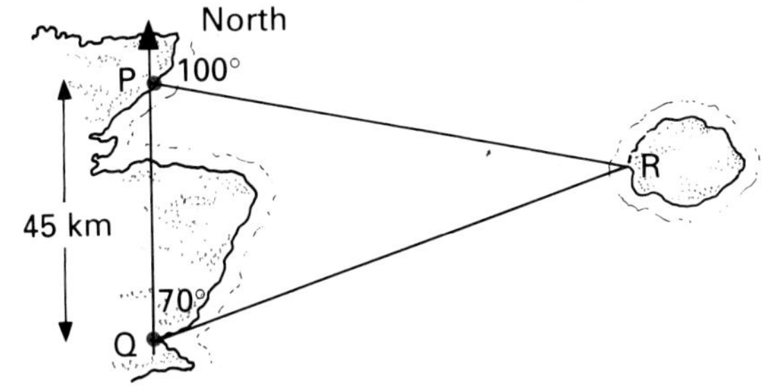

4 Calculate the height of the hill.

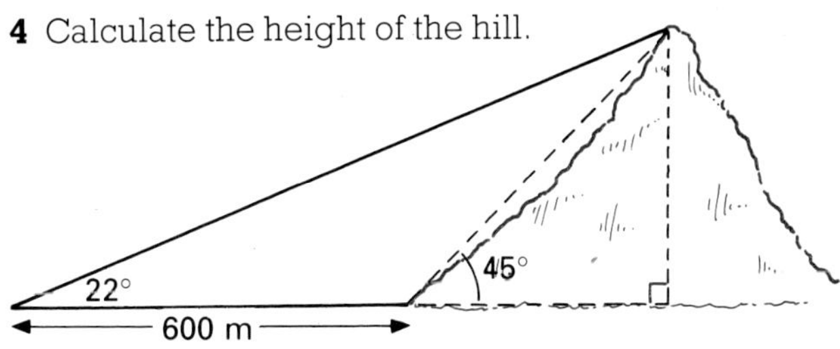

PRACTICAL PROJECT

Use an anglemeter and long measuring tape to enable you to find the height of the tallest part of the school, the nearest church spire, or a tall tree.

CHALLENGE

Make a scale-model house

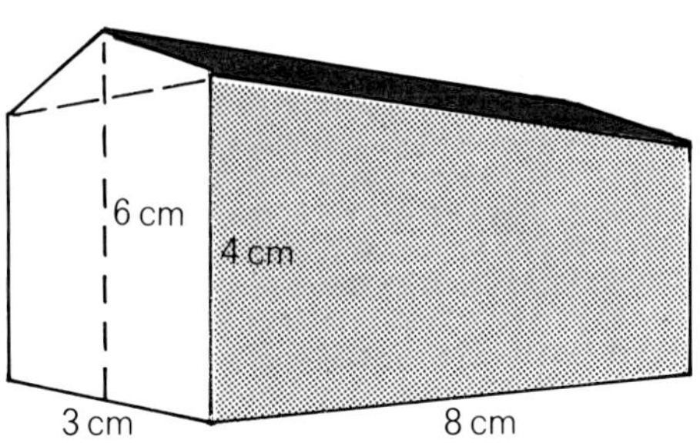

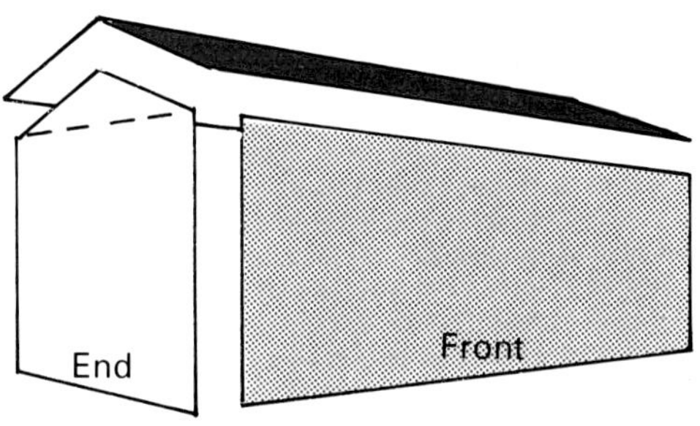

Dave has a model railway, and he has decided to make several cardboard houses like the one shown above. Before cutting the cardboard he has to make accurate drawings.

1 *What shape is the front of the house?*

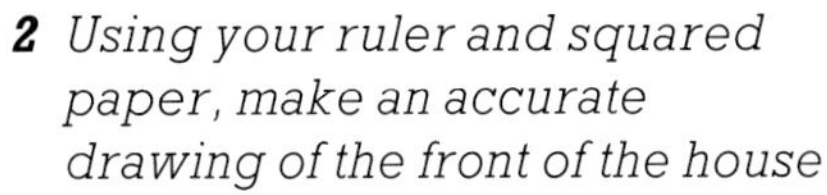

2 *Using your ruler and squared paper, make an accurate drawing of the front of the house.*

3 *How would you describe the shape of the end of the house? Make an accurate drawing of the end.*

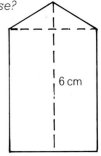

4 *Measure the length of the sloping sides of the triangular part of your drawing.*

5 *The roof is made from a rectangle which is creased and folded as shown. Make an accurate drawing of this rectangle.*

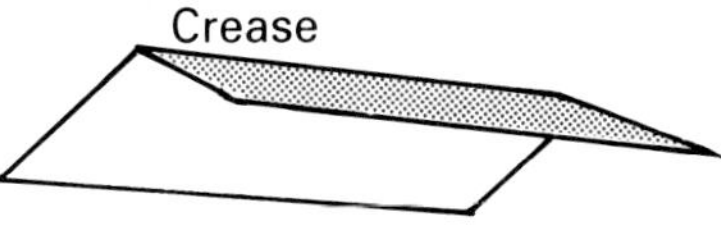

6 *You can now make the model house if you wish.*

COMPASS BEARINGS

EXERCISE 2A

1 Copy the diagram, and write out the names of the eight points of the compass in full.

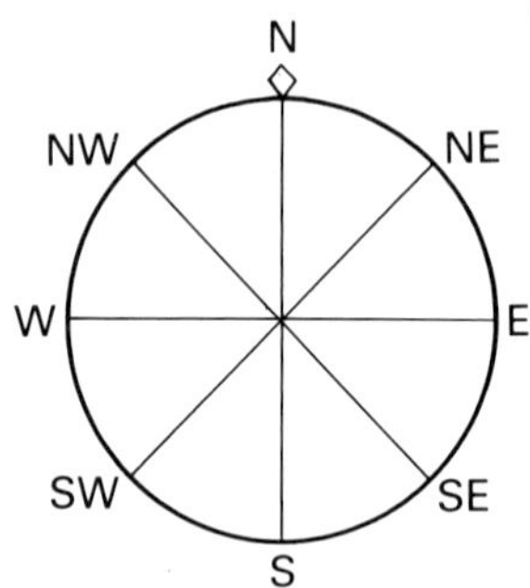

2 How many degrees are in the angles between:
a N and NE **b** N and E **c** N and SE
d SE and SW?

3 If you start facing N each time, what direction do you face after turning:
a 90° clockwise **b** 180° clockwise
c 45° anti-clockwise?

4

a From Portree, which place is approximately:
(i) N (ii) S (iii) W (iv) SW (v) SE?
b Which two places would be most affected by winds from the south?

5 Sketch this football pitch.
a Draw, and letter, new stands for spectators to the N, S, E and W of the pitch.
b Draw, and letter, four corner stands.

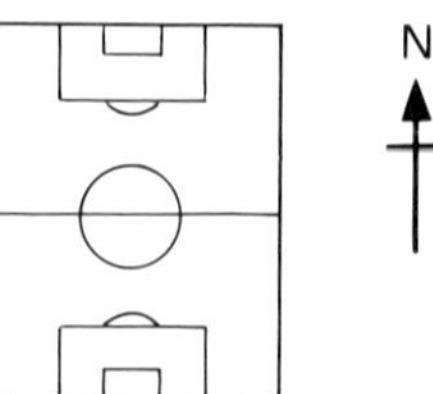

EXERCISE 2B/C

1

a *Silver Spray* sails due east, then turns 90° clockwise.
In which direction is she sailing now?
b Find the entries for the third row of the table.
(C = clockwise, AC = anti-clockwise.)

Course	S	SE	SW	NW	W
Turn	90° AC	90° C	90° C	45° AC	135° C
New course					

2 *Silver Spray* leaves her mooring and sails 5 km NE, then 12 km SE.
a Make a rough sketch of the voyage.
b Using a scale of 1 cm to 1 km, make a scale drawing.
c Find the direct distance between the yacht and its mooring.

3 Here is a scale drawing of *Silver Spray's* next trip.

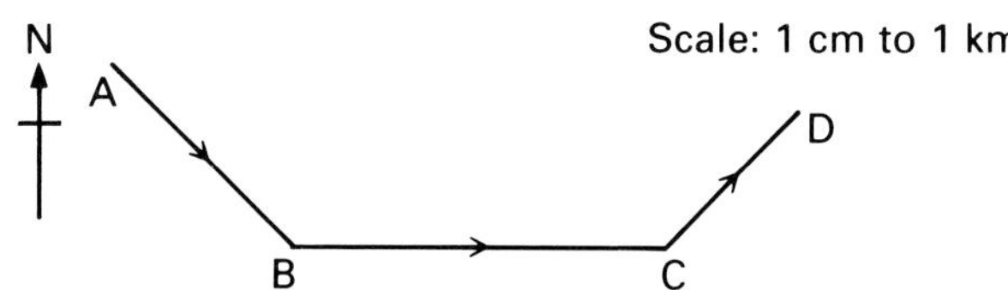

Copy and complete the table, to describe the voyage.

	Distance	**Direction**
A to B		
B to C		
C to D		

4 Make scale drawings of these journeys, and then find the direct distance from start to finish in each.

a 30 km NW → 60 km E → 25 km S → 35 km SW

b 40 km NE → 25 km E → 32 km SE → 48 km SW → 13 km W

5 A ship's radio has a range of 100 km. The ship sails 80 km north from its port, P, then south-east until it loses radio contact with P.
Use a scale drawing to find how far:

a west the ship has to sail to be due south of P

b north it will then have to sail to return to port.

PUZZLE

A tourist leaves his camp and travels 10 km south, then 10 km west, where he meets a bear. He then travels 10 km north and finds that he is back at his camp. What colour is the bear?

CHALLENGE

The knight's move in chess can be made in two stages—a diagonal move, in any direction, followed by a vertical or a horizontal move. The move can be described by using the eight points of the compass. For example:

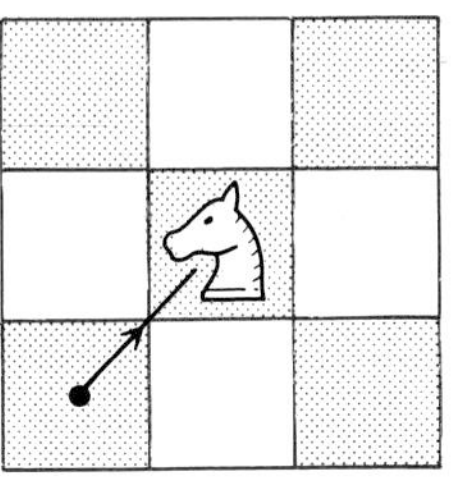

Stage 1 (NE)

then

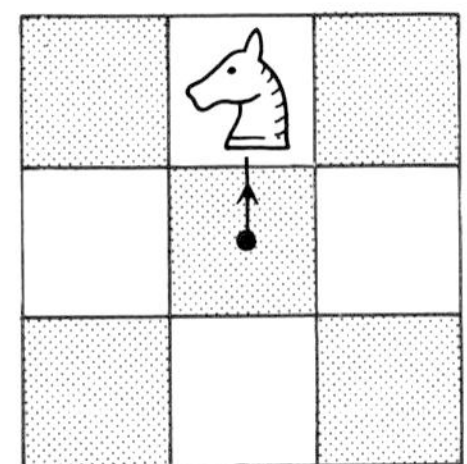

or

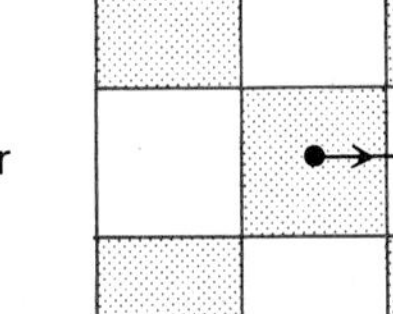

Stage 2 (N, or E)

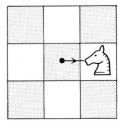

This example shows NE, N or NE, E.
Use the eight compass directions to describe all the possible moves from a square near the centre of the board.

INVESTIGATION

Work out the names of all 32 points on the mariner's compass.

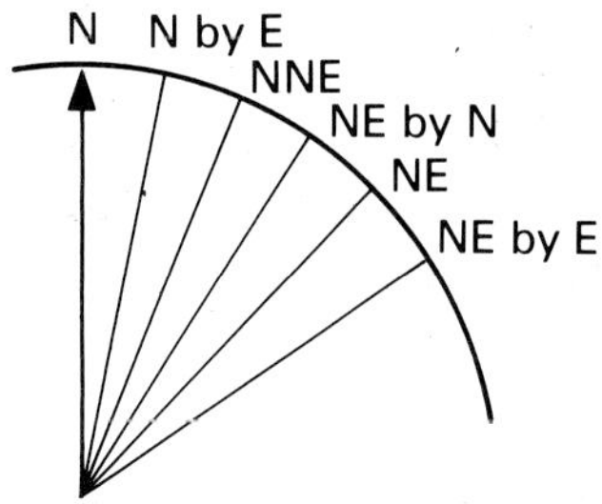

THREE FIGURE BEARINGS

Three figure bearings are angles measured from north in a clockwise direction, always given in three figures.

(i) Face north at Glasgow.
(ii) Turn clockwise through 80° to face Edinburgh.
(iii) The bearing of Edinburgh from Glasgow is 080° (three figures).

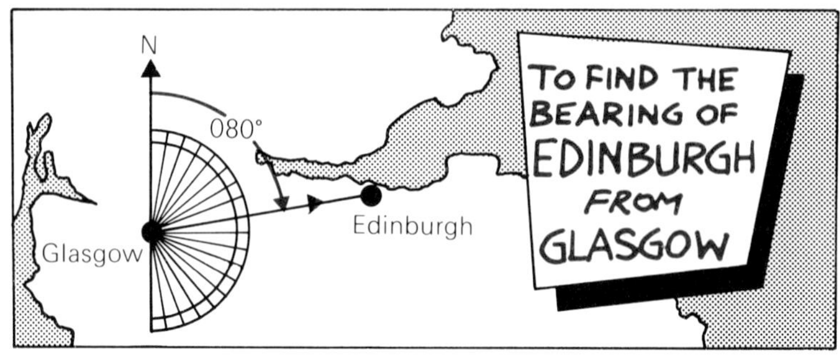

EXERCISE 3A

1 List these towns in alphabetical order, and give their 3-figure bearings from Kirkcaldy.

Town	Bearing
Anstruther	070°
Cupar	
Dollar	290°
Edinburgh	
...	...

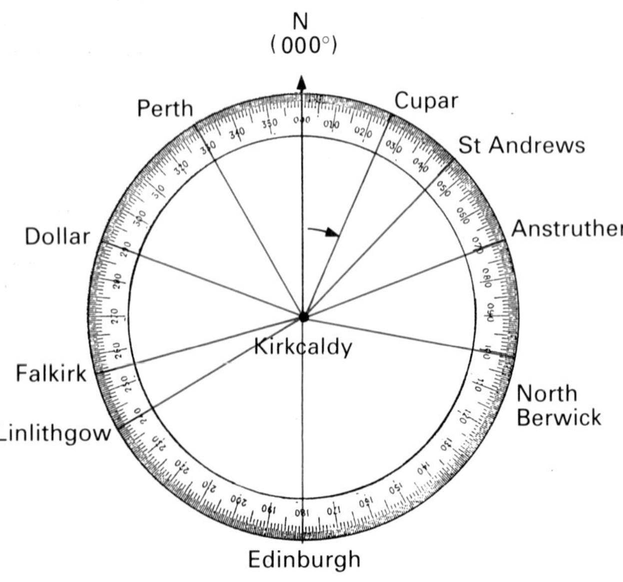

2 a Copy the diagram.
b Write the correct 3-figure bearing against each of the eight compass directions.

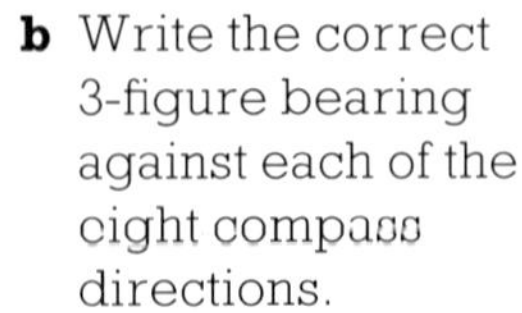

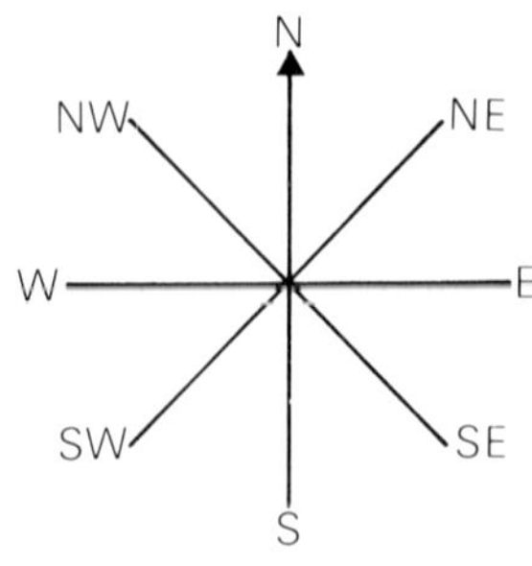

3

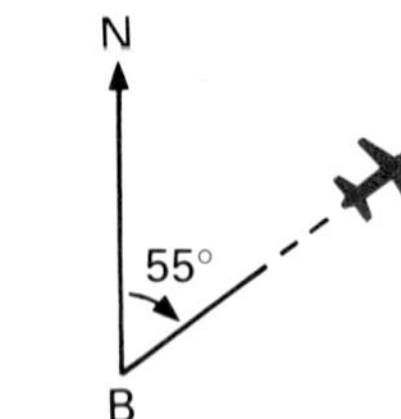

The bearing of this plane from B is 055°. Write down the bearings of all the planes shown below from B.

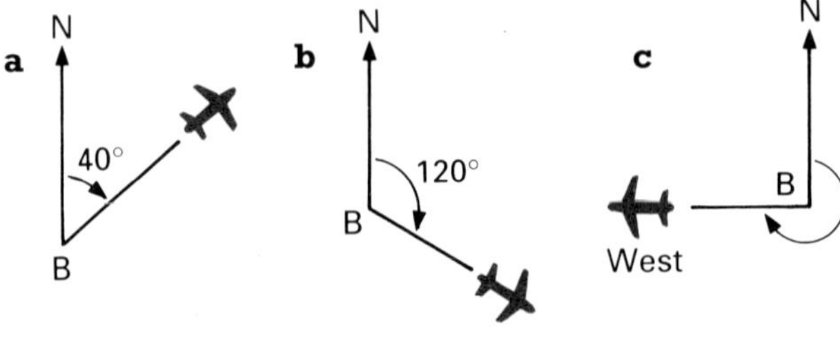

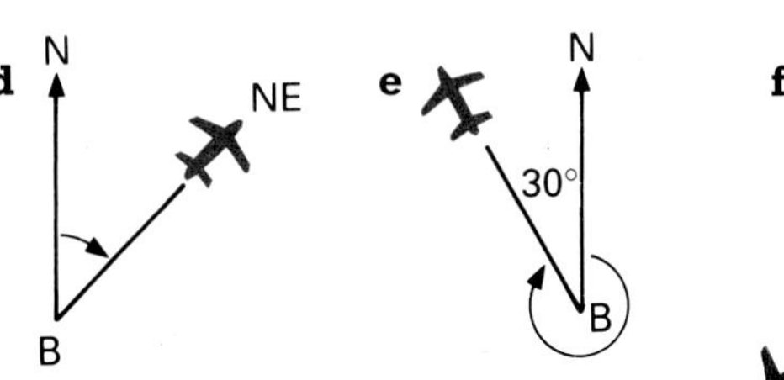

4 Make accurate drawings showing planes at these bearings from a point O. Start with the north (000°) line each time.
a 030° **b** 125° **c** 200° (180° + 20°) **d** 300°

5 Follow the leader! Start at A, and find the point which bears 050°. Move to this point, and find the point which bears 090°. Move on, finding the points bearing 240°, 100° and 260° from the previous ones. List the points in the order in which you visit them.

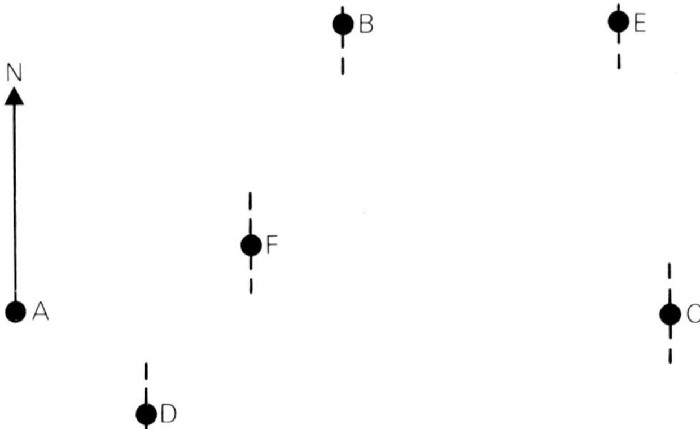

6 a Make sketches to show where you estimate these bearings to be:
(i) 020° (ii) 090° (iii) 135° (iv) 220° (v) 350°
b Check your accuracy, using a protractor.

7 Measure the bearings from Plymouth of each place on the map. List the places with their bearings.

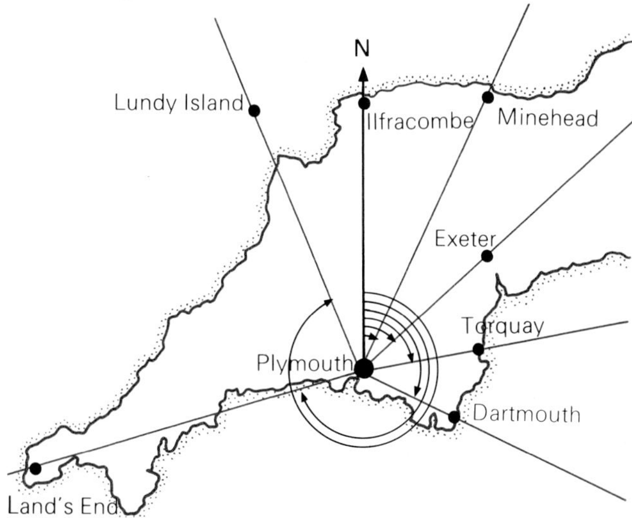

EXERCISE 3B/C

1 Calculate the bearings of A, B, C, D and E from O.

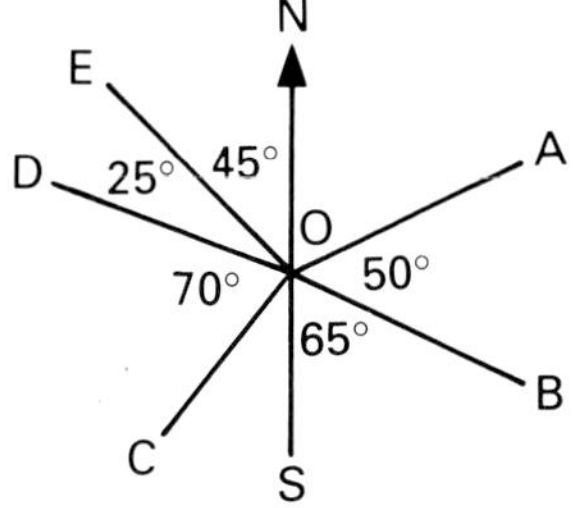

2 Copy and complete this table. A rough sketch should help you in each part.

Direction	Angle turned through	New direction
070°	140° clockwise	
175°	55° clockwise	
240°	360° anti-clockwise	
East	75° anti-clockwise	
215°	140° anti-clockwise	
175°	200° clockwise	

3 A plane changes its course from 085° to 120°. Show in a diagram the angle it has to turn through. What size is the angle?

4 Copy and complete this table.

First direction	Second direction	Smaller angle turned through
040°	070°	
105°	170°	
160°	210°	
100°	350°	
070°	300°	
295°	010°	

5 Measure the bearing of:
(i) B from A (ii) A from B.
How can you find (ii) from (i)?

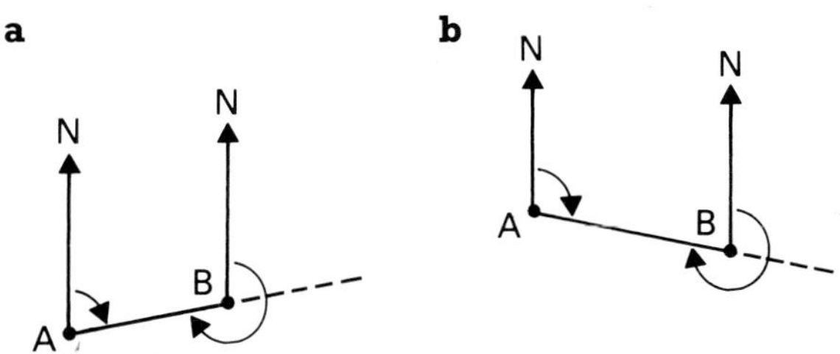

6 Calculate the bearing of A from B in each diagram.

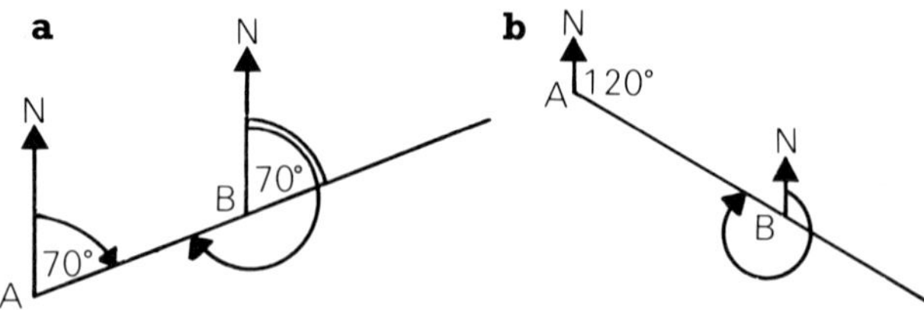

7 Using the map in question **7** of Exercise 3A on page 71, **calculate** the bearing of Plymouth from:
a Minehead **b** Land's End **c** Lundy Island.

8 A pilot flies on a course of:
a 160° from Nottingham to London
b 330° from Birmingham to Manchester.
Calculate his course for each return flight.

AIRSTRIKE

Draw two targets with circles of radii 1 cm, 2 cm and 3 cm, and draw in radii every 45°. Ask an opponent to do the same.
On one of your targets place your three airports at any three points where the circles and lines cross.
On your other target, record shots at your opponent, for example X (2,225°), and circle them if you score a hit.

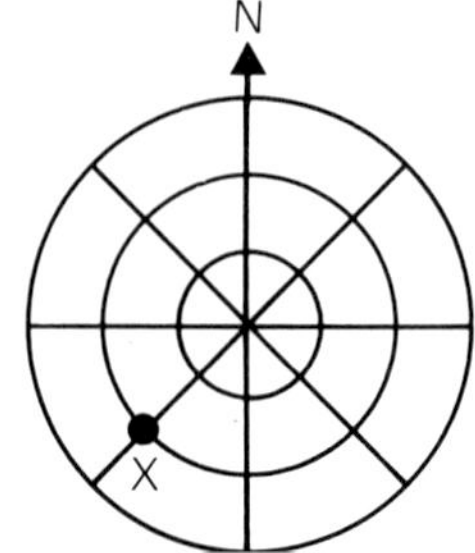

Take turns, three shots each, until one person's airports have all been hit.

BEARINGS AND SCALE DRAWINGS

EXERCISE 4A

1 Use a scale of 1 cm to 1 km to make scale drawings of these journeys from O to A.

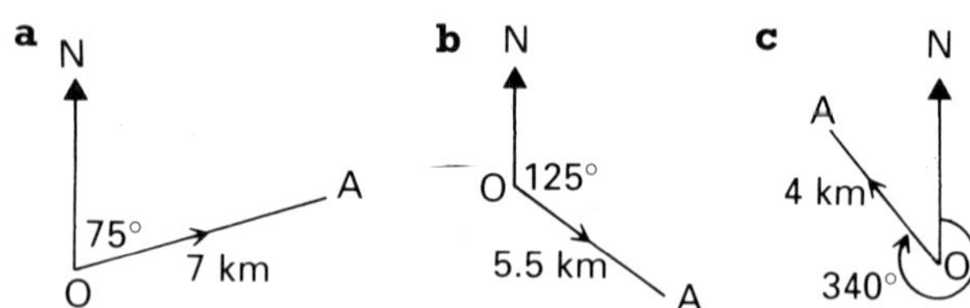

2 This scale drawing shows a yacht's voyage from pier P, to marker buoy B, to its mooring M. Use your ruler and protractor to find the distance and bearing of:
a B from P **b** M from B **c** P from M.

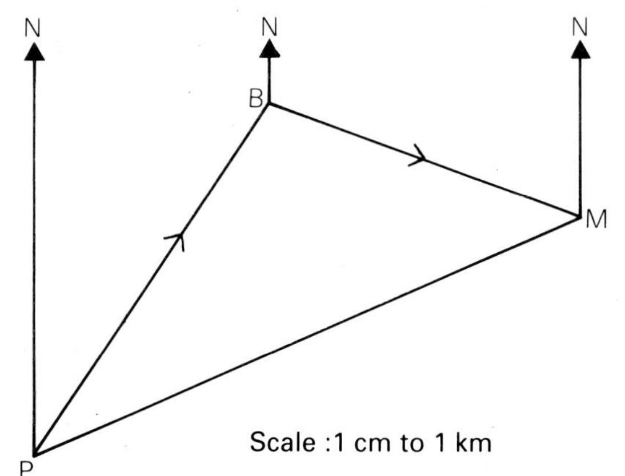

3 The *Flying Swan* sails from A to B to C.
Make a scale drawing, and find the distance and bearing of C from A.

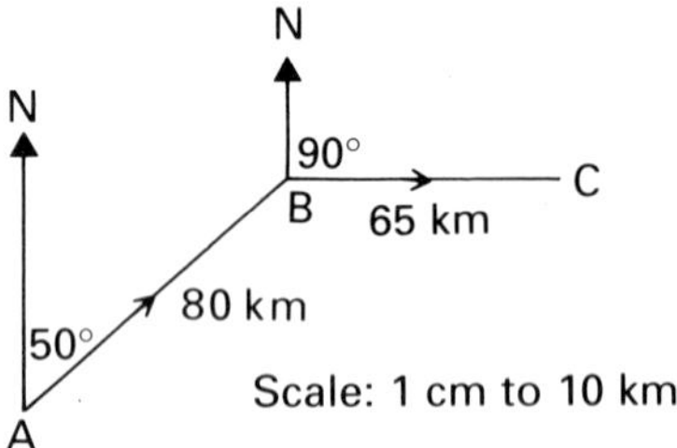

4 A helicopter sets off in search of a ship in distress. The diagram shows the helicopter's course from A to B to the ship's position at C.

a Make a scale drawing, taking 1 cm to 5 km.
b What is the ship's distance and bearing from A?

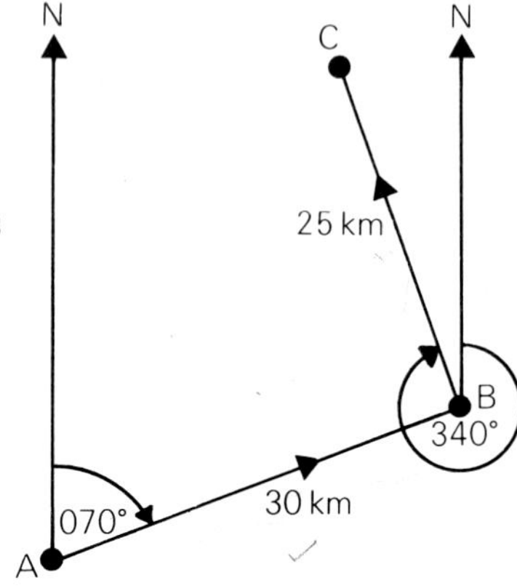

EXERCISE 4B/C

Make a rough sketch first in each question.

1 

The *Jolly Jack* sailed for 10 km on a course of 060°, and then sailed for 10 km on a course of 180°.

a Make a scale drawing of the boat's course so far, taking 1 cm to 1 km.

b What is the boat's distance and bearing from the starting point?

2 From Londonderry, Belfast is 100 km away on a bearing of 115°, and Armagh is 80 km away on a bearing of 150°. Using a scale of 1 cm to 10 km make a scale drawing, and find the distance and bearing of Belfast from Armagh.

3 The bearing of the lighthouse from the ship is 025°.

The ship sails 7 km east, and the bearing of the lighthouse is then 300°.
Make a scale drawing to find how close the ship passed to the lighthouse.

4 Treasure Hunt clues: 'From the barn, go 600 m on a bearing of 155°; then 400 m east; finally 600 m on a bearing of 045°. Find the treasure and return to the barn.' Make a scale drawing, and find the distance and bearing from the treasure to the barn.

5 A picnic to Mystery Island: 20 km on a bearing 030°. But then, disaster—the engine breaks down halfway there. The boat drifts 6 km on a bearing of 300° until, finally, the engine restarts.

a Make a scale drawing. Remember to state your scale.

b Picnic abandoned! What is:
(i) the bearing for home
(ii) the distance home?

CHALLENGE

Working in groups, and using a compass and trundle wheel, either:

a *measure and record distances and bearings from one place in school to another, or*

b *make a map of the perimeter of the school grounds.*

Exchange results with other groups, and write a brief report on the project.

BRAINSTORMER

List the distances and bearings for the flight plan of a pilot who wishes to fly over the capitals of Wales, Northern Ireland, Scotland and England in turn.

INVESTIGATION

The scale of this map is 1 : 10 million.

a *Copy and complete:*
1 : 10 million = 1 cm : 10 000 000 cm
= 1 cm : ______ m
= 1 cm : ____ km

b *How far are the actual distances in a straight line between:*
(i) Bristol and London
(ii) London and Glasgow
(iii) Land's End and John O'Groats?

c *Manchester is on a straight line between Birmingham and Edinburgh, and is about 100 km from Birmingham. How far is it from Manchester to Edinburgh?*

d *Investigate the kind of scales used in maps of countries, towns, parks, etc.*

CHECK-UP ON DISTANCES AND DIRECTIONS

1 a Make a scale drawing of this metal plate, using a scale of 1 cm to 1 m.

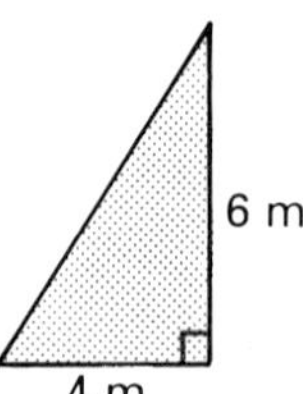

b Measure the length of the sloping edge, and the sizes of the angles.

2 a Use a scale of 1 cm to 4 km to make a scale drawing of the plane's flight from A to B to C.
b Find the direct distance from A to C.

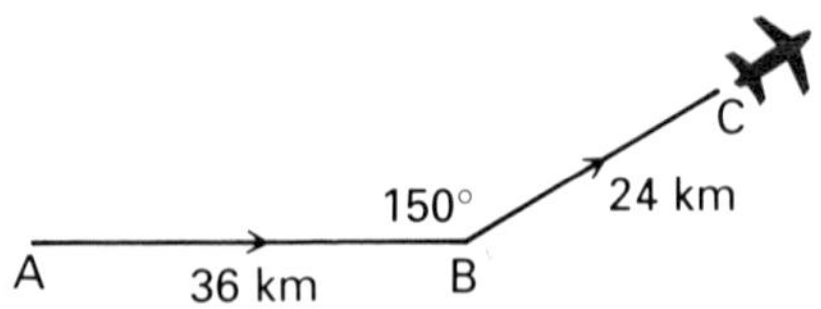

3 During the Ice Age, glaciers flowed over Britain. Use the map to give the direction of flow at the places marked A, B, C, . . . , G.

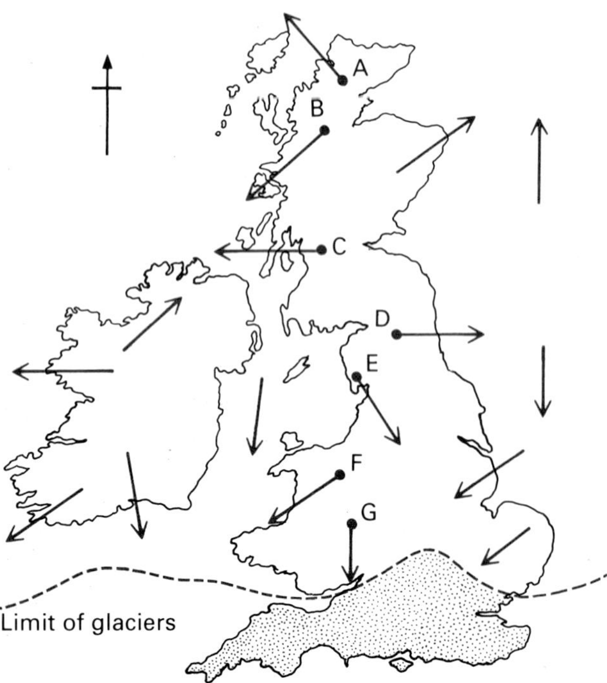

4 Write down the 3-figure bearings for these directions from O.

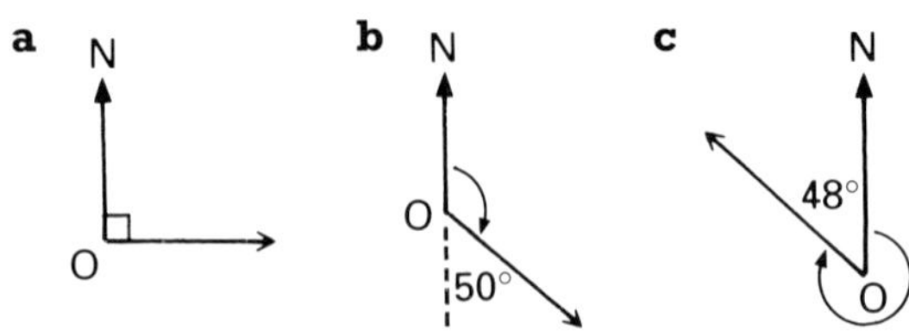

5 a Measure the bearing of each town on the Isle of Wight from Newport.

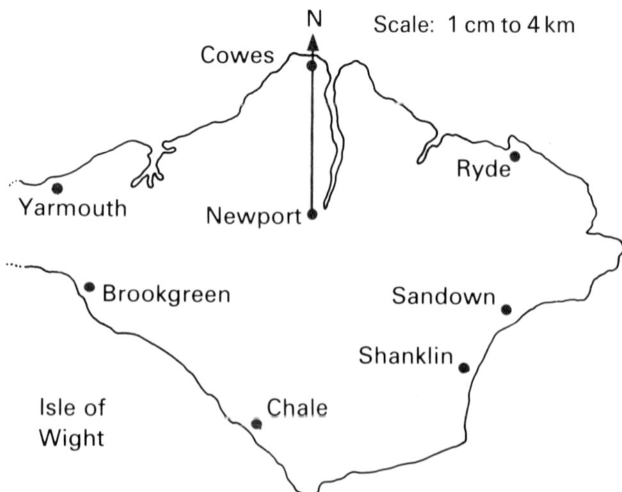

b What is the bearing of Newport from:
(i) Brookgreen (ii) Ryde?
c Find the direct distance from Newport to each town.

6 The bearing of A from B is 125°. What is the bearing of B from A?

7 A life-boat sails 16 km from harbour on a bearing of 135°, then 20 km on a bearing of 075°. Make a scale drawing to find the distance and bearing from the life-boat to the harbour.

7 POSITIVE AND NEGATIVE NUMBERS

LOOKING BACK

1 Copy this number line, and fill in all the numbers.

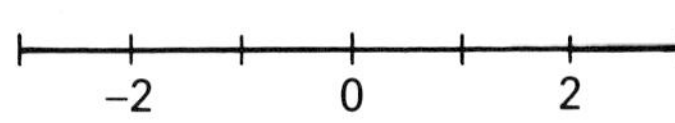

2 On the map,
a which temperature is:
(i) highest (ii) lowest?
b which temperatures are:
(i) positive (ii) negative?

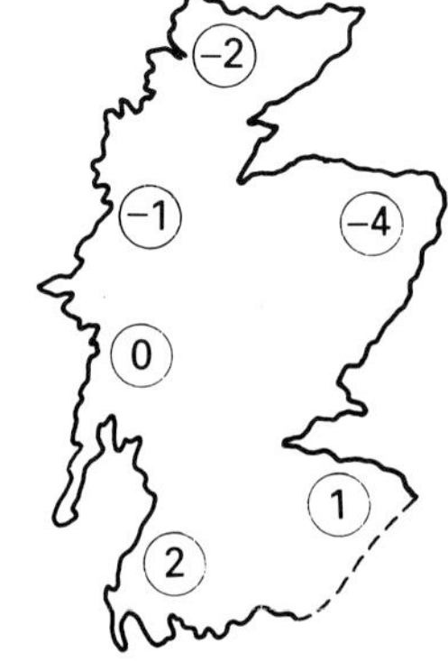

3 What number is:
a 5 more than 5 **b** 9 less than 9
c 10 more than 0?

4 Write down the greatest and least numbers in each set.
a 4, 9, 3, 6 **b** 1, 0, 3, 2 **c** 0, −1, 1

5 a Write down the coordinates of A, B, C and D.

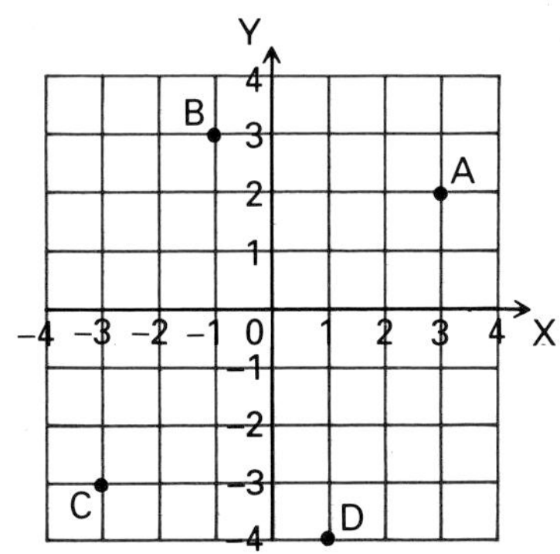

b (i) Is A to the left or right of B?
(ii) Is C above or below D?
c What are the coordinates of the image of A under reflection in:
(i) the x-axis (ii) the y-axis?

6 a What are the temperatures on the thermometer at P, Q and R?

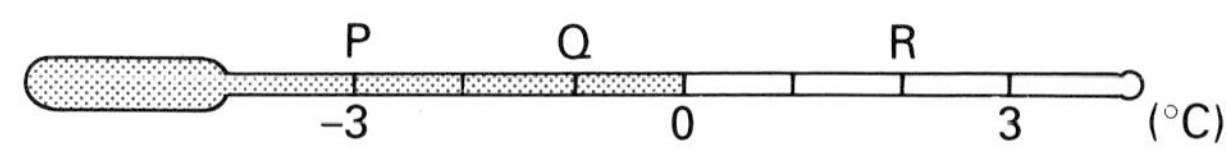

b If the temperature falls 2° from zero, what is it then?

7 Find the numbers that go in the spaces.

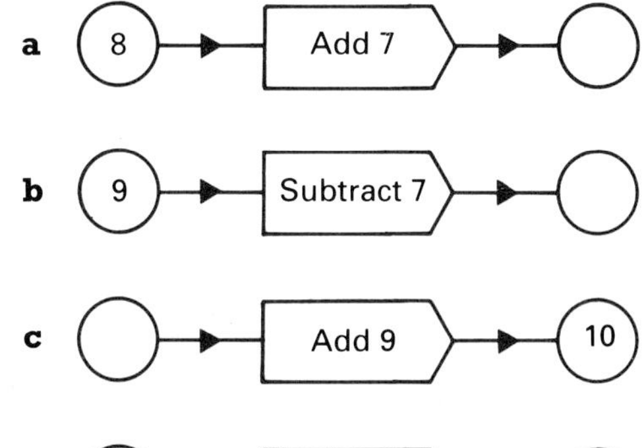

8 What number does x stand for in each equation?
a $x+3=5$ **b** $4+x=9$ **c** $4-x=2$
d $x-7=3$ **e** $x-10=0$ **f** $1-x=0$

9

How many steps is Sophie above or below the platform if she takes:
a 3 steps up, then 7 up
b 5 steps up, then 6 down
c 2 steps down, then 4 down
d 2 steps down, then 4 up?

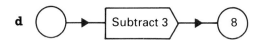

EXPLORING THE NUMBER LINE

Here is the number linesman preparing for action!

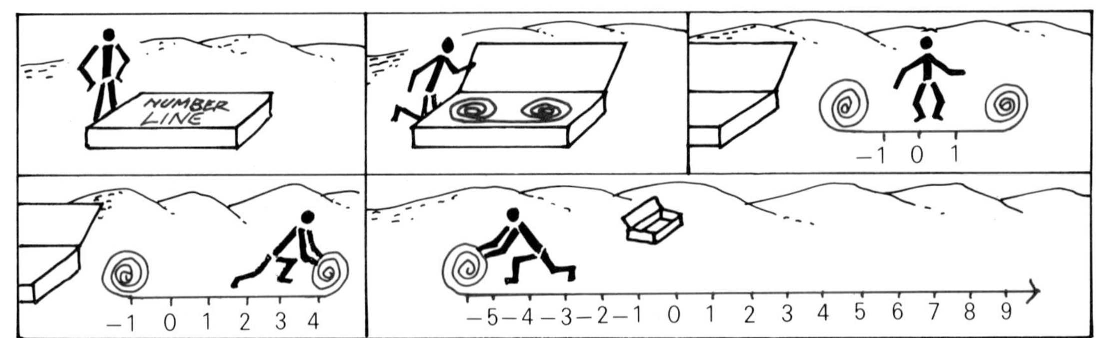

If he starts at 0 and walks to the **right**, he is among **positive** numbers.
If he starts at 0 and walks to the **left**, he is among **negative** numbers.

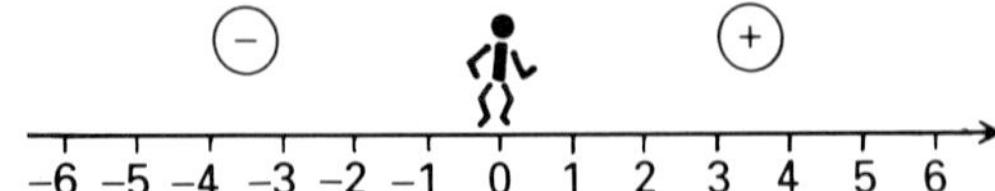

Zero is neither positive nor negative.

If one number is **greater** than another, it lies to the **right** of it.
If one number is **smaller** than another, it lies to the **left** of it.

EXERCISE 1A

1 a

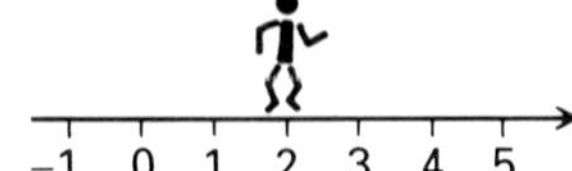

Name three numbers on the line:
(i) greater than 2 (ii) less than 2.

b

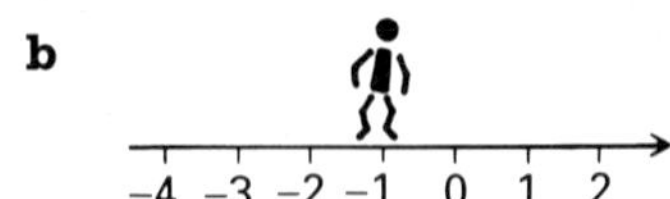

Name three numbers on the line:
(i) less than −1 (ii) greater than −1.

2 Write down the smaller number in each pair.
a 10, 9 **b** 0, 1 **c** 0, −1 **d** 5, −5
e 9, 11 **f** 9, −1 **g** −1, −2 **h** −5, 0

3 Write down the larger number in each pair.
a 1, 10 **b** 1, −10 **c** −1, 1 **d** −2, 0
e 0, −1 **f** −1, −2 **g** 4, −5 **h** −3, −4

4 a

Which number is:
(i) 1 greater than 1 (ii) 2 less than 1
(iii) 3 greater than 1 (iv) 3 less than 1?

b

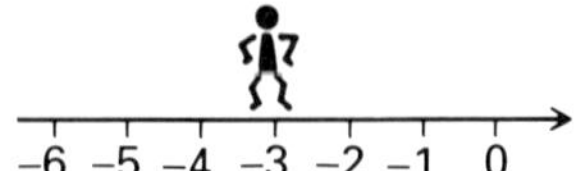

Which number is:
(i) 2 less than −3 (ii) 2 greater than −3
(iii) 3 greater than −3 (iv) 1 less than −3?

5 Write down the highest and lowest temperatures in each group.
a Berlin 1°C, Chicago −4°C, Copenhagen −2°C
b London −3°C, Moscow −4°C, Zurich −2°C

6 Write down the temperatures (°C) at the points A to E on the thermometer.

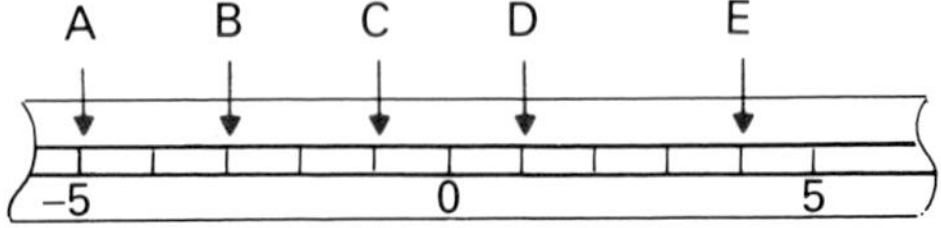

7 Using the thermometer in question **6** to help you, if you wish, write down temperatures:
a 2°C lower than (i) 2°C (ii) 0°C (iii) −2°C
b 3°C higher than
(i) −5°C (ii) −1°C (iii) 0°C.

8 In a treasure hunt, Ahmed guesses the treasure is at the point A(−3, −2). Copy and complete the points for:
a Becky, B(1, . . .) **b** Carole, C(−2, . . .)
c Denzil, D **d** Eve, E **e** Finlay, F.

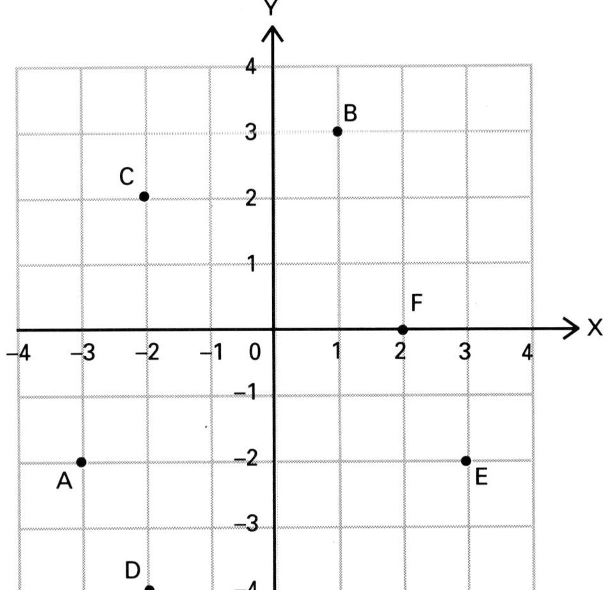

9 A signwriter is painting letters for the name of her shop. Plot each set of points on squared paper, with the x- and y-axes numbered −5 to 5. Join the points as shown by the arrows to find the letters.
a (3, 2) → (3, 5) → (5, 2) → (5, 5)
b (−4, 3) → (−3, 0) → (−2, 3)
c (−5, −2) → (−5, −5) → (−3, −5)
d (1, 0) → (2, −3) → (3, −1) → (4, −3) → (5, 0)

10 a Use negative numbers to give the positions of the whale and the submarine from sea-level.

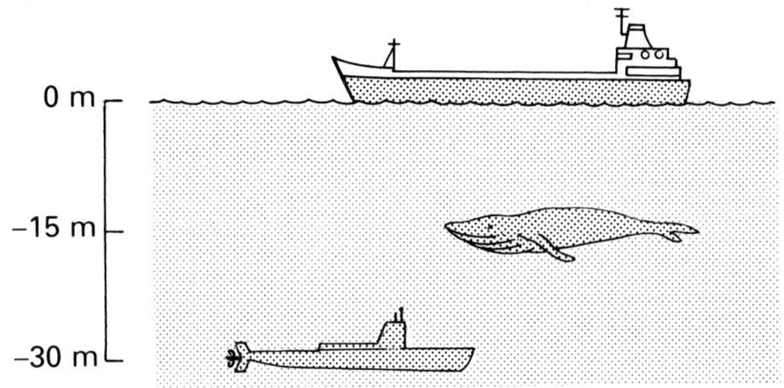

b The whale goes down 10 m, and the submarine rises 20 m. Describe their new depths, using negative numbers.

EXERCISE 1B/C

1 Copy and complete the 'Balance' column in the bank statement. Pay-ins are marked '+', and pay-outs are marked '−'.

Date	Pay in/out (£)	Balance (£)
1 March	+20.00	20.00
5 March	−10.00	10.00
8 March	−5.00	
12 March	−10.00	
20 March	+15.00	
25 March	−10.00	

2

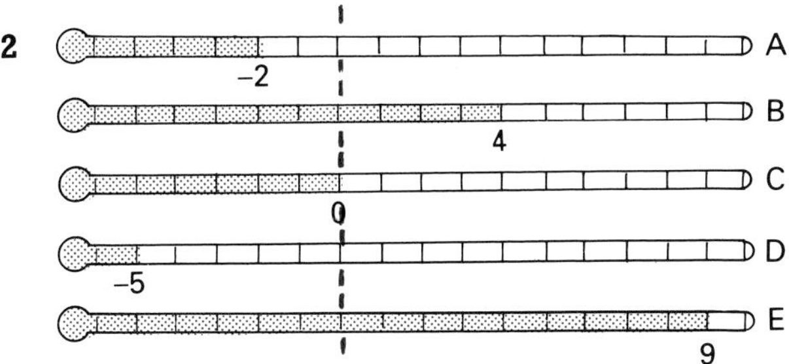

a Calculate the rise in temperature from:
(i) A to B (ii) C to E (iii) D to E
b Calculate the fall in temperature from:
(i) B to C (ii) B to D (iii) A to D
c Calculate the change in temperature (+ or −) from: (i) A to D (ii) D to A (iii) E to A.

3 Here are some temperatures from places around the world:

Paris	Aberdeen	Calgary	Rome	Hawaii
3°C	−5°C	5°C	−1°C	20°C

a (i) Which is the warmest place and which is the coldest place?
(ii) What is the difference between their temperatures?
b Write down the change in temperature from place to place, working from left to right.

4 a Plot these points, and join each one to the next to make an interesting shape.
(4, 0) → (1, 1) → (0, 4) → (−1, 1) → (−4, 0) → (−1, −1) → (0, −4) → (1, −1) → (4, 0).
b Is the x-axis a line of symmetry for the shape?
c Name another line of symmetry.
d Has the star-shape half-turn symmetry about O?
e Has it quarter-turn symmetry about O?
f What is its order of symmetry?

5 Over the centuries, as sea-level fell, raised beaches appeared like this:

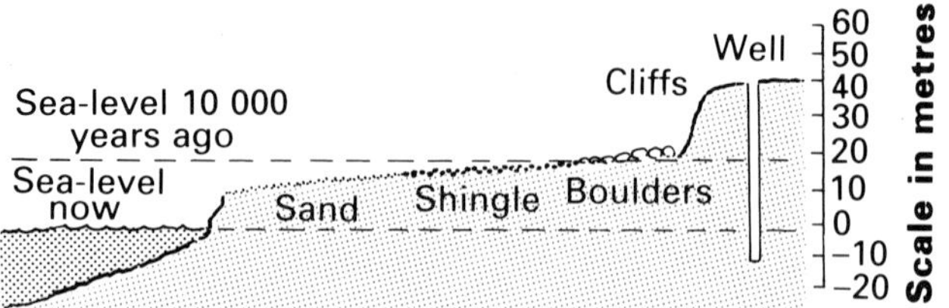

a Relative to sea-level now, what is the height of:
(i) the top of the well
(ii) sea-level 10 000 years ago
(iii) the lowest part of the sand
(iv) the bottom of the well?

b Answer part **a** again, relative to sea-level 10 000 years ago.

PRACTICAL PROJECT

a *In a newspaper, or on television, find a list of temperatures at places throughout the world. List them in order, name the warmest and coolest, and calculate the difference between the temperatures at various places.*
b *If you list the temperatures regularly, you can draw line graphs to illustrate the changes.*

ADDING POSITIVE AND NEGATIVE NUMBERS

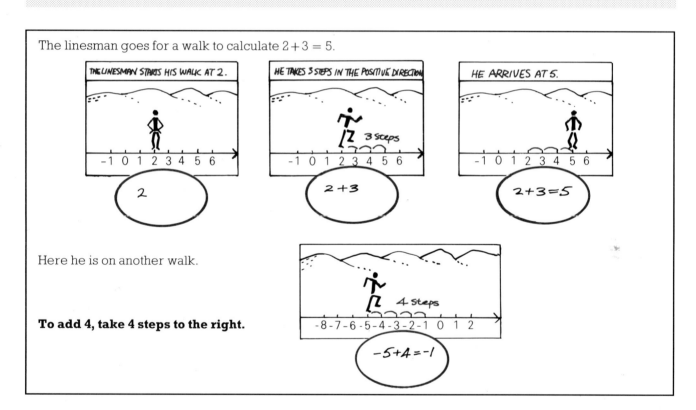

The linesman goes for a walk to calculate $2+3=5$.

THE LINESMAN STARTS HIS WALK AT 2.
-1 0 1 2 3 4 5 6
2

HE TAKES 3 STEPS IN THE POSITIVE DIRECTION
3 steps
-1 0 1 2 3 4 5 6
2+3

HE ARRIVES AT 5.
-1 0 1 2 3 4 5 6
2+3=5

Here he is on another walk.

To add 4, take 4 steps to the right.

4 steps
-8 -7 -6 -5 -4 -3 -2 -1 0 1 2
−5+4=−1

EXERCISE 2

1 Write down the calculations for these walks.

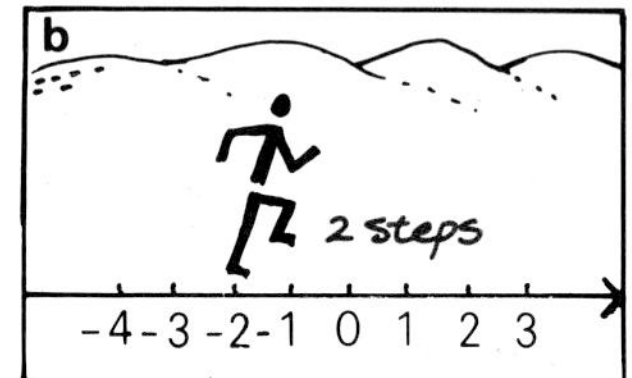

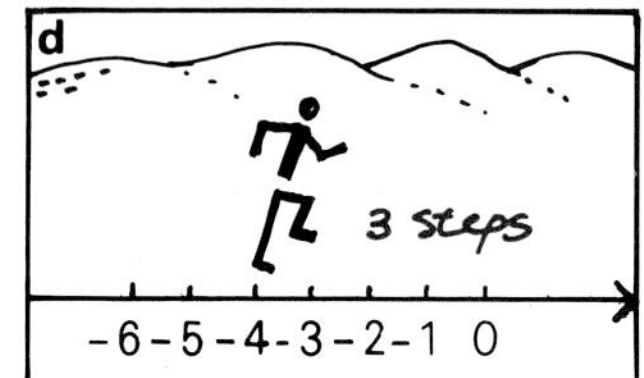

2 Write down the calculations for these walks.

a

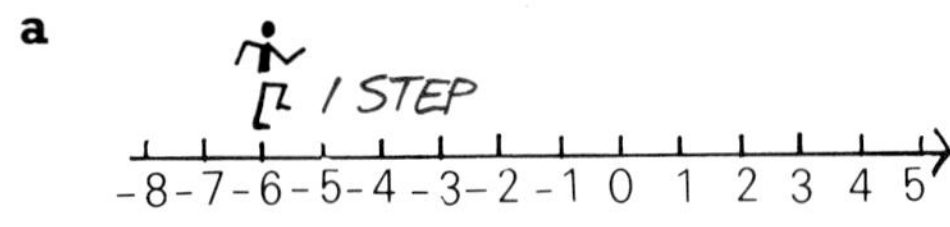

b

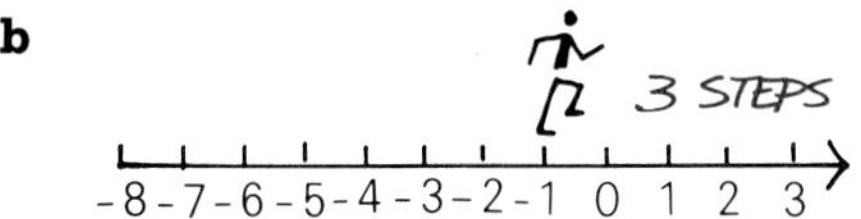

c

d

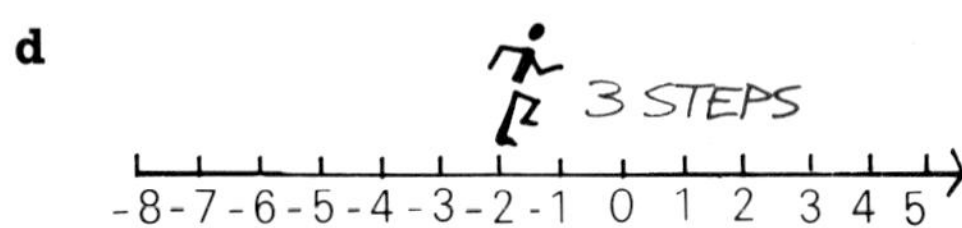

3 Copy and complete these calculations.

a

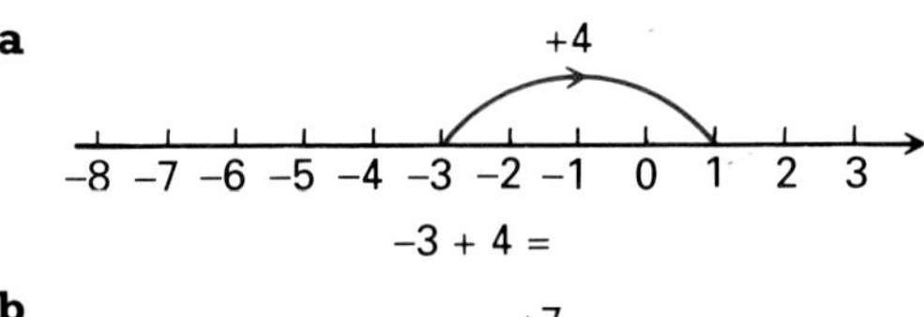

b

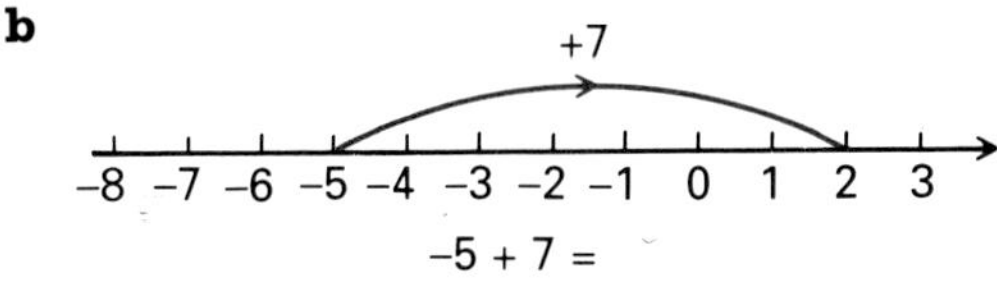

−12 −11 −10 −9 −8 −7 −6 −5 −4 −3 −2 −1 0 1 2 3 4 5 6 7 8 9 10 11 12

4 Use the number line above to calculate:

a $-5+4$	**b** $-1+6$	**c** $-3+3$	**d** $-1+8$
e $-7+4$	**f** $-10+2$	**g** $-6+6$	**h** $-1+11$
i $0+2$	**j** $-7+8$	**k** $-1+4$	**l** $-7+7$
m $-4+0$	**n** $-2+8$	**o** $-10+11$	**p** $-7+12$
q $-10+17$	**r** $-7+15$	**s** $-4+13$	**t** $-6+16$

5 Check your answers to question **4**, using the [+/−] key on your calculator for negative numbers.

6 Temperatures shown on the map are in °C.

a Find the highest and lowest temperatures.

b If the temperatures in Scotland rise by 1°, and in England and Wales by 2°, what are they all now, from north to south?

7 The controller counts down the seconds to launch-time: $-10, -9, -8, \ldots$
What number does he call:

a 5 seconds after '−5' **b** 10 seconds after '−3'
c 7 seconds after '−9' **d** 9 seconds after '−6'?

8 An aircraft A is at $(-30, -20)$ on a coordinate map. The units are kilometres. Find its position after it flies:

a 50 km east
b 100 km north, from A.

Y (North)
X (East)
0
A (−30, −20)

9 Try to calculate these without using the number line or a calculator.

a $-1+4$	**b** $-3+3$	**c** $-2+0$
d $-2+7$	**e** $-1+1$	**f** $5+8$
g $-7+6$	**h** $-1+9$	**i** $-8+8$
j $-9+7$	**k** $-1+0$	**l** $-1+8$

10 Calculate the temperatures after the increases shown:

a 6°C, +2° **b** 0°C, +4° **c** −2°C, +4°
d −4°C, +2° **e** −8°C, +8° **f** −3°C, +7°

POSINEGI GAME 1—FOR TWO PLAYERS

Each player has a set of cards like this:

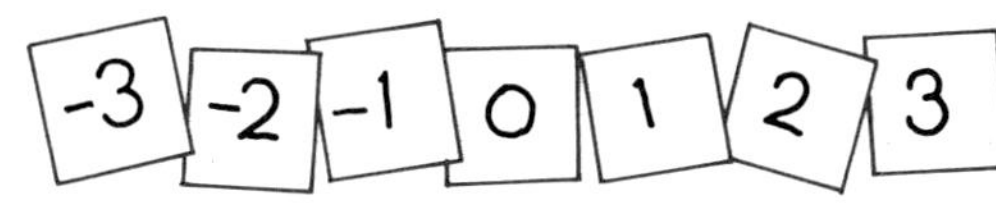

The players each choose one of their cards, and place it face down on the table. The cards are then turned over. The player with the greater number scores the sum of the two numbers, and the other player scores zero. The two cards are then put aside. If the two cards are equal they are laid aside, and both players score zero. The winner is the player with the greater score when all the cards have been played.
A larger set of cards, from −9 to 9, makes the game last longer!

SUBTRACTING POSITIVE AND NEGATIVE NUMBERS

To subtract, the linesman takes a walk in the negative direction.

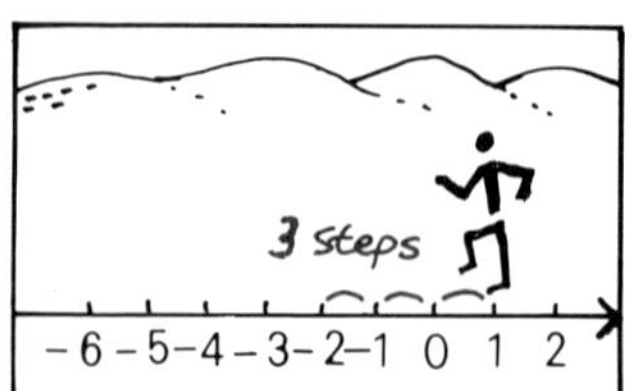

He finds that **$1-3=-2$**

Also, using the sequence
3, 2, 1, 0, −1, −2, . . . ,

$$1+2=3$$
$$1+1=2$$
$$1+0=1$$
$$1+(-1)=0$$
$$1+(-2)=-1$$
$$\mathbf{1+(-3)=-2}$$

It seems that $1-3=1+(-3)=-2$. Subtracting 3 is the same as adding -3.

Here he is on another walk.
$-2-3=-5$, or $-2+(-3)=-5$.

To subtract 3, take 3 steps to the left.

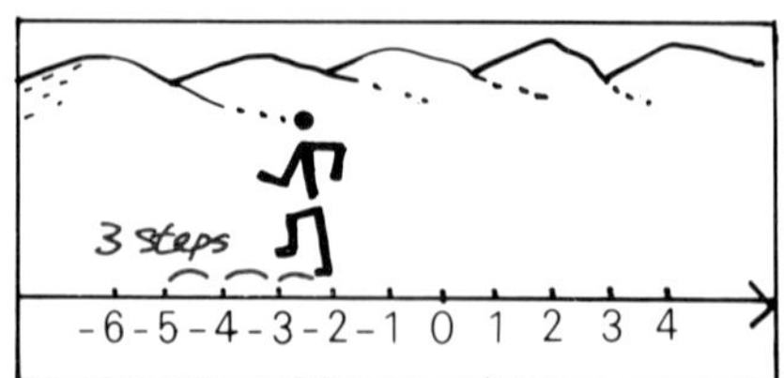

EXERCISE 3

1 Describe each of these walks in two ways. For example, $1-2=-1$ and $1+(-2)=-1$.

a **b**

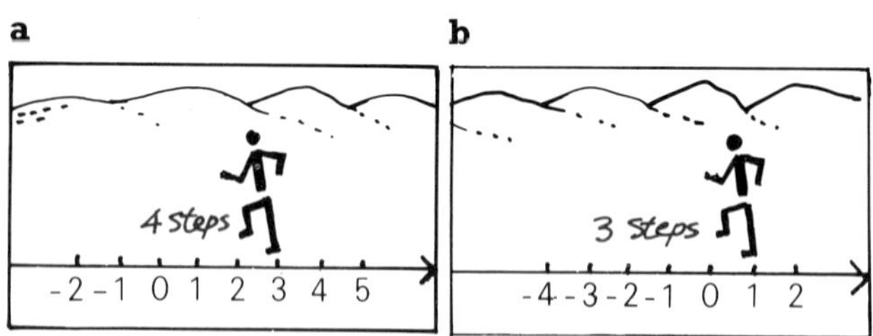

c **d**

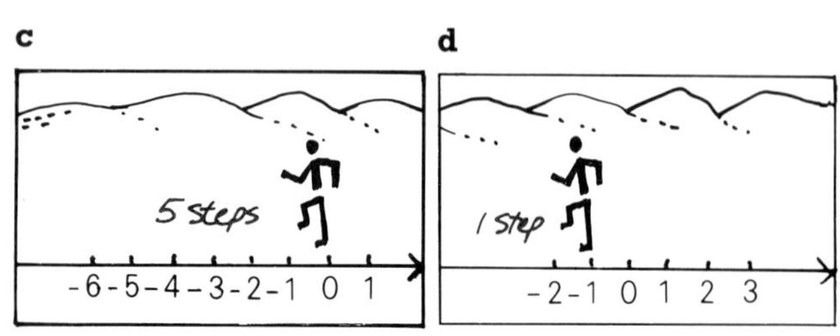

e **f**

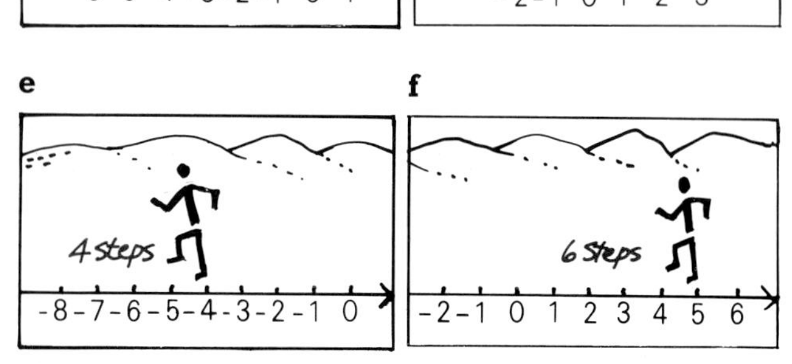

2 Describe each of these walks in two ways.

a

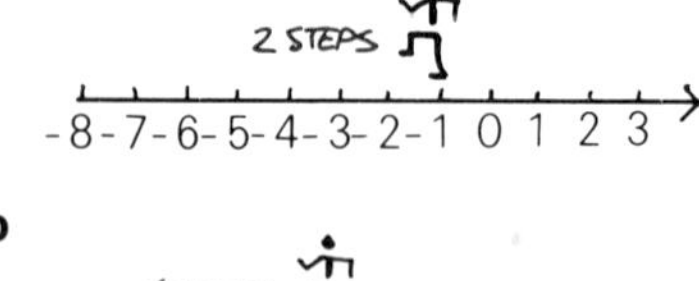

b

4 STEPS
-8 -7 -6 -5 -4 -3 -2 -1 0 1 2 3

c

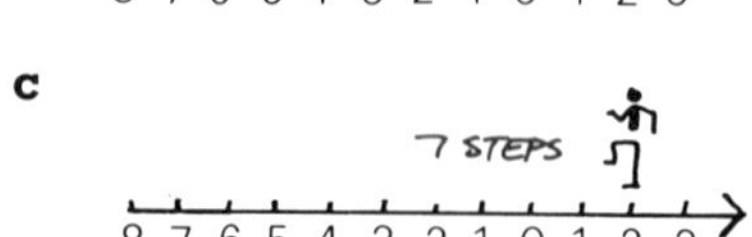

3 Copy and complete these calculations.

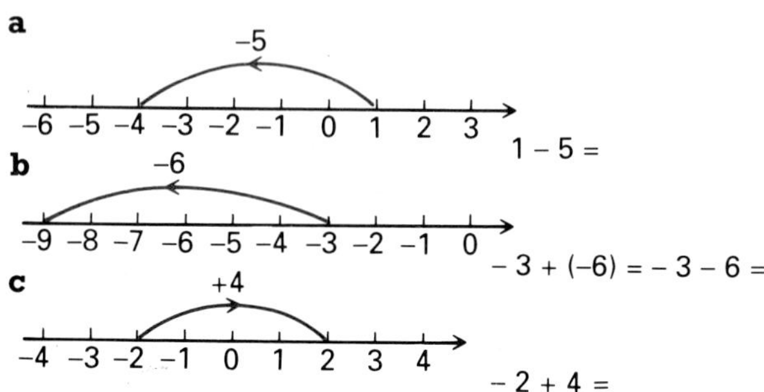

a $1-5=$

b $-3+(-6)=-3-6=$

c $-2+4=$

−12 −11 −10 −9 −8 −7 −6 −5 −4 −3 −2 −1 0 1 2 3 4 5 6 7 8 9 10 11 12

4 Use this number line to calculate:

a $4-6$ **b** $4+(-6)$ **c** $3+(-2)$
d $-1-4$ **e** $-1+(-4)$ **f** $-3-6$
g $8+(-3)$ **h** $5+(-5)$ **i** $2+(-3)$
j $-4+(-2)$ **k** $2+(-9)$ **l** $4-8$
m $0-10$ **n** $0+(-8)$ **o** $-4-5$
p $-9-2$ **q** $-8+(-2)$ **r** $-3+(-3)$
s $3+(-3)$ **t** $-3-3$

5 Check your answers to question **4** with your calculator.

6 Calculate the midnight temperatures. The first calculation is $15+(-5)=10$.

Midday	15°C	6°C	7°C	−2°C	0°C	−5°C
Change	−5°	−6°	−9°	−3°	−10°	−1°
Midnight						

7

In a television quiz you score 2 for a correct answer and −2 for a wrong answer. Calculate these scores:

a Sally: 10 correct, 5 wrong
b Sean: 6 correct, 9 wrong.

8 Kevin counts money paid into his bank account as positive, and money he takes out as negative. How much did he have in his account after each pay-in or pay-out shown below?

+£10; +£5; −£20; −£5; +£10; −£5; +£30

$2-(-3)$. The linesman is puzzled.
For $2-3$, he goes **left** to -1.
So for $2-(-3)$ should he go **right** to 5?

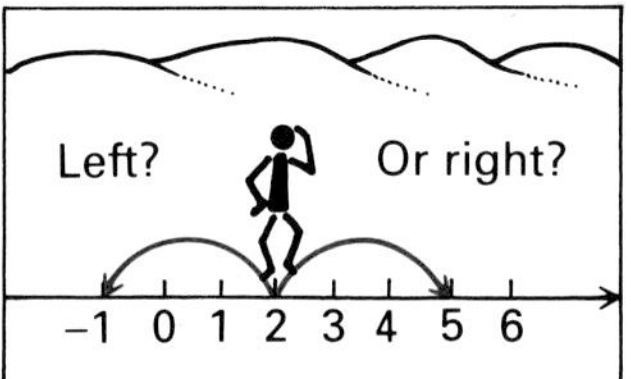

Try sequences again.

$2-2=0$
$2-1=1$
$2-0=2$
$2-(-1)=3$
$2-(-2)=4$
$\mathbf{2-(-3)=5}$

It seems that $2-(-3)=2+3=5$.
Subtracting −3 is the same as adding 3.

Examples

a $8-(-3)$ $=8+3$ $=11$

b $-2-(-4)$ $=-2+4$ $=2$

c $0-(-5)$ $=0+5$ $=5$

9 Copy and complete the right-hand columns:

a		**b**		**c**	
$4-3$	$=1$	$6-2$	$=4$	$5-1$	$=$
$4-2$	$=$	$6-1$	$=$	$5-0$	$=$
$4-1$	$=$	$6-0$	$=$	$5-(-1)$	$=$
$4-0$	$=$	$6-(-1)$	$=7$	$5-(-2)$	$=$
$4-(-1)$	$=5$	$6-(-2)$	$=$	$5-(-3)$	$=$
$4-(-2)$	$=$	$6-(-3)$	$=$	$5-(-4)$	$=$

10 Calculate, without using a calculator:

a $5-(-2)$ **b** $4-(-3)$ **c** $8-(-8)$
d $1-(-5)$ **e** $2-(-4)$ **f** $-1-(-2)$
g $-2-(-6)$ **h** $-8-(-3)$ **i** $1-(-1)$
j $-6-(-5)$ **k** $0-(-2)$ **l** $7-(-8)$
m $-7-(-8)$ **n** $5-(-5)$ **o** $-2-(-2)$
p $-1-(-1)$ **q** $11-(-1)$ **r** $-1-(-9)$
s $0-(-1)$ **t** $-9-(-1)$ **u** $4-(-4)$

11 Check your answers to question **10** with your calculator.

12 In North America, one of the highest temperatures ever recorded was 56.7°C. One of the lowest was −62.8°C, in the Yukon. Calculate the difference between these temperatures.

13 Calculate each rise or fall in temperature.

a −2°C to 2°C **b** −1°C to −5°C
c 7°C to 0°C **d** −9°C to −2°C

CHALLENGE

Replace $*$ by $+$ or $-$. For example, for $2*3=-1$ the answer is $2-3=-1$.

a $2*1=3$
$2*1=1$
$1*2=-1$
$-1*(-2)=1$
$-1*(-2)=-3$

b $3*2*1=4$
$3*2*1=0$
$1*2*3=-4$
$1*2*3=2$
$-1*2*3=-6$

c $5*4*3*2*1=-1$
$-5*4*3*2*1=-7$
$-3*2*1*2*3=-5$

POSINEGI GAME 2

Play this like the first Posinegi game, but the system of scoring is different. Two score cards are now included and the player with the greater score chooses one of these cards and calculates his or her score like this:

SMALLER NUMBER

Summary

$2-3 = -1$	$2+3 = 5$	$-4-5 = -9$	$-4+5 = 1$
$2+(-3)=2-3=-1$	$2-(-3)=2+3=5$	$-4+(-5)=-4-5=-9$	$-4-(-5)=-4+5=1$

EXERCISE 4A/B

1 Calculate:
a $6+5$ **b** $-2+5$ **c** $0+3$
d $-7+4$ **e** $-3+0$ **f** $-1+1$

2 Calculate:
a $4+(-2)$ **b** $-3+(-2)$ **c** $0+(-2)$
d $-5+(-5)$ **e** $-1+(-6)$ **f** $1+(-1)$

3 Calculate:
a $5-2$ **b** $2-5$ **c** $-4-1$
d $-4-4$ **e** $0-10$ **f** $-1-1$

4 Calculate:
a $5-(-1)$ **b** $2-(-3)$ **c** $-1-(-6)$
d $-1-(-1)$ **e** $0-(-4)$ **f** $1-(-1)$

5 What is the score in each game of dice? The numbers have to be added or subtracted in each pair.

a **b** **c**

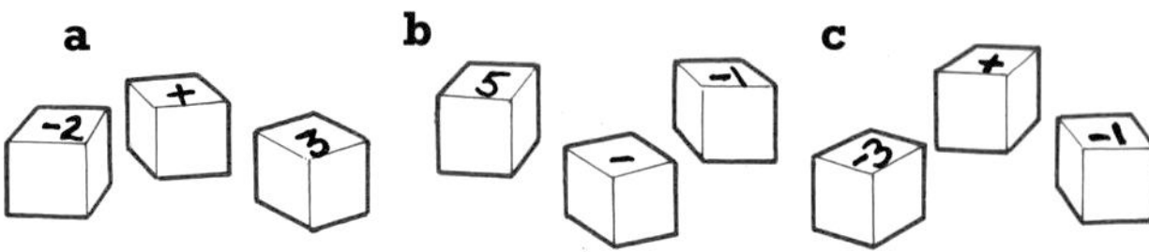

d **e** **f**

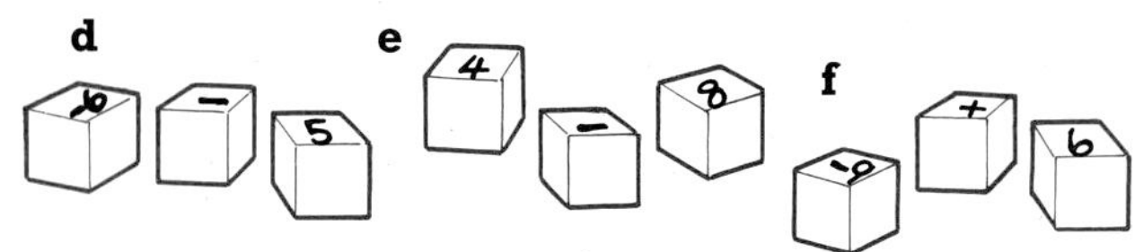

6 $x=2$, $y=-1$ and $z=-3$. Calculate the values of **a–h**. Write out your answers like this:
$y-z=-1-(-3)=-1+3=2$
a $x+y$ **b** $x+z$ **c** $y+z$ **d** $x-y$
e $y-z$ **f** $z-x$ **g** $z-y$ **h** $y-x$

7 If you are turning a telescope, '+' means anti-clockwise and '−' means clockwise. Starting from north, to which directions would these turns point it?
a $+90°$
b $-45°$
c $-30°$ then $-60°$
d $-160°$ then $+25°$

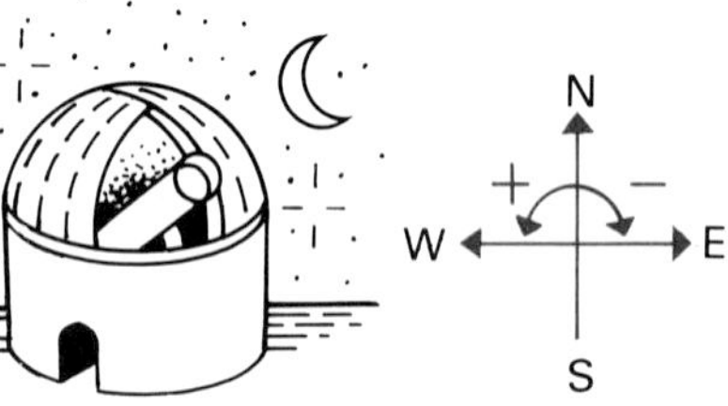

8 Copy and complete these 'crossadds'. Add across, then add down. For example, in **a**, $-2+1=-1$, $-2+3=1$

a

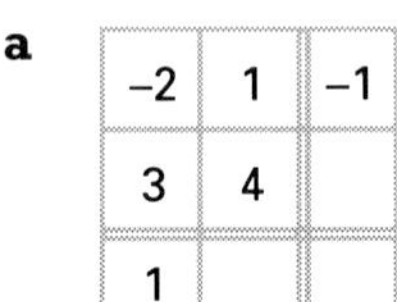

−2	1	−1
3	4	
1		

b

−5	2	
4	1	

c

−6	1	
2	1	

d

5		0
	2	−1

9 Use the 'cover up' method to solve these equations.
a $x+2=3$ **b** $x+2=1$ **c** $x-2=1$
d $x-2=-1$ **e** $-3+x=2$ **f** $2-x=-5$.

10 Write an equation for each diagram below, and solve it.

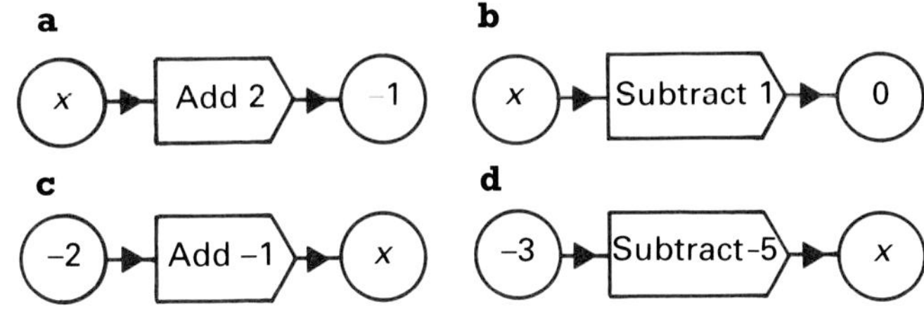

11 Calculate:
a $8-5$ **b** $5-8$ **c** $4-(-1)$
d $-2+(-3)$ **e** $0-7$ **f** $6+(-4)$
g $2-(-3)$ **h** $-1+(-4)$ **i** $-7-(-1)$
j $-9+6$ **k** $-1-3$ **l** $7-(-4)$
m $-6-6$ **n** $-9-(-9)$ **o** $0-(-1)$
p $9-(-1)$ **q** $9-1$ **r** $-9-1$
s $-9+1$ **t** $-9-(-1)$ **u** $10-10$

12 Simplify:
a $3x+2x$ **b** $3x-2x$ **c** $2x-3x$
d $-2x-3x$ **e** $3x-(-2x)$ **f** $-6x+x$
g $4x+(-x)$ **h** $-x-(-x)$

13 True or false?
a $1-3=3+(-1)$ **b** $-3+2=-3-1$
c $4+(-2)=-2+4$ **d** $5+(-1)=14-10$
e $11+(-14)=8-10$ **f** $-3+(-7)=-7-3$

14 Copy and complete these 'crossoffs'. Subtract across, then down. For example, in **a**, $2-3=-1$, $2-4=-2$.

a

2	3	−1
4	1	
−2		

b

−3	4	
8	3	

c

6	8	
7	2	

d

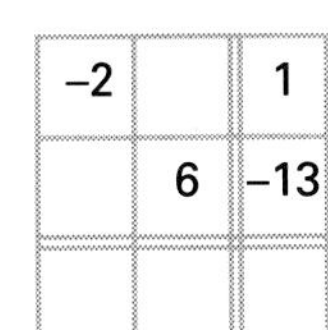

−2		1
	6	−13

EXERCISE 4C

1 Calculate:
a $-12-8$ **b** $17+(-9)$ **c** $-15+7$
d $14-(-6)$ **e** $0-(-11)$ **f** $0+(-11)$
g $-11+9$ **h** $-11-9$ **i** $-11-(-9)$

2 What number does x stand for in each of these equations?
a $x+4=3$ **b** $-5+x=0$
c $-6+x=1$ **d** $x+1=-2$
e $x+1=-3$ **f** $-8+x=-8$
g $-7+x=-1$ **h** $-10+3x=-4$
i $x+(-3)=0$ **j** $x+(-1)=-4$
k $x+(-2)=6$ **l** $x+(-5)=-1$
m $2+x=-1$ **n** $-3+x=5$

3 A triangular tile has its vertices at A$(-60, -30)$, B$(-30, -30)$ and C$(-30, -60)$. Find the new coordinates of the vertices if the tile slides 50 units parallel to:
a the x-axis, in the direction OX
b the y-axis, in the direction OY.

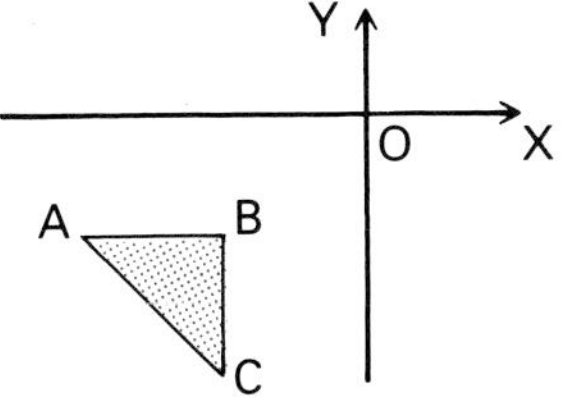

4 Calculate:
a $4+(-3)+(-1)$ **b** $6-(-2)-(-1)$
c $-2+3-4$ **d** $1-(-1)-1$
e $-3-(-2)+(-5)$ **f** $1+(-2)-(-4)$

5 Which numbers go in the circles?

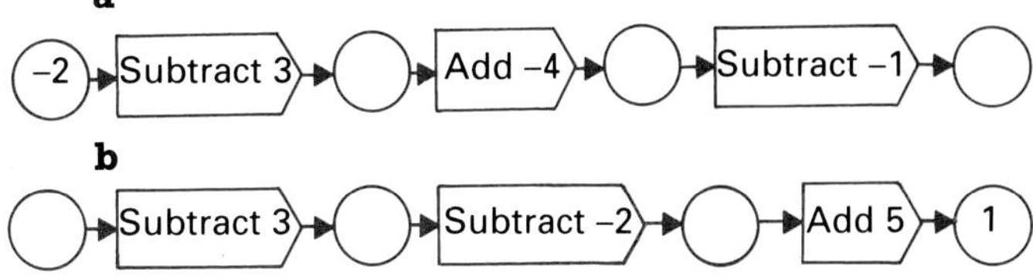

6 Simplify:
a $5t-6t-t$ **b** $-u-2u-3u$
c $-4v-(-v)+2v$ **d** $a-5a-a$
e $-b+(-b)-b$ **f** $-2c+c-3c$

7 Copy and complete these magic squares. (Remember—all rows, columns and the two diagonals should add up to the same number.)

a

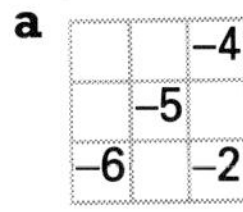

		−4
	−5	
−6		−2

b

−6		−1	3
7	−5	2	
		8	
0			5

c

8			4	−4
	6		10	2
−5		−1		3
1	−12		−3	
7	−6	11		

CHECK-UP ON POSITIVE AND NEGATIVE NUMBERS

1 Copy and complete these scales:

a

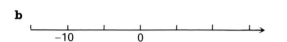

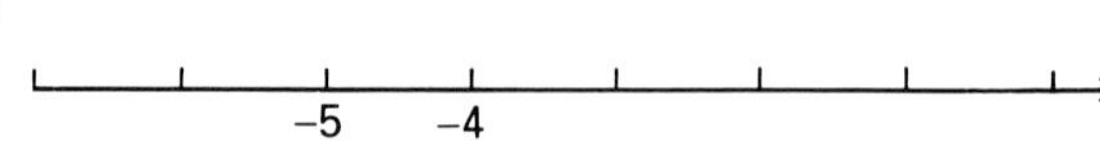

b

–10 0

2

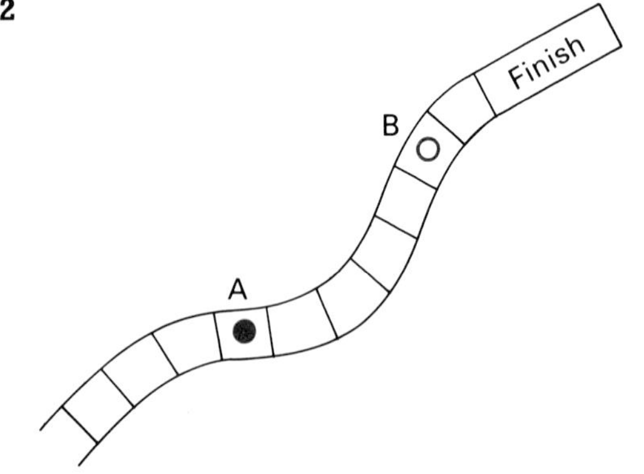

In this game a [+3] card moves your counter three spaces forward, and a [–4] card moves it four spaces back.

Amy plays [+2] [–1] [–3],

and Beatrice [–8] [+1] [+1].

Who is now nearer the finish, and by how many squares?

3 Using a number line if you wish, find the number which is:

a 3 greater than -4 **b** 2 less than 1
c 7 greater than -6 **d** 13 less than 6.

4 Arrange each set of numbers in order, smallest number first in each.

a

1 –3
–1 3
5 –2

b

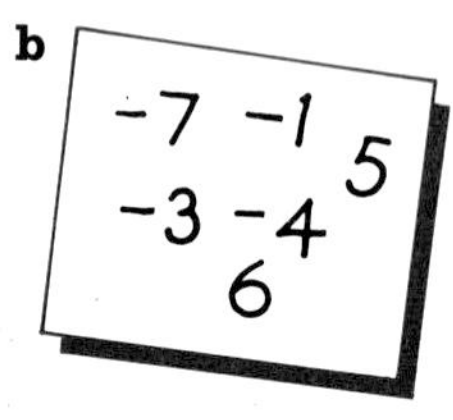

5 Write out one calculation for each walk, **a** and **b**, and two calculations for **c** and **d**.

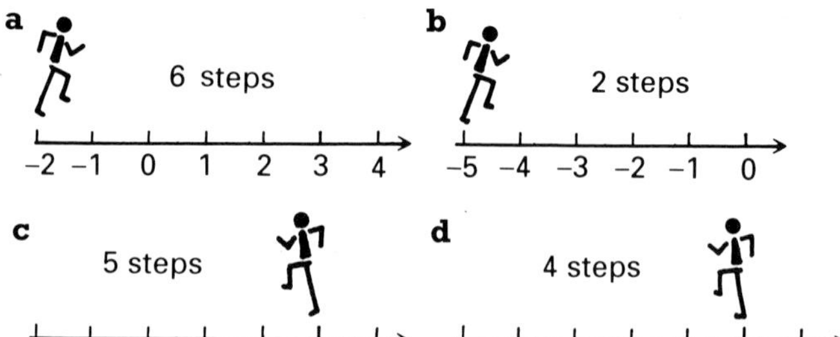

6 Write down the next five terms in this number pattern, and calculate the values of all ten terms.
$4-3, 4-2, 4-1, 4-0, 4-(-1), \ldots$

7 Calculate:

a $2-(-3)$ **b** $3-4$ **c** $-3-7$
d $8+(-4)$ **e** $7+(-9)$ **f** $12-13$
g $3-7$ **h** $6-(-2)$ **i** $-3+(-2)$
j $0-8$ **k** $4+(-2)$ **l** $7-(-5)$
m $-3-(-5)$ **n** $-7-(-7)$ **o** $8+(-8)$

8 Which numbers go in the spaces?

a

b

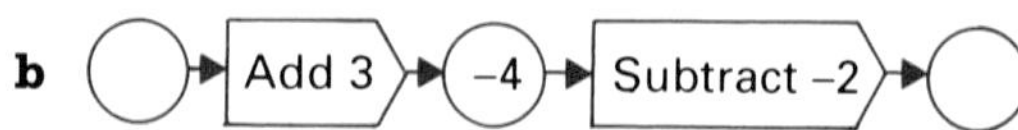

9 Solve these equations:

a $x+2=-1$ **b** $5-x=-3$ **c** $-3+x=-1$
d $4+x=2$ **e** $-1+x=-5$ **f** $4-x=6$.

10 A rectangular tile has vertices at S(1, -2), T(14, -2), U(14, 13) and V.

a Find the coordinates of V.
b The tile is moved, parallel to the y-axis, until VU lies on the x-axis. Find the coordinates of S, T, U and V now.

11 $p=-3$, $q=4$ and $r=-5$. Find the value of:

a $p+q+r$ **b** $p+q-r$ **c** $p-q+r$
d $p-q-r$ **e** $q-p+r$ **f** $q-p-r$

8 ROUND IN CIRCLES

LOOKING BACK

1 Name the parts of these circles marked by heavy lines.

a **b** **c**

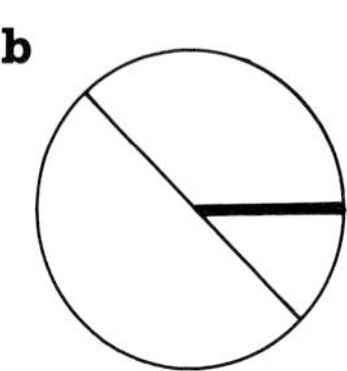
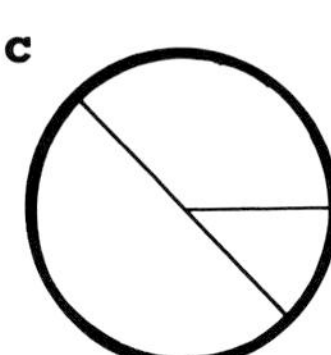

2 Make a list of six objects that are circular.

3 Measure the diameters (widths) of these circles, in millimetres.

a **b** **c**

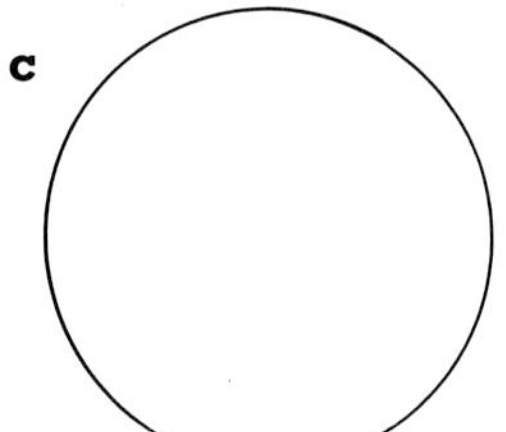

4 Trace the circles in question **3**. By folding them, mark:

a the centres of symmetry

b some lines of symmetry.

5 Use a ruler to measure the perimeter of the:

a rectangle

b triangle

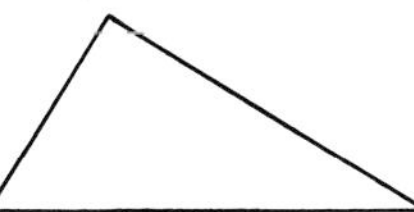

c circle in question **3c** (as best you can).

6 a Draw three circles with radii (compass gaps):
(i) 2 cm (ii) 3 cm (iii) 2.5 cm

b What lengths are their diameters?

7 By counting squares, estimate the areas of these circles. Include parts that are $\frac{1}{2}$ squares or more in area.

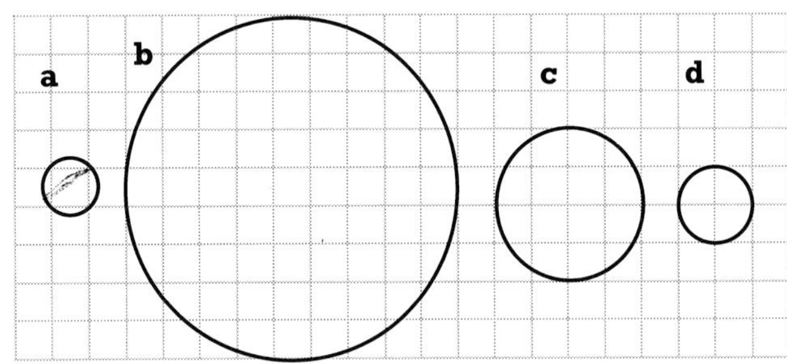

8 a Which of these are cylinders?

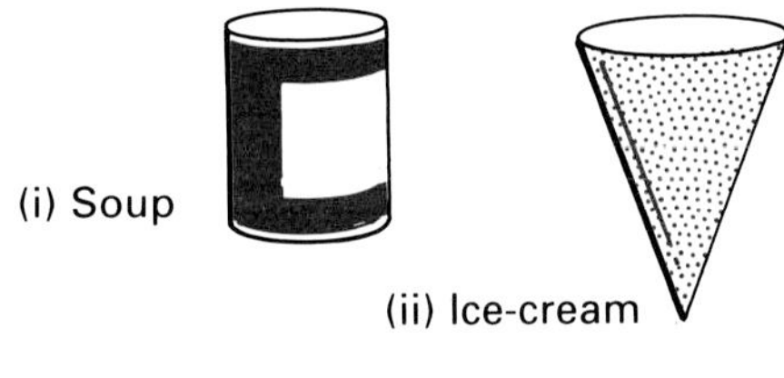

(i) Soup

(ii) Ice-cream

(iii) Oil storage tank

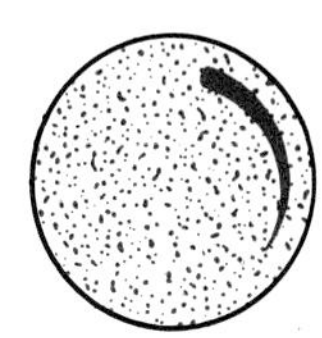

(iv) Football

(v) Plant pots

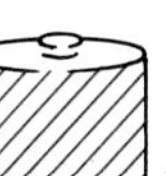

(vi) Torch battery

(vii) Thimble

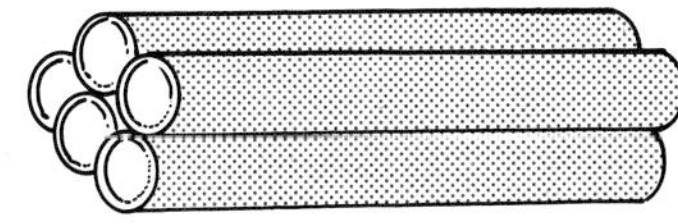

(viii) Water pipes

b What other shapes do you recognise?

PARTS OF A CIRCLE

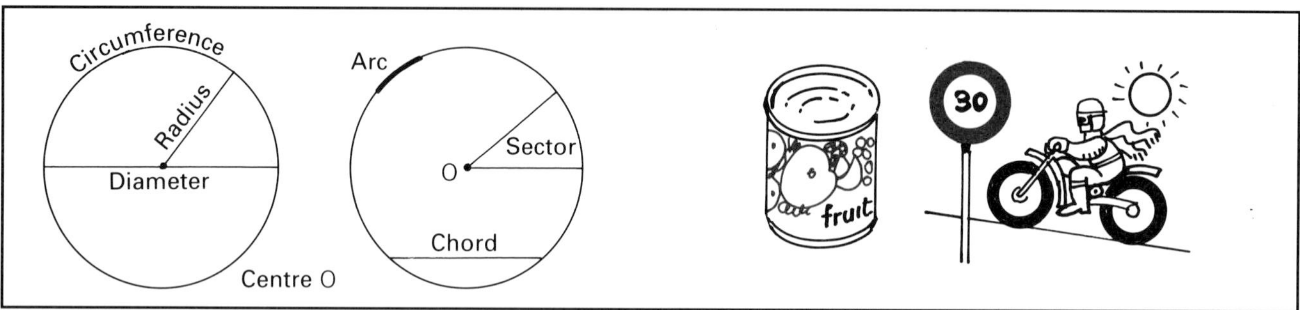

EXERCISE 1

1 In the diagrams below, find examples of: circles, centres, diameters, radii, circumferences, arcs and chords.

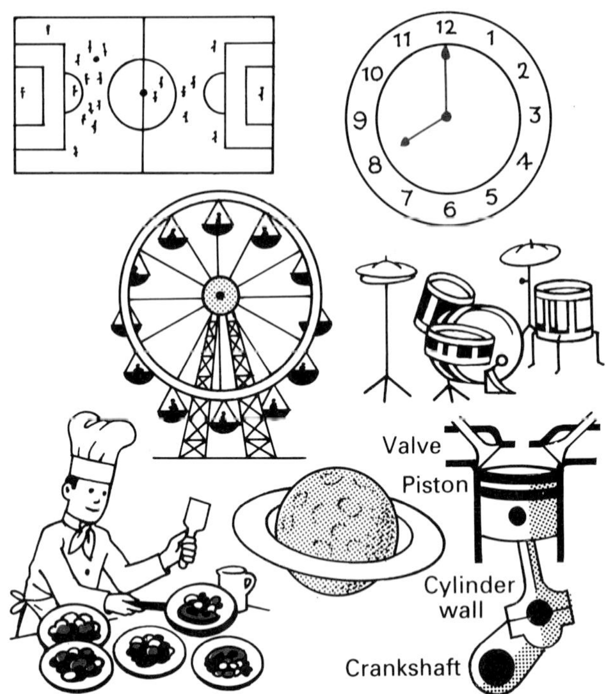

2 a How many radii make a diameter?
b What is the length of the diameter of each of these circles?

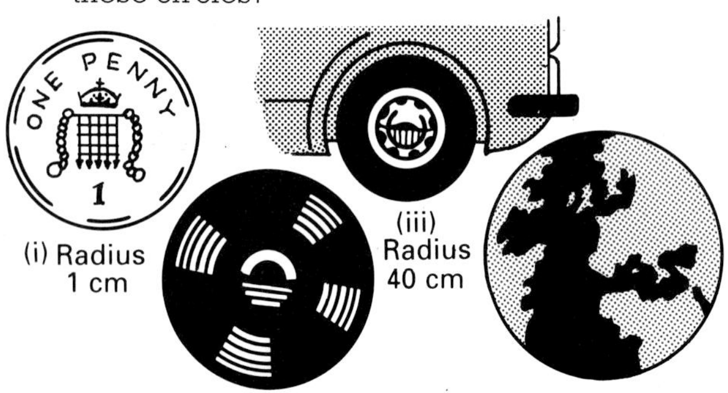

(i) Radius 1 cm (ii) Radius 6 inches (iii) Radius 40 cm (iv) Radius 4000 miles

3 Calculate the radius of each of the circles shown below.

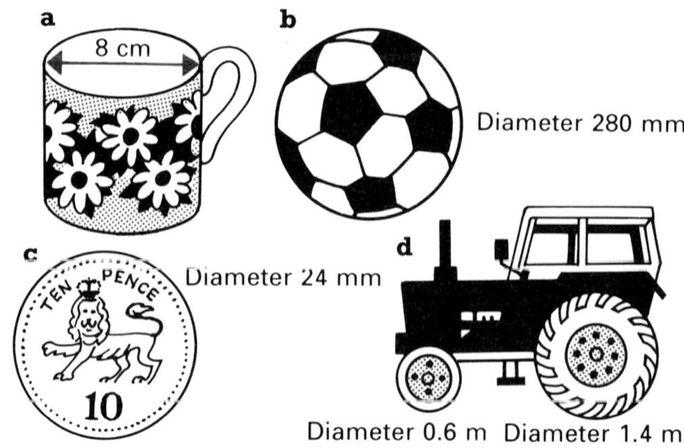

4 Make a full size drawing of this compact disc.

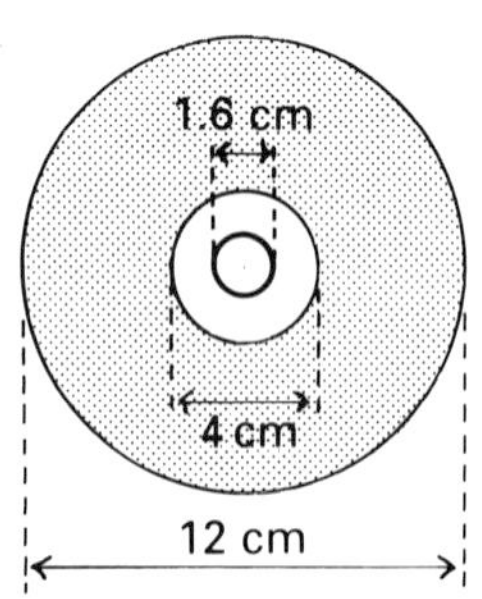

5 Copy and complete this table.

Radius	16 mm	3 cm	1.5 m			
Diameter				36 cm	18 cm	5.6 m

6 a Draw a circle, centre O, radius 3 cm.
b In it, draw:
(i) radius OA (ii) diameter BOC
(iii) arc AB (iv) sector OAB
(v) chord DE.

CIRCUMFERENCE AND DIAMETER

Question: Is the circumference of a circle twice as long as the diameter, three times as long, four times . . .? What do you think?

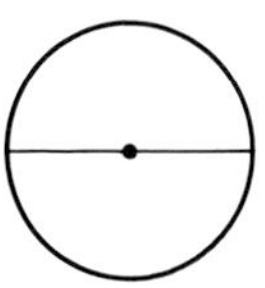

1 *Either:*

a measure the circumferences and diameters of a variety of circular objects such as tins of juice or soup, cotton reels, plates, LPs, wheels, etc.

or:

b measure the circumferences and diameters of the red circles on this page, using a ruler for the diameters, and string or a coil of paper for the circumferences.

2 Fill in your measurements in a table like this:

Object (or circle)	Circumference (C mm)	Diameter (D mm)	$C \div D$

Calculate $C \div D$, correct to 2 decimal places, for: **a** your table **b** the whole class.

About how many times longer than the diameter *is* the circumference?

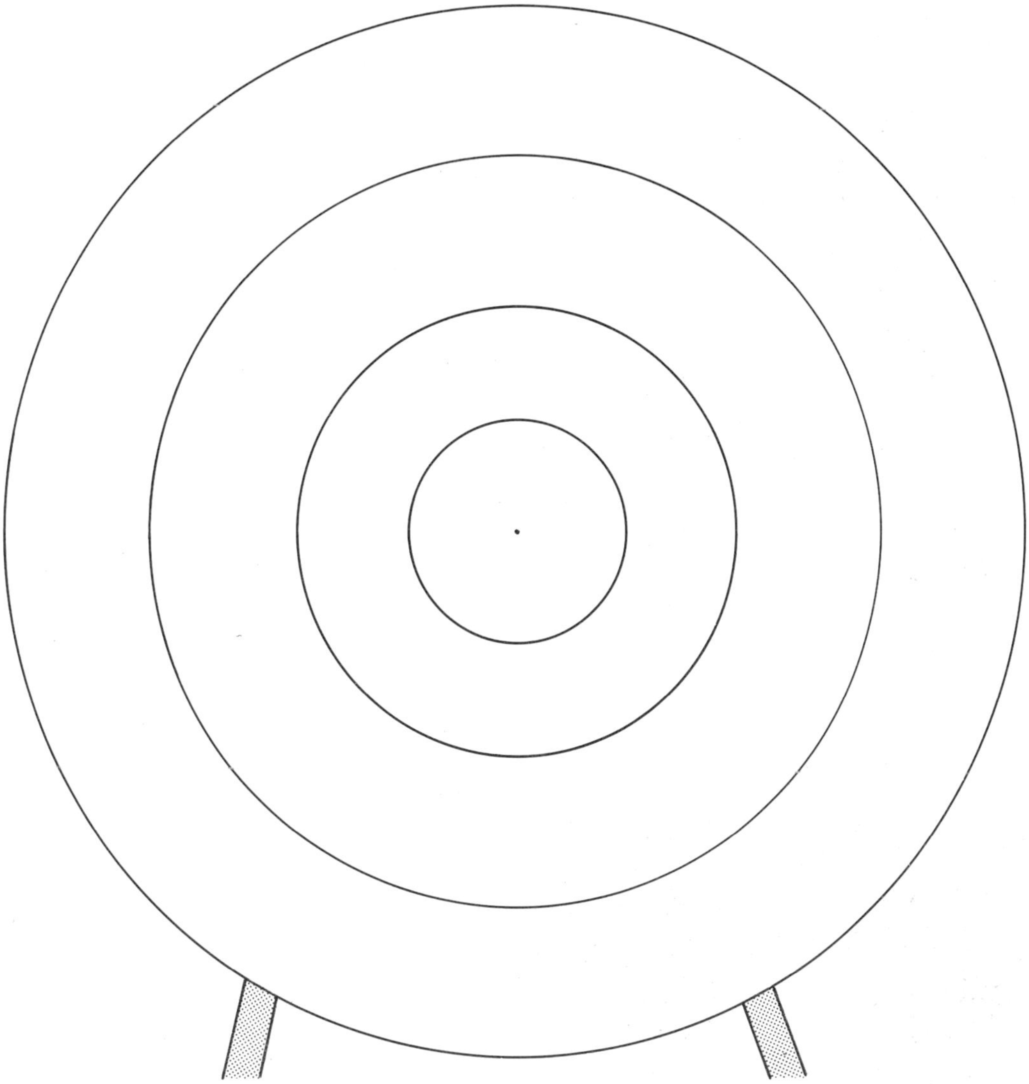

ESTIMATING THE CIRCUMFERENCE OF A CIRCLE

For all circles, the circumference is about '3 × diameter'. $C \doteqdot 3 \times D$.

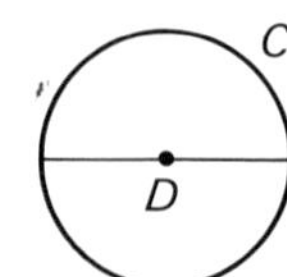

EXERCISE 2A

1 The diameter (D cm) of a compact disc is 12 cm.

For its circumference (C cm),
$C = 3 \times D = 3 \times \ldots$ cm $= \ldots$ cm.

Copy and complete the calculation.

2 Calculate the circumference of this roll of packing tape.

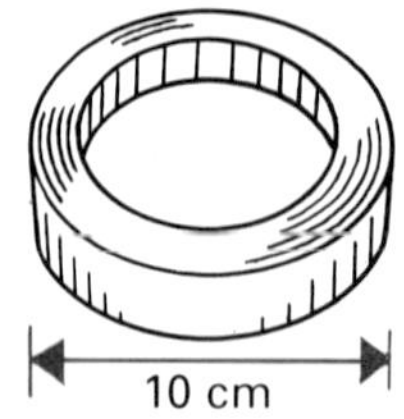

3 The diameter of the roundabout is 20 m. Estimate its circumference.

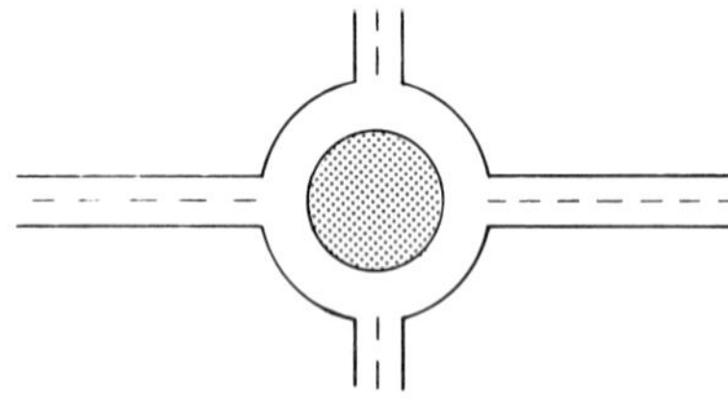

4 Estimate the circumference of each tyre shown below.

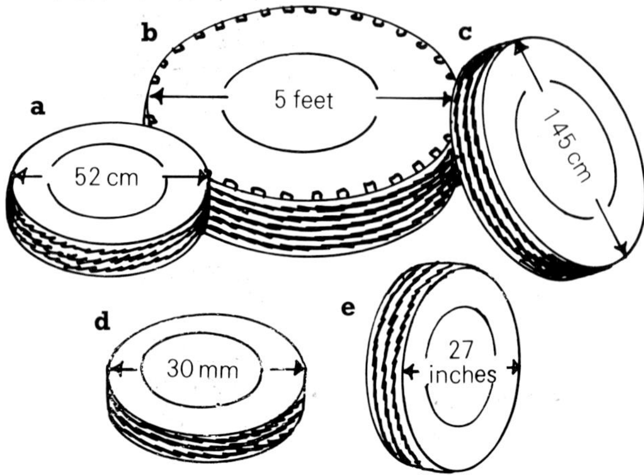

5 The radius of the top of the basketball net is 22.5 cm.
Calculate:
a its diameter
b its circumference.

6 The radius of a cartwheel is 60 cm. Calculate:
a its diameter, in cm
b its circumference in: (i) cm (ii) m.

7 A bicycle wheel has a diameter of 50 cm.
a Calculate its circumference, in metres.
b How far will the bicycle travel in 20 turns of the wheel?

8 Measure again the diameter of each circle in question **3** on page 85, and calculate the three circumferences.

9

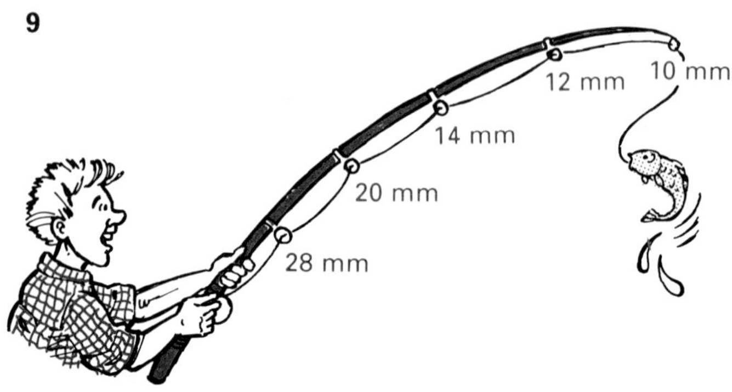

The diameter of each ring in Kevin's fishing rod is shown. Calculate:
a the circumference of each ring
b the radius of each ring.

CALCULATING THE CIRCUMFERENCE MORE ACCURATELY

In the last section you probably found that $C \div D$ was between 3.1 and 3.2.
If you measured carefully, you may have found that the value was about 3.14, or 3.142.

In fact the ratio $\frac{C}{D}$ is the same for all circles and is a never-ending decimal, 3.141 592 653 589 793 . . . .

It is impossible to write it out in full. Because of this we use a special symbol, π (pronounced **pi**), to represent it. $\pi = 3.14$ to *3 significant figures*, so answers cannot be given with greater accuracy than this when you use this approximation for π.

For Exercise 2A you used $C \doteqdot 3 \times D$ but you can now be more accurate as you know that $C = \pi D$.

If your calculator has a π key, compare its value with the one given above.

Example

The diameter of the tin of fruit is 6 cm. Calculate its circumference.

$C = \pi D$

$= \pi \times 6$ *or* 3.14×6 (without π key)

$= 18.8$ (using π key) $= 18.8$

The circumference is 18.8 cm, correct to 3 significant figures.

EXERCISE 2B

Use the π key, or 3.14, for this exercise. In questions **1–3**, round the answers to 2 significant figures.

1 Calculate the circumference of each object.

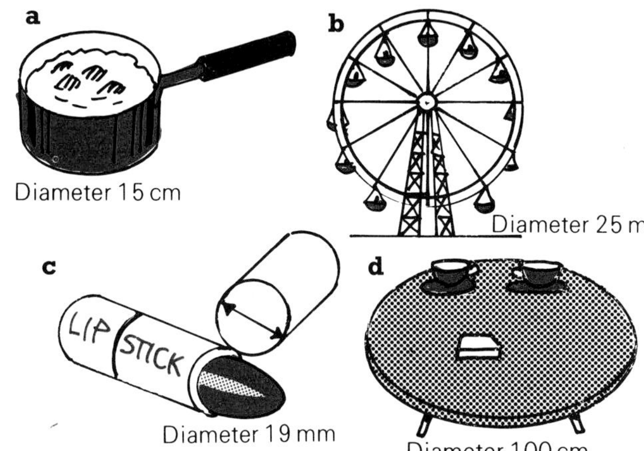

a Diameter 15 cm

b Diameter 25 m

c Diameter 19 mm

d Diameter 100 cm

2 Calculate the diameter, and then the circumference, of each of the following:

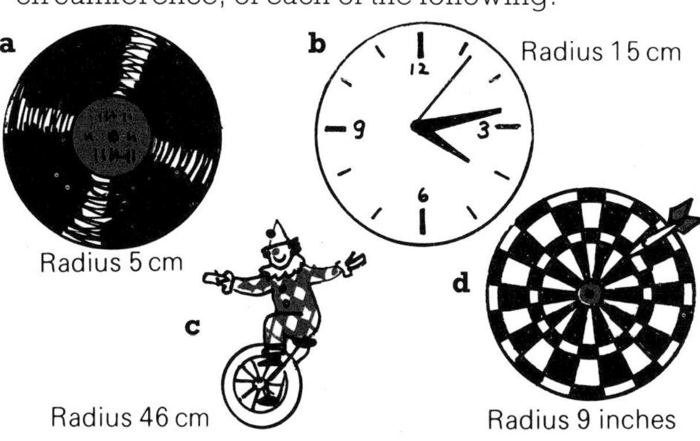

a Radius 5 cm

b Radius 15 cm

c Radius 46 cm

d Radius 9 inches

3 Here are some knitting needle sizes:

Size	0	2	4	6	8	10
Diameter (mm)	8	7	6	5	4	3.25

a Calculate the circumference of each needle.

b Which size of needle would be used to knit a chunky sweater?

Round your answers to questions **4–5** to 3 significant figures.

4 Calculate the circumference of:

a the ribbon round the hat

b the flex-holder

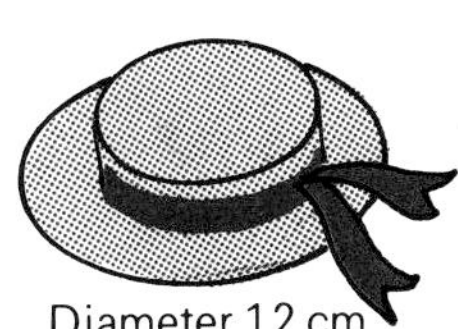

Diameter 12 cm

Diameter 16 cm

c the tape spool

d the reel.

Diameter 9 cm

Diameter 30 mm

5 Calculate the lengths of the circular parts of these objects.

a The edges of the biscuits being coated with brown sugar.

Diameter 6.5 cm

b The frill round the edge of the cushion.

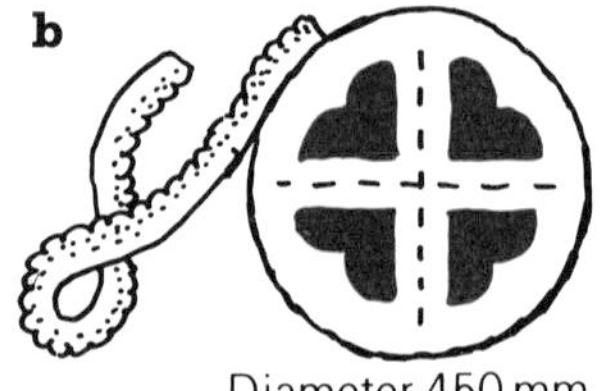

Diameter 450 mm

c The rubber seal round the washing machine door.

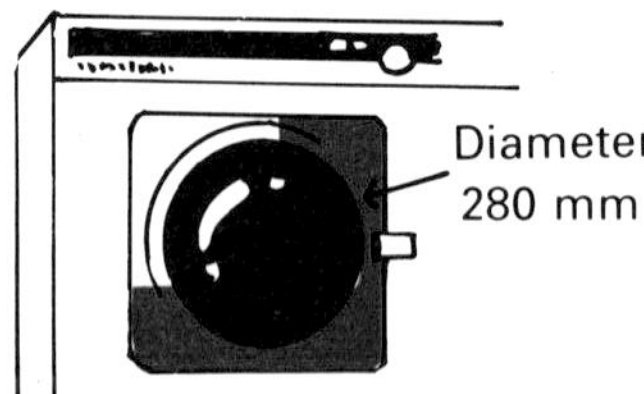

Diameter 280 mm

d The edging on this lace mat.

Diameter 162 mm

e The rims of the sugar bowl and the cup.

Diameter 110 mm

Diameter 80 mm

f The fancy trim round the three tiers of cake.

Diameters 18 cm, 27 cm, 40 cm

EXERCISE 2C

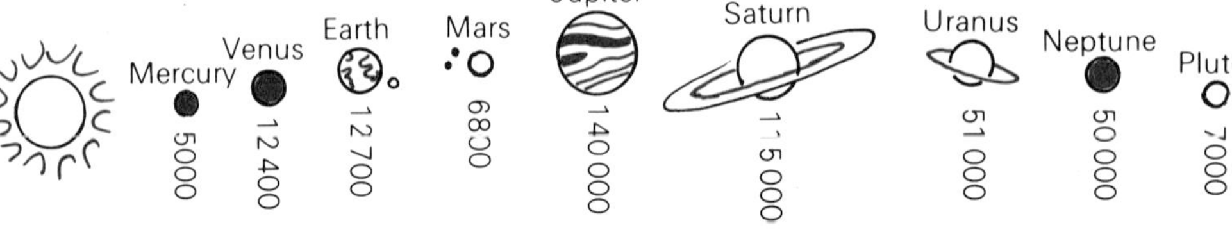

1 This picture shows the planets in our solar system in order from the sun outwards. The figure beside each planet gives its diameter in kilometres. Calculate the circumferences of the planets, correct to the number of significant figures given in the diameter of each.

2 a The radius of the moon is 1740 km, and the radius of the sun is 696 000 km. Calculate the circumference of each, correct to 3 significant figures.

b There is a big difference between the circumferences of the sun and moon. Why do they appear to be the same size during an eclipse?

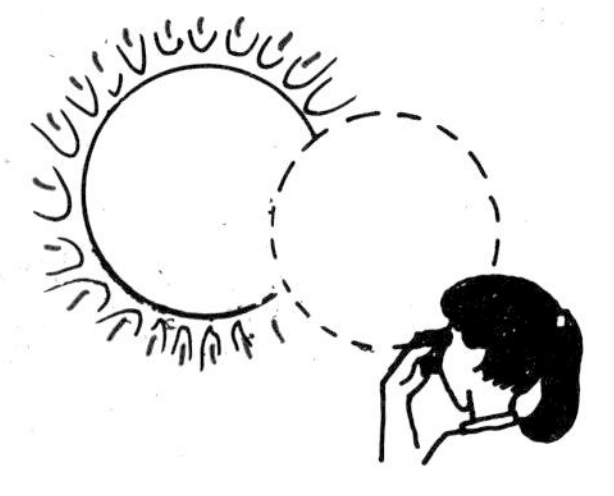

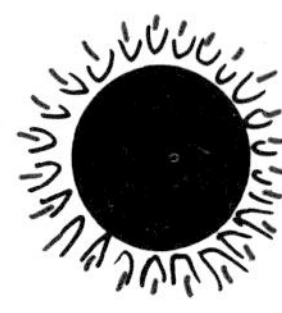

3 The diameter of the Earth across the equator is 12 760 km, and from pole to pole is 12 710 km. Calculate the Earth's circumference to 4 significant figures:

a round the equator **b** round the poles.

What does this suggest about the shape of the Earth?

4 NOAA9 is a satellite which orbits the earth 850 km above the surface. Taking the radius of the Earth to be 6350 km, calculate, correct to 2 significant figures:

a the radius of the orbit

b how far the satellite travels in one orbit.

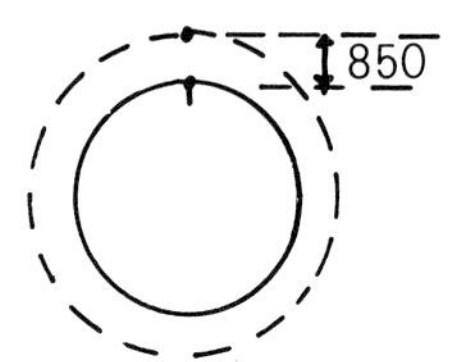

5

James finds that the diameter of the wheels on his bicycle is 68 cm, to the nearest cm. He knows that this means that the diameter lies between 67.5 cm and 68.5 cm. Calculate the upper and lower limits of the circumference, correct to 3 significant figures.

6 Taking the Earth's radius to be 6400 km, correct to 2 significant figures, calculate the lower and upper limits of its circumference, also to 2 significant figures.

BRAINSTORMER

Here are some historical approximations for π. Put them in order, from most accurate to least accurate. (π is given to 15 decimal places on page 89.)

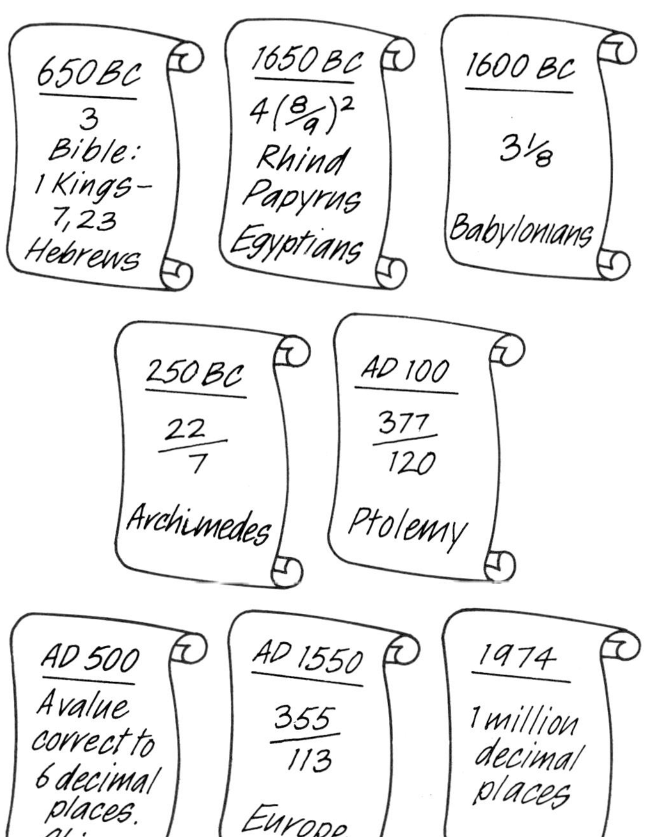

INVESTIGATION

Try this program in a computer, and watch it slowly calculating π.

```
10 LET N = 1: LET P = 2
20 LET D = 4 * N * N
30 LET T = D/(D - 1)
40 LET P = P * T
50 PRINT N; "2 spaces"; P
60 LET N = N + 1
70 GOTO 20
```

CHALLENGES

1 *Copy the first pattern, which consists of circles and arcs. In it colour the second pattern.*

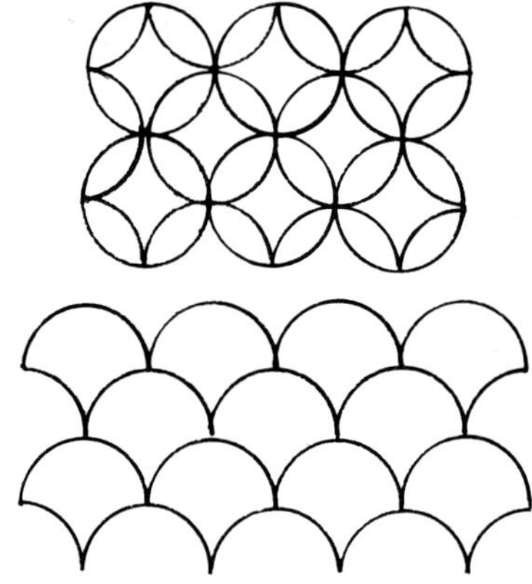

2 *The Moors used parts of circles in designing arches and domes in their buildings. Experiment with your own designs.*

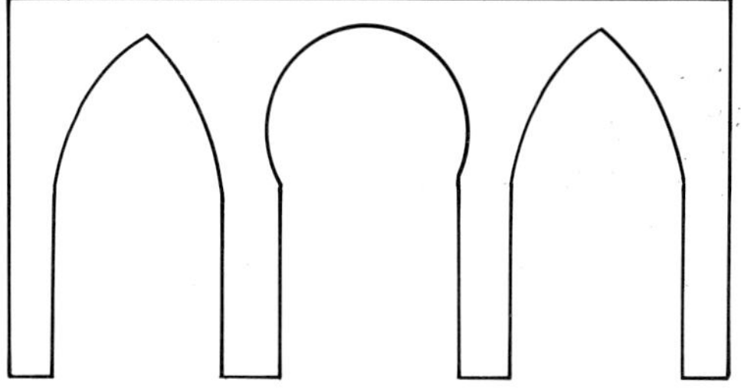

3 *How could you find the diameter of:*

a *a 2p coin, by rolling it*

b *a roll of kitchen paper by unrolling it?*

ESTIMATING THE DIAMETER OF A CIRCLE

$C \doteqdot 3 \times D$. For example, if diameter = 2 cm, circumference = 3 × 2 cm = 6 cm.
If circumference = 6 cm, diameter = 6 ÷ 3 cm = 2 cm

$$D = C \div 3$$

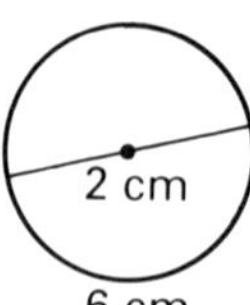

EXERCISE 3A

1 The circumference of the bin is 75 cm.
To estimate its diameter, copy and complete:
$D = C \div 3 = 75\text{ cm} \div 3 = \ldots\text{ cm}$

2

The circumference of the stone column is 120 cm. Calculate its diameter.

3 The circumference of the roundabout is 54 m. Find its diameter.

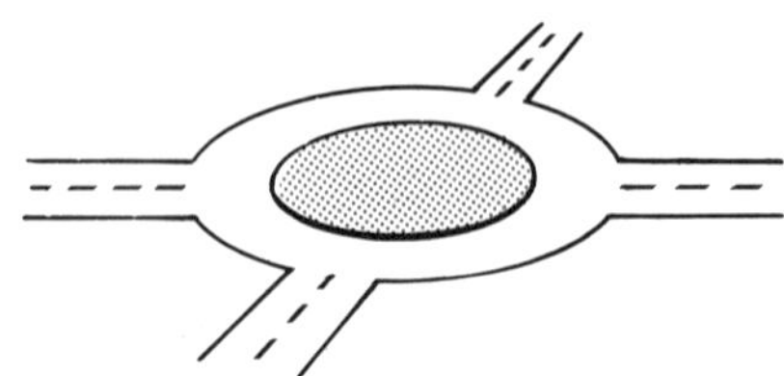

4 Copy and complete this table for circles.

Circumference	30 cm	24 m	6 m	15 cm		
Diameter					10 m	
Radius						12 cm

5 Sam puts a tape measure round the tree trunk, and finds that the circumference is 144 cm.

Calculate the tree's:
a diameter **b** radius.

6 The trundle wheel measures 1 metre for each turn.

a Calculate its diameter in cm.
b How many turns does the wheel make in 100 m?

CALCULATING THE DIAMETER

Example

This circular floor has a circumference of 25 m. Calculate its diameter.

$C = \pi D$, so $25 = \pi \times D$
$\pi \times D = 25$, so $D = 25 \div \pi = 8.0$, to 2 significant figures

Using the π key, or 3.14, the diameter is 8 m.

EXERCISE 3B/C

Give answers to 2 significant figures.

1 If you wanted to draw a circle with circumference 24 cm:
a what would the diameter of the circle be
b what distance apart would you set the points of your compasses?

2 Draw circles with circumferences:
a 27 cm **b** 18 cm **c** 12 cm.

3

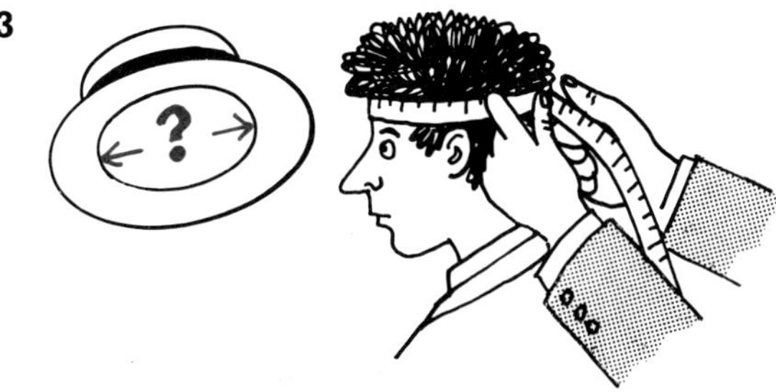

A hatter is measuring a customer's head. If he assumes the head to be circular, what diameter of hat should he make for a customer whose head measures:
a 54 cm **b** 57 cm **c** 60 cm?

4 Collar sizes in a catalogue are recorded as $14\frac{1}{2}$ inch, 15 inch, $15\frac{1}{2}$ inch, 16 inch and $16\frac{1}{2}$ inch. If the collars are circular, what diameter of neck is each of the above sizes meant for?

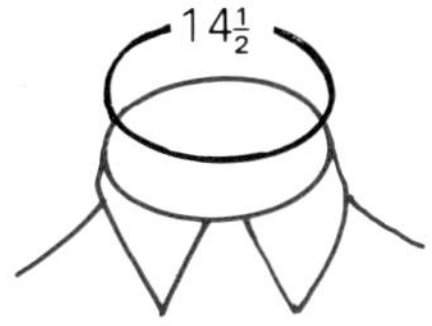

5

The well is 18 m deep, and it takes 40 complete turns of the handle to raise the bucket from the bottom to the top. Calculate the diameter of the roller, in cm.

6

After a bank robbery, the police found a track left by the getaway car.

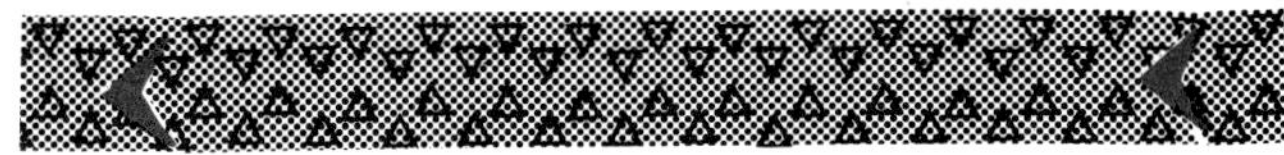

They measured the distance between two marks on the tyre track.
The distance was 226 cm. What was the diameter of the wheel?

ESTIMATING THE AREA OF A CIRCLE

Examples

a This circle has a radius of 2 cm.
Estimate its area like this:
count half a square or more as one square, and don't count the others.
Check that the area of the circle is approximately 12 cm^2.

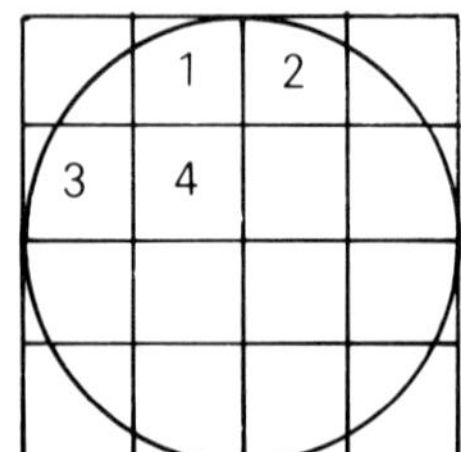

b On 1 cm squared paper draw circles with radii 2, 3, 4 and 5 cm. Estimate their areas by counting squares. Copy and complete the table.

Radius (r cm)	Area (cm^2)	r^2	$3r^2$
2	12	$2 \times 2 = 4$	12
3		$3 \times 3 = 9$	27
4			
5			

Check that in each case the area is equal to, or slightly more than, $3r^2$.

The area of a circle is about 3 × radius × radius, or $3r^2$.

EXERCISE 4A

1 The minute hand of the clock is 10 cm long. To estimate the area of the clock, copy and complete:

area $= 3r^2$
$= 3 \times 10 \times 10$ cm^2
$= \ldots$ cm^2

2

The radius of one penny is 1 cm. Copy and complete:

area of coin $= 3r^2$
$= 3 \times \ldots \times \ldots$ cm^2
$= \ldots$ cm^2

3 The radius of the compact disc is 6 cm. Calculate its area.

4 Estimate the areas of these circular shapes.

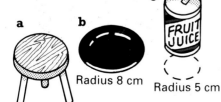

a **b**

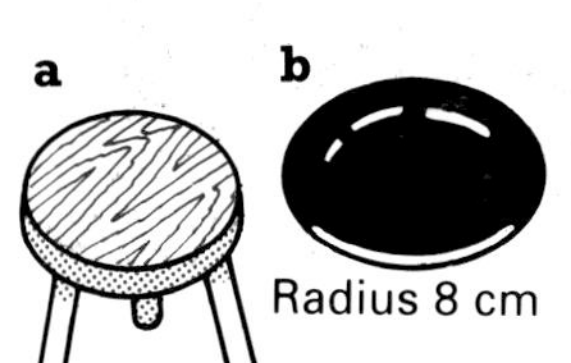

Radius 16 cm

Radius 8 cm

c

Radius 5 cm

d

Radius 12 mm

5 a The radius of an LP is 15 cm. Estimate the area of the LP (one side).

b What shape is the record sleeve? What area of cardboard is needed to make it?

6 A single has a radius of 9 cm. Repeat question **5** for a single.

7 Estimate the areas of the rug, which has a radius of 2 m, the table, with diameter 80 cm and the table mat, with diameter 66 mm.

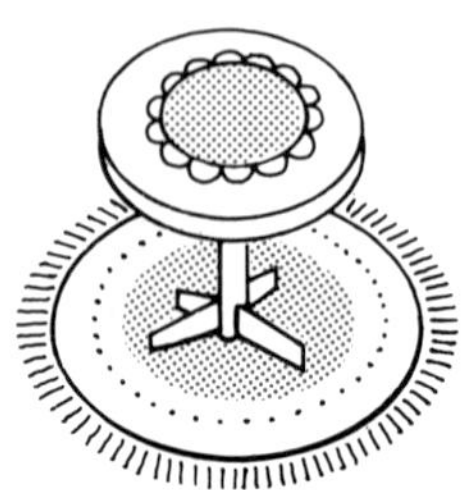

8 Copy and complete this table for circles.

Radius	6 cm	9 m	20 mm	1 km	2.5 m
Area					

9 Find the radius of each circle in question **3** on page 85, and calculate the area of each circle.

CALCULATING THE AREA MORE ACCURATELY

Here is the circular end of a roll of kitchen paper.

Using a sharp knife we cut along the top of the roll, down to the centre:

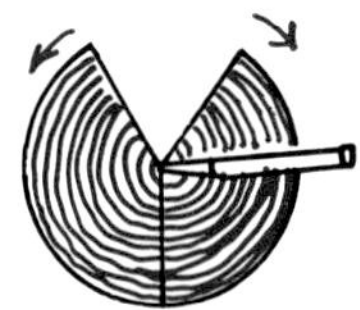

We straighten out the layers of paper:

The layers of paper form an isosceles triangle.
The base of the triangle is the circumference of the roll of paper.
Why can we say its length is $2\pi r$?
What is the area of the triangle? (Remember that the area of a triangle $= \frac{1}{2}$ base × height.)
So what is the area of the circle?
The investigation above suggests that the area is πr^2. This is, in fact, true.

The area of a circle is $A = \pi r^2$

Example Calculate the area of the circular cover for the speaker.
Its diameter is 16 cm.

$D = 16$
So $r = 8$
$A = \pi r^2$
$= \pi \times 8 \times 8$
$= 201$, to 3 significant figures, using the π key, or 3.14 for π.

The area of the cover is 201 cm^2.

EXERCISE 4B

Give answers correct to 3 significant figures.

1 Calculate the areas of these circular shapes.

a

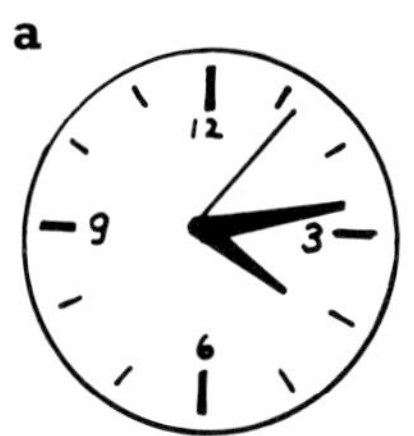

Radius 12 cm

b

Radius 17.5 cm

c

Radius 4 cm

d

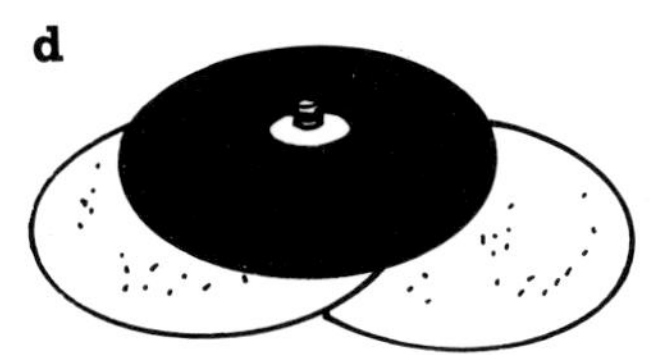

Sanding discs

Radius 8.5 cm

2 This circular table has a glass centre.
The diameter of the table is 1.5 m, and the diameter of the glass circle is 1 m.

a Write down the radius of:
(i) the table
(ii) the glass.

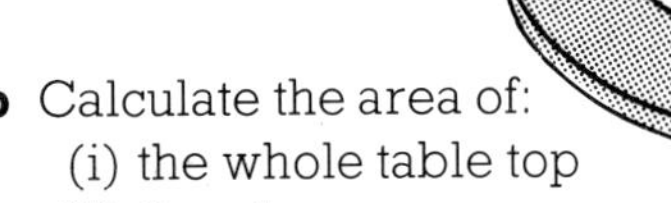

b Calculate the area of:
(i) the whole table top
(ii) the glass
(iii) the wooden surround.

3 Calculate:

a the area of the label

b the area available for recording on each side of the record.
(Remember to find the radii first.)

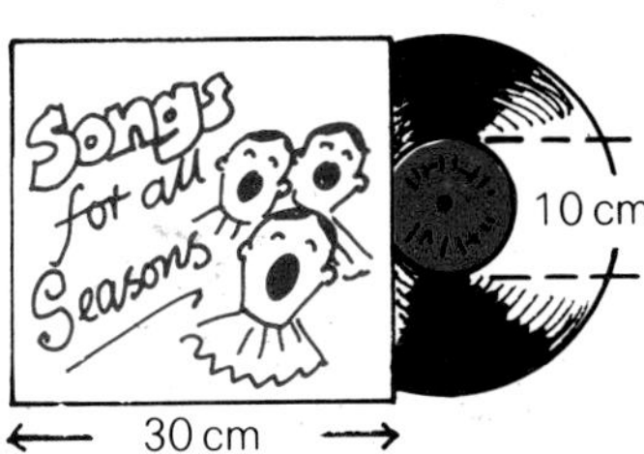

4 The rug is a semi-circle. Calculate the area of:

a a circle with radius 80 cm

b the rug.

5 a What is the radius of the large semi-circular part of the doorway?

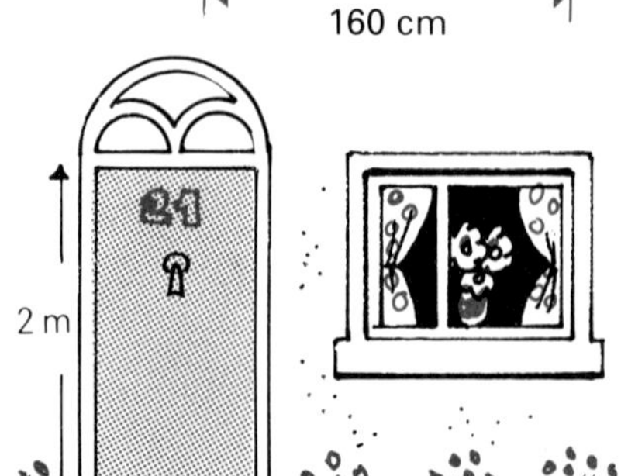

b Calculate the area, in m^2, of:

(i) the rectangular door

(ii) the large semi-circle at the top

(iii) the whole doorway.

6 Diagram (i) below shows a square lawn in a park, with a circular flower bed. Calculate the area of:

a the flower bed **b** the lawn.

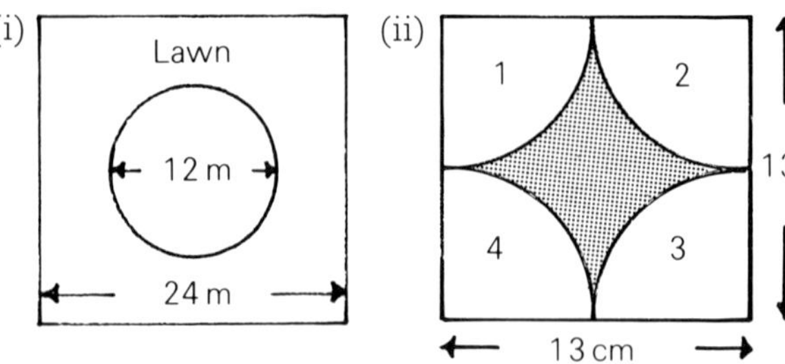

7 In diagram (ii) above, the arcs are parts of equal circles with centres at the corners of the square.

a What shape would parts 1, 2, 3 and 4 make if they were cut out and fitted together?

b Calculate the area of the shaded part (called an asteroid).

EXERCISE 4C

Give answers correct to 3 significant figures.

1 The two most popular sizes of floppy disc have diameters of $5\frac{1}{4}$ inches and $3\frac{1}{2}$ inches. The black part of the disc is coated with magnetic material, and the disc is in a protective square sleeve. For each size of disc, calculate the area of:

a the outer circle

b the hole (diameter 1 inch)

c the coated part

d the sleeve not occupied by the disc.

2 The end of an Underground train looks like this. Calculate the area of:

a the circle

b the door

c the part of the end that is not door.

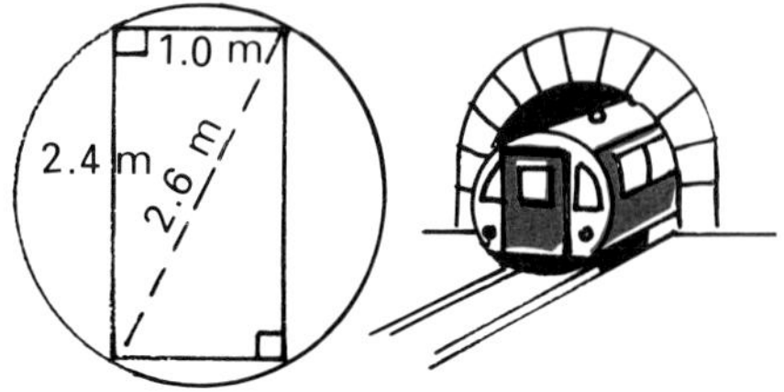

3 Calculate the area of the top of the corner table.

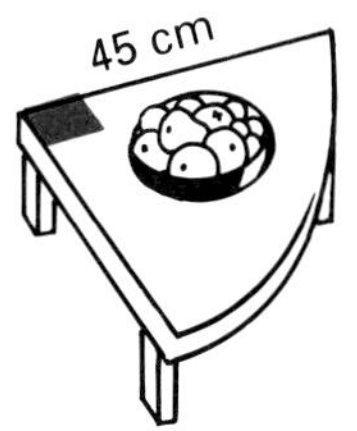

4 Calculate the area of the end of the shelf below. The curves are quarter-circles.

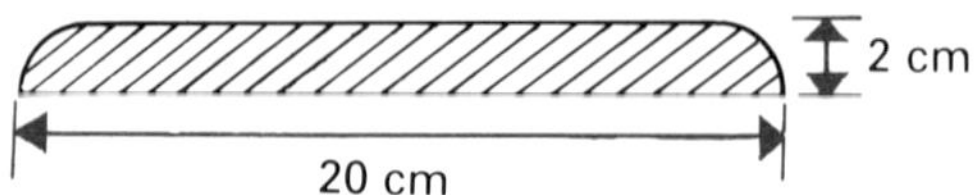

5 Calculate the goal area on this hockey pitch, bounded by the straight lines and quarter-circles.

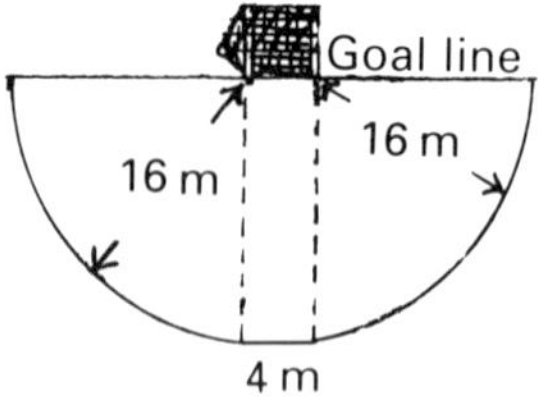

6 An archery target has a gold centre 24.4 cm in diameter, ringed by four concentric bands, each 12.2 cm wide. Calculate the area of each of the five colours on the target.

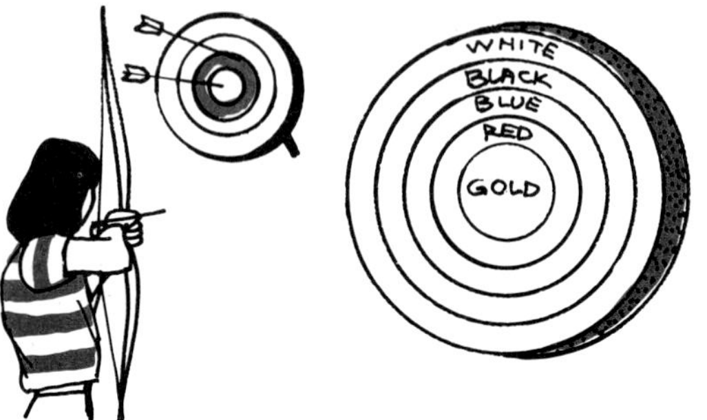

INVESTIGATIONS

1 *A farmer has 100 m of flexible fencing.*
  a *Which would give him the larger enclosed area—a square pen or a circular pen?*
  b *Calculate the difference between the two areas.*

2 **a** *Which of the two squares do you think has the larger coloured area in it? Check, by calculation.*

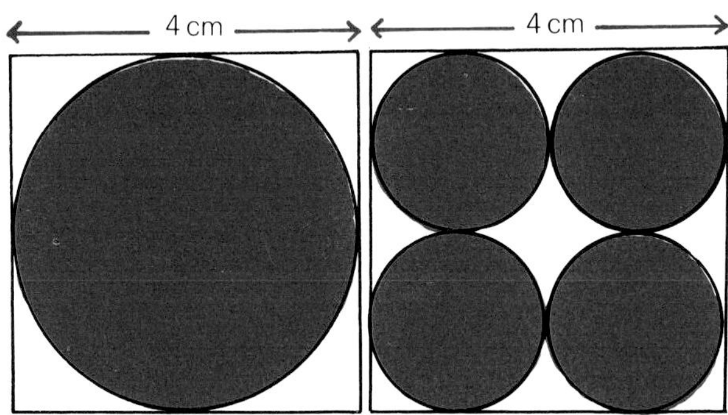

  b *Investigate the result for squares of side 6 cm.*
  c *Find out what happens if a square has 9 or 16 circles in it.*

CHALLENGES

1 *O is the centre of circle (i) below, which has a radius of 5 cm.*
  a *Draw the design, which is based on semi-circles.*
  b *Explain why the dark and light areas are equal.*
  c *Calculate the dark area.*

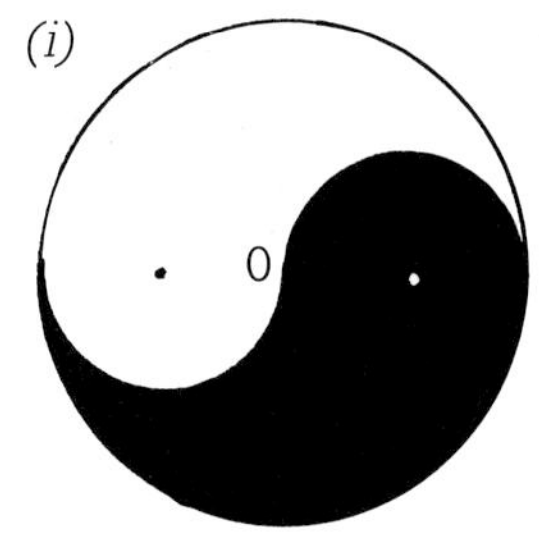

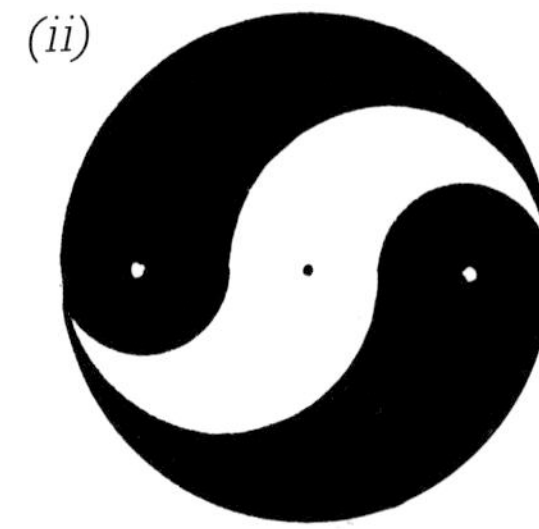

2 *The diameter of circle (ii) has been divided into six parts, each 2 cm long.*
  a *Draw the design, which is based on semi-circles.*
  b *Calculate the area of every semi-circle you can find in the design.*
  c *Can you extend the pattern to draw circles with four equal parts and five equal parts?*

BRAINSTORMERS

1 *Billy the goat has a 5 m chain which is fixed to a post.*
  a *Make scale drawings to show where he can graze if the post is:*
    (i) in the centre of a square field of side 10 m
    (ii) at a corner of the field
    (iii) half-way along one side of the field.
  b *Calculate each grazing area.*

2 *Michelle is puzzled. In her calculation she has the same number of units in the area and circumference of a circle. Can you find the diameter of her circle?*

CHECK-UP ON ROUND IN CIRCLES

1 a Measure the diameter of:
(i) the clockface
(ii) the inner and outer edges of the Olympic rings, in millimetres.

b Write down the radius of each circle.

2 a The TV aerial has a diameter of 3 m. What is its radius?

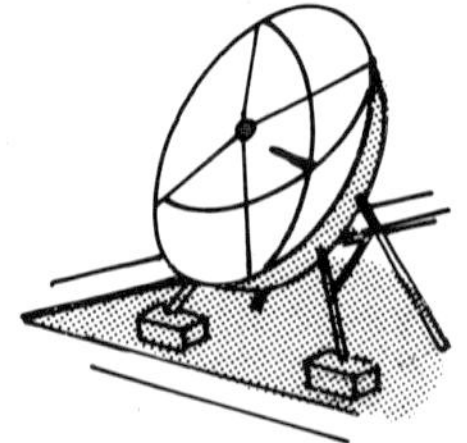

b The radius of the cymbal is 12 inches. What is its diameter?

3 Copy and complete: $C = \ldots D$ for every circle.

4 Calculate the circumference of:
a the fishing net, diameter 40 cm
b the tin of shoe polish, radius 3.5 cm.

5 A 2p coin makes one complete turn as it rolls from A to B.

A ——————————————— B

a What is the length of its circumference, in mm?

b Calculate its diameter.

6 Copy and complete: $A = \ldots r^2$ for every circle.

7 Calculate the area of:
a the glass on the barometer, radius 8 cm
b the tin lid, diameter 24 cm.

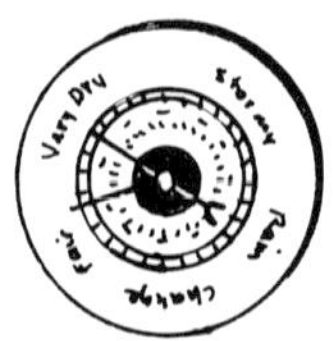

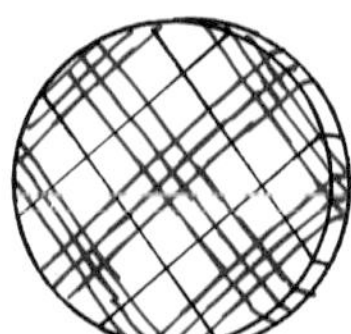

8 Calculate the circumference (in mm) and the area (in mm^2) of the clockface in question **1**.

9 Calculate the area of the base of the tin of shoe polish in question **4b**.

9 TYPES OF TRIANGLE

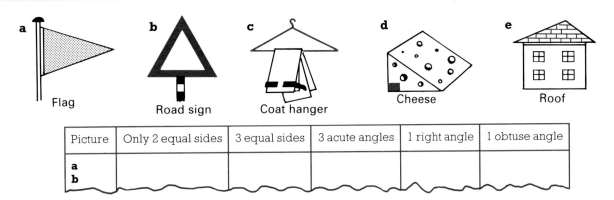

LOOKING BACK

a 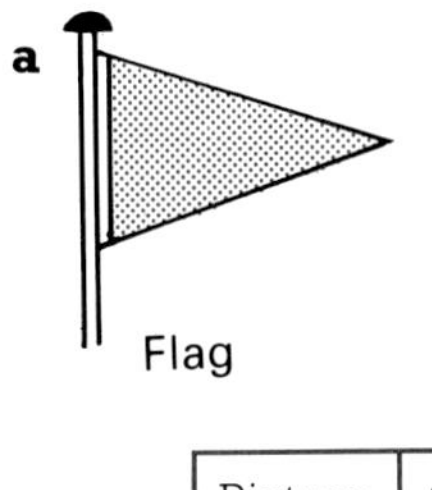
Flag

b 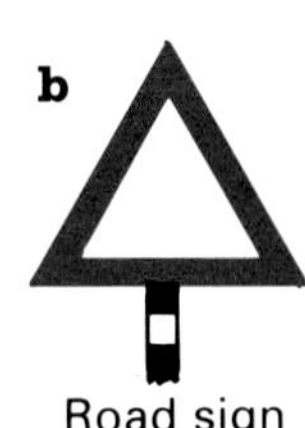
Road sign

c 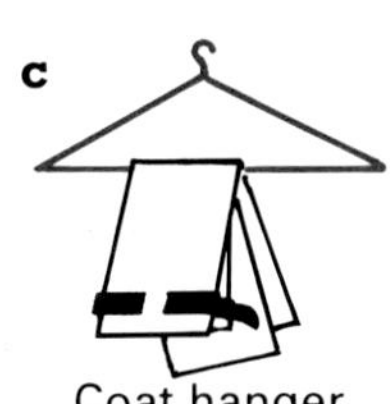
Coat hanger

d 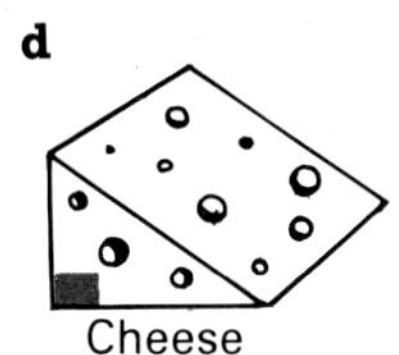
Cheese

e 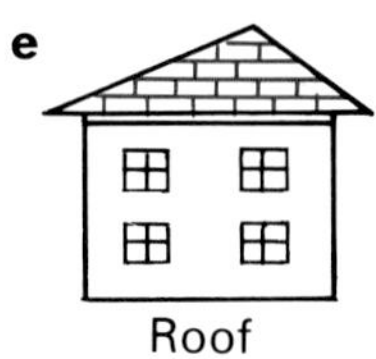
Roof

Picture	Only 2 equal sides	3 equal sides	3 acute angles	1 right angle	1 obtuse angle
a **b**					

1 Copy and complete the table, ticking all the headings which apply to each triangle pictured above.

2 Sketch this wooden gate, and show how you could make it much stronger.

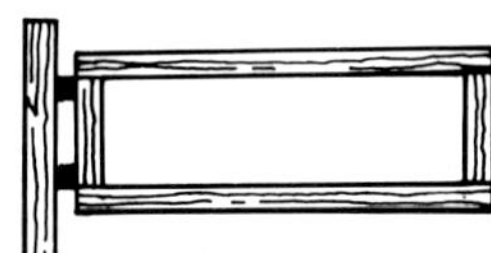

3 BD is a line of symmetry in the shape ABCD.

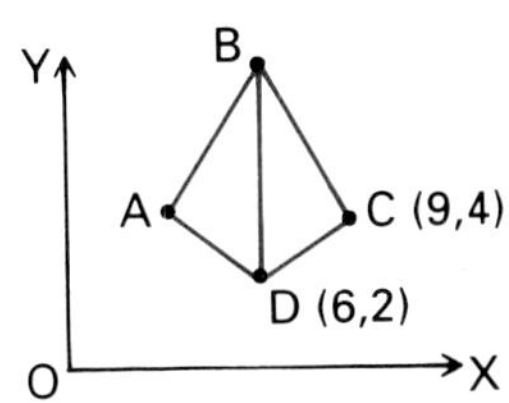

a What are the coordinates of A?
b What do you call triangles like ABD and CBD, which have the same shape and size?

4 a $x^\circ + y^\circ = 90^\circ$, and $x^\circ = 66^\circ$. Calculate y°.
b $p^\circ + q^\circ + r^\circ = 180^\circ$, $p^\circ = 45^\circ$ and $q^\circ = 75^\circ$. Calculate r°.

5 a How many triangles can you see in the drawing?
b Name the triangle which has:
(i) a right angle
(ii) an obtuse angle.

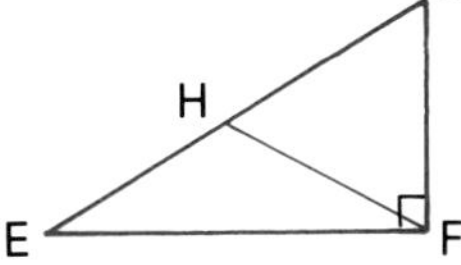

6 a Write down the number of the angle:
(i) corresponding to 1
(ii) alternate to 4
(iii) supplementary to 5.
b Copy the diagram, and fill in the sizes of all the angles, if angle 1 is 108°.

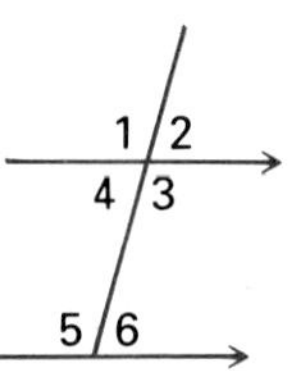

7 Calculate the areas of these triangles.

a
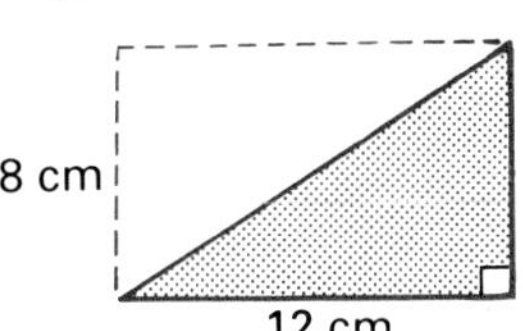

b
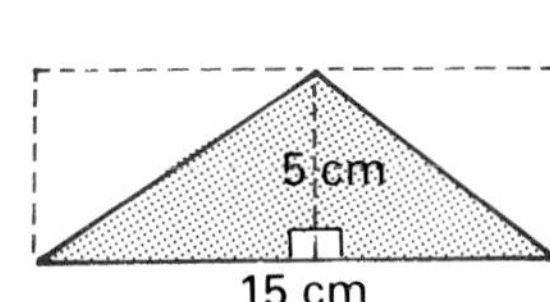

PUZZLE

Why is the triangle on the road longer than the one on the road sign? Keep a look-out for these and other road signs with triangles round them.

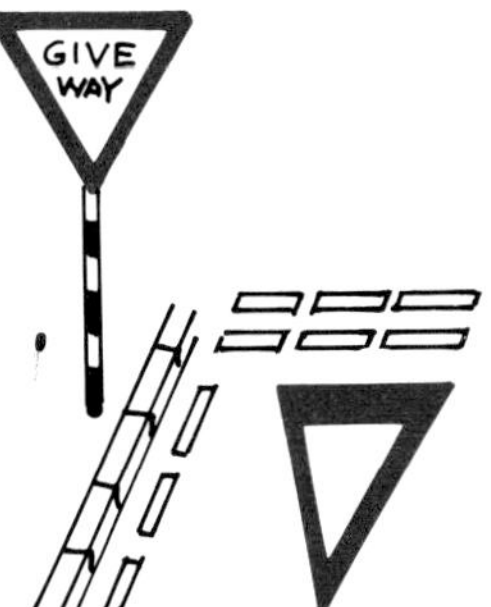

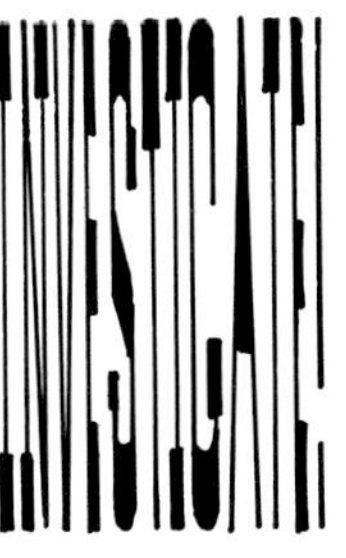

THE SUM OF THE ANGLES OF A TRIANGLE

Try this.
a Draw a triangle, and number its angles 1, 2 and 3.
b Cut out the triangle.
c Tear off the angles, and fit them together as shown.
d Do the three angles together make a straight angle? How many degrees is this?

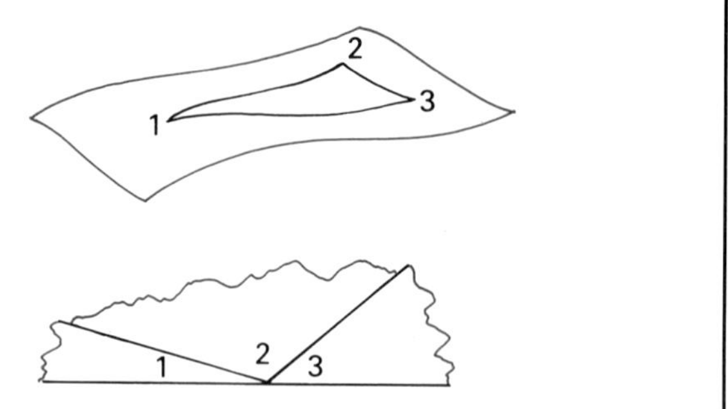

EXERCISE 1A

1 a In △ABC, write down the sizes of the angles at A and B. (Remember to measure from 0° on one arm.)

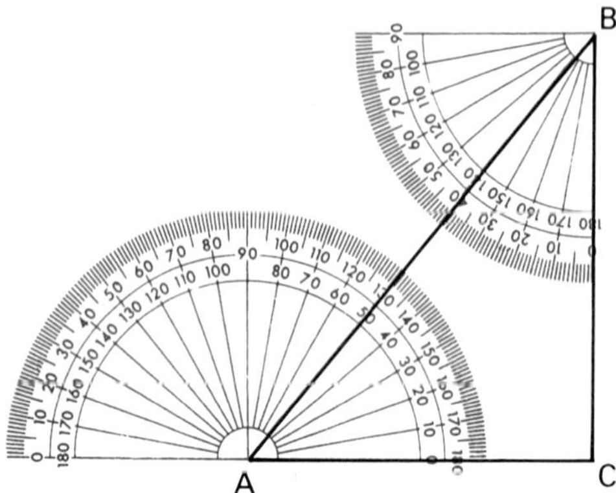

b Measure ∠ACB with a protractor.
c Calculate the sum of the three angles.

2 a Draw a large triangle, and measure each angle.
b Add the three angles together. Compare your answer with your neighbour's.

3 Repeat question **2** for a different size and shape of triangle.

These results illustrate the fact that: **the sum of the angles of a triangle is always 180°.**

4 Calculate the size of the third angle in each triangle shown below.

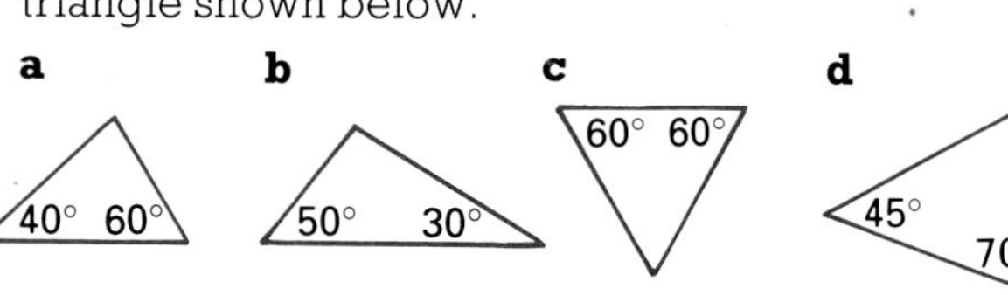

5 Calculate the sizes of the third angle in each of these triangles.

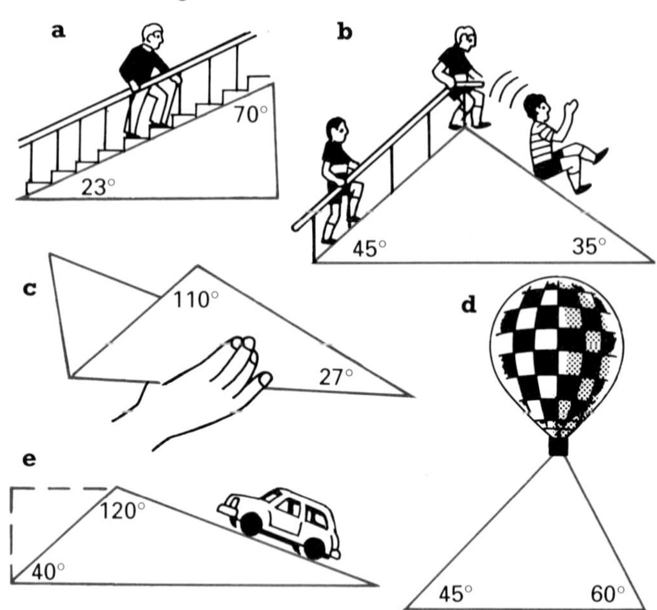

6 Calculate the sizes of the angles marked with letters. For example, $x° = 30°$.

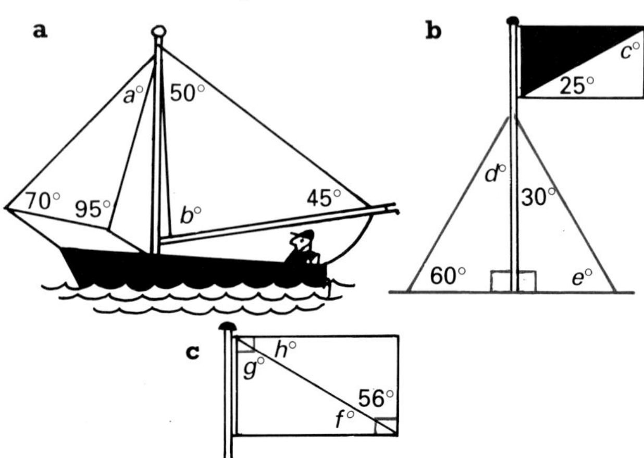

7 Calculate the size of the third angle in each triangle, and say whether or not the triangles are right-angled.

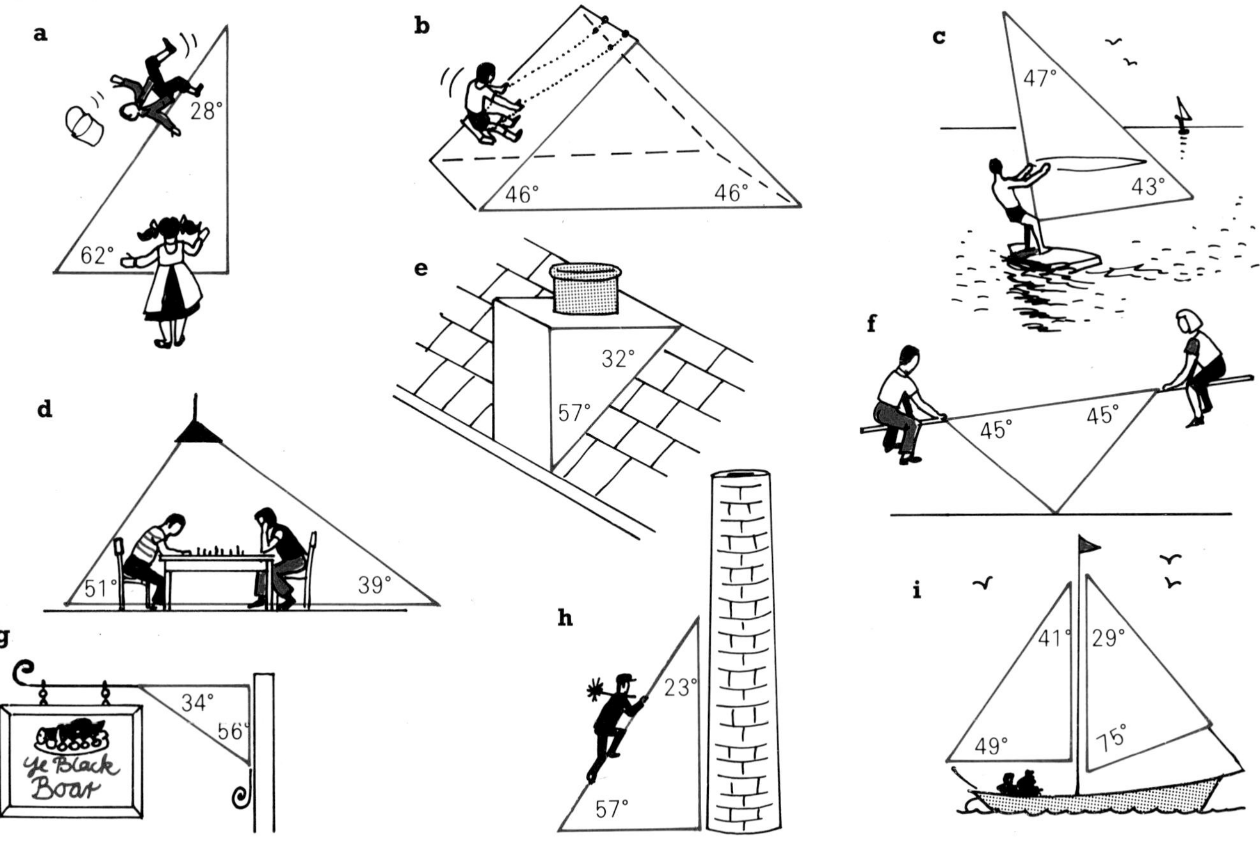

EXERCISE 1B/C

1 Calculate:
- **a** ∠BAC
- **b** ∠s ABC and ACB (Remember alternate angles?)

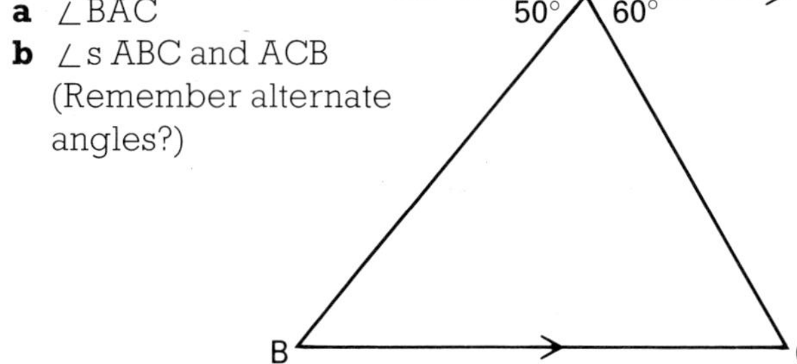

- **c** ∠ABC + ∠ACB + ∠BAC

2 a What is the value of $a^\circ + b^\circ + c^\circ$?

b Why does:
(i) ∠ABC = a°
(ii) ∠ACB = c°?

c What is the sum of the angles of △ABC?

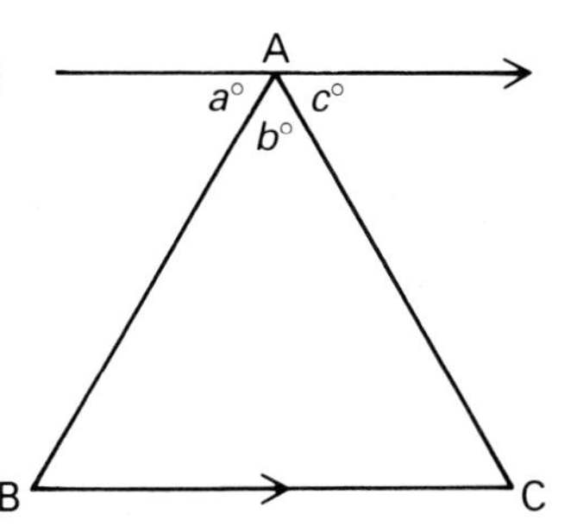

3 Marion asks her computer to make up sets of three angles that can be used as the angles of a triangle. She finds that the program has a fault. Which sets *are* correct?

4 a Make an equation for each triangle, and solve it. For example, $x° + 2x° + 3x° = 180°$.

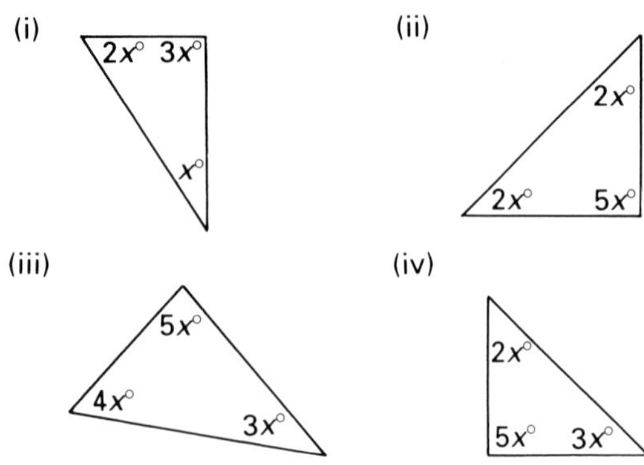

b Which of the triangles are right-angled?

5 Make equations for these diagrams, and find the values of x and y.
Then write down the sizes of all the angles.

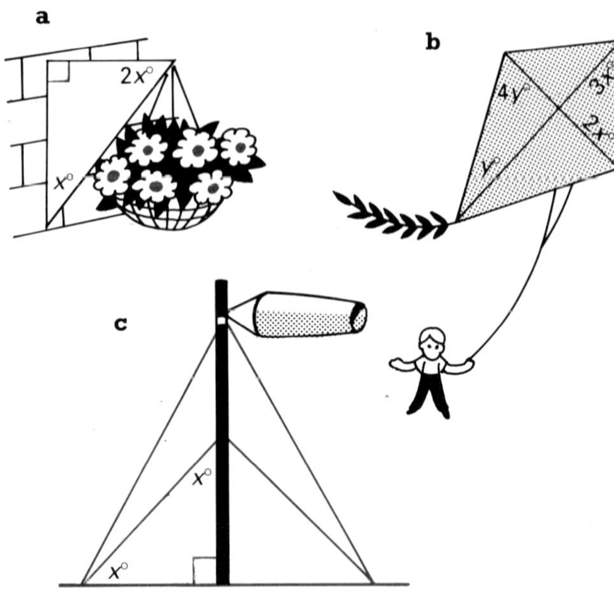

6 Prove that △RSU is right-angled. Give a reason for each step in your proof.

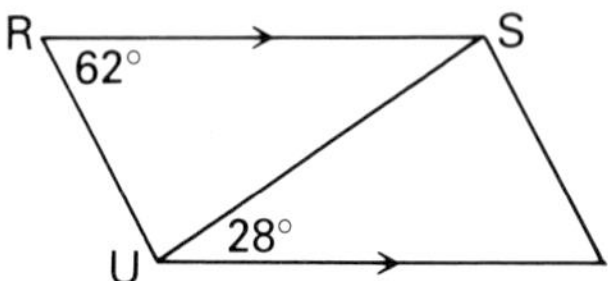

7 Find an equation with x and the other letters in each triangle. Solve the equations when $a = 33$, $b = 54$, $c = 67$, $d = 38$.

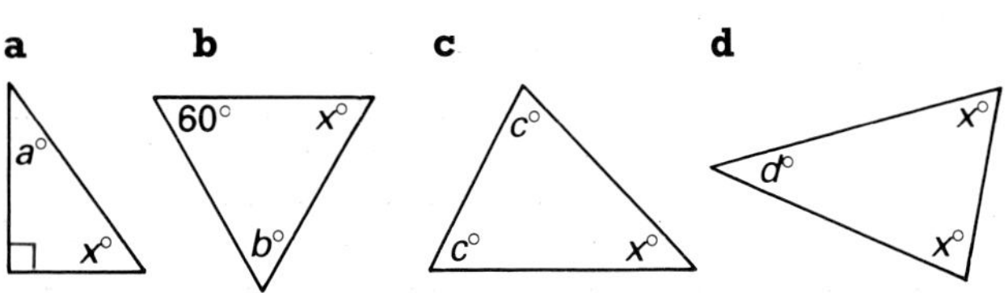

PRACTICAL PROJECT

Use card or plastic strips and fasteners to make some three, four, five and six-sided shapes.

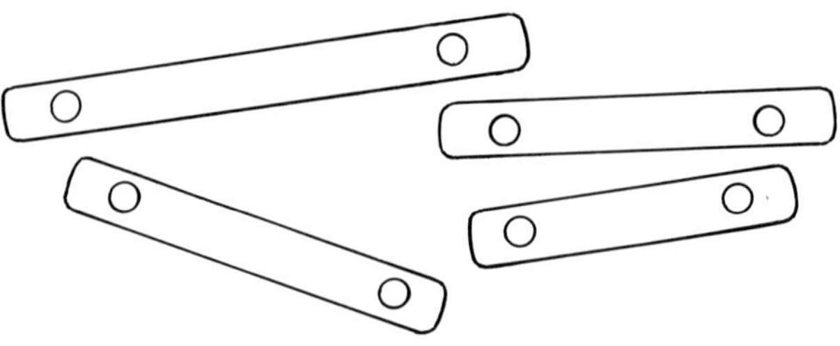

Investigate ways of fixing the shapes, so that they cannot be altered. Write a sentence about your findings and illustrate it with diagrams.

BRAINSTORMER

How many triangles can you find in this pentagon? It may help if you list them methodically: ABC, ABD, . . .

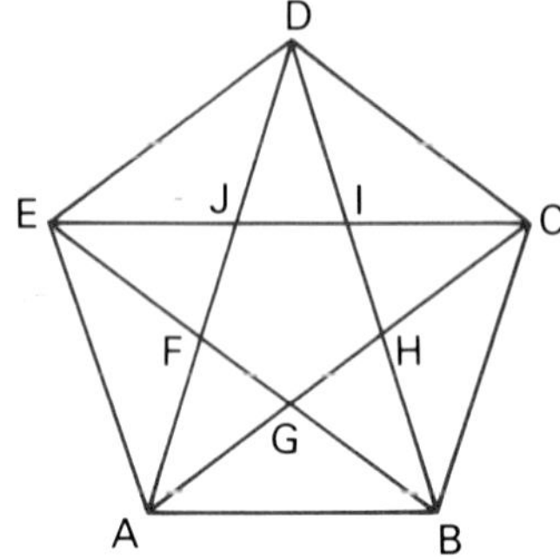

INVESTIGATION

a *Complete the sequences to find the sum of the angles of a pentagon, hexagon, octagon and n-sided polygon.*

Shape:

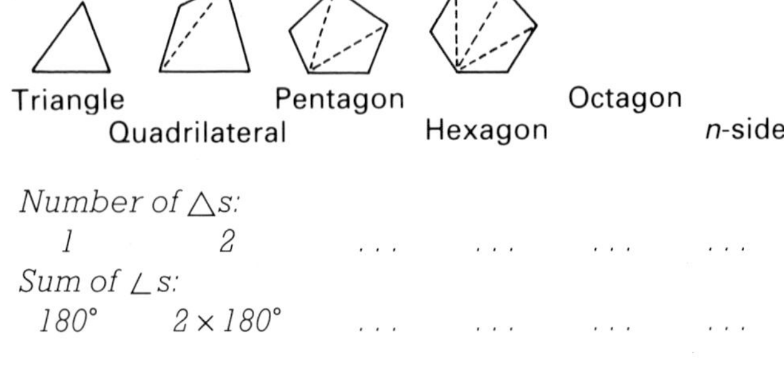

Number of △s:
1 2 . . . .

Sum of ∠s:
180° 2 × 180° . . . .

b *If each shape is regular, having all its sides and angles equal, calculate the size of an angle in each.*

THE AREA OF A TRIANGLE

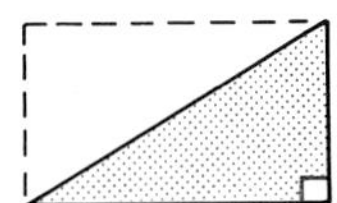

Area of triangle
$= \frac{1}{2}$ area of surrounding rectangle
$= \frac{1}{2}$ base × height

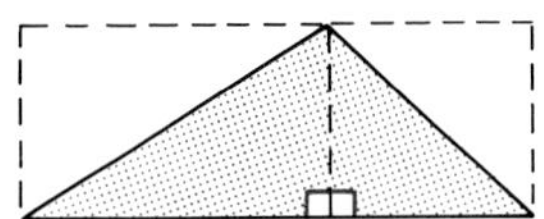

Area of whole triangle
$= \frac{1}{2}$ area of each rectangle
$= \frac{1}{2}$ area of whole rectangle
$= \frac{1}{2}$ base × height

The area of a triangle $= \frac{1}{2}$ base × height. Formula: $A = \frac{1}{2}bh$

EXERCISE 2A

1 Calculate the area of each shaded right-angled triangle.

a

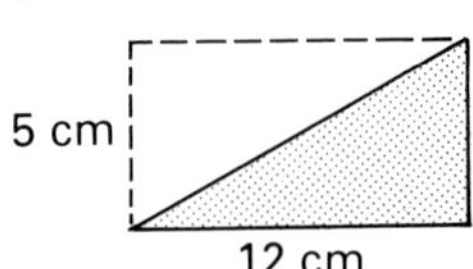

b

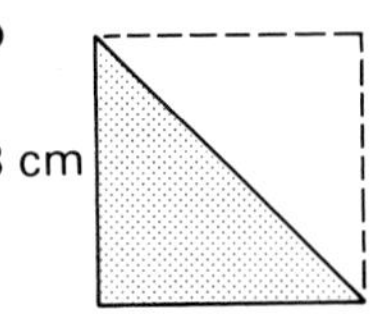

c

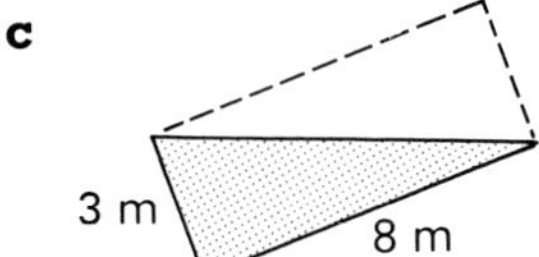

d

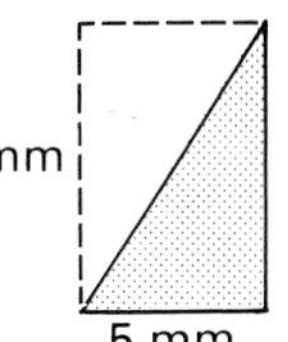

2 Calculate the area, in squares, of each triangle by:
(i) first calculating the area of the surrounding rectangle
(ii) using the formula 'Area of triangle $= \frac{1}{2}$ base × height'.

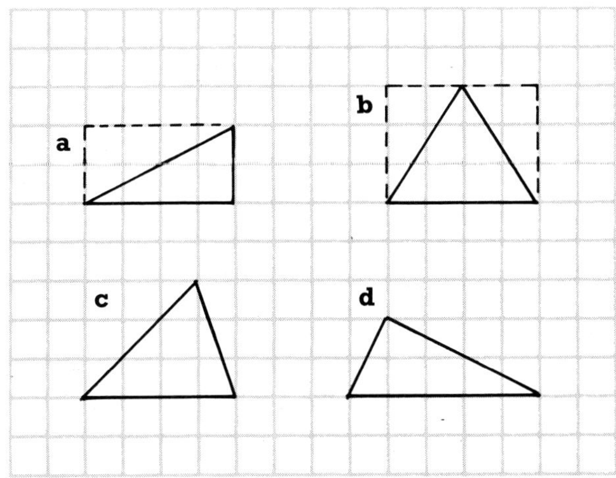

3 Calculate the areas of these triangles. The lengths are in centimetres.

a

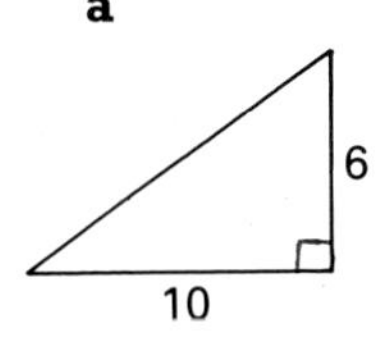

b

c

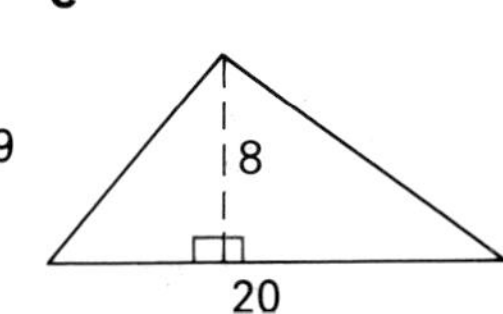

d

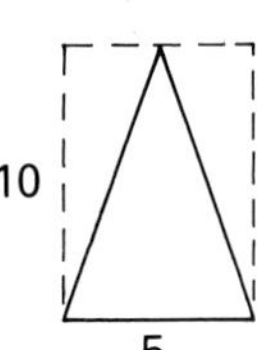

e

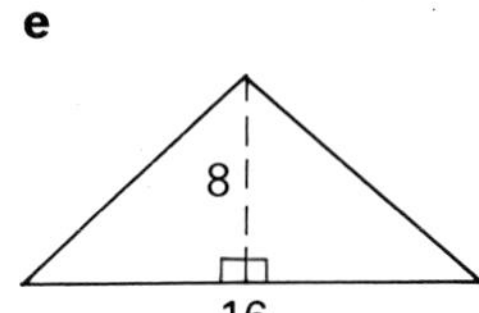

f

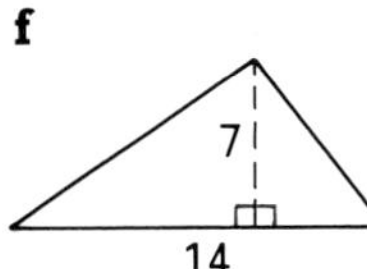

g

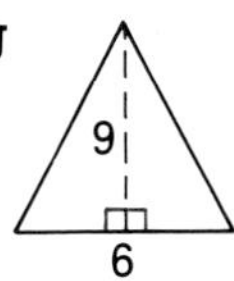

h

4 Find the areas of these triangular shapes.

a

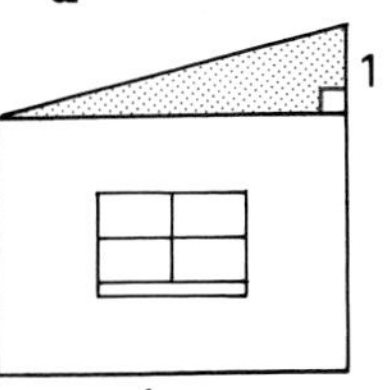
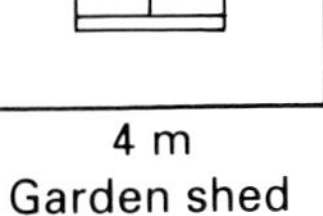

Garden shed

b

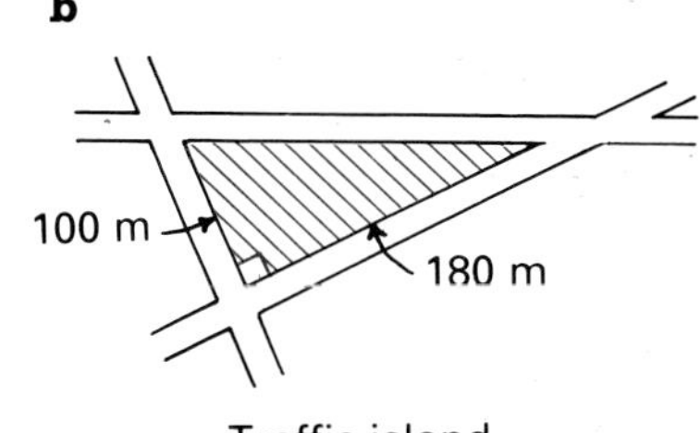

Traffic island

5 Calculate the areas of these triangles.

Base	10 m	12 cm	20 cm	6 m	70 mm	25 cm
Height	8 m	4 cm	9 cm	6 m	30 mm	15 cm

EXERCISE 2B/C

1 Calculate the areas of these triangles. The lengths are in centimetres.

a

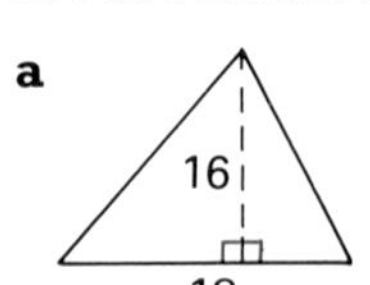

b

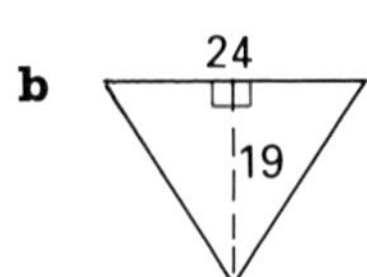

c

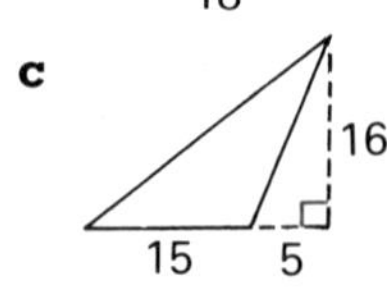

d

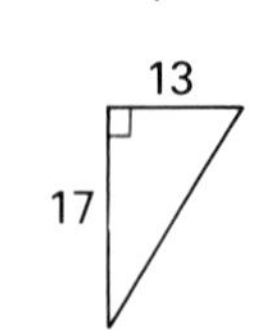

2 Find the area of each shape. The units are metres.

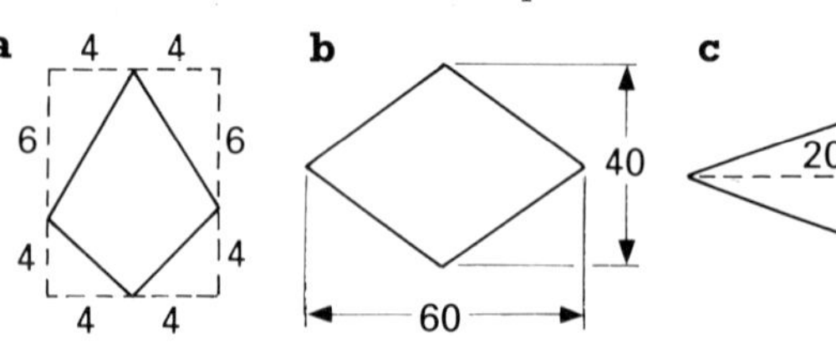

3 a Sketch a rectangle, and mark its length 12 cm and its breadth 7 cm.

b Draw its diagonals, crossing at O, and calculate the areas of the triangles which have O as one vertex.

4 The diagonals of a square are 12 cm long. Find the area of the square.

5 Copy and complete this table of triangle data.

Base	80 m	4.5 cm	16 m		12 m
Height	60 m	1.2 cm		6 m	
Area			64 m²	75 m²	14.4 m²

6 On squared paper draw the triangle with vertices A(2, 2), B(6, 8) and C(8, 4), and calculate its area.

ISOSCELES TRIANGLES

The steps used by ABC Airways have sides which are congruent right-angled triangles. When they are not in use, two sets of steps are parked back-to-back.

Put together, the two right-angled triangles make an **isosceles triangle**. The line joining the two right-angled triangles is an **axis of symmetry**.

EXERCISE 3A

1 a Copy right-angled triangles (i)–(v) on squared paper, or trace them.

b Draw their reflections in the dotted lines, to make isosceles triangles.

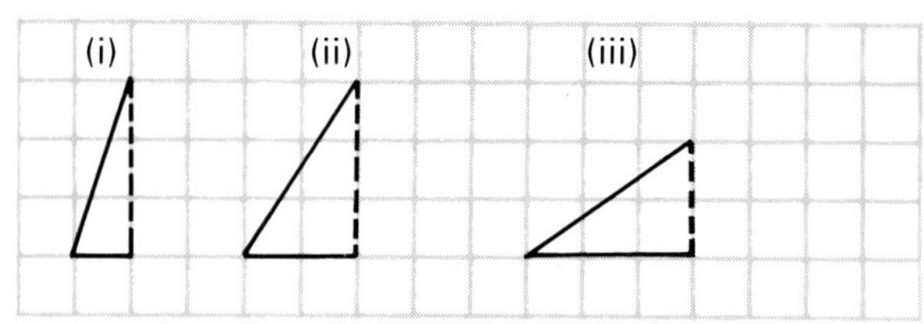

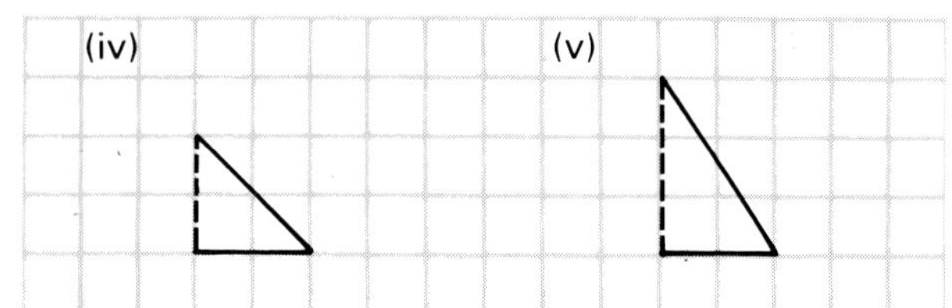

2 a Draw a large isosceles triangle on squared paper.
b Cut it out, and fold it along its axis of symmetry.
c Check that it has two equal sides and two equal angles.

An isosceles triangle has one axis of symmetry. It has two equal sides, and two equal angles.

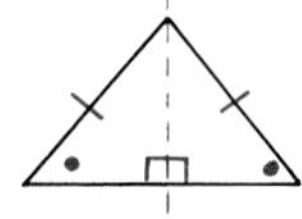

3 a Copy this figure, and fill in as many lengths and angles as you can. Remember that BD is an axis of symmetry.

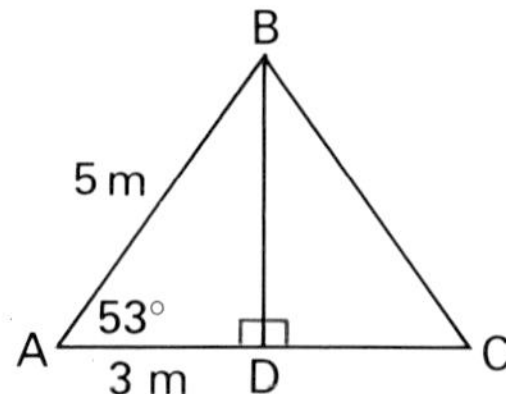

b Draw the triangle again, this time without its axis of symmetry BD. Mark in the lengths of the sides and the sizes of the angles.

4 Sketch these isosceles triangles, and fill in as many lengths and angles as you can.

a
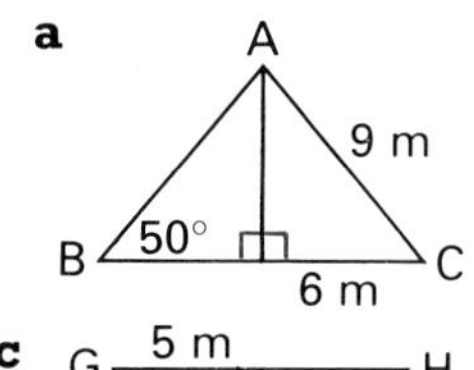

b
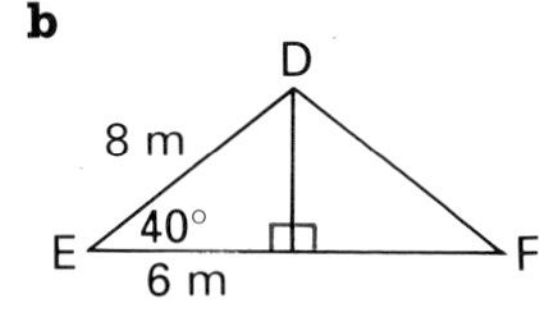

c
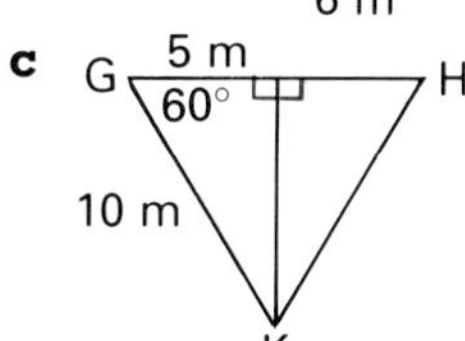

d
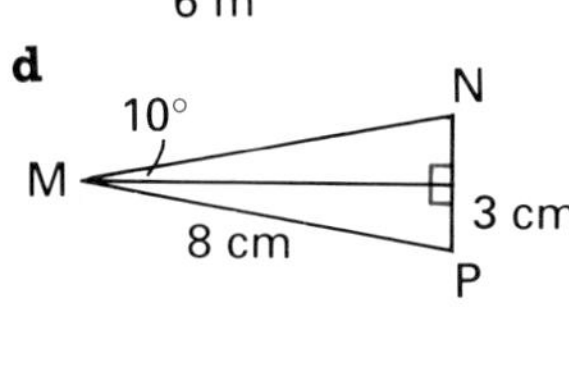

5 Sketch these isosceles triangles, and fill in as many lengths and angles as you can.

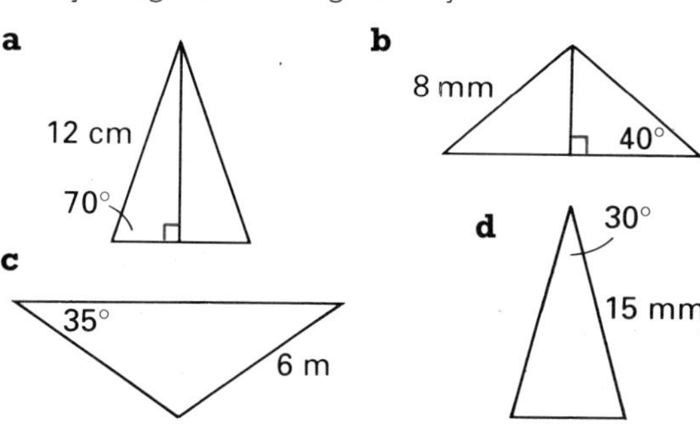

6 Sketch these isosceles triangles, and fill in as many lengths and angles as you can.

7 a On squared paper plot the points A(2, 2), B(8, 2) and C(5, 4).
What kind of triangle is ABC?
b Plot point D so that △ABD is congruent to △ABC, and AB is an axis of symmetry of ACBD. Write down the coordinates of D.

8 Repeat question **7**, for A(−4, 0), B(4, 0), C(0, 3).

9 Copy these diagrams, and fill in the sizes of all the angles. Each triangle is isosceles.

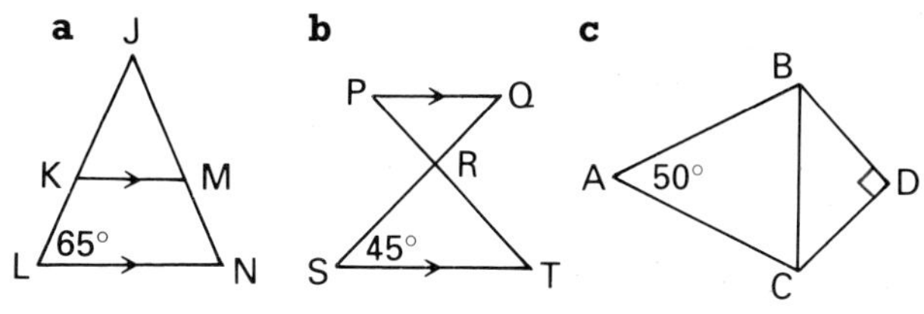

10 Calculate the area of each rectangle, then the area of each isosceles triangle.

a
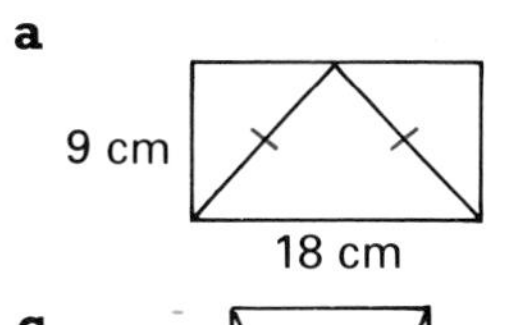

b
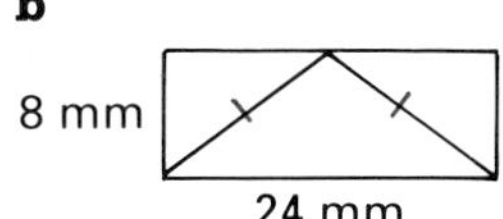

c
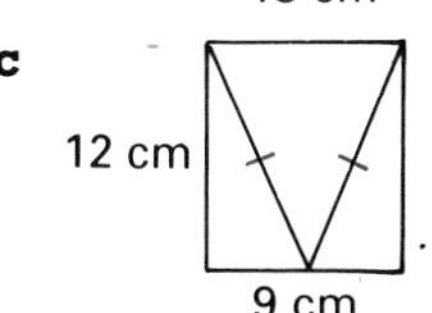

d
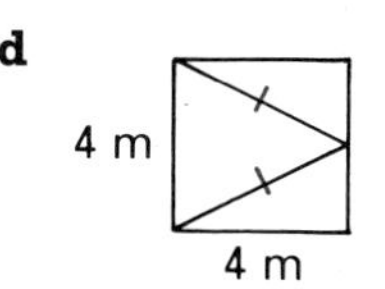

EXERCISE 3B/C

1 Draw a right-angled isosceles triangle, and fill in the sizes of all its angles.

2 This is part of a tiling of isosceles triangles. Calculate x and y.

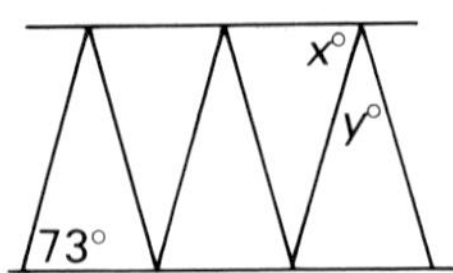

3 B is the point $(-2, 0)$ and C is $(2, 0)$. A is a point (x, y) so that AB = AC.

a Name:
(i) the axis of symmetry of triangle ABC
(ii) two equal angles in the triangle.

b What can you say about the numbers x and y stand for?

4 A is the point $(5, 5)$ and B is $(2, 1)$. AB and AC are equal sides of an isosceles triangle ABC. Find the coordinates of C if:

a BC is parallel to OX
b BC is parallel to OY.

5 a A is the point $(2, 1)$, B is $(6, 1)$ and C is $(4, 7)$.
(i) What kind of triangle is ABC?
(ii) Calculate the area of the triangle.

b D is the point $(4, 0)$. Calculate the area of:
(i) triangle ABD (ii) the shape ADBC.

6 Calculate the areas of these shapes:

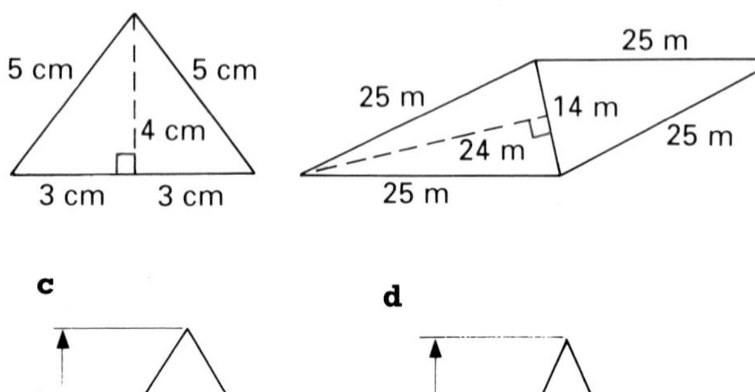

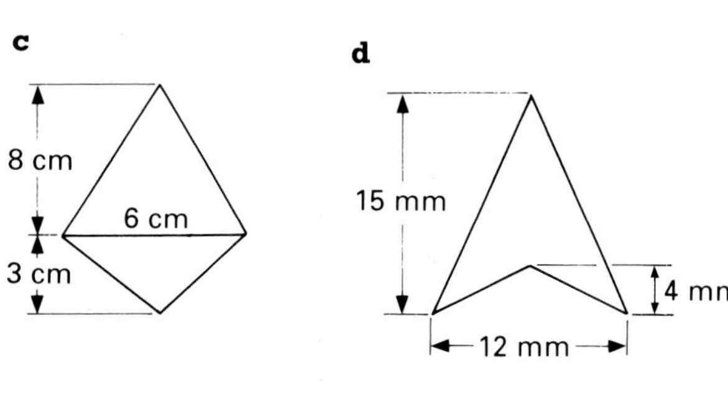

7 Prove that triangles ABC and CDE are isosceles. Set the proof down clearly, giving a reason for each step.

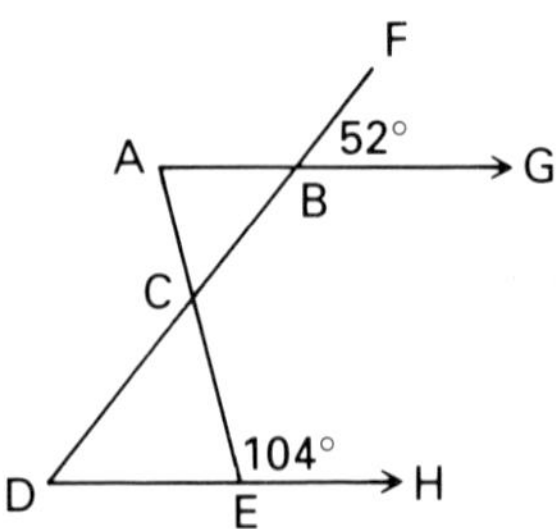

PRACTICAL

1 a *Draw the outlines of your two set squares.*
b *Measure all the angles with a protractor.*
c *Which set square has an isosceles triangle outline?*

2 *Draw a line AB across your page. Slide a set square along it so that you can draw parallel lines at:*
a *45°* **b** *30°* **c** *60° to AB.*
What kind of angles are made by AB and the parallel lines?

INVESTIGATION

1 *A, B, C are equally-spaced points on the circle, centre O.*

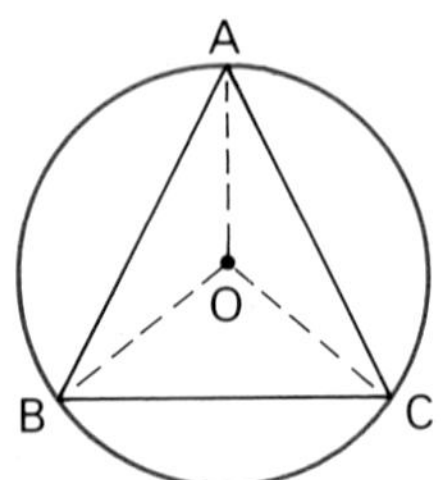

a *What is the size of ∠BOC?*
b *Why is OB = OC? What kind of triangle is OBC?*
c *Calculate: (i) ∠OBC (ii) ∠ABC.*

2 *Use the same kind of diagram and calculation for 4, 5, 6, 8 and 10 equally-spaced points on a circle, and so find the size of each angle of a square, regular pentagon, hexagon, octagon and decagon.*

EQUILATERAL TRIANGLES

The road-sign triangle is an **equilateral triangle**.
This is a special kind of isosceles triangle with **all its sides equal**.

EXERCISE 4

1

Equilateral   Isosceles   Isosceles

Isosceles   Equilateral

a How many axes of symmetry has an equilateral triangle?
b What can you say about the angles in an equilateral triangle?
c What size is each angle in an equilateral triangle?

2 a Draw an equilateral triangle with sides 8 cm long. You may use compasses . . . or a protractor.

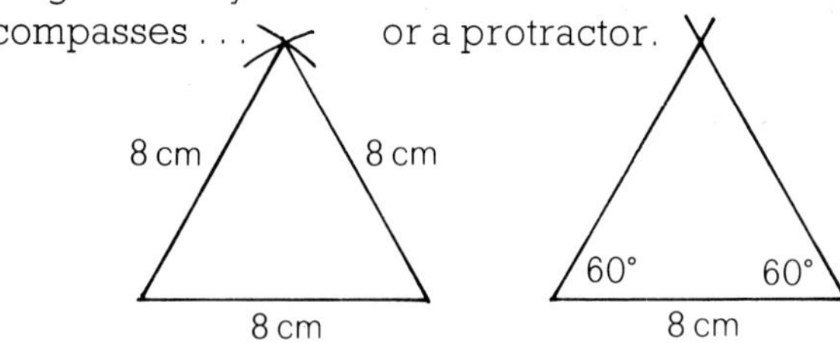

b Cut it out, and fold it to check the three axes of symmetry.
c In how many ways will it fit into its space as you turn it round?

An equilateral triangle has three equal sides and three equal angles.
Each angle is 60°.
It has three axes of symmetry.

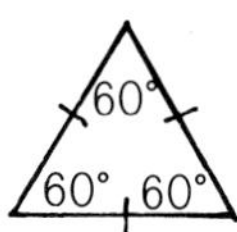

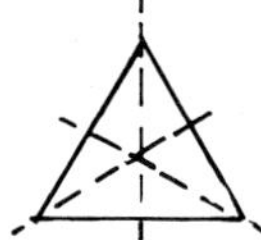

3

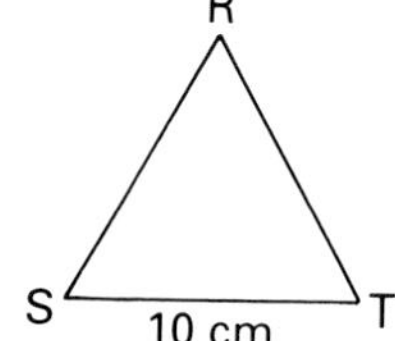

△RST is equilateral. Copy it, and fill in the size of each angle and the length of each side.

4 Elsa's kite is in the shape of a right-angled isosceles triangle and an equilateral triangle put together. Sketch it, and fill in all the angles.

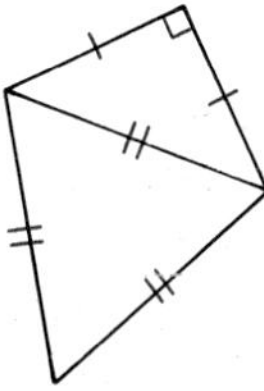

5 Which of these triangles:
a are equilateral **b** might be equilateral
c are not equilateral?

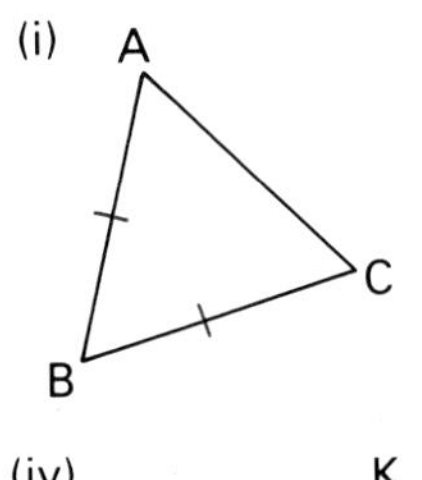

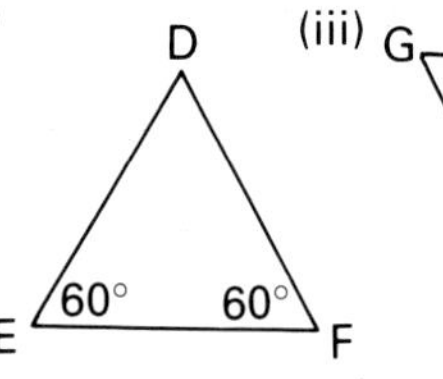

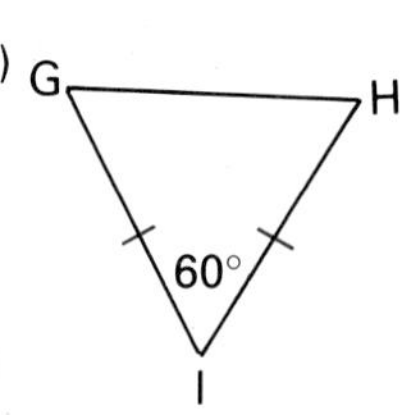

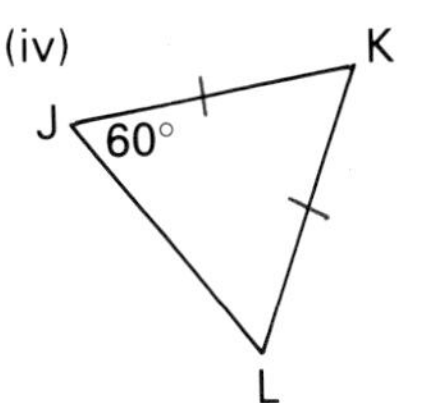

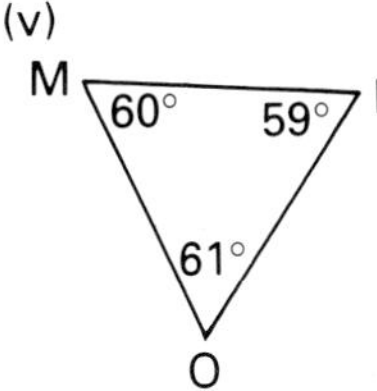

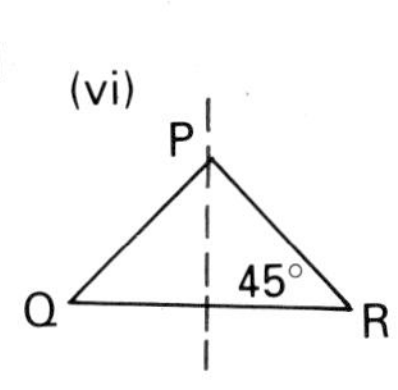

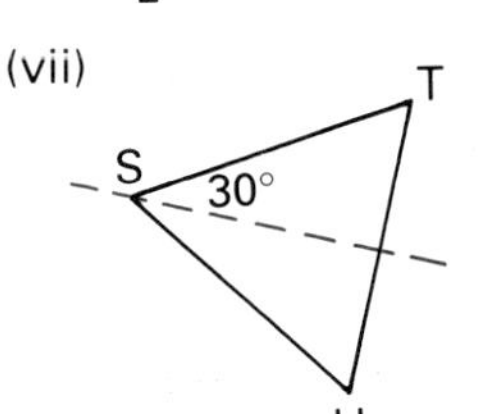

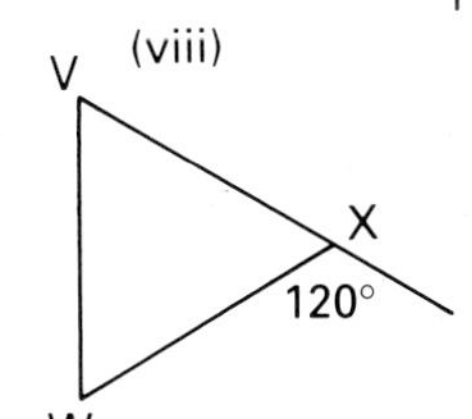

6 This diagram shows part of a tiling of equilateral triangles.

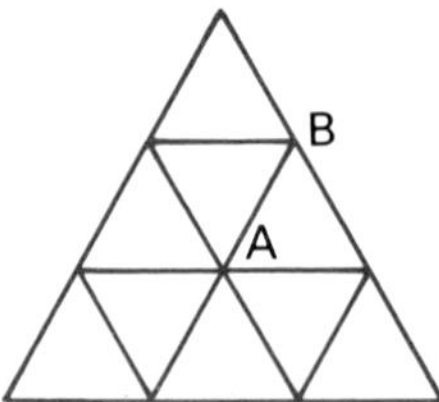

a Why do the triangles fit together at A?
b Why do the sides of the triangles make a straight line at B?
c AB = 1 cm. How many triangles are there with sides of length:
(i) 1 cm (ii) 2 cm (iii) 3 cm?

7 A record is playing on a turntable. The record label has an equilateral triangle printed on it.

a If the record spins once, how many times will the triangle be in a position like the one shown in the picture?
b This tells you that an equilateral triangle has another kind of symmetry.
(i) What kind of symmetry?
(ii) What is the order of symmetry?

8 Matthew is setting up the balls for a game of snooker.

a Without measuring, he knows that the triangular frame is equilateral. How does he know this?
b What is the size of each angle?
c How many different ways can the frame fit over the triangular arrangement of snooker balls?

A PUZZLE

Trace this equilateral triangle pattern. Cut along the red lines, then fit the four parts together to make a square.

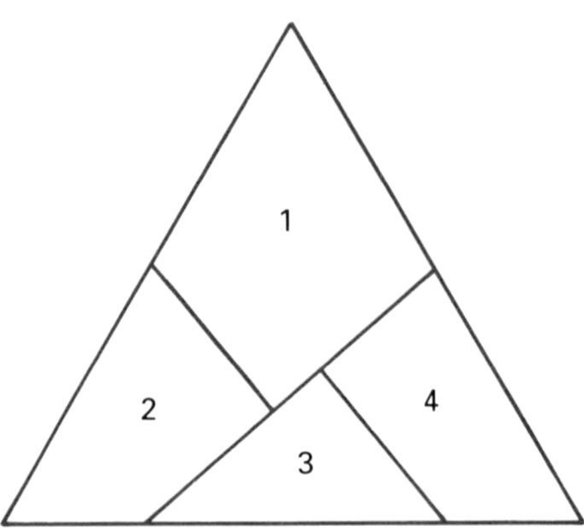

CHALLENGE

Draw and cut out larger versions of these nets of equilateral triangles, and make the 'solids'.

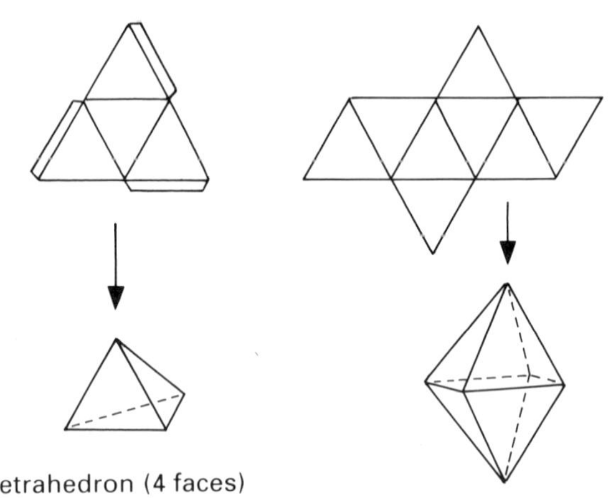

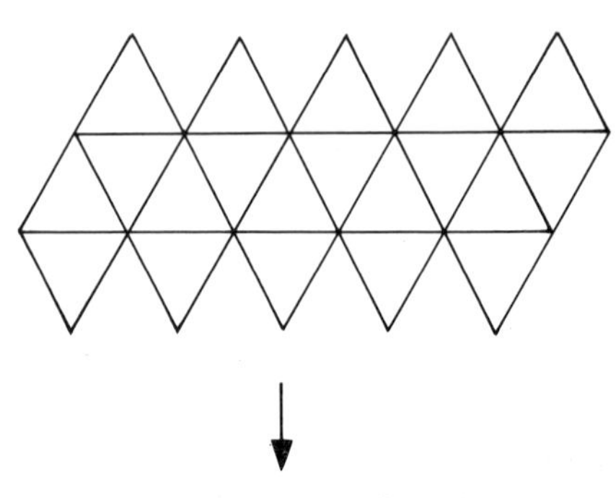

ALL SORTS OF TRIANGLES

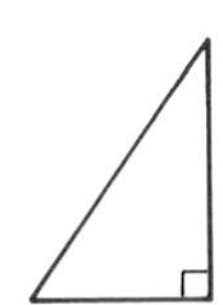
Right-angled

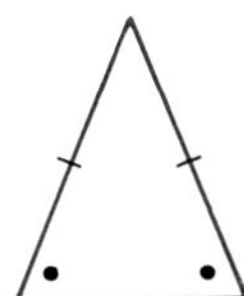
Isosceles

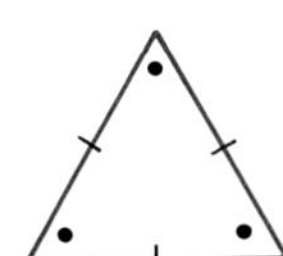
Equilateral

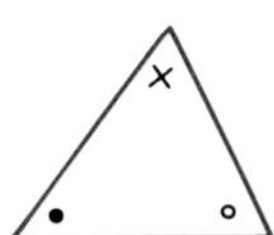
Acute-angled

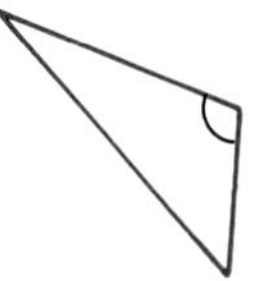
Obtuse-angled

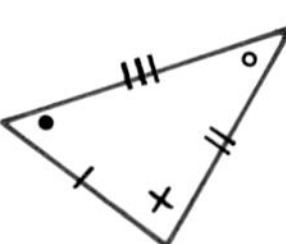
Scalene (all sides unequal)

EXERCISE 5

1 Calculate the missing angle in each of these triangles. Then say whether the triangle is right-angled, isosceles, equilateral, acute-angled, obtuse-angled (or more than one of these).

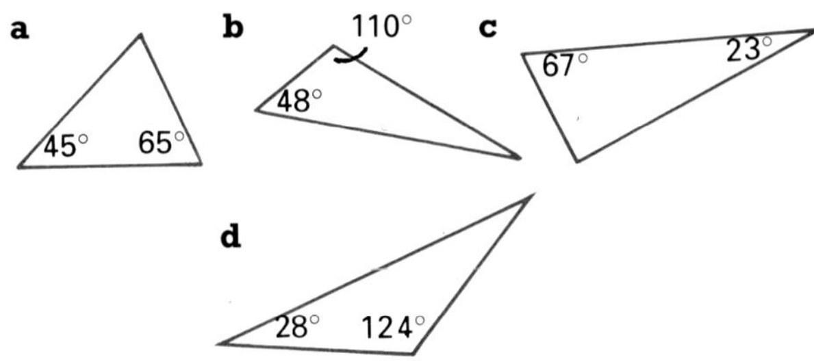

2 Calculate the missing angles in the triangles below. What type is each large triangle?

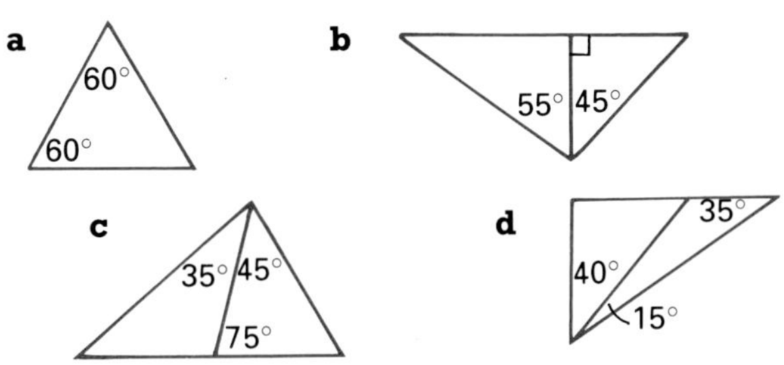

3 Draw these triangles on squared paper, and say which type each one is.

a $\triangle O(0, 0), A(6, 0), B(2, 2)$
b $\triangle C(0, 3), D(-4, 3), E(-4, 0)$
c $\triangle F(1, -1), G(0, -4), H(-1, -1)$
d $\triangle K(3, -1), L(5, -2), M(3, -4)$

4 **a** Copy the rectangle, and fill in all the angles.
b Name:
(i) two acute-angled triangles
(ii) two obtuse-angled triangles
(iii) as many right-angled triangles as possible.

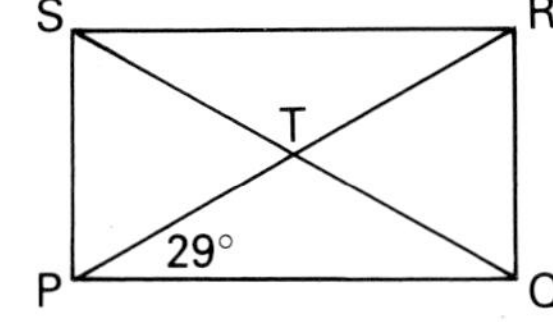

5 Which types of triangle are these?

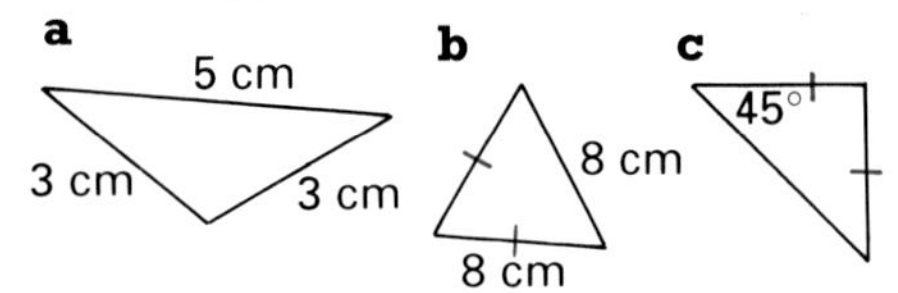

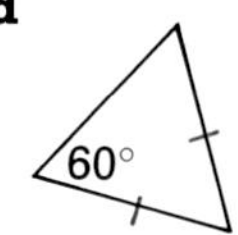

6

a An equilateral triangle can be obtuse-angled.
b A right-angled triangle can be isosceles.
c A triangle can have two right angles.
d The sum of the angles in a triangle is equal to two right angles.
e The sum of two angles in a triangle is always greater than the third angle.

CHALLENGE

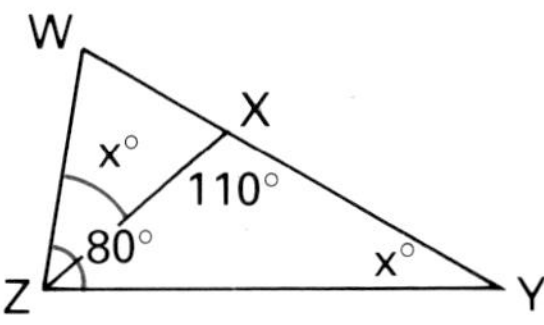

This is Ewan's diagram, but Nicola says that $\angle WZX$ cannot be equal to $\angle XYZ$. Who is right?

INVESTIGATION

Investigate all the different shapes and sizes of triangles you can make by joining three points on this seven dot pattern on 'dotty' paper or a geo-board.

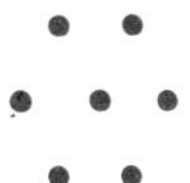

DRAWING TRIANGLES, USING A RULER, PROTRACTOR AND COMPASSES

EXERCISE 6

Given 2 angles and a side

1 Follow these diagrams. Use a ruler and protractor to draw the triangles. Measure the third angle in triangle (iii), then check by calculation.

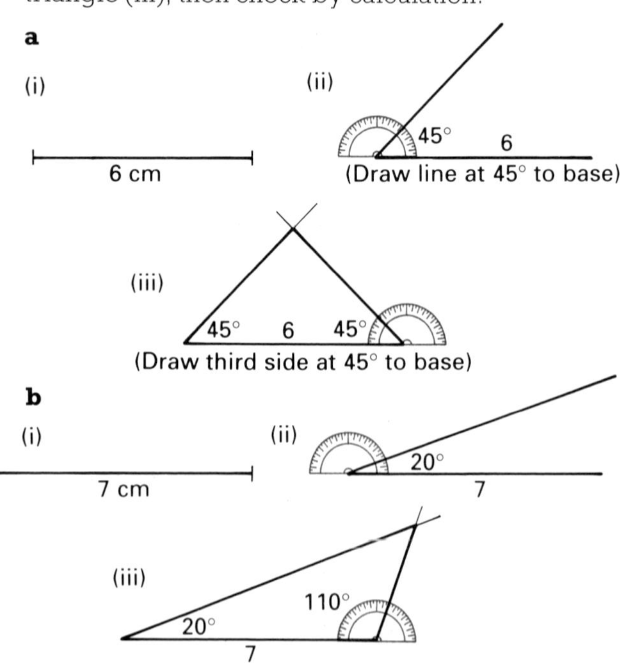

Given 2 sides and the included angle

2 Construct these triangles. Measure the third side in triangle (iv), correct to 1 decimal place.

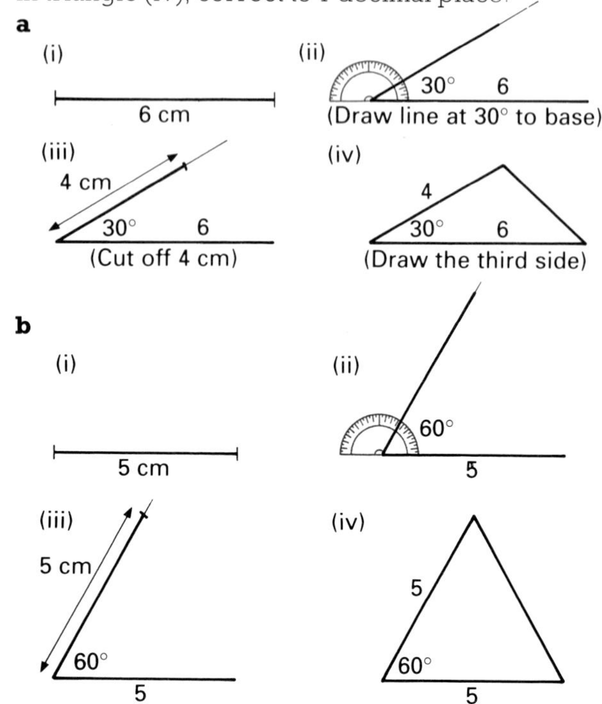

Given 3 sides

3 Use a ruler and compasses to draw these triangles. Measure the angles in triangle (iv), to the nearest degree.

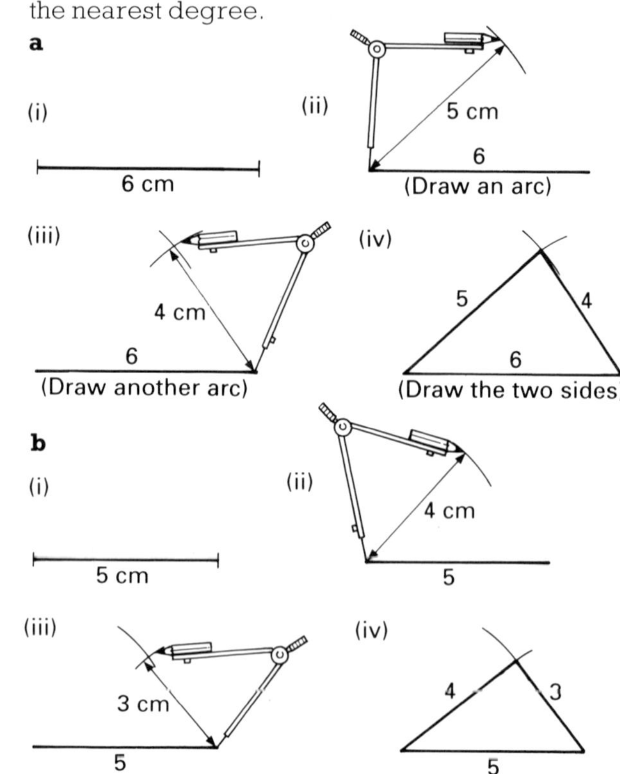

4 Make accurate drawings from these sketches.

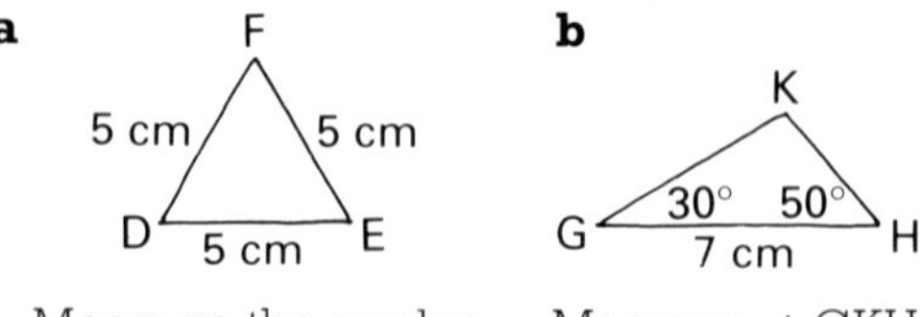

Measure the angles. Measure ∠GKH.

5 Draw these triangles accurately, then measure the unknown sides and angles.

a △ABC: AB = 6 cm, AC = 5 cm, BC = 7 cm

b △DEF: DE = 5 cm, ∠FDE = 35°, ∠DEF = 65°

c △PQR: PQ = 5 cm, ∠QPR = 120°, PR = 5 cm

INVESTIGATION

You have a collection of strips 1 cm, 2 cm, 3 cm and 4 cm long. Investigate the number of different shaped triangles you can make with:

a *1 cm strips only* ***b*** *1 and 2 cm strips*

c *1, 2 and 3 cm strips* ***d*** *1, 2, 3 and 4 cm strips.*

COMPUTER GRAPHICS

When Jeff types TRIANGLE(7, 6) the flashing dot moves to the point (7, 6) and colours in the triangle formed by this point and the last two points visited by the dot.

For example,

MOVE(1, 6)
MOVE(1, 1)
TRIANGLE(7, 6)

produces

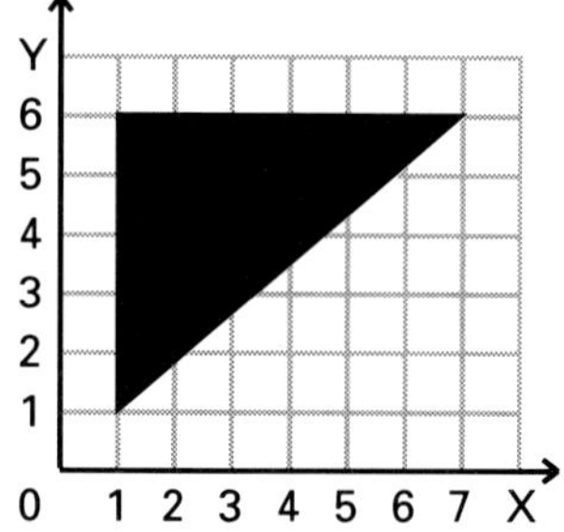

EXERCISE 7

1 a Copy Jeff's diagram. Label the corners A, B and C to show the order in which they were visited by the flashing dot.

b What command could Jeff add to his list to change his diagram to a coloured-in rectangle?

2

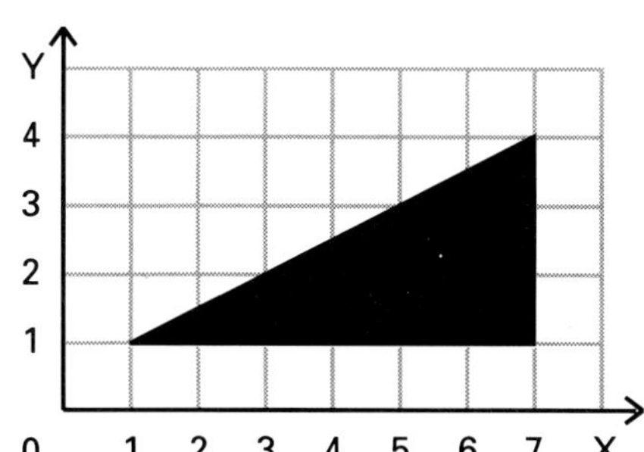

For his next attempt, Jeff typed

MOVE(7, 1)
MOVE(7, 4)
TRIANGLE(1, 1).

Repeat question **1** for the triangle he got this time.

3 a On a computer, or on squared paper, show the result of:

MOVE(2, 3)
MOVE(7, 1)
TRIANGLE(7, 5).

b Which type of triangle have you drawn?

4 a List the commands needed to draw this triangle. Visit the points in the order $1 \rightarrow 2 \rightarrow 3$.

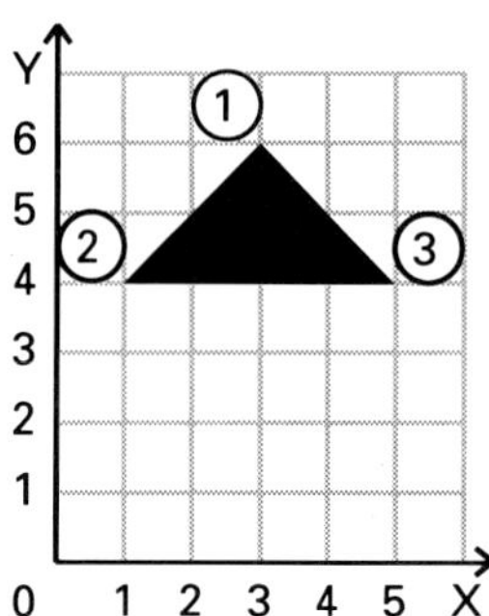

b What shape will it become if you add the command TRIANGLE(3, 2)?

5 Anita thought she was colouring in a square when she gave the commands:
MOVE(2, 1) MOVE(2, 4) TRIANGLE(5, 4)
TRIANGLE(5, 1).

a Follow the steps carefully to see what shape she *did* get. (Remember 'TRIANGLE' uses the last two points visited.)

b Can you rearrange Anita's list to make it do what she wants?

6 List instructions for your own diagram, using some or all of the commands MOVE, TRIANGLE and DRAW.

CHECK-UP ON TYPES OF TRIANGLE

1 Which of these could be sets of angles in a triangle?
a 30°, 60°, 90° **b** 10°, 70°, 120° **c** 45°, 55°, 80°

2 Calculate the sizes of the unmarked angles in these triangles.

a
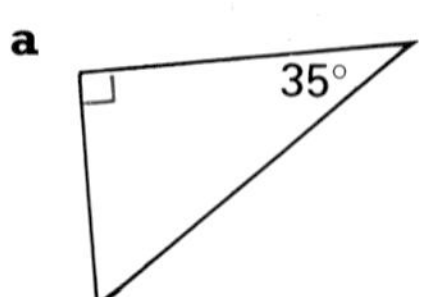

b
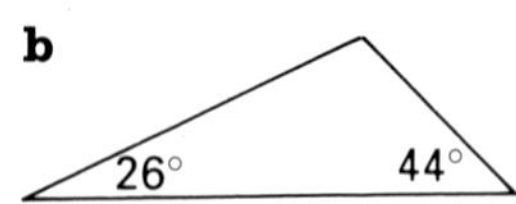

c
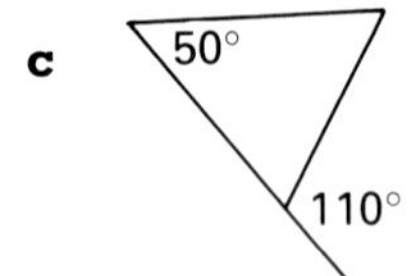

d
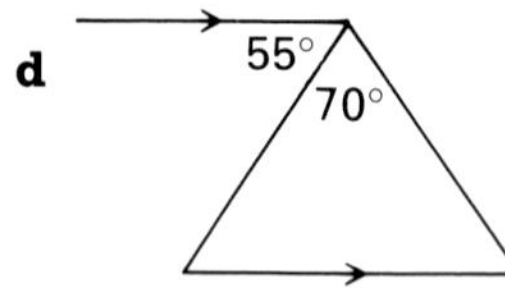

3 Calculate the shaded areas.

a
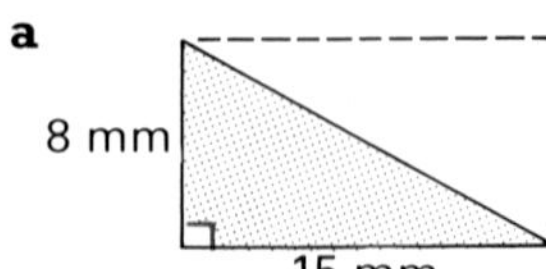

b
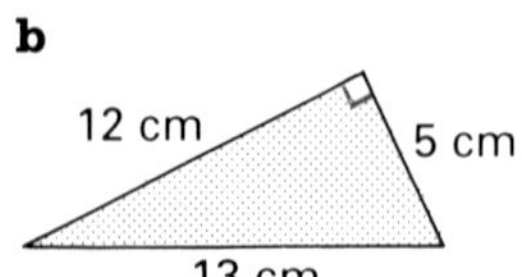

c
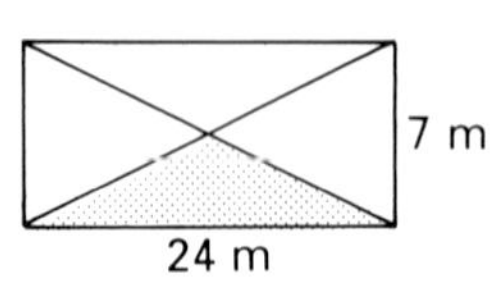

d
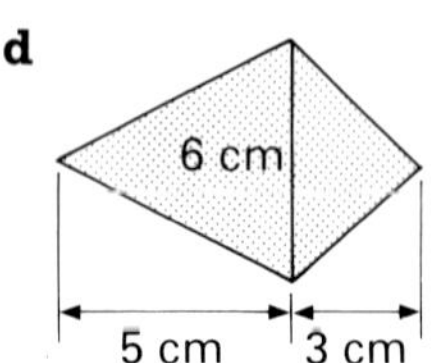

4 Describe each of the triangles below: right-angled, isosceles, equilateral, acute-angled or obtuse-angled.

a

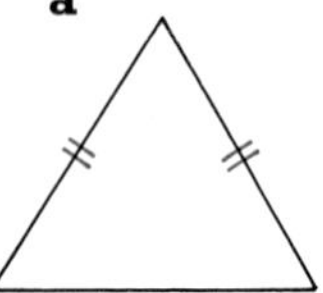

b **c**
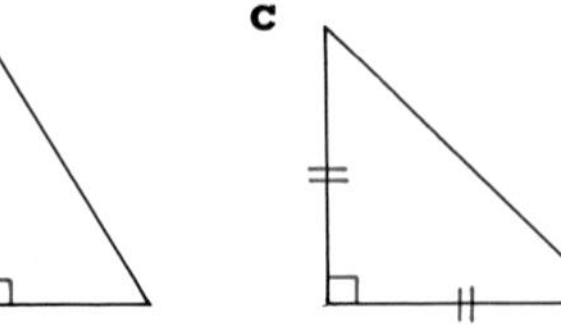

d
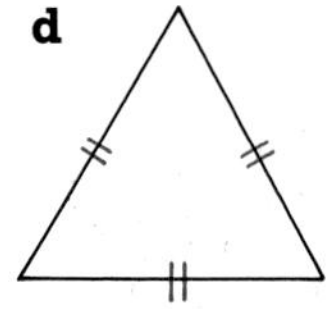

e
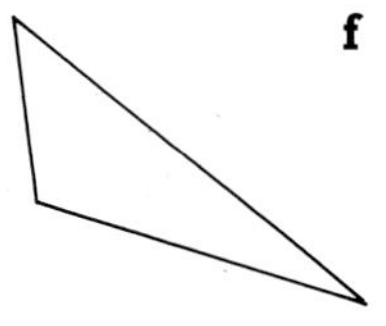

f
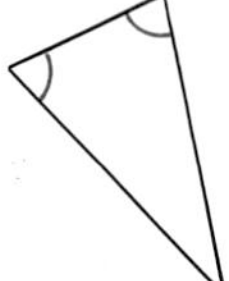

5 Copy the triangles in question **4** that have axes of symmetry, and draw in these axes.

6 What numbers do these letters stand for?

a **b** **c**
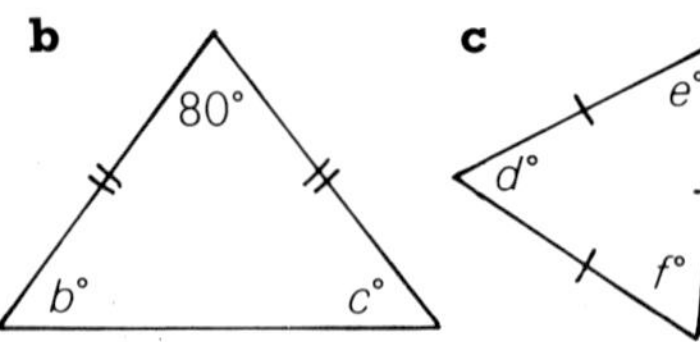

d **e**
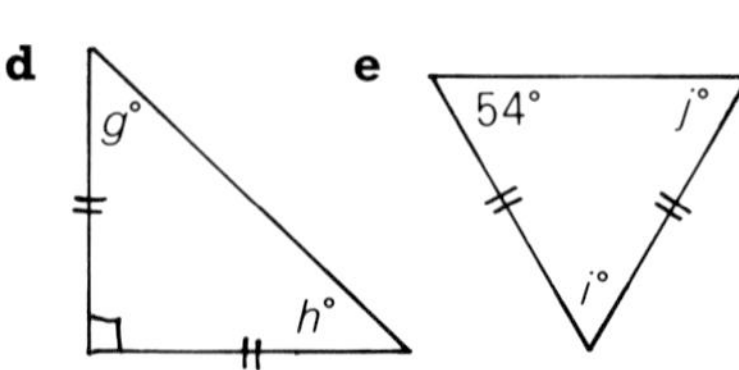

7 a Sketch this diagram, in which SR = 5 cm. Fill in the sizes of as many angles and the lengths of as many lines as you can.

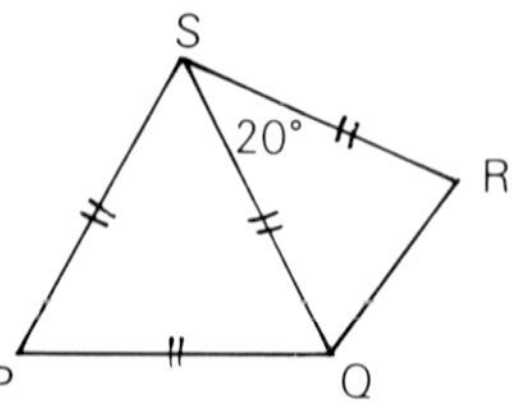

b What is the sum of the angles of the quadrilateral PQRS?

8 a Calculate the area of the triangle ABC where A is the point (1, 3), B is (6, 1) and C is (6, 5).
b D is the point (4, 3). Calculate the area of ABDC.

9 Copy this diagram, and fill in the sizes of all the angles.

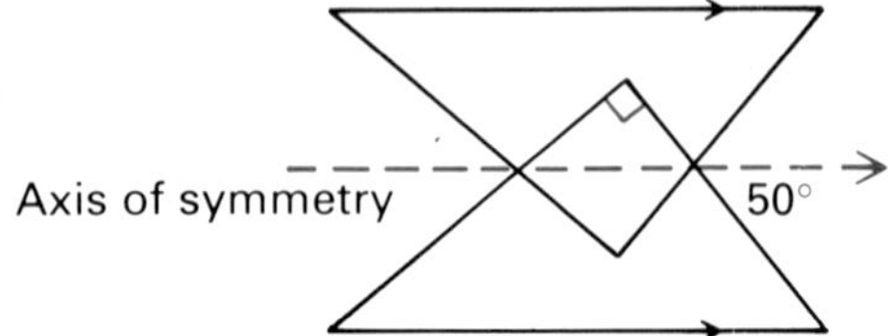

10 METRIC MEASURE

LOOKING BACK

1 Which unit of length would you use to measure:
a the length of a pencil
b the height of a door
c the thickness of a matchstick
d the distance between towns
e the length of a playing field?

2 Estimate:
a the length of the classroom
b the width of your desk or table
c the height of a double-decker bus
d the thickness of a penny.

3 Arrange these lengths in order, smallest first:
8.9 cm, 9.1 cm, 8.7 cm, 10.5 cm, 9.6 cm

4 a Which of the areas shown below is greatest, do you think?

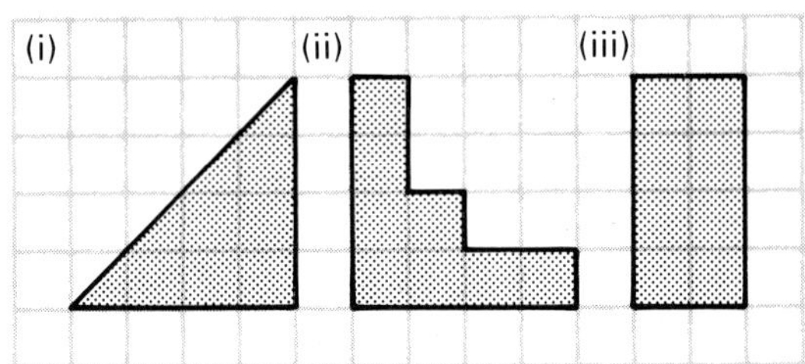

b Check by counting squares to find the three areas.

5 A rectangular carpet is 6 m long and 3 m broad. Calculate:
a its area **b** its cost, at £22 per square metre.

6 Each carton contains one litre of juice. How much juice is in each of these shop displays?

a

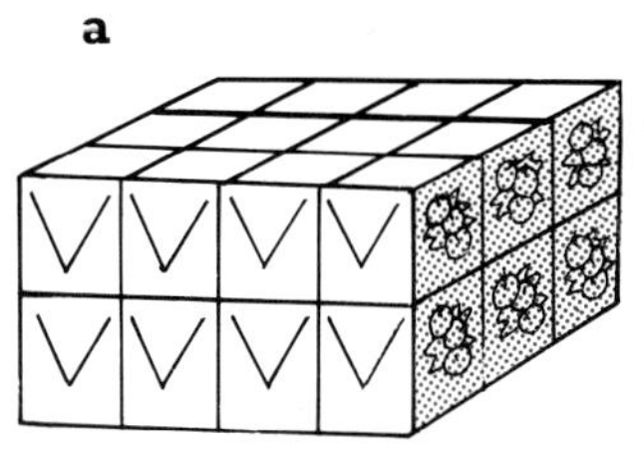

b

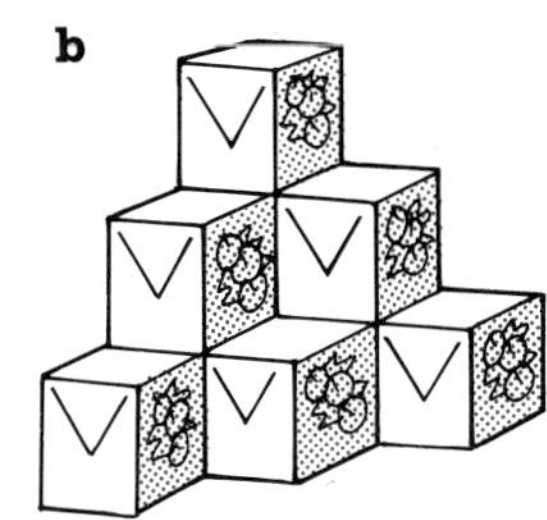

7 This carton, in the shape of a cuboid, has sides 5 cm, 8 cm and 20 cm long. Calculate its volume:
a in cm^3 **b** in litres.

8 (i)

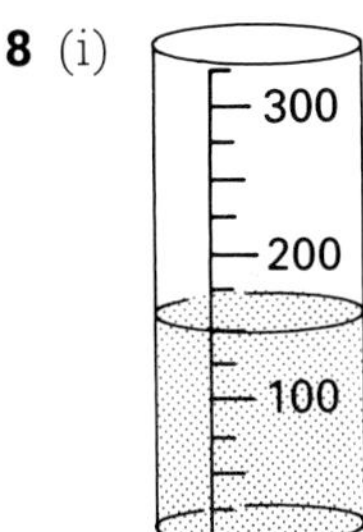

(ii)

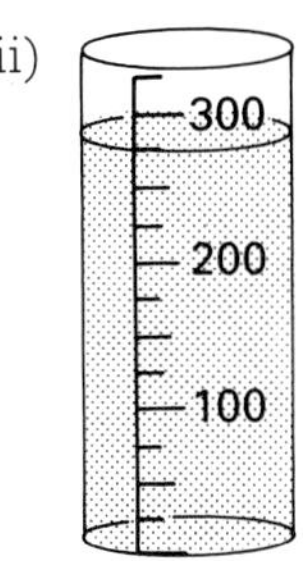

The scales in these jars are marked in millilitres. How much liquid:
a is in jar (i)
b was added to (i) to give (ii)?

9 Without using a calculator, write down the value of:
a 2.5×10 **b** 13.1×10 **c** 1.43×100
d 1.23×1000 **e** $57 \div 10$ **f** $8.5 \div 10$
g $365 \div 100$ **h** $4781 \div 1000$

10 Tom is always making up stories. Here is one of them:
'On my way home I passed a boy 170 centimetres tall. He was standing beside a car 100 metres long. Then I saw a man jump from a building 3 kilometres high. He gave me such a fright that I jumped 1 millimetre into the air.'

Which parts of his story are hard to believe?

MEASURING LENGTH

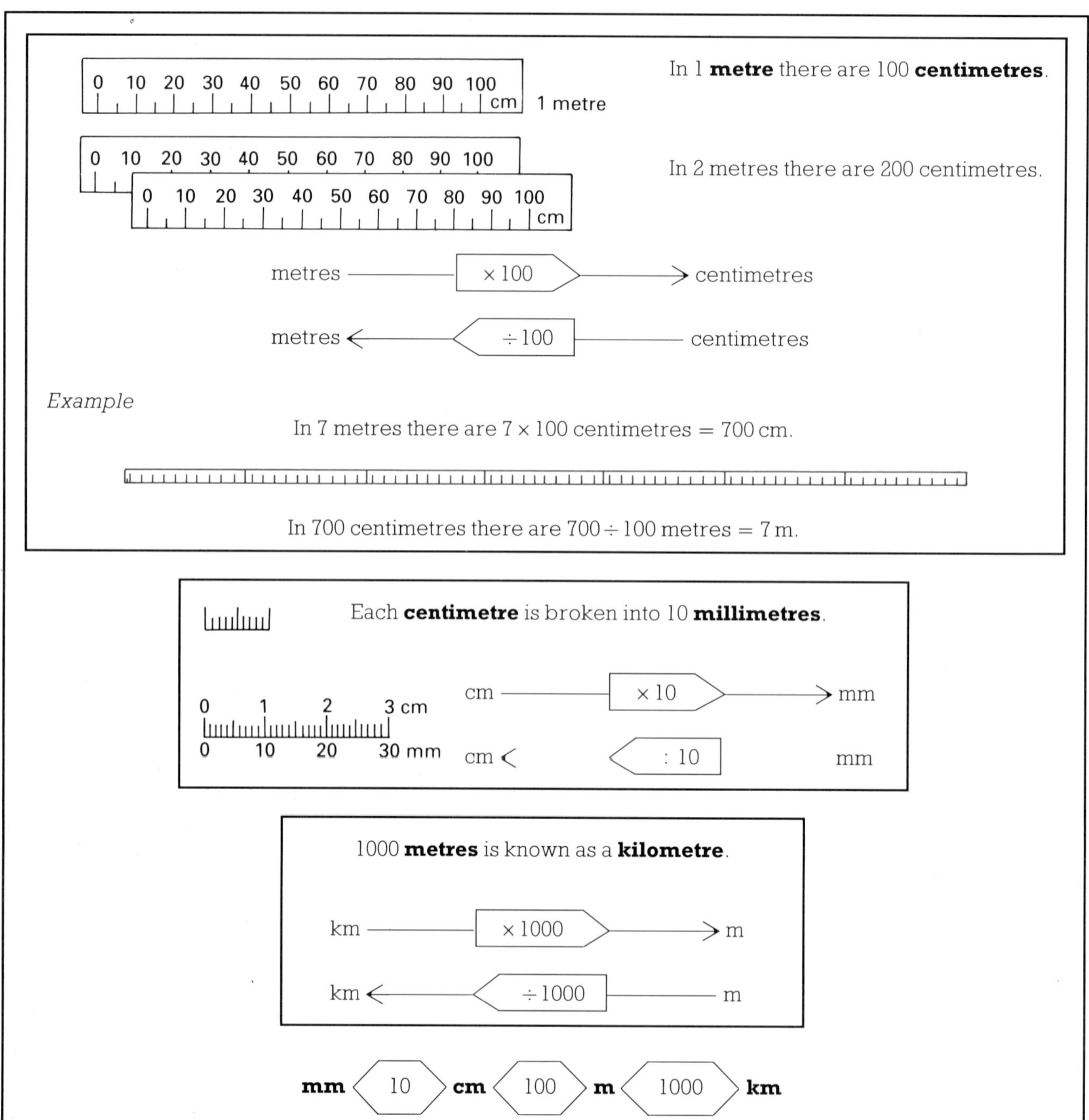

0 10 20 30 40 50 60 70 80 90 100 cm
1 metre
In 1 **metre** there are 100 **centimetres**.
0 10 20 30 40 50 60 70 80 90 100
0 10 20 30 40 50 60 70 80 90 100 cm
In 2 metres there are 200 centimetres.
metres
× 100
centimetres
metres
÷ 100
centimetres
Example
In 7 metres there are 7 × 100 centimetres = 700 cm.
In 700 centimetres there are 700 ÷ 100 metres = 7 m.
Each **centimetre** is broken into 10 **millimetres**.
0 1 2 3 cm
0 10 20 30 mm
cm
× 10
mm
cm
: 10
mm
1000 **metres** is known as a **kilometre**.
km
× 1000
m
km
÷ 1000
m
mm
10
cm
100
m
1000
km

EXERCISE 1A

The 'number machines' above can help you in questions **1–6**.

1 Change to millimetres:
a 5 cm **b** 12 cm **c** 4.8 cm **d** 0.4 cm

2 Change to centimetres:
a 400 mm **b** 60 mm **c** 72 mm **d** 9 mm

3 Change to centimetres:
a 4 m **b** 2.5 m **c** 0.37 m

4 Change to metres:
a 800 cm **b** 480 cm **c** 75 cm

5 Change to metres:
a 5 km **b** 2.6 km **c** 0.65 km

6 Change to kilometres:
a 6000 m **b** 8400 m **c** 120 m

7 On squared paper draw these rectangles, and measure the length of a diagonal in each one.

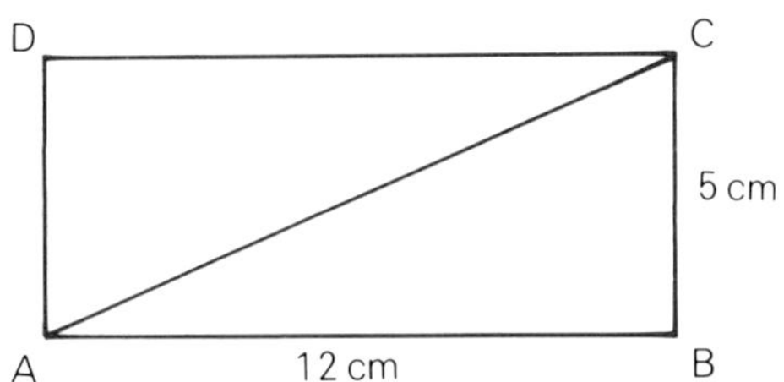

a ABCD, with AB = 12 cm and BC = 5 cm
b PQRS, with PQ = 60 mm and QR = 45 mm.

8 **a** Measure the width and height of the rectangle forming the outside edge of the stamp, in mm.
b Calculate its perimeter in: (i) mm (ii) cm.

9

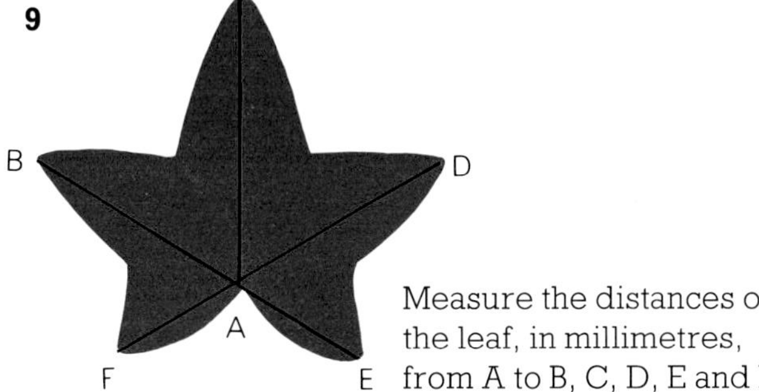

Measure the distances on the leaf, in millimetres, from A to B, C, D, E and F.

10

Sandy boasted that he could cover 1 km with four drives on the golf course. His first three shots went 205 m, 217 m and 196 m towards the hole. How far would he need to hit his fourth shot? Do you think he will do this?

11 A piece of elastic is 47 cm long. How far must it be stretched to be 1 m long?

12 A pile of 12 tiles is 6 cm high. Calculate the thickness of each tile, in millimetres.

13 A swimming pool is 50 m long. How many lengths would be swum:
a in the men's 1500 m freestyle race
b in the 4 by 100 m medley event?

EXERCISE 1B/C

1 Change the following measurements to the units given in brackets:
a 12 mm (cm) **b** 3 m (cm) **c** 7 cm (mm)
d 6.5 km (m) **e** 189 cm (m) **f** 96 mm (cm)
g 1200 m (km) **h** 1.5 m (cm)

2 Copy and complete:
a 4 km + 200 m = ... m
b 1 km − 400 m = ... m
c 5 cm + 20 mm = ... mm
d 8 cm − 30 mm = ... mm
e 2 m + 40 cm = ... cm
f 4 m − 20 cm = ... cm

3 Calculate the perimeters of these shapes:

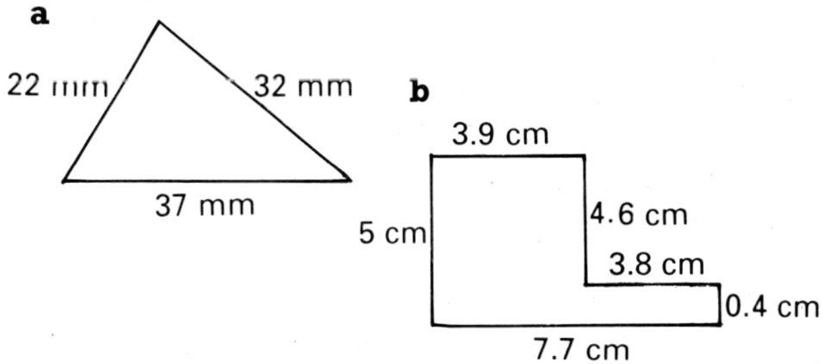

4 Taking 1 mile to be the same as 1.6 km, calculate the speed limits of 30 mph and 70 mph in km/h.

5 How many blocks of wood 40 cm long can be cut from the length shown?

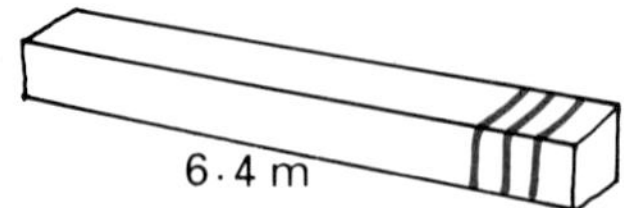

6 Lorna walks to and from school each day, a distance of 2.25 km each way. How far does she walk in a week?

7 In a roll of photographic film, each frame is 35 mm long. There is a 2 mm gap between each frame and 5 cm is provided at each end of the film for loading.
What is the length of a film with:
a 10 frames **b** 24 frames?

8 The longest track event at the Olympic Games is the 10 000 m race.
a How many kilometres is this?
b How many laps of a 400 m track would have to be run?

9 This table lists the lengths and breadths in metres of some popular cars.

Car	Length	Breadth
Metro	3.40	1.55
Maestro	4.05	1.69
Fiesta	3.65	1.57
Mondeo	4.47	1.76
Corsa	3.73	1.59
Astra	4.00	1.67
Volvo	4.80	1.71
Panda	3.38	1.45
Rover	4.74	1.77
Renault	3.99	1.66

a What is the difference in length, in metres, between the longest and shortest cars?
b What is the difference in width, in centimetres, between the widest and narrowest cars?
c Miss Allen has decided to buy a new car. The space in her garage for a car is 3.50 m by 1.60 m. Which of the cars could she choose?

10 a Sound travels at 341 metres per second. How far does it travel in an hour:
(i) in metres
(ii) in kilometres (to 3 significant figures)?
b Light travels at 299 800 kilometres per second. How long, to the nearest second, does it take to travel from the sun, a distance of 149 675 000 kilometres?

CHALLENGE

Estimating length can be difficult. Which line is equal in length to **a***?*

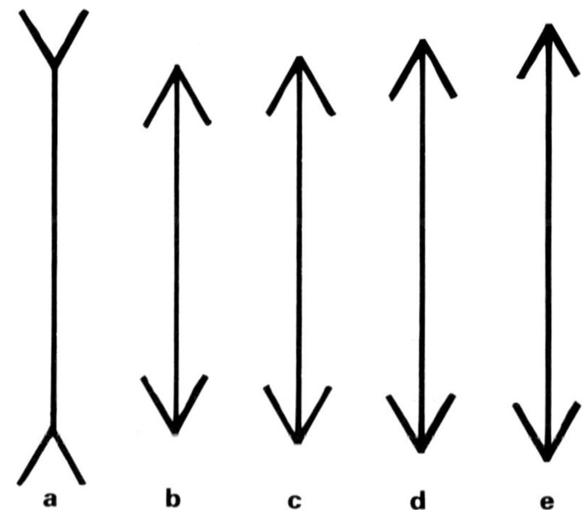

Did you choose **d***? Measure each line in millimetres to find out if you were correct!*

BRAINSTORMER

Each encyclopaedia is 4.8 cm thick.

A bookworm eats its way from the front cover of volume 1 to the back cover of volume 5. What distance does it travel?

INVESTIGATION

Investigate ways of finding the thickness of a page of this book.

MEASURING AREA

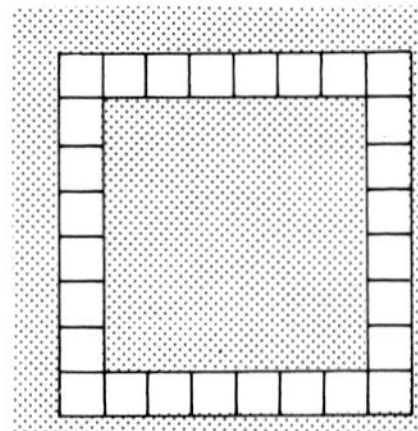

Frank's garden has grass on both sides of the path. Which grass plot is larger, the inner or outer area in the picture on the left?
Count the squares in the picture on the right to find out. Does the result surprise you?

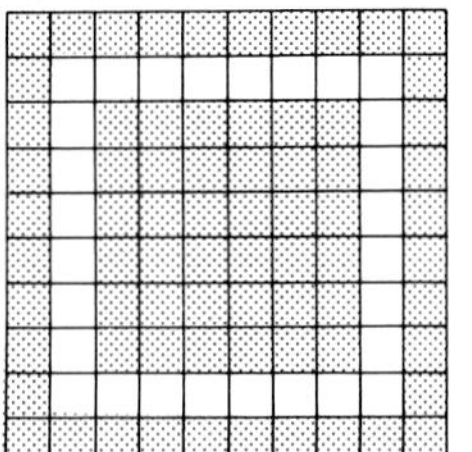

This is how we compare or calculate areas, by breaking them into equal squares.

The **square millimetre** (mm^2) is used for small areas.
Each small square on this coin is $1\ mm^2$.

The **square centimetre** (cm^2) is used for larger areas.
The area of the stamp is about $4\ cm^2$.

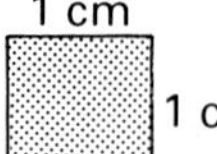

In 1 square centimetre there are 100 square millimetres.

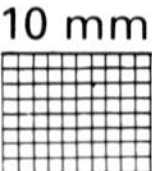

For larger areas we use the **square metre** (m^2). The area of this table is about $1\ m^2$.

The area of the football pitch is about 1 hectare.
1 **hectare**
$= 100 \times 100\ m^2$
$= 10\,000\ m^2$

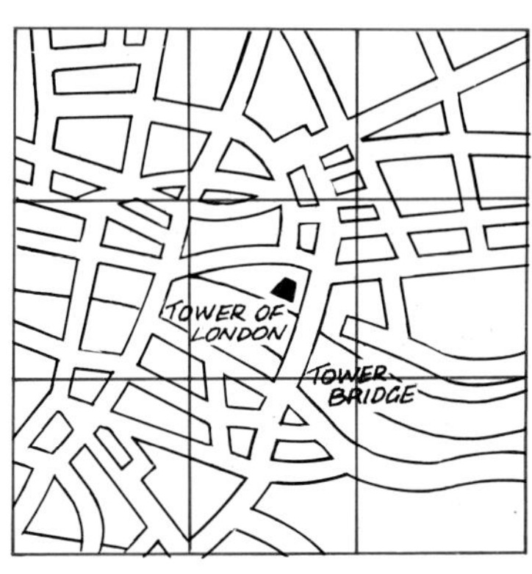

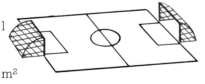

Each square on this map is 1 **square kilometre**.

Reminder
The area of a rectangle = length × breadth. $A = lb$

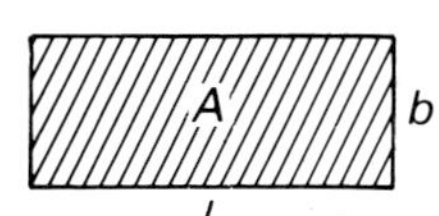

EXERCISE 2A

1 Which units would you use to measure the area of:
- **a** the classroom floor
- **b** the cover of your book
- **c** the top of your desk
- **d** one of your fingernails
- **e** a farmer's field
- **f** Wales?

2 Estimate the area of:
- **a** the classroom door
- **b** a 10p coin
- **c** the blackboard
- **d** the top of your pen
- **e** this page
- **f** your thumbnail.

3 Find the areas of these shapes by counting the number of squares inside them. Count squares that are half-squares or more—don't count smaller parts.

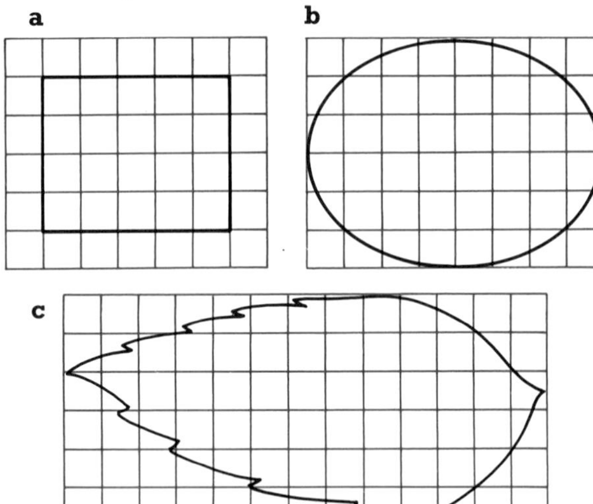

4 How could you calculate the area of the rectangle in question **3a** without adding up the squares? Could you do this for the other shapes?

5 Calculate the areas of these rectangles and squares:

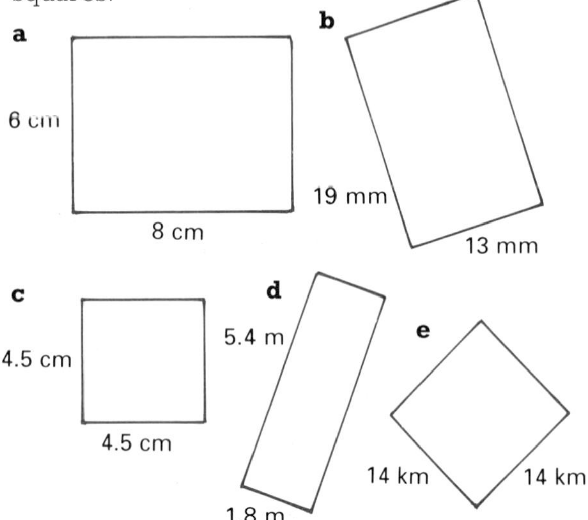

6 a Write down the length and breadth of the stamp in mm.

b Calculate the area of the stamp in: (i) cm² (ii) mm².

c How many mm² make 1 cm²?

7 a Write down the length and breadth of this notice board in: (i) cm (ii) m.

b Calculate its area in: (i) cm² (ii) m².

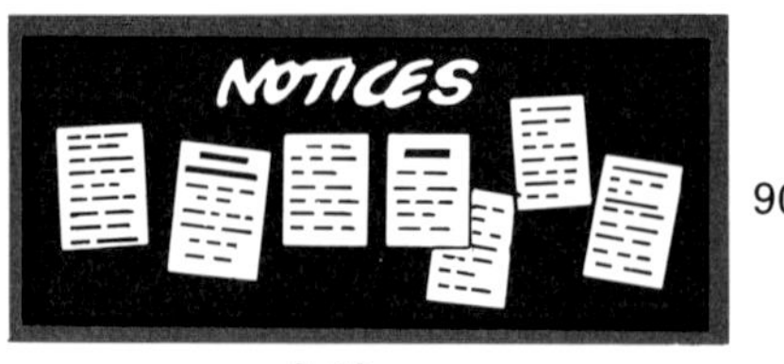

8 Notices on the board must be squares, with each edge 15 cm long. How many notices can be fitted:

a side by side along the bottom of the board

b up the left side of the board

c over the whole board?

9 Tom has another story to tell.
'My Dad painted the garage door. It took all day because its area was 100 square centimetres. Next day he cut the grass, which covered 10 hectares. After that he washed the windows which each had 2 square metres of glass, and swept the garage floor which had an area of 9 square millimetres.'

Which parts of this story do you not believe?

10 Daniel's small sister Lucy was playing with two wooden rectangles, making different shapes. An edge of one rectangle was always wholly or partly in contact with an edge of the other.

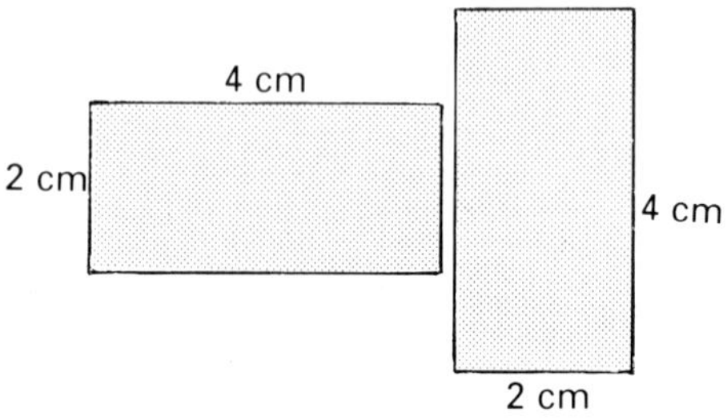

a Sketch six different shapes she could make.

b What can you say about the area of each shape?

EXERCISE 2B

1 a A toyshop window had been broken at the weekend, so Denis was sent to mend it. He measured the window, and found that it was 4 m by 3 m. Calculate the area of glass he would need.

b Change the units of the length and breadth to centimetres. Calculate the area in cm². Why is the number in your answer much greater?

2 The runway at an airport has to be resurfaced. It is 2 km long and 20 m wide.

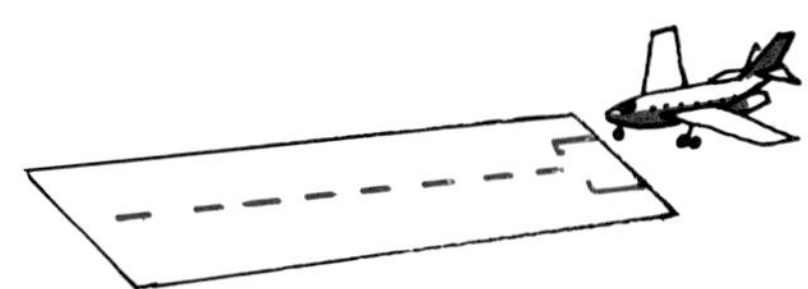

a Calculate: (i) its length in m (ii) its area in m².

b Divide your answer by 10 000 to find the area in hectares.

3 Each of these shapes can be cut along some of the lines, and put together to make a rectangle.
(i) Count the number of squares in each shape.
(ii) Find the lengths and breadths of possible rectangles that can be made from each shape.

a
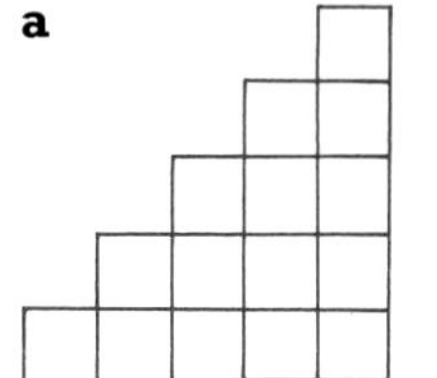
b
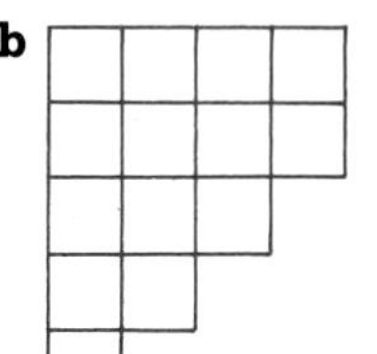
c
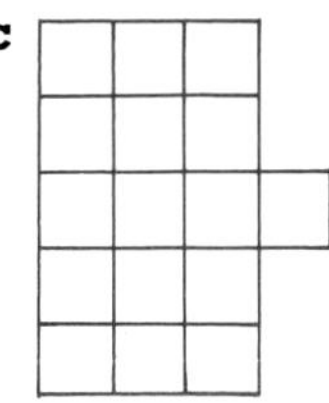

4 A maths classroom measures 8 m by 6 m. There is a school rule which says that before a class can use a room, there must be at least 1.5 m² of floor space for each pupil.

a What is the largest number of pupils allowed in the classroom?

b What floor area would your class need, using this rule?

5 An estate agent advertises a rectangular plot of ground for sale. It measures 250 m by 180 m. What is the area of the plot in hectares?

6 An office floor is 4 m by 3 m. It has a fixed cupboard with a square base 50 cm long at each corner. What area of floor is left to be carpeted?

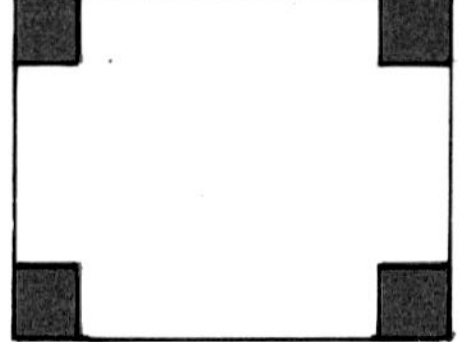

7 A rectangle is 5.3 cm long and 3.5 cm broad. Calculate its area to 2 significant figures (the same number as in the given data).

8 Calculate the area of a square of side 5.18 m to an appropriate number of significant figures.

EXERCISE 2C

1 Each shape below can be cut up, and made into a square. Find the length of the side of each square.

a **b**
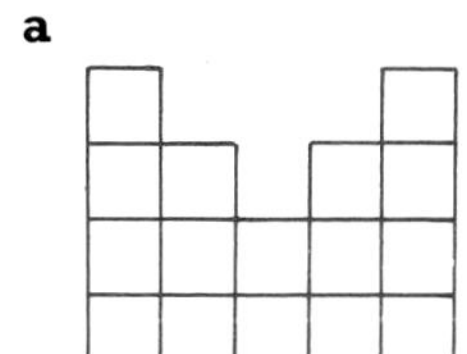

2 a How many: (i) mm in 1 cm (ii) mm² in 1 cm² (iii) mm² in 5 cm²?

b How many: (i) cm in 1 m (ii) cm² in 1 m² (iii) cm² in 8 m²?

3 Find the number of hectares in 1 km² of land.

4 A park is laid out for four football pitches. It is in the shape of a rectangle 400 m long and 150 m broad. Calculate the area of the park in square metres, and in hectares.

5 Calculate the area of this L-shape in three different ways. The dotted lines should help you.

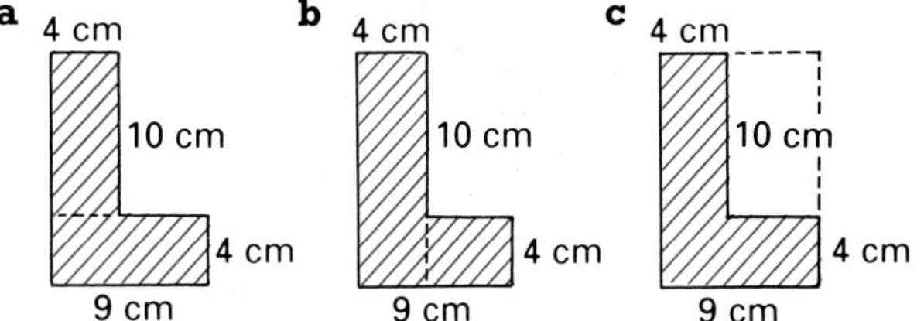

6 Calculate the areas of these shapes:

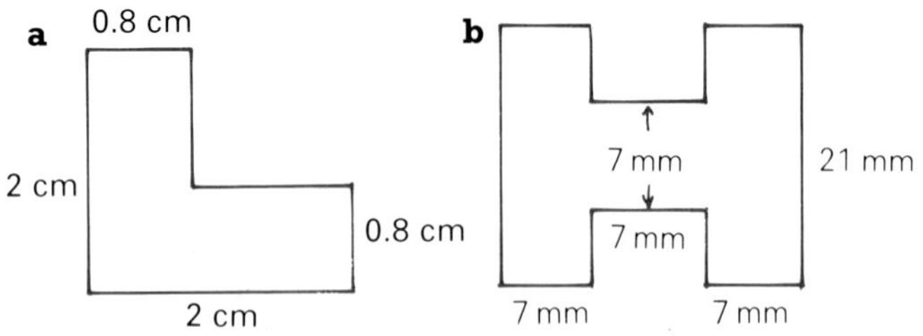

BRAINSTORMER

With only two cuts, this rectangle can be turned into a square. What is the length of the side of the square, and where would you make the cuts?

INVESTIGATION

Investigate all the different rectangles that have an area of 36 cm², and sides with lengths that are whole numbers of centimetres.
Calculate all their perimeters.

PRACTICAL PROJECTS

1 *Draw a rough sketch of the school grounds, or of your garden or street.*
Make the necessary measurements, and calculate the area in square metres and in hectares.

2 *Make the measurements, and find the prices needed to calculate the cost of painting the walls and covering the floor with carpet tiles, in the classroom or in a room at home. Write a report on your project.*

MEASURING VOLUME

Volume can be measured by using cubes.
In one layer of the cuboid there are 4×3 cubes.
There are two layers, so the number of cubes $= 4 \times 3 \times 2 = 24$.

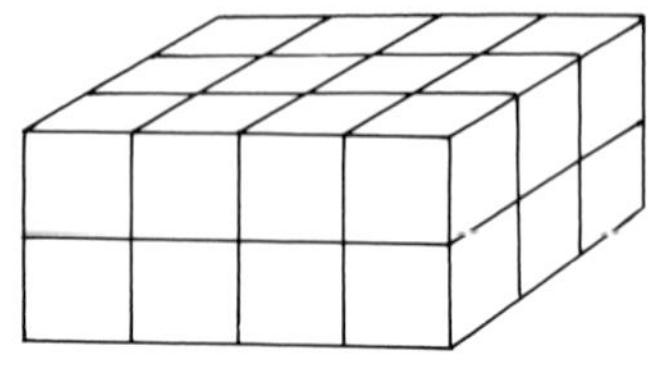

Reminder
The volume of a cuboid = length × breadth × height. $V = lbh$

A **cubic millimetre** (mm^3) is like a grain of sugar, very small.

A **cubic centimetre** contains 1000 mm^3.
Some dice are about 1 cm^3.

Larger sizes are measured in **cubic metres** (m^3).
A typical washing machine takes up about $\frac{1}{4}$ m^3 of space.

1 m^3 = 1 000 000 cm^3, but this is such a large number that a **litre**, 1000 cm^3, was introduced.

Milk is often sold in litres.

1 litre = 1000 cm³ = 1000 ml (millilitres). So 1 cm³ = 1 ml.

EXERCISE 3A

1 Which units would you use to measure the volume, or capacity, of:
a a carton of fruit juice **b** a thimble
c a classroom **d** a bottle of perfume
e this book **f** a car's petrol tank?

2 Estimate the volume of water in each jug below.

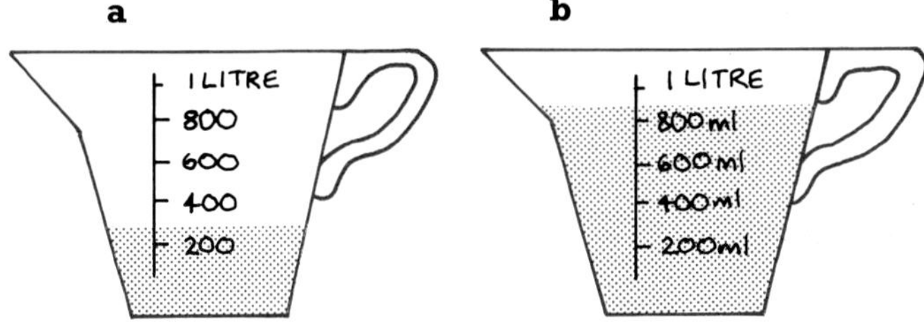

3 Change to millilitres:
a 5 litres **b** 25 litres **c** 0.2 litre

4 Change to litres:
a 7000 ml **b** 12 000 ml **c** 4500 ml

5 Calculate the capacity of each of these, in millilitres.

a 50 litres **b** 0.8 litre **c** 10 litres

6 Calculate the volume of each box of breakfast cereal in cm³; the measurements are in cm.

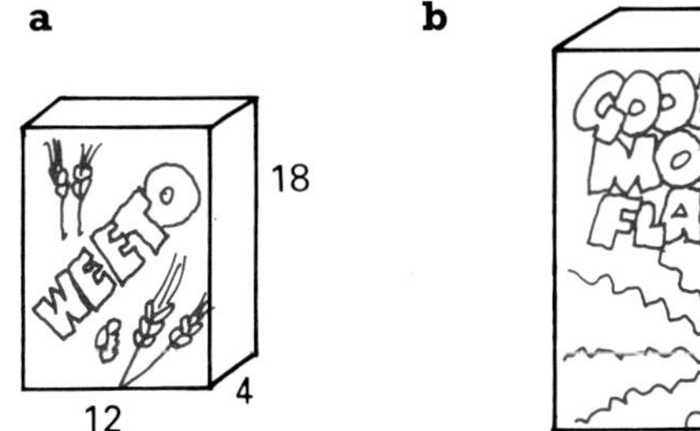

7 a Calculate the volume of this packet of Fruitos.

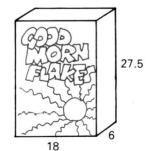

b The packet contains 12 sweets. What is the volume of each one?

8

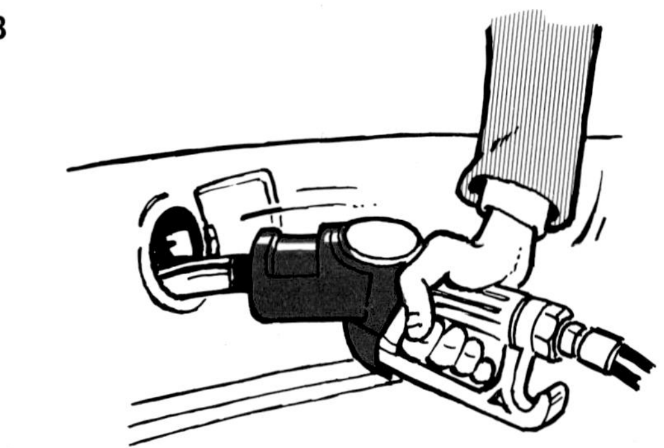

The petrol tank in Mr Thomas' car measures 58 cm by 50 cm by 12 cm. Calculate:
a the volume of the tank in
(i) cm³ (ii) ml (iii) litres
b the cost of a tankful of petrol at 48p per litre.

9 1 centilitre (cl) = 10 ml.
(Compare 1 cm = 10 mm.)
How many ml are in each bottle below?

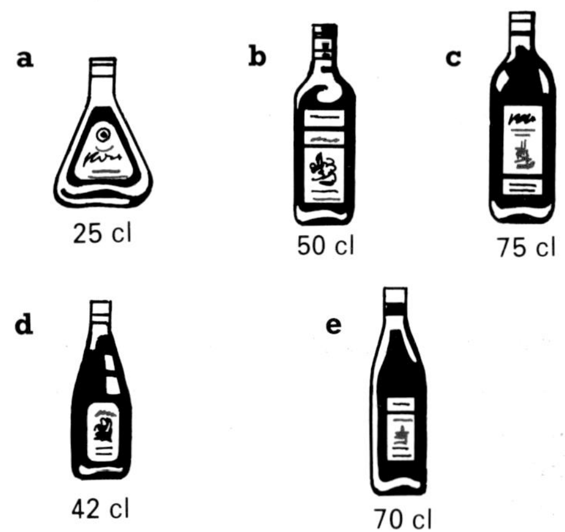

10 List the capacities of these tins in order, largest first.

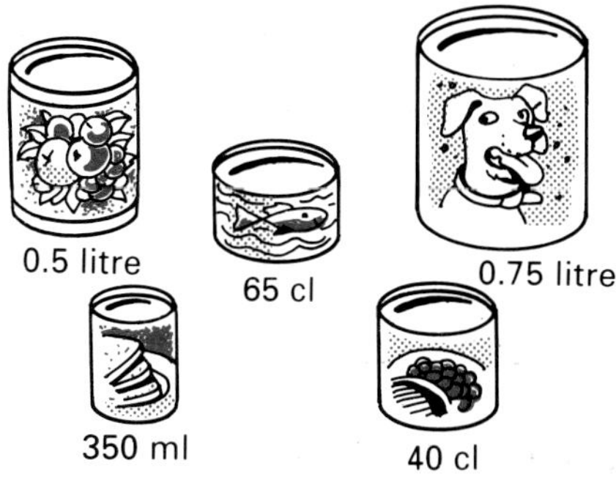

EXERCISE 3B

1 Ian got a 250 millilitre bottle of medicine from the chemist, and with it a 5 ml spoon. The label said 'one spoonful, three times a day'. How long will the bottle last?

2 The school fish tank is 2 metres long, 1.5 metres broad and 50 cm high. How many litres of water does it hold?

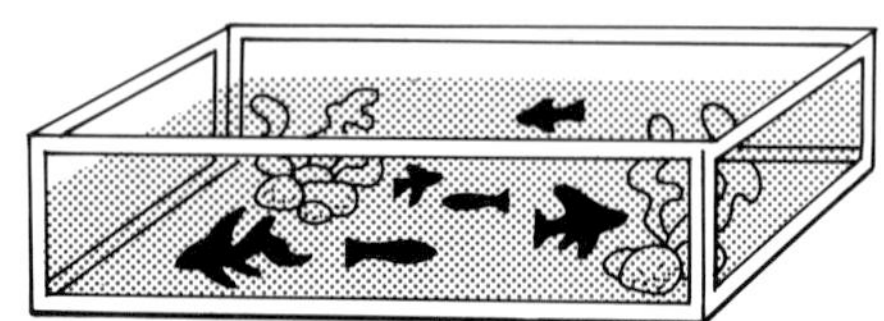

3 Which flat-roofed garden shed has most space in it?
Shed A: 4.5 m by 3.5 m by 2.6 m.
Shed B: 3.8 m by 3.8 m by 2.8 m.

4 Sandra was planning a party for 24 friends.
- **a** 'A glass of juice holds 150 millilitres. How many millilitres will I need—not forgetting myself—for one glass each?'
- **b** 'Say each person drinks two glasses. How many millilitres now?'
- **c** 'The juice comes in 2-litre bottles. How many will I need?'
- **d** At the party her friend Mark is very thirsty. Would there be enough juice for him to have another glass?

5 A rectangular tank is 2 m long, 1 m broad and 50 cm high. It is open at the top to collect rainwater.
- **a** How many litres of rainwater can it hold?
- **b** If the tank is quarter full of rainwater, what is the depth of water in it?

EXERCISE 3C

1 Beef stock cubes of edge 2 cm have to be packed in cubical boxes, each holding 27 cubes.
- **a** What is the volume of the box?
- **b** What are the dimensions of the box?

2 A central heating tank is being designed to hold 1250 litres of oil. It has to be fitted into a space 1.8 m long and 1.5 m broad. What height should the tank be, to the nearest 0.1 m?

3 Calculate the volumes of these solids; all angles are right angles.

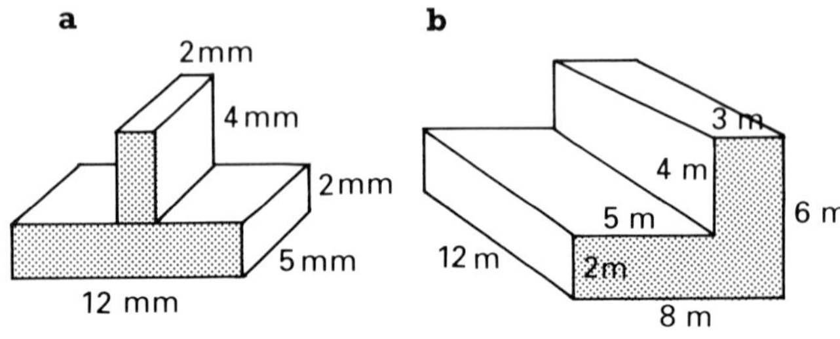

4 A firm is choosing between two different boxes for its chocolates. Each is a cuboid; one is 20 cm by 20 cm by 10 cm, and the other is 40 cm by 20 cm by 5 cm.
- **a** Which box has:
  (i) the greater volume (ii) the greater surface area?
- **b** Which do you think the firm will choose? Why?

5 Find all possible ways of packing boxes 5 cm by 5 cm by 5 cm into cartons which hold 20 boxes. Which carton would be cheapest to manufacture?

INVESTIGATION

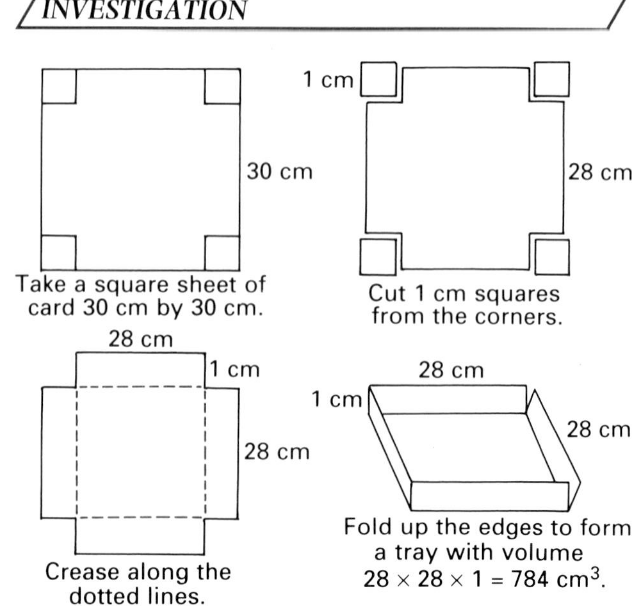

a *What is the volume of the tray formed when 2 cm squares are cut from the corners?*

b *Find the depth of the tray with greatest volume. (A graph of depths against volumes might help.)*

c *Investigate for other sizes of square sheets.*

MEASURING WEIGHT

A **gram** is the weight of 1 cm^3 of water. A stock cube weighs about 6 g. 	The **milligram** is used for lighter weights. 1000 mg = 1 g. 
A **kilogram** is for heavier objects. 1000 g = 1 kg 1 litre of water weighs 1 kilogram. 	Very heavy objects are weighed in **tonnes**. 1000 kg = 1 tonne A small car weighs about 1 tonne.

mg ⟨ 1000 ⟩ **g** ⟨ 1000 ⟩ **kg** ⟨ 1000 ⟩ **tonne**

EXERCISE 4A

1 Which units would you use to measure the weights of:

a a bag of sugar **b** a key
c a bus **d** a sheet of paper
e a letter for posting **f** a baby
g a chestnut leaf **h** an aeroplane?

2 Estimate the weight of:

a this book **b** a sack of coal
c a calculator **d** a stick of chalk
e a feather **f** a brick.

3 Here's Tom again.

'I put my 2 kilogram pencil in my pocket, and on my way to school I stopped to weigh myself. I was glad to see I only weighed 5 tonnes. I posted a letter weighing 100 grams, and bought a 5 milligram box of chocolates for my Mum.'

Which parts of this story are hard to believe?

4 a Write down the readings on these scales.

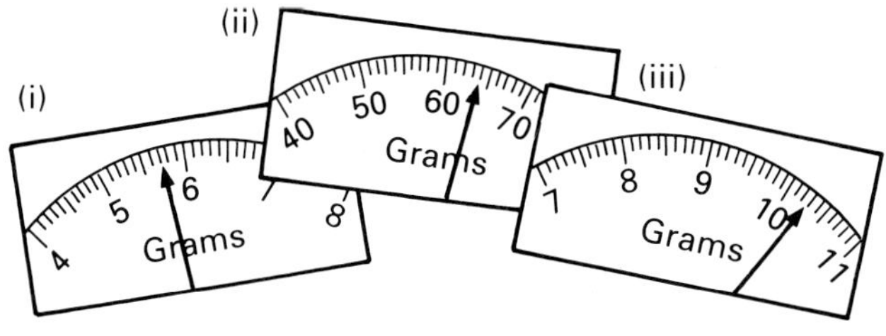

b What is the difference between the readings on scales (i) and (ii)?

5 Adam and Bethan had a good day's fishing. The scales show the weights of the fish they caught, in kilograms.

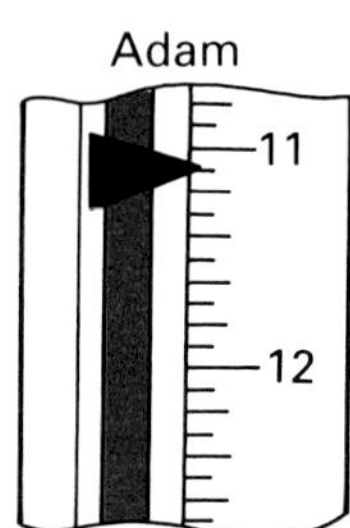

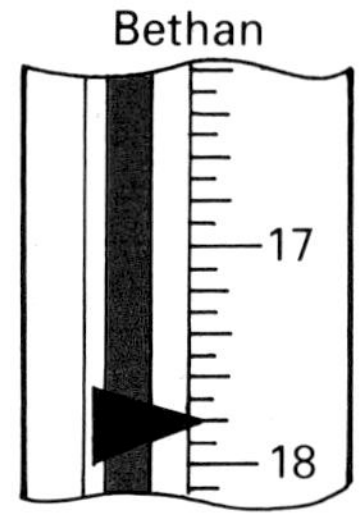

a How much did each catch?
b What was their total catch?

6 Lorries are weighed as they enter and leave a factory. Calculate the weight of:

a each lorry
b each load collected at the factory.
(*Note* 100 kg scale.)

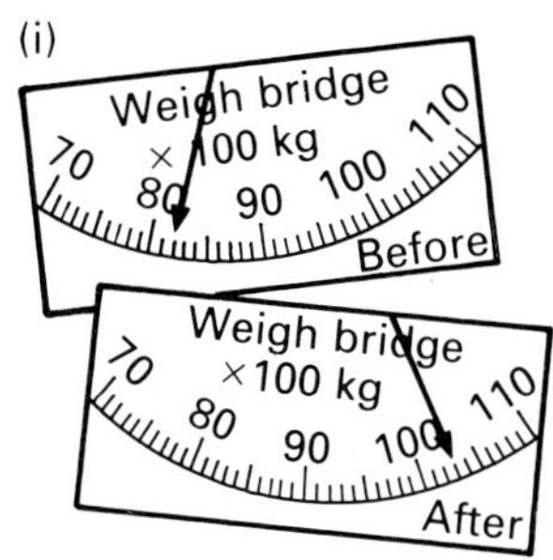

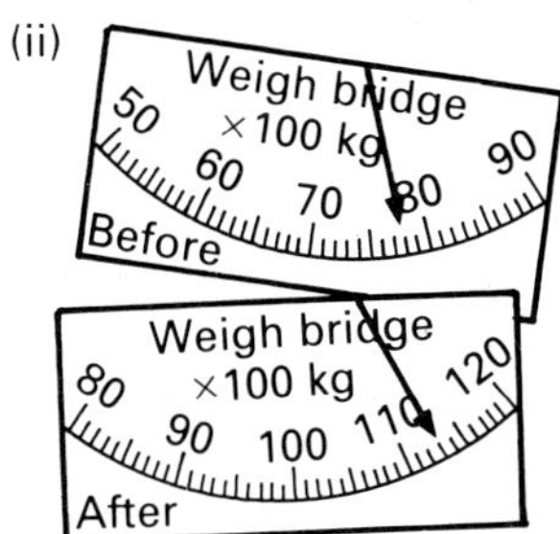

EXERCISE 4B/C

1 Change to grams:
a 7000 mg **b** 15 000 mg **c** 800 mg

2 Change to grams:
a 1 kg **b** 6 kg **c** 2.3 kg

3 Change to kilograms:
a 3000 g **b** 300 g **c** 10 000 g

4 Change to kilograms:
a 1 tonne **b** 5 tonnes **c** 6.4 tonnes

5 A recipe for blackberry crumble lists these ingredients:
750 g blackberries
225 g brown sugar
75 g butter
175 g self-raising flour
50 g granulated sugar
What is the total weight of the ingredients?

6 An empty lorry weighs 8750 kg. When it is filled with sand it weighs 13 320 kg. What is the weight of the sand?

7 An empty box weighing 300 g is filled with 50 packets of crisps, each weighing 25 g. What is the total weight of the box and crisps:
a in grams **b** in kilograms?

8 Mohammed lost weight while training for a marathon. His starting weight was 65.3 kg, and he lost 1500 g. What was his new weight?

9

Mr Johnson laid four rolls of this felt. Calculate:
a the total length of felt laid
b the total weight of felt laid
c the total cost of the felt.

10 Calculate the following, giving your answers in kilograms:
a 1.2 kg + 800 g **b** 1100 g + 900 g
c 4 tonnes + 500 kg

11 **a** 1 litre of water weighs 1 kilogram. Calculate the weight of 1 cm^3 of water.
b 20 cm^3 of material weighs 60 g. Why will this material sink in water?
c 20 cm^3 of another material weighs 15 g. Will it sink or float?

12 A sugar refinery has 6.5 tonnes of sugar to pack in 1 kg bags.
How many bags will be filled?

13 A postman is struggling with a big load of mail. In his bag there are 75 letters weighing 50 g each, 43 weighing 120 g each, and 18 weighing 175 g each. How much does the mail weigh, in kg?

14 The inland letter postal rates were:

	Weights up to—			
	60 g	100 g	150 g	200 g
1st class	24p	31p	38p	45p
2nd class	18p	22p	28p	34p

For the postman's load in question **13**, calculate:
a the least possible total postage paid
b the greatest possible total postage paid.

PUZZLE

Which is heavier, a tonne of bricks or a tonne of feathers, and by how much?

PRACTICAL PROJECTS

1 *Make a list of items that have the weights (g, kg, etc), volumes (cm^3, etc), or capacities (litre, ml) of their contents marked on them.*

2 *Compare different sizes and prices of items to find the 'best buys'.*

INVESTIGATIONS

1 *Find out the weights of the heaviest bird, the heaviest land creature and the heaviest sea creature. Discuss why they are so different.*

2 *A market trader wants a set of weights which will weigh articles from 1 kg to 40 kg, in steps of 1 kg. What is the least number of weights that he requires, and which ones are they?*

CHECK-UP ON METRIC MEASURE

1 Copy and complete these 'number machines'.

For example, mm —[÷10]— cm.

a 

b 

c 

d 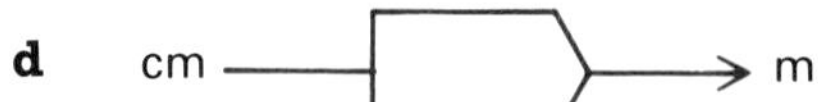

e 

f 

2 Find the perimeter of this shape in:
a cm **b** m.

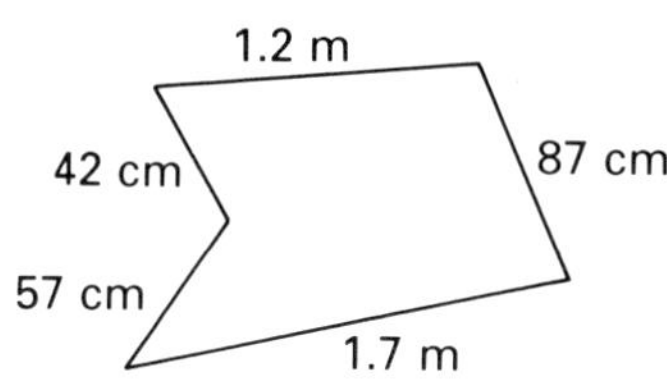

3 Find the areas of these fields in:
(i) m^2 (ii) hectares.

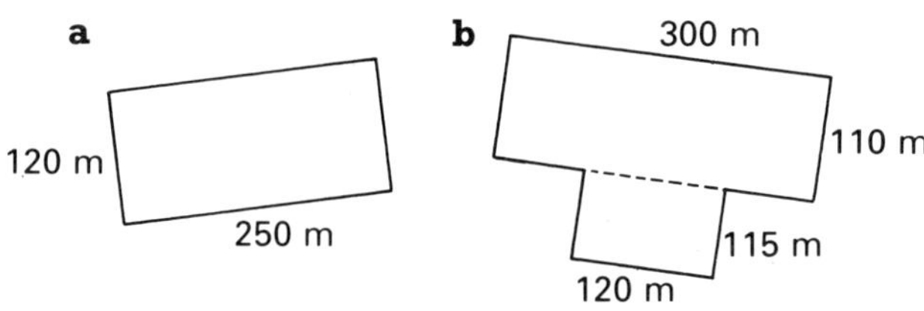

4 How many 10 cm by 10 cm square tiles are needed to cover a wall 5 m long and 2 m high?

5 A baking bowl is weighed, then an egg is added. What is the weight of:
a the bowl **b** the egg?

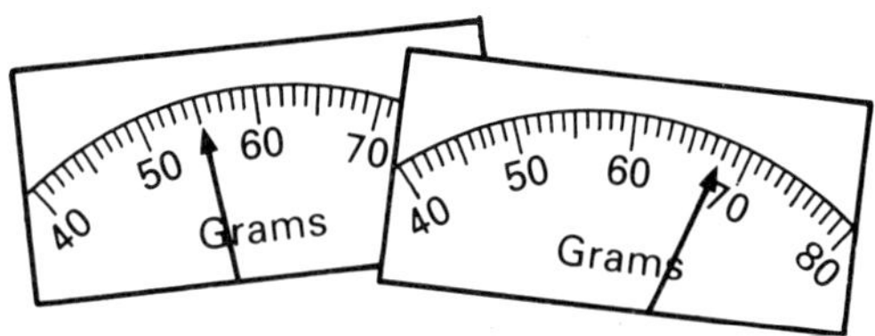

6 A boxer dieted to reduce his weight by 720 g, to 85.6 kg. What was his original weight?

7 **a** 375 cm. How many m?
b 2304 ml. How many litres?
c 1 m^2. How many cm^2?
d 1 km^2. How many m^2?

8 **a** What is the volume of this carton, in ml?

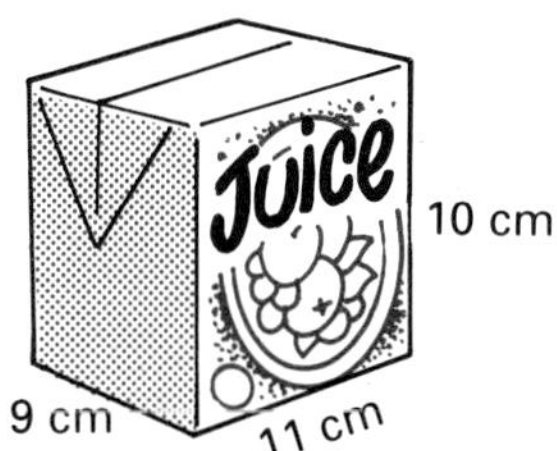

b Which measurement would you change to make the carton hold 1 litre of juice (with the least amount of empty space in it)?

9 By cutting along the lines you can put all the small squares together to make a rectangle.

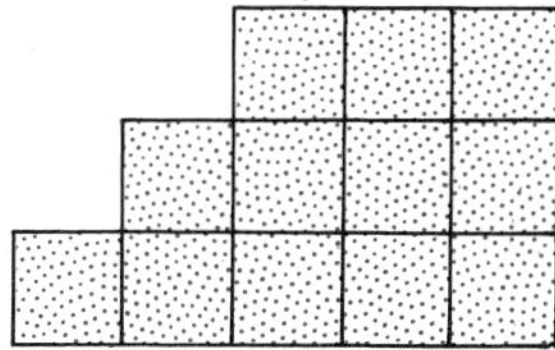

What are the lengths and breadths of the possible rectangles?

10 A toy box is 1 m long, 30 cm wide and 20 cm deep. How many bricks, each a cube of side 5 cm, can it hold?

11 EQUATIONS AND INEQUATIONS

LOOKING BACK

1 Simplify:
a $2n+3n$ **b** $5k-k$ **c** $9a-9a$
d $4t+t-t$ **e** $2d+3d+4$ **f** $3n+1-2n$

2 Write without brackets:
a $2(x+1)$ **b** $2(y-3)$ **c** $3(5+x)$
d $2(2x-1)$ **e** $4(3w+4)$ **f** $5(1-2t)$

3 Find an expression for the perimeter of each rectangle. Lengths are in cm.

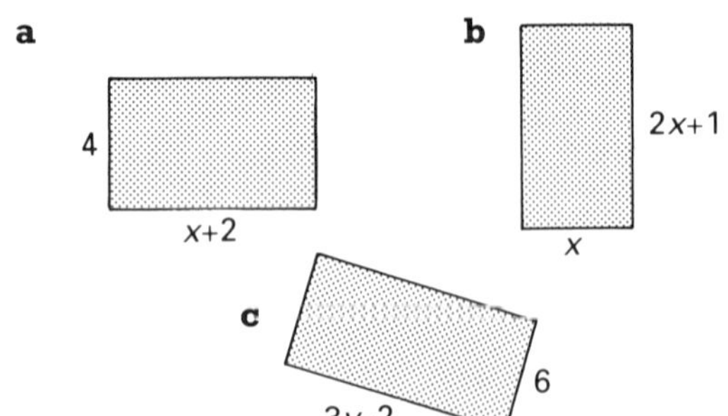

4 Solve these equations, and check your solutions.
a $y+2=3$ **b** $5u=15$ **c** $t-8=10$
d $3x+1=16$ **e** $2m-3=7$ **f** $18-2w=8$

5 Write down:
(i) the number of weights on each tray
(ii) the number left after the action shown.

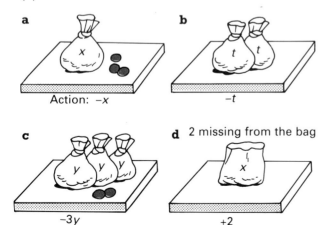

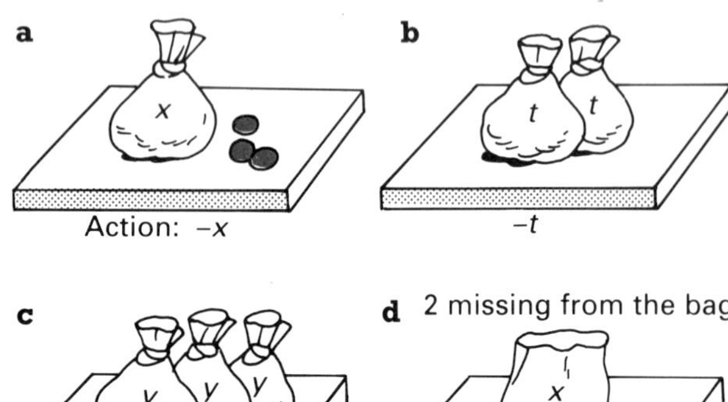

6 In **a** and **b** below find x, and check your solution.

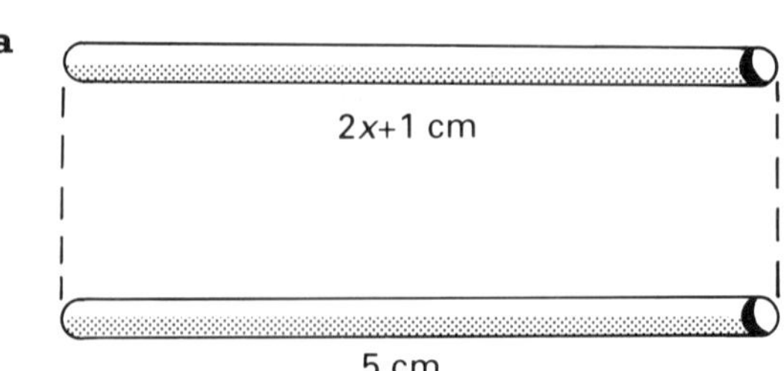

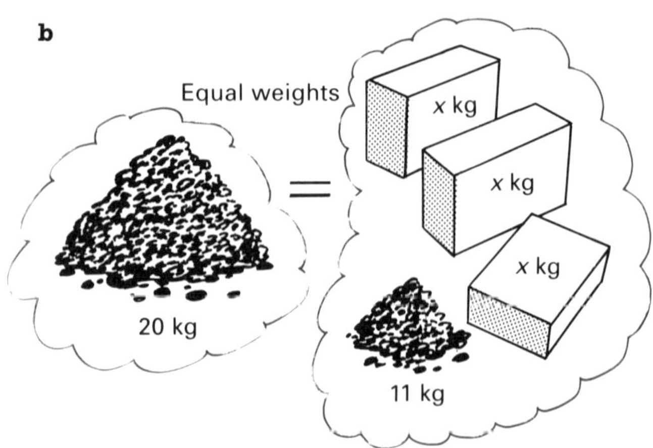

7 Solve these equations, and check your solutions.
a $2x=x+5$ **b** $3y=y+8$
c $8w=5w+15$ **d** $5m+4=2m+13$
e $2x-3=x+4$

8 The numbers in row n in columns **A** and **B** are equal. Find n.

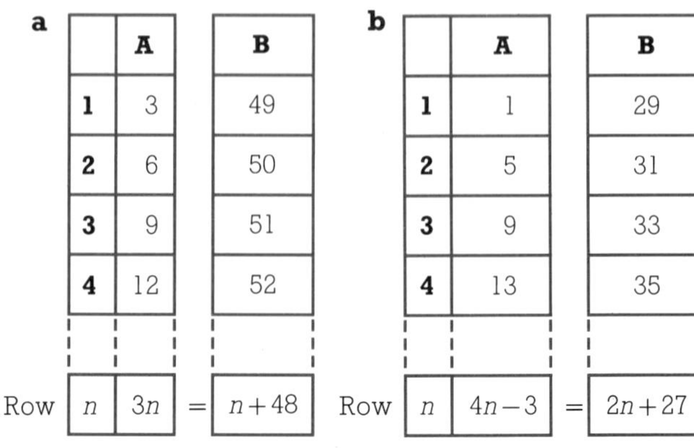

a

	A		B
1	3		49
2	6		50
3	9		51
4	12		52
Row n	$3n$	=	$n+48$

b

	A		B
1	1		29
2	5		31
3	9		33
4	13		35
Row n	$4n-3$	=	$2n+27$

SOLVING EQUATIONS

EXERCISE 1A

Example Solve $2x+3 = 11$

$2x+3 = 11$

$2x = 8$

$x = 4$

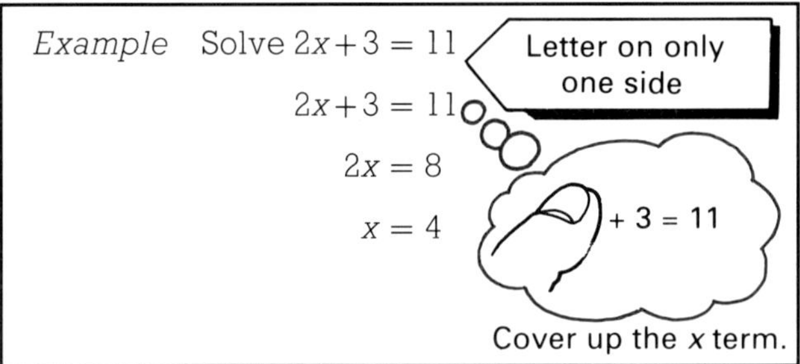

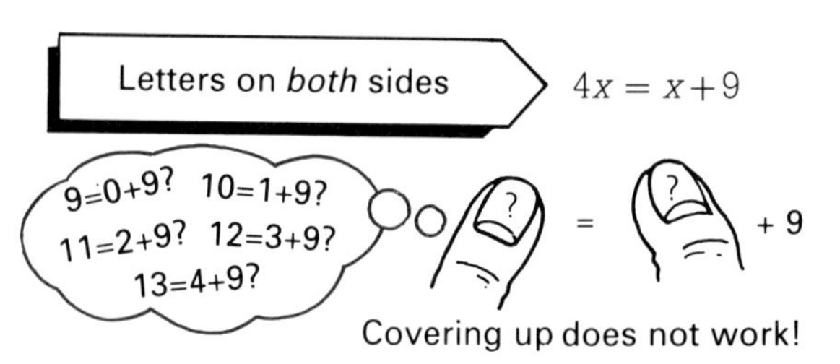

$4x = x+9$

The balance	The equation
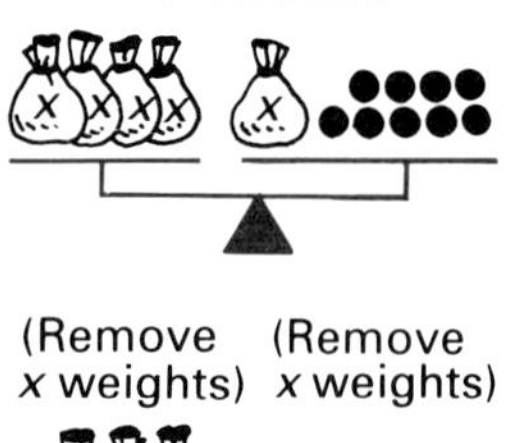	$4x = \quad x+9$
(Remove x weights) (Remove x weights)	$-x \quad -x$
	$3x = 9$
Each bag contains 3 weights.	$x = 3$

Rule
Eliminate an x term from one side by adding or subtracting the same term on each side of the equation.

Examples

a $3x+1 = \quad x+7$
$-x \qquad -x$
$2x+1 = 7$
$x = 3$

b $x = 12-x$
$+x \qquad +x$
$2x = 12$
$x = 6$

Solve the equations in questions **1–6**.

1 **a** $x+6 = 10$ **b** $t+4 = 7$ **c** $e+2 = 8$

2 **a** $y-1 = 5$ **b** $p-2 = 4$ **c** $n-10 = 1$

3 **a** $5-k = 4$ **b** $9-m = 2$ **c** $6-t = 0$

4 **a** $2n = 14$ **b** $3u = 21$ **c** $9v = 36$

5 **a** $2x+1 = 15$ **b** $3x+2 = 17$ **c** $5y-1 = 9$

6 **a** $4w+2 = 18$ **b** $6y-1 = 29$ **c** $7t-7 = 0$

7 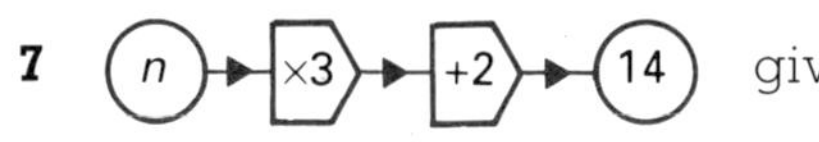 gives $3n+2 = 14$.

What is the value of n?

8 Make an equation for each of these, and solve it.

a 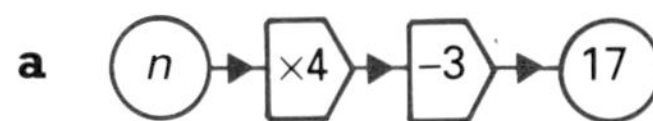

b 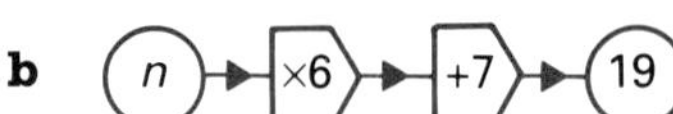

9 **a**

Eve thought of a number . . . doubled it . . . added 6.
If she got 20, make an equation and solve it to find the number she thought of.

b Jemma thought of another number, trebled it and subtracted 1. She got 14. Find the number she thought of.

10 Copy these equations, and solve them by carrying out the given action.

a $2x = \quad x+5$
$-x \qquad -x$

b $4a = \quad 2a+6$
$-2a \qquad -2a$

c $3x+4 = \quad x+6$
$-x \qquad -x$

Solve the equations in questions **11–13**.

11 **a** $3x = 2x+4$ **b** $5y = y+8$ **c** $2x = x+1$

12 **a** $3w = w+4$ **b** $9t = t+24$ **c** $5x = x+16$

13 **a** $3t = t+2$ **b** $6u = u+20$ **c** $4v+5 = v+8$

14 Copy these equations, and solve them using the given action.

a $x = 6 - x$ $+x$ $+x$ **b** $2y = 9 - y$ $+y$ $+y$ **c** $3t = 8 - t$ $+t$ $+t$

15 Solve:

a $x = 8 - x$ **b** $2y = 3 - y$ **c** $4k = 12 - 2k$

16 Make an equation for each pair of equal straws below, and solve it. Lengths are in cm.

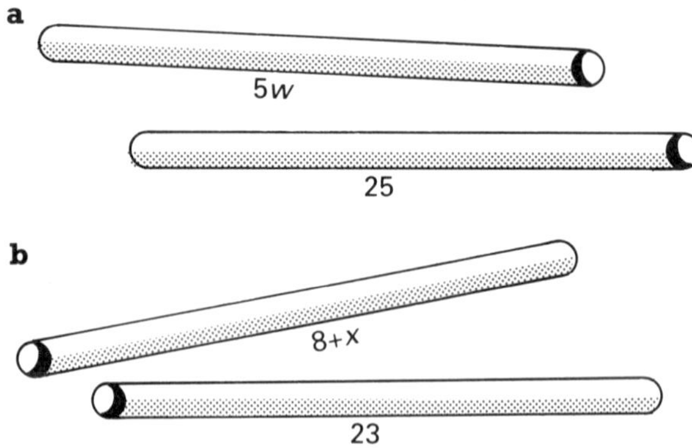

17 (i) Make an equation for each picture below, and solve it.
(ii) Write down the lengths (all in metres) of one car and one house.

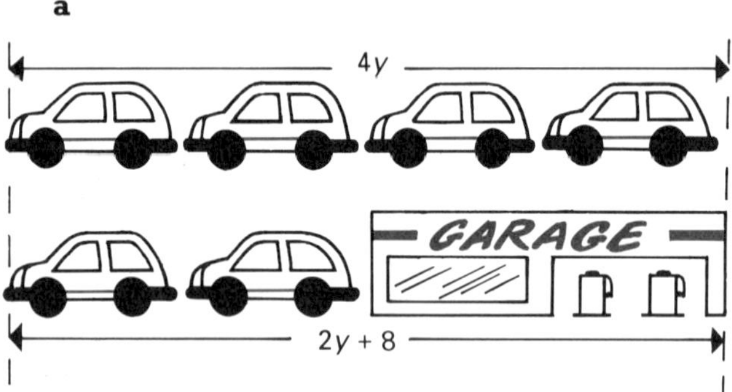

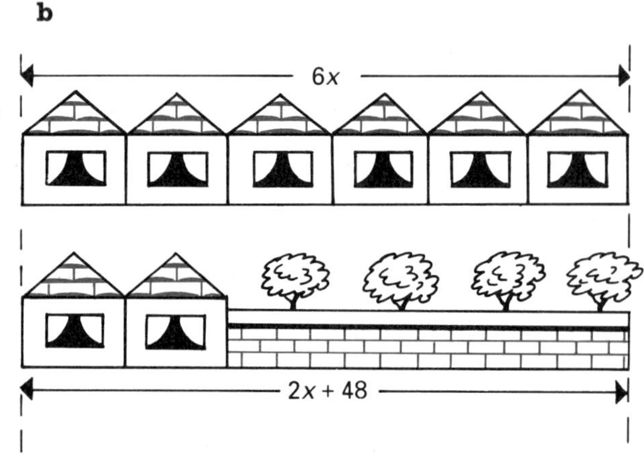

18

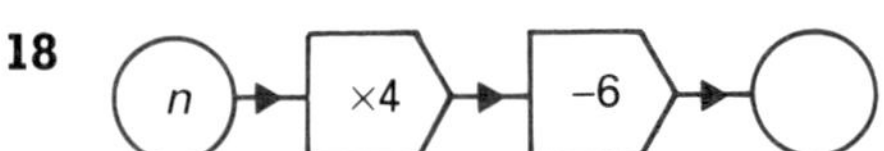

gives the same result as

Make an equation, and solve it.

CHALLENGE

Copy and complete this cross-number puzzle.

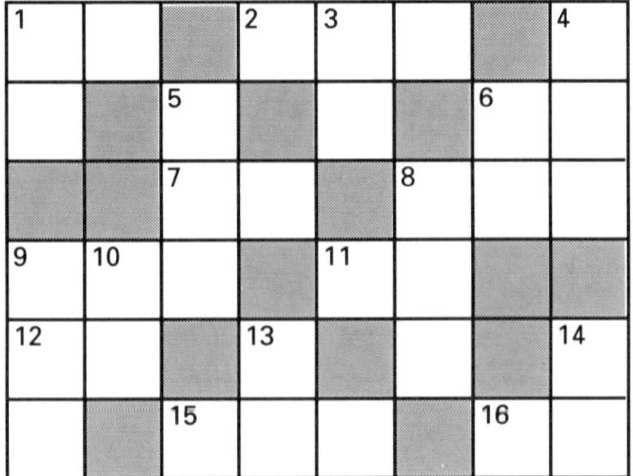

Across

1 $8x = 80$
2 $x + 750 = 1000$
6 $2x = x + 11$
7 $4x + 2 = 50$
8 $5x = 500$
9 $x - 25 = 100$
11 $3x = x + 24$
12 $x = 110 - x$
15 $x + x = 468$
16 $3x - 2 = 100$

Down

1 $x = 24 - x$
3 $5x = 275$
4 $x - 10 = 100$
5 $x + 85 = 200$
6 $3x + 5 = 35$
8 $2x = x + 120$
9 $8x = 1200$
10 $3x = x + 50$
13 $3x + 1 = 100$
14 $2x + 2 = 50$

EXERCISE 1B

You can eliminate an x-term **or** a number from one side of an equation by adding or subtracting the same x-term or number on each side.

Examples

a
$$\begin{array}{rcl} x+4 &=& -1 \\ -4 & & -4 \\ x &=& -5 \end{array}$$

b
$$\begin{array}{rcl} x &=& 2x+5 \\ -x & & -x \\ 0 &=& x+5 \\ -5 & & -5 \\ -5 &=& x, \\ \text{or} \quad x &=& -5 \end{array}$$

c
$$\begin{array}{rcl} 2x+9 &=& 4x-15 \\ -2x & & -2x \\ 9 &=& 2x-15 \\ +15 & & +15 \\ 24 &=& 2x \\ x &=& 12 \end{array}$$

Solve these equations (**1–9**).

1 a $x-3=-1$ **b** $x-1=-2$ **c** $x+1=-2$

2 a $x=2x-3$ **b** $y=3y-6$ **c** $2t=3t-1$

3 a $4u=5u+3$ **b** $3v=4v+1$ **c** $3w=4w+6$

4 a $t+7=2t$ **b** $3m+5=4m$ **c** $3k+10=5k$

5 a $4n=5n-4$ **b** $7p=8p-6$ **c** $3t=6t-9$

6 a $2y=15-y$ **b** $10-t=4t$ **c** $12-g=2g$

7 a $3x+7=2x+9$ **b** $8t-1=5t+8$

8 a $x+8=2x+5$ **b** $x+11=4+2x$

9 a $y-7=3y-11$ **b** $2t+7=5t-2$

10 Make an equation for each pair of equal straws below. Solve it, and find the length of each straw. All lengths are in centimetres.

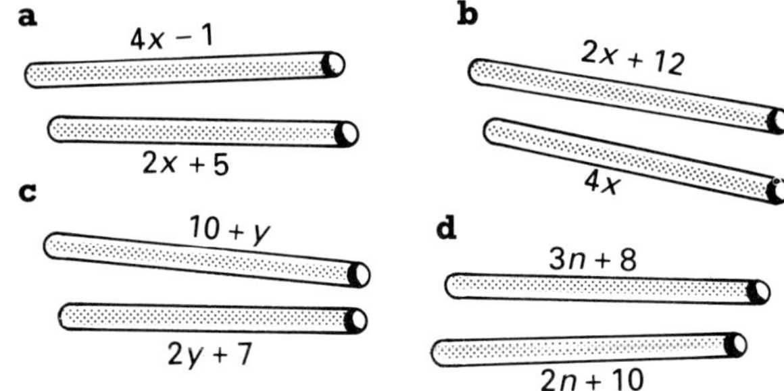

11 Make equations, and find the number of weights in each bag.

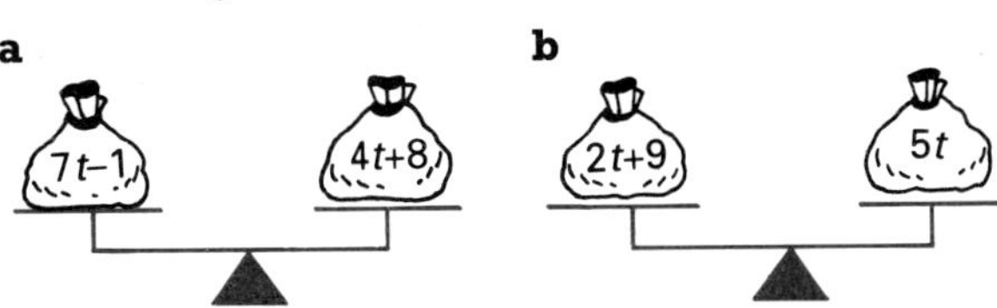

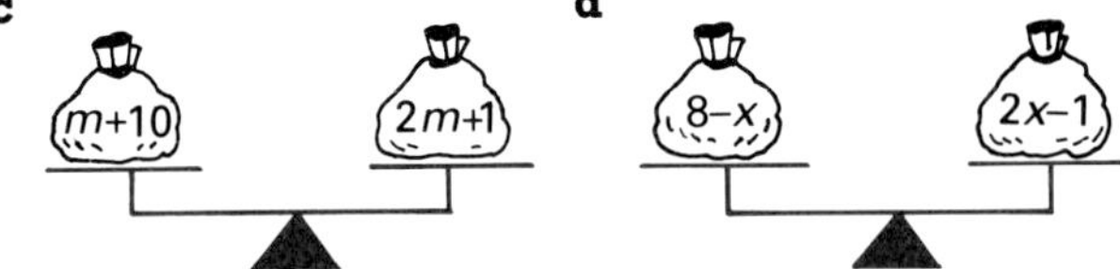

12 The two numbers in row n of **A** and **B** are equal.
(i) Form an equation, and find n.
(ii) Use this value of n to find the number. (Check each one.)

a

	A		B
1	3	**1**	101
2	6	**2**	102
3	9	**3**	103
4	12	**4**	104
n	$3n$	n	$n+100$

b

	A		B
1	1	**1**	20
2	3	**2**	21
3	5	**3**	22
4	7	**4**	23
n	$2n-1$	n	$n+19$

c

	A		B
1	5	**1**	53
2	10	**2**	54
3	15	**3**	55
4	20	**4**	56
n		n	

d

	A		B
1	14	**1**	24
2	18	**2**	26
3	22	**3**	28
4	26	**4**	30
n		n	

EXERCISE 1C

The problem

Two of these bags together contain the same number of coins as the third bag.
Which two, and how many coins are in each bag? Try all possible arrangements.

Understand the problem

	First arrangement	Second arrangement	Third arrangement
Understand the problem	x+1, 2x–1 	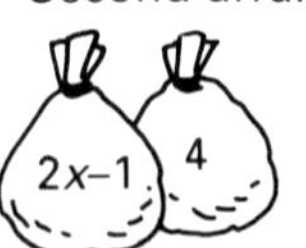 	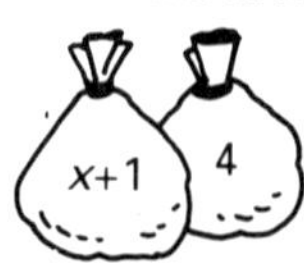 2x–1
Form an equation	$x+1+2x-1=4$	$2x-1+4=x+1$	$x+1+4=2x-1$
Solve the equation	$3x=4$ x is not a whole number.	$2x+3=x+1$ $-x \quad -x$ $x+3=1$ $x=-2$	$x+5=2x-1$ $-x \quad -x$ $5=x-1$ $x=6$
Understand the solution	This arrangement is not possible.	This arrangement is not possible.	The bags contain 7, 4 and 11 coins.

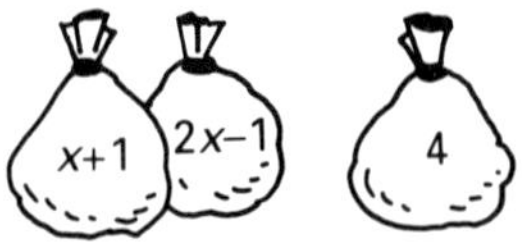

In each picture below (**1–14**), two of the bags together contain the same number of coins as the third one. Find the number of coins in each bag by checking all possible arrangements.

1 x+1, 6, 4

2

3 x, 2x, 4

4

5 4x, 6, x

6

7

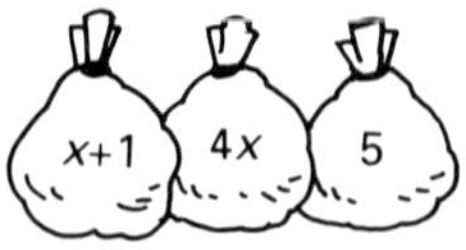

8

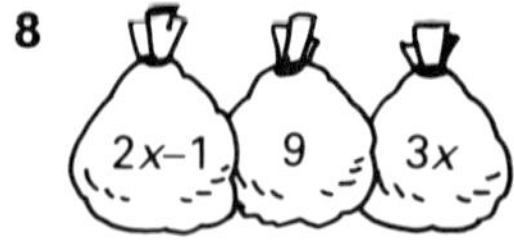

9

10

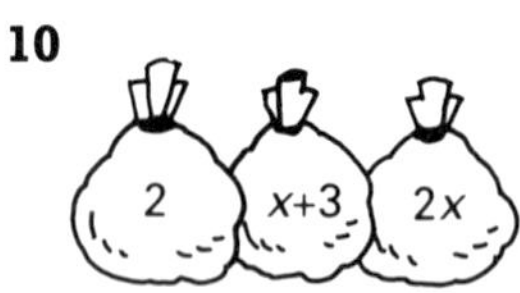

11

12

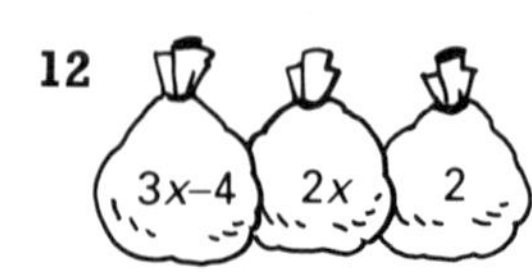

13

14

EQUATIONS WITH BRACKETS

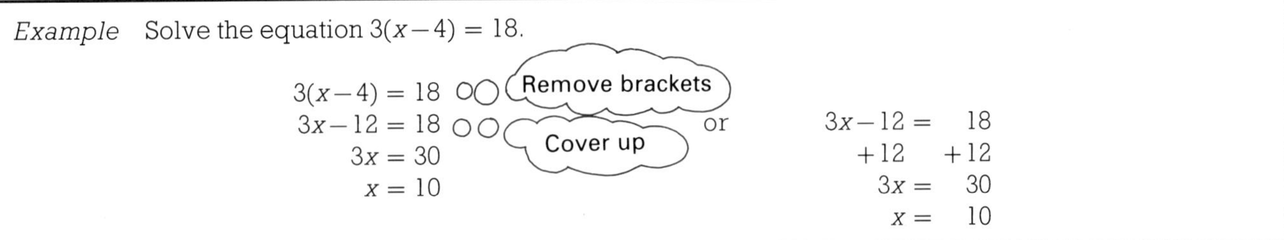

Example Solve the equation $3(x-4) = 18$.

$3(x-4) = 18$ (Remove brackets)
$3x-12 = 18$ (Cover up)
$3x = 30$
$x = 10$

or

$3x-12 = 18$
$+12 \quad +12$
$3x = 30$
$x = 10$

EXERCISE 2A

Solve these equations, by first removing the brackets.

1 **a** $2(x+1) = 8$ **b** $3(x+2) = 6$ **c** $4(x+3) = 16$

2 **a** $2(x-1) = 10$ **b** $3(x-2) = 3$ **c** $5(x-3) = 10$

3 **a** $2(x+3) = 10$ **b** $3(x+1) = 9$ **c** $3(x-1) = 12$

4 **a** $4(y-2) = 20$ **b** $6(y+1) = 6$ **c** $7(y+2) = 14$

5 **a** $3(x+1)+2 = 11$ **b** $4(y+1)-3 = 13$

6 **a** $5(t-1)+7 = 22$ **b** $2(m-5)-1 = 1$

7 Make an equation for the area of each rectangle, as shown in **a**. Solve it, and check that 'length times breadth' gives the area.

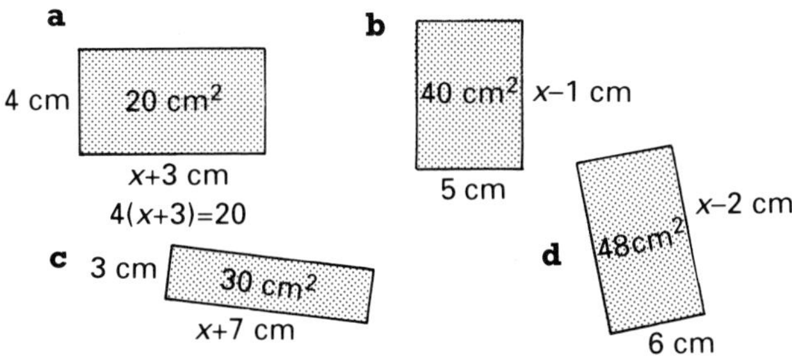

8 Use the diagrams to make equations, as in **a**, then solve the equations.

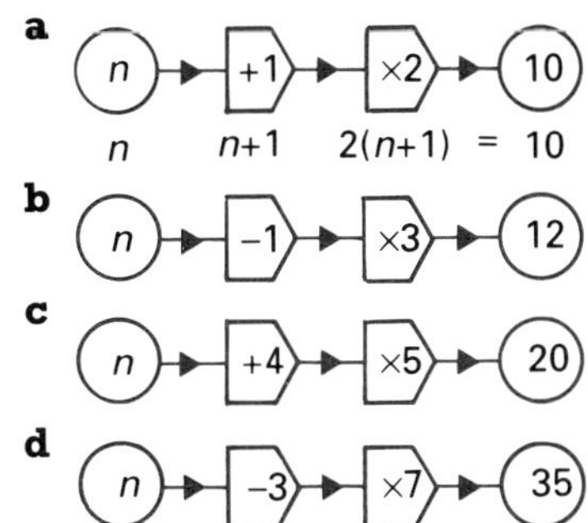

9

'Easy,' said Rose, 'I'll make an equation, and find out . . . I think it was 12.'
Make your own equation, and find out if Rose was correct.

10 'My turn,' said Rose. 'I've thought of a number, taken away 10, and multiplied the answer by 5. This gives me 120. What number did I start with?'
Make an equation, solve it and write down Rose's number.

CHALLENGE

Ask some friends to think of a number, multiply it by 4, add 4, halve the answer, and subtract 2. Ask them for their answer, then tell them the number they thought of. (It is half the number in their answer—why?)

EXERCISE 2B

Solve these equations.

1 **a** $2(3x+1) = 14$ **b** $3(2x-1) = 9$
c $5(4x+3) = 15$

2 **a** $2(x-1) = x$ **b** $3(x+1) = 4x$
c $5x = 6(x-2)$

3 **a** $4x+1 = 3(x+1)$ **b** $5(x-1) = 4x+11$
c $2(x+1) = 3(x-3)$

4 **a** $4(x-1) = 3(x+2)$ **b** $2(x+1) = 6(x-1)$
c $3(3+x) = 2(7+x)$

5 Make an equation for the area of each rectangle below. Solve it, and then calculate the perimeter of the rectangle.

a 8 cm; 40 cm²; 2x+3 cm

b 190 cm²; 3x–2 cm; 10 cm

c 63 m²; 7 m; 2x+1 m

d 8x m²; 18 m; x–5 m

6 The seesaw is balanced, because the pull down on the left, 3×4, is equal to the pull down on the right, 2×6.

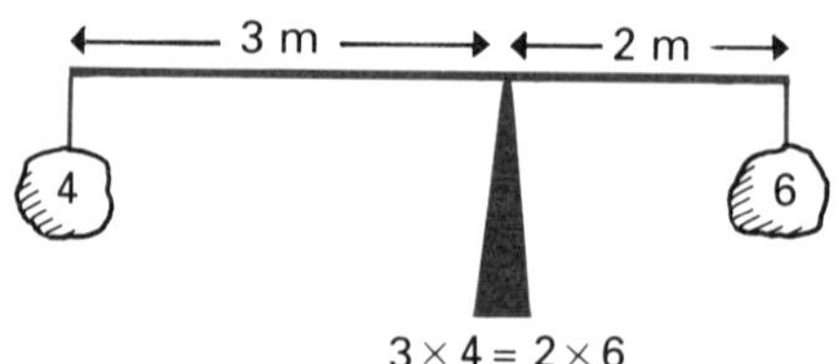

Make an equation for each seesaw below, and solve it. The weights are in kg.

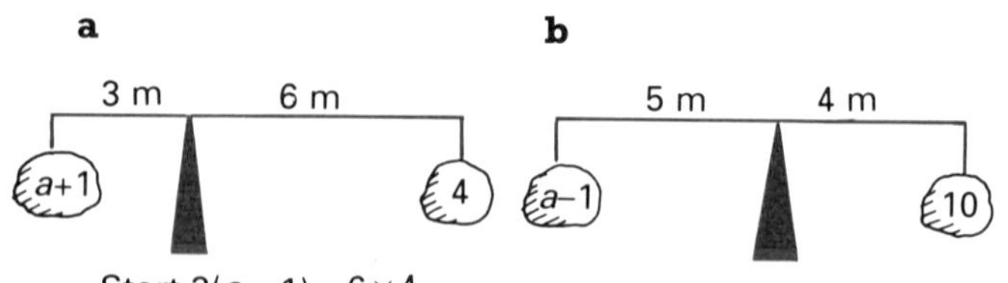

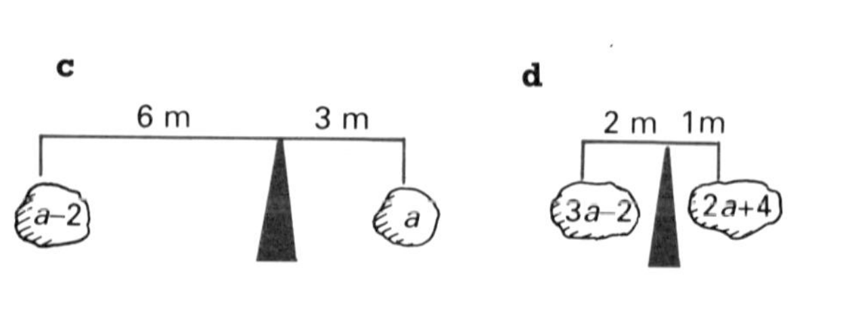

EXERCISE 2C

1 Mrs Kelly is organising a charity run for the x pupils in her class and herself. Each person has to run 6 km.
a How many people will be running?
b What will be the total distance covered by the runners?
c If this total distance is 144 km, form an equation and solve it to find the number of pupils in the class.

2 Ross can save £6 a week for his holiday. He has been saving for three weeks and plans to save for another x weeks.
a For how many weeks altogether does he save?
b How much will he save in this time?
c He needs £210 for his holiday. Make an equation, and find x.

3 **a** Find expressions for:
(i) the perimeter P cm
(ii) the area A cm²
of the rectangle.
b (i) If $P = 30$, find A.
(ii) If $A = 40$, find P.

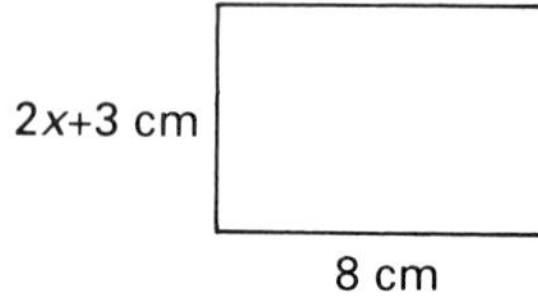

4 **a** Repeat question **3a** for this rectangle.
b (i) If $P = 46$, find A.
(ii) If $A = 190$, find P.

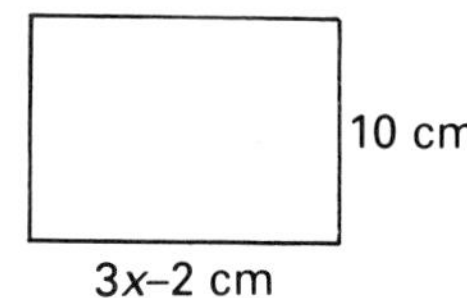

5 Sean had been thinking about number problems. He made this one up:
'Fred's father is 30 years older than Fred. In 20 years, three times Fred's age will be 12 years more than double his father's present age. How old are Fred and his father at present?'
Can you solve this? Take Fred's age to be x years, and make an equation.

6 Lisa's mum works in a bank, so she gave her a money puzzle:
'A bag is full of 10p, 5p and 2p coins, which are worth, in all, £36. There are the same number of 5ps as 2ps, and there are twenty more 10ps than 5ps. How many coins are in the bag?'

PUTTING NUMBERS IN ORDER

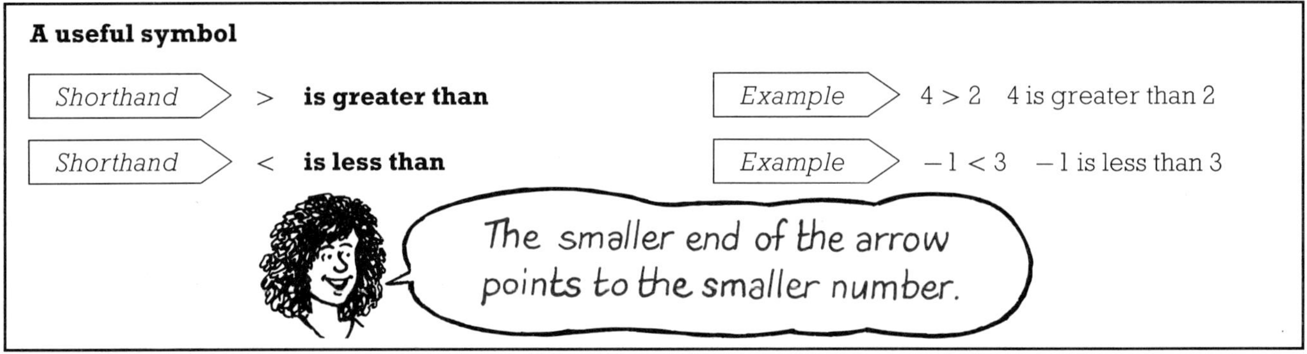

A useful symbol

Shorthand $>$ **is greater than** *Example* $4 > 2$ 4 is greater than 2

Shorthand $<$ **is less than** *Example* $-1 < 3$ -1 is less than 3

The smaller end of the arrow points to the smaller number.

EXERCISE 3A

1 Put the symbol $>$ (is greater than) or $<$ (is less than) between each pair of numbers given below. For example, for 2 and 3, $2 < 3$.
a 4 and 3 **b** 1 and 2 **c** 3 and 0
d 8 and 7 **e** -1 and 1 **f** 5 and -5
g 9 and 11 **h** 0 and 2 **i** 0 and -2
j 3 and -3 **k** -2 and -1 **l** -3 and -6

2 x and y are weights, in kilograms. Say whether $x = y$, or $x > y$, or $x < y$ for each pair of scales.

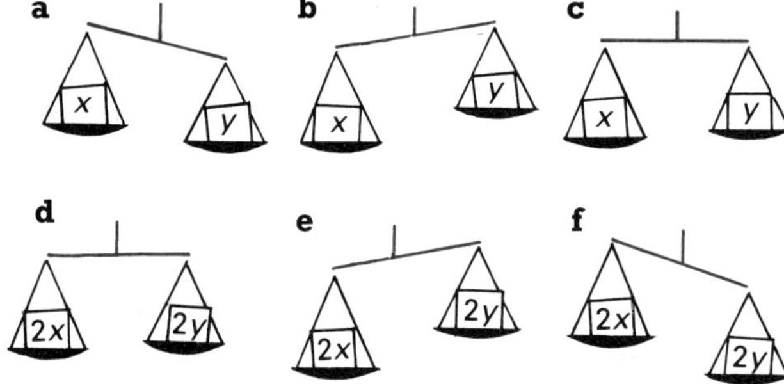

3 Write out the meanings of these. For example, $-1 < 0$ means -1 is less than 0.
a $4 > 2$ **b** $3 < 5$ **c** $-1 > -3$ **d** $-2 < 0$

4 p, q, r, s and t are the lengths of the canes, in cm. Copy these pairs, and fill in $>$ or $<$.
a $p \ldots q$ **b** $q \ldots r$ **c** $r \ldots s$ **d** $s \ldots t$
e $p \ldots r$ **f** $r \ldots t$ **g** $q \ldots s$ **h** $p \ldots t$

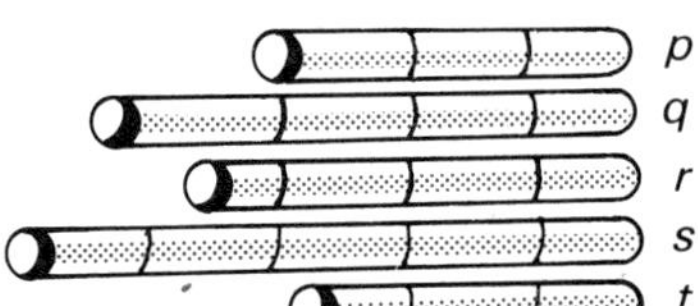

5 Find the value of n, a whole number, in each of these:
a $n > 4$ and $n < 6$ **b** $n > 0$ and $n < 2$
c $n < 10$ and $n > 8$ **d** $n > -1$ and $n < 1$

6 Draw a number line from -2 to 2, and use it to help you find out whether each of these is true or false.
a $1 > 2$ **b** $-1 < 2$ **c** $1 < 2$
d $2 > 1$ **e** $-2 > 1$ **f** $-2 > -1$
g $-2 < 1$ **h** $1 > -2$ **i** $-1 < -2$
j $1 < -2$ **k** $2 < -1$ **l** $-1 > -2$
m $-2 < -1$ **n** $2 < 1$ **o** $-1 > 2$

INEQUATIONS

More useful symbols

$\geqslant$ 'is greater than or equal to' . . . — Price (P) at least £80 000 — $P \geqslant 80\,000$

$\leqslant$ 'is less than or equal to' . . . — Maximum speed (S) 30 mph — $S \leqslant 30$

Example

All police constables must have a height of 170 cm or more.
PC Jones is H cm in height.

So $H \geqslant 170$

$H \geqslant 170$ is an inequation.

EXERCISE 3B

Read each story carefully, and then make an inequation.

1

Alan's lorry weighs W tonnes. Its weight must be less than, or equal to, 40 tonnes.

2 Bluebell Rovers' ground can hold up to 20 000 people. The attendance at the cup replay was S spectators.

3 Voters must be 18 or over. Mr Ndabe is Y years old.

4 'Pass mark 50%', says Mrs Alexander. Anna scored T%, and passed.

5 The minimum cost of a meal after 7 pm is £5. The Bertrands' bill came to £B.

6 The speed limit on the motorway is 70 mph. Tom was stopped by the police for speeding. His speed was S mph.

7 'Sorry, madam, the lift is full. It has more than 11 people in it.'
There were P passengers in the lift.

8 A plane can seat 245. P passengers are aboard.

9 Jack is washing a window. The pole is 2 m long, and has an extension 1.8 m long. For this window, the pole and extension come to D m.

10 An indoor clothes drier can stretch to 4 m. The length of line in use is L m.

11 The bag on the left weighs less than the one on the right, so $x < 12$, or $12 > x$.
Write down two inequations for each picture below.

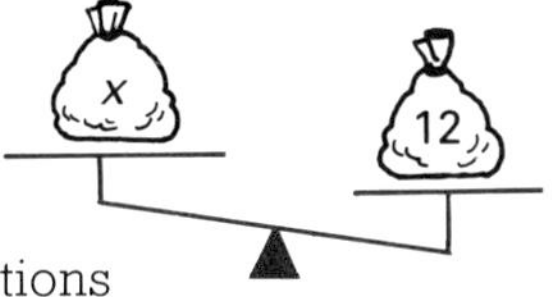

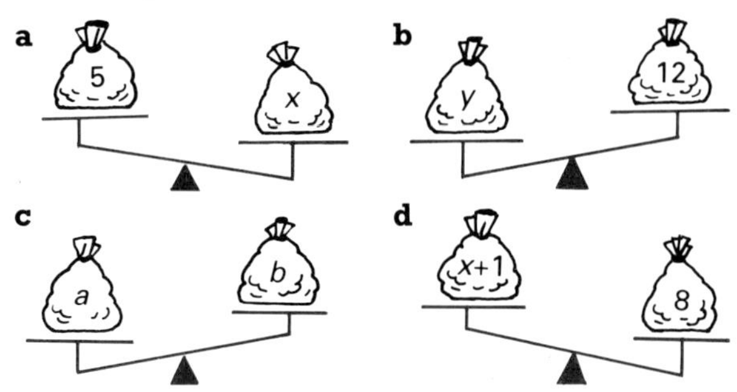

12

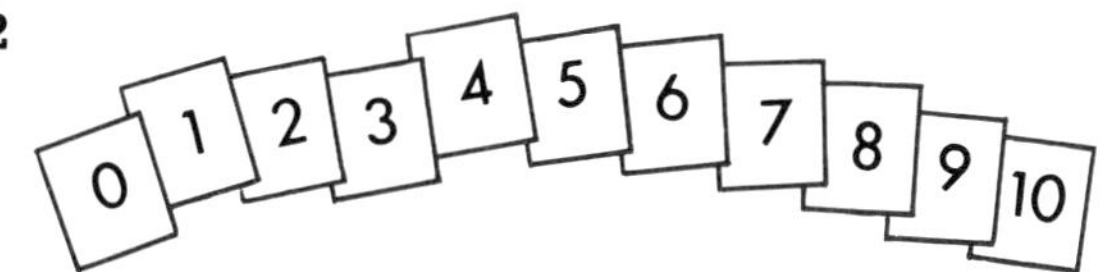

a Which numbers in this collection might n stand for in each inequation below?
(i) $n > 9$ (ii) $n < 4$ (iii) $n \geqslant 7$ (iv) $n \leqslant 1$

b Write an inequation which describes each of these sets of numbers from the collection:
(i) 0, 1, 2, 3, 4, 5 (ii) 8, 9, 10 (iii) 0

EXERCISE 3C

Example
Solve $p+1 \leqslant 3$, choosing replacements for p from $\{0, 1, 2, 3, 4, 5\}$.
Use a table to check each replacement in turn.

p	0	1	2	3	4	5
$p+1$	1	2	3	4	5	6
Is $p+1 \leqslant 3$?	Yes	Yes	Yes	No	No	No

So $p = 0, 1$ or 2

HOW MANY SOLUTIONS? NONE? ONE? MANY?

1 Solve $y+5 > 7$, using replacements from $\{0, 1, 2, 3, 4, 5\}$. Copy and complete this table to help you.

y	0	1	2	3	4	5
$y+5$						
Is $y+5 > 7$?						

2 Make a table to help you to solve $10-y > 8$, for replacements from $\{0, 1, 2, 3, 4, 5\}$.

3 Choosing replacements from $\{0, 1, 2, 3, 4, 5\}$, find solutions for these inequations.

a $x+2 \geqslant 6$ **b** $y+1 > 3$ **c** $5-t < 1$
d $w+5 < 10$ **e** $2x \geqslant 6$ **f** $5m < 5$
g $2x+1 > 9$ **h** $15-2x \leqslant 11$ **i** $x^2 > 10$

4 (i) Choose replacements from $\{-2, -1, 0, 1, 2\}$ to solve these inequations.
(ii) Mark the solutions as dots on a number line.
For example, the solution of $x > 0$ is 1, 2, which can be shown on the number line like this: −1 0 1 2 3

a $x+2 > 1$ **b** $x-3 < -2$ **c** $1-x > 1$
d $-4 < -2+x$ **e** $-1+x > 0$ **f** $-3-x \leqslant -4$

5 Escape-stair regulations state that the length of the landing (L) must be at least equal to the door width (W).
The number of steps (N) between landings should not exceed 16.

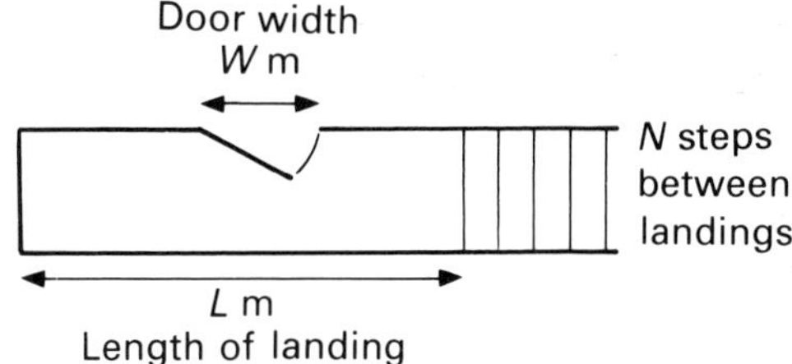

a Write these regulations as inequations.

b The regulations are changed, requiring the landing to be at least 2 m longer than the door width. Write this as an inequation.

c If $W = 1$, and you must choose replacements from $\{2, 3, 4, 5\}$, find solutions for this new inequation to give possible lengths for the landing.

CHECK-UP ON EQUATIONS AND INEQUATIONS

Solve the equations in questions **1** and **2**.

1 a $4x = 20$ **b** $y+5 = 8$ **c** $t-1 = 9$
d $12-x = 6$ **e** $2n+6 = 10$ **f** $14-k = 0$

2 a $2x = x+6$ **b** $3y = 16-y$ **c** $5t+2 = 2t+8$

3 Make an equation for each of these, and solve it.

a

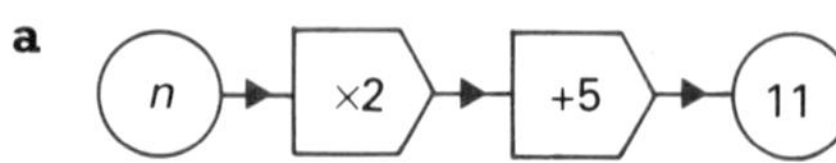

b

4 Solve these equations:
a $4(x+2) = 8$ **b** $2(x-1) = 16$ **c** $5(x-4) = 5$

5 a **b**

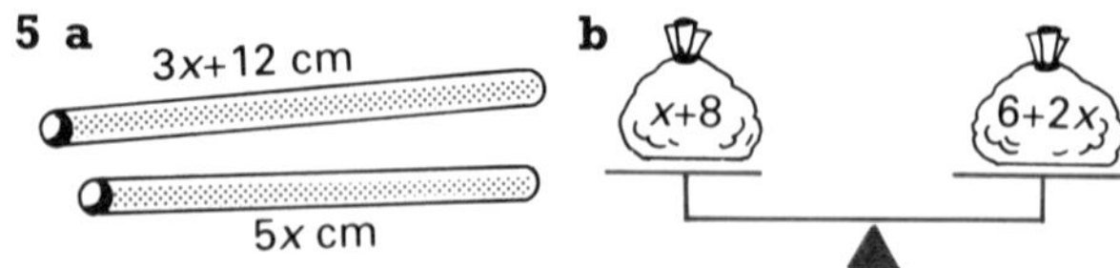

Find the length of the equal straws.

How many weights are in each bag?

6 Make equations for these, and solve them:
a **b**

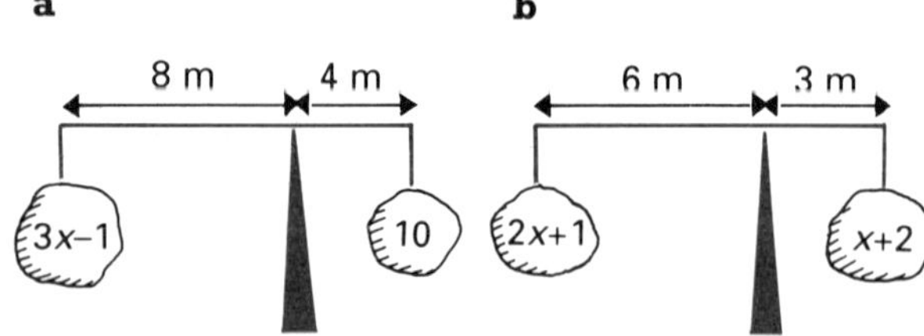

7 a Find the nth terms of these sequences:
(i) 1, 4, 7, 10, . . . (ii) 41, 42, 43, . . .
b If the nth terms are equal, make an equation and solve it. Check that the two nth terms have the same value.

8 a (i) Write down an expression for the area of the rectangle.

(ii) If the area is $2x$ cm^2, make an equation, and find x.
b (i) Find an expression for the perimeter of the rectangle.
(ii) If the perimeter is 50 cm, make an equation, and find x.

9 Write down an inequation for each picture below.

a This sack weighs less than 10 kg.

b

This bottle contains more than 8 litres.

c This straw is at least 20 cm long.

d y kg of sand

The pile of sand weighs less than 3 kg.

10 Choose replacements from {0, 1, 2, 3, 4, 5} and solve:
a $x+1 \geqslant 4$ **b** $x-3 < 1$ **c** $3y \geqslant 6$
d $4-x < 0$ **e** $-3+x \leqslant 0$ **f** $4(x+1) \geqslant 16$

11

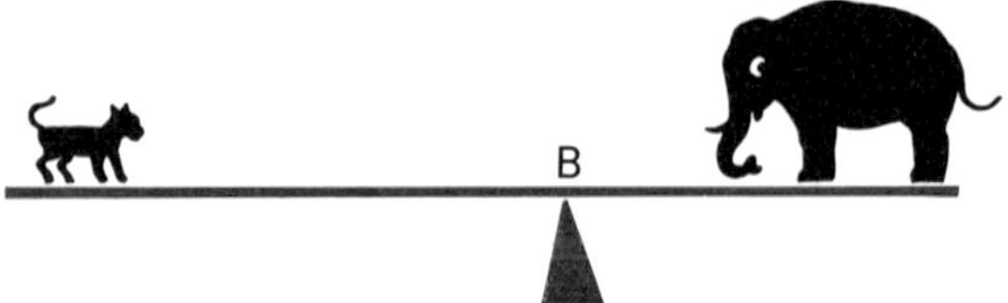

The cat and the elephant are balancing on a seesaw 50 m long. The cat is x m from the point of balance, B. If the cat weighs 1 kg and the elephant weighs 800 kg, find how far each is from B, to the nearest tenth of a metre.

12 $5(x-3)$ metres of rope are used to fence this rectangular enclosure. Find x.

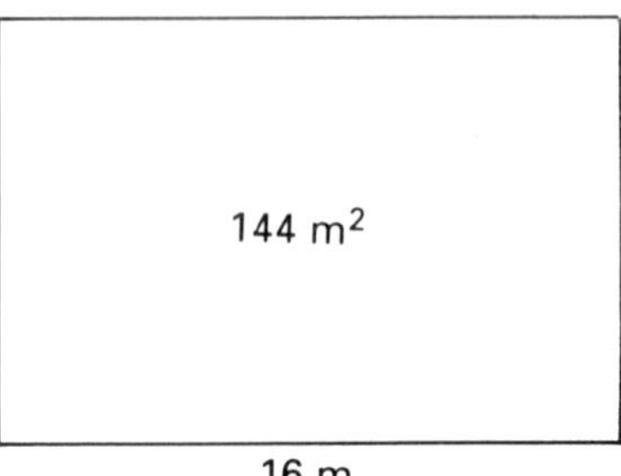

12 RATIO AND PROPORTION

LOOKING BACK

1 What fraction is the shaded part of each whole shape shown below?

a

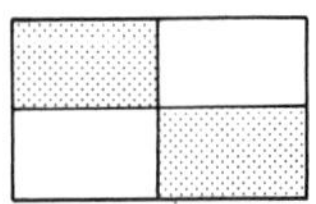

b

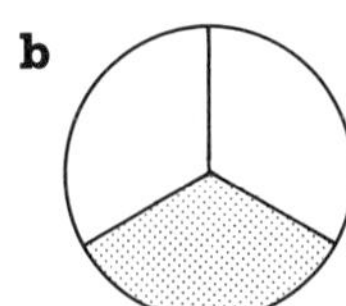

c

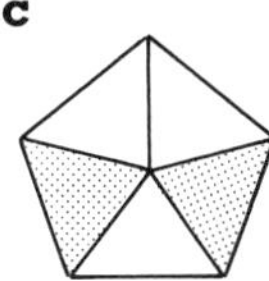

d

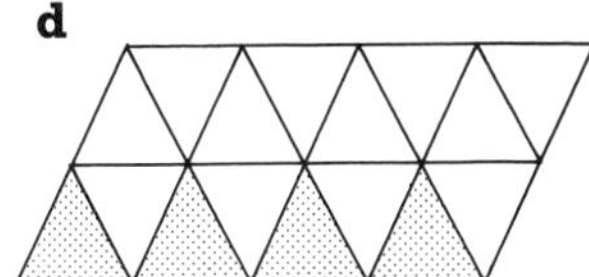

2 What fraction of £1 is:
a 50p **b** 25p **c** 10p **d** 75p?

3 Calculate:
a $\frac{1}{2}$ of £10 **b** $\frac{1}{4}$ of £12 **c** $\frac{1}{3}$ of £6 **d** $\frac{1}{5}$ of £50.

4 How many:
a millimetres are in a centimetre
b centimetres are in a metre
c metres are in a kilometre
d grams are in a kilogram
e millilitres are in a litre?

5 Measure the lengths of these lines.

a ——————————

b ——————

c ————————

d —

6 Karl and Grace weeded the garden. If Grace did three quarters of the work, what is her fair share of the £2 their Dad gives them?

7 Write each fraction in its simplest form.
a $\frac{3}{6}$ **b** $\frac{3}{9}$ **c** $\frac{6}{8}$ **d** $\frac{2}{10}$ **e** $\frac{12}{16}$ **f** $\frac{16}{24}$ **g** $\frac{15}{30}$

8 What fraction of one hour, in its simplest form, is:
a 30 minutes **b** 15 minutes **c** 20 minutes
d 50 minutes?

9 Six books cost £18. What is the cost of:
a 1 book **b** 2 books **c** 10 books
d 25 books?

10 True or false?
a If two coins weigh 50 g, four of the same coins will weigh 100 g.
b The more money you spend, the more you will have left.
c The faster you run, the farther you'll go in one minute.
d The higher the sun in the sky, the longer the shadow cast by a vertical flagpole.

RATIO

Alison and Robert share a bar of chocolate.
The ratio of Alison's share to Robert's share is $\frac{2}{4}$, or $\frac{1}{2}$.

$$\frac{\text{Alison's share}}{\text{Robert's share}} = \frac{1}{2}, \text{ or } 1:2.$$

In calculations, the quantities in the ratio must be in the same units.

Example $\frac{\text{Value of £1}}{\text{Value of 20p}} = \frac{100\text{p}}{20\text{p}} = \frac{100}{20} = \frac{5}{1}$, or 5:1

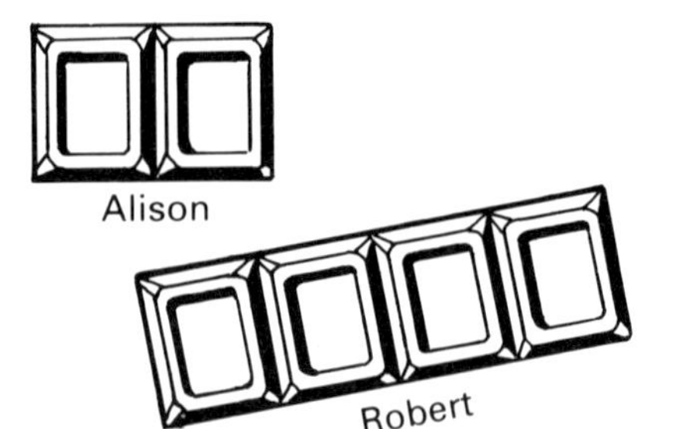

EXERCISE 1A

1 Simplify these fractions. For example, $\frac{6}{8} = \frac{3}{4}$ (dividing 'top' and 'bottom' by 2).
a $\frac{4}{6}$ **b** $\frac{8}{10}$ **c** $\frac{6}{9}$ **d** $\frac{9}{12}$ **e** $\frac{2}{6}$ **f** $\frac{8}{6}$ **g** $\frac{10}{5}$

2

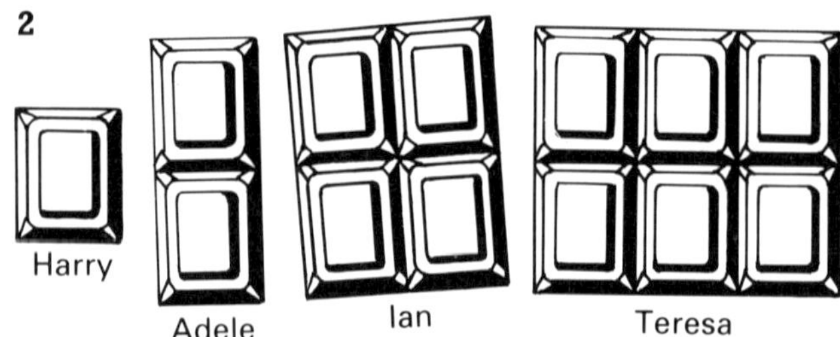

Calculate these ratios of pupils' shares of a bar of chocolate, in simplest form.

a $\frac{\text{Harry's}}{\text{Adele's}}$ **b** $\frac{\text{Adele's}}{\text{Ian's}}$ **c** $\frac{\text{Ian's}}{\text{Teresa's}}$ **d** $\frac{\text{Teresa's}}{\text{Adele's}}$

3 Calculate these ratios of lengths of fish.
a perch's:pike's **b** perch's:tench's
c perch's:salmon's **d** salmon's:pike's
e tench's:pike's **f** pike's:perch's

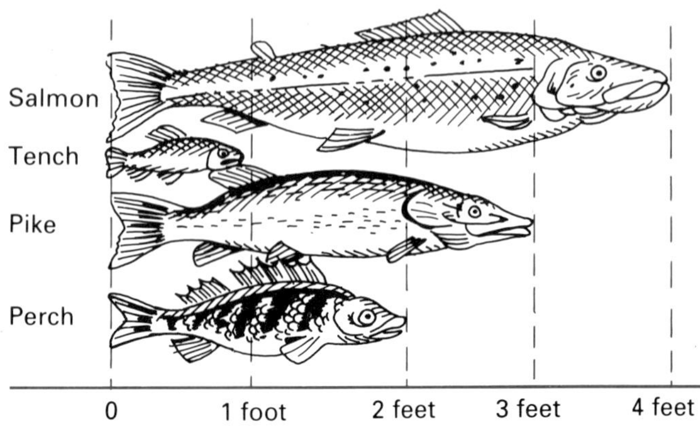

4 The order of the numbers in a ratio is important.
Is $\frac{2}{3} = \frac{3}{2}$?
Which of these could be correct?
a age of father:age of son = 1:3
b height of car:height of bus = 1:4
c population of England:population of Scotland = 10:1
d weight of elephant:weight of mouse = 2:1

5

Calculate these ratios of values of coins, in simplest form.

a $\frac{B}{D}$, **b** $\frac{B}{E}$ **c** $\frac{F}{D}$ **d** $\frac{E}{C}$ **e** $\frac{C}{F}$ **f** $\frac{F}{G}$ **g** $\frac{C}{G}$

EXERCISE 1B

1 Shirts and blouses are often made from a mixture of cotton and polyester. Write these ratios in their simplest form:
a cotton 70% : polyester 30%
b cotton 60% : polyester 40%
c cotton 65% : polyester 35%
d cotton 50% : polyester 50%

2

Pocket radio £10

Radio cassette £30

Stereo radio cassette £75

Write these price ratios in their simplest form:
a pocket radio : radio cassette
b pocket radio : stereo cassette
c stereo cassette : radio cassette
d stereo cassette : pocket radio

3 The ratio of the length of the snake to the length of the beetle is not 1 : 2! The units must be the same. What is the correct ratio?

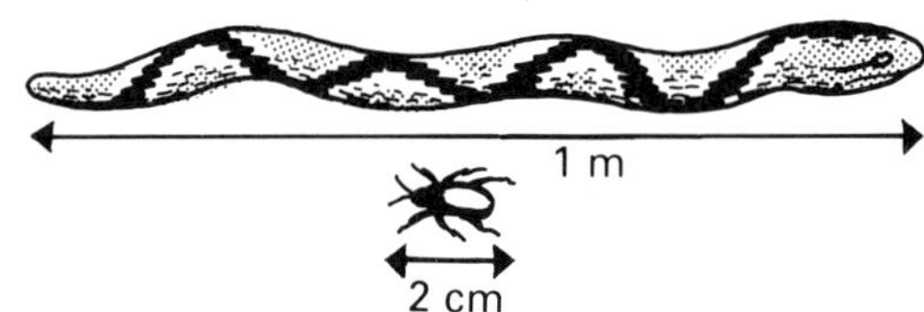

4 In the examples below find the ratios, A:B, of lengths, weights, etc, in simplest form.

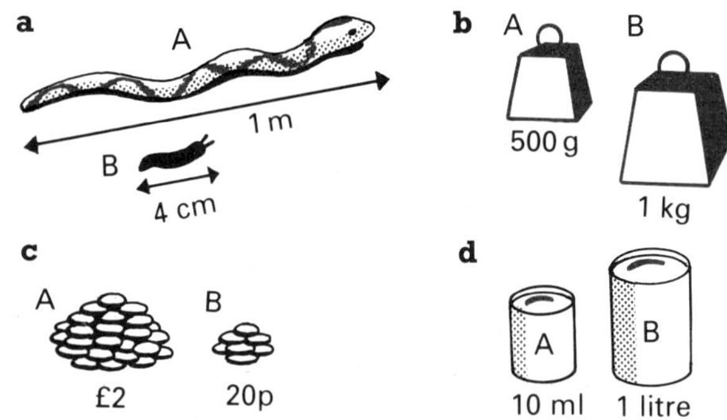

5 Write each of these ratios in its simplest form.
a 1 mm : 1 cm **b** 1 m : 1 km
c 1 kg : 100 g **d** 15 seconds : 1 minute
e 60p : £2 **f** 2 days : 2 weeks
g 1 ml : 1 litre **h** 25 cm : 1 metre

EXERCISE 1C

It is useful to be able to write a ratio as $n:1$, or $1:n$.

For example, $\frac{5}{2} = \frac{5 \div 2}{2 \div 2} = \frac{2.5}{1} = 2.5:1$, or

$\frac{5}{2} = \frac{5 \div 5}{2 \div 5} = \frac{1}{0.4} = 1:0.4$.

1 Express each ratio below in the form $n:1$.
a 6:2 **b** 3:2 **c** 10:5 **d** 700:200

2 Express each ratio in the form $1:n$.
a 3:9 **b** 4:6 **c** 5:6 **d** 10:85

3 The ratio of the width to the height of a normal television screen is 4:3, and the ratio for the newer wide screen is 16:9. Express each ratio in the form $n:1$, with n correct to 2 decimal places.

4

     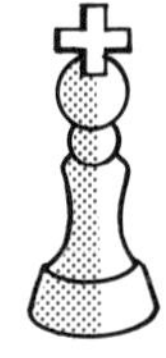

In chess, as a rough guide:
1 bishop is worth 3 pawns,
1 knight is worth 3 pawns,
1 rook is worth 5 pawns,
and 1 queen is worth 9 pawns.
Choose either $1:n$ or $n:1$ to compare the values of:
a 1 knight : 1 queen
b 1 queen : 1 rook
c 1 bishop : 1 rook
d 2 rooks : 3 bishops and 1 pawn.

5 Some approximate populations and areas:
United Kingdom............ 56 million, 240 000 km^2
United States of America 228 million, 9 300 000 km^2
China .... 975 million, 9 500 000 km^2
Use the form $1:m:n$ to compare, correct to 2 decimal places:
a the populations of the three countries
b their areas.
Which is most densely populated (most people per km^2)?

INVESTIGATIONS

1

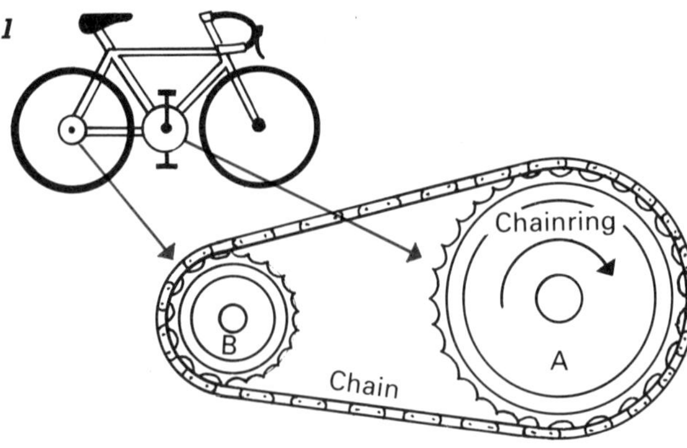

A has 48 teeth and B has 24.
a *Calculate, in simplest form, the ratio of the number of teeth on A to the number on B.*
b *If A makes one full turn clockwise what happens to B, and to the bicycle?*
Two more gears are added at B; C with 20 teeth and D with 16.
c *Calculate the ratios of the number of teeth A:C and A:D.*
d *One full turn of the rear wheel moves the bicycle two metres. Investigate the distances travelled for one turn of the pedals in gears B, C and D.*

2 *Work out the ratios for your own, or a friend's, bicycle.*

PROPORTIONAL DIVISION—FAIR SHARES FOR ALL

How can Brian and Norma share £32 in the ratio 3:5?

3:5 means that there are $3+5 = 8$ shares.
8 shares = £32, so 1 share = £32 ÷ 8 = £4.
Brian takes $3 \times$ £4 = £12.
Norma takes $5 \times$ £4 = £20.
(*Check* £12 + £20 = £32.)

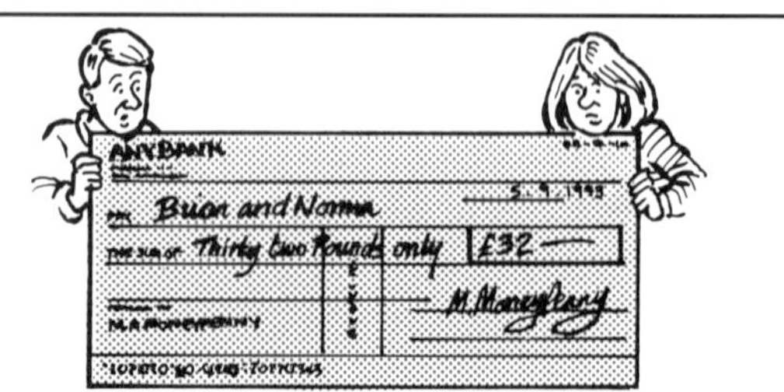

EXERCISE 2A

1 Ravi and Sonia share £48 in the ratio 5:7. Copy and complete this calculation to find how much each receives.
Number of shares = 5 + 7 = . . .
12 shares = £48, so 1 share = £ . . .
Ravi gets $5 \times$ £4 = £ . . .
Sonia gets $7 \times$ £4 = £ . . .
(*Check* £20 + £28 = £ . . .)

2 Nick and Emma share £12 in the ratio 1:2.
a How many shares?
b What is each worth?
c How much does each of them get?

3 Tamsin and Hugh share £36 in the ratio 5:4. How much does each get?

4 Roger and Ranjith share £100 in the ratio 2:3. How much does each get?

5 Divide:
a 20 cm of wire in the ratio 1:3
b 5 litres of juice in the ratio 2:3
c 24 kg of sand in the ratio 5:1
d 1 hour in the ratio 7:5.

6 a

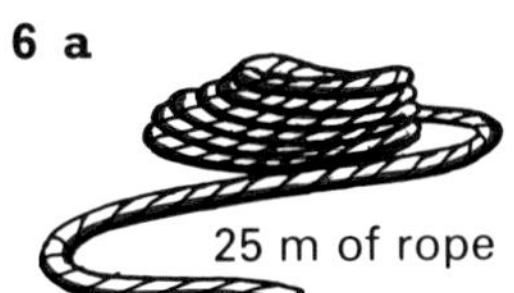

Coiled length : uncoiled = 3:2.
What length is coiled?

b

Length of program : adverts = 10:3.
How long are the adverts?

c

Weight of Brazilian beans : Zambian beans = 3:4.
What weight is Brazilian?

7 How would you share out this bar of chocolate in the ratio:
a 2:1 **b** 3:1
c 1:5 **d** 1:1?

8 A coin weighs 12 g. It contains three parts copper and one part nickel.
What weight of each metal was used for it?

9 Kim and Curtis are delivering leaflets. Kim hands out three bundles, and Curtis two. How should they share the £20 they are paid?

10 Sally makes 10 litres of orange drink for a party, using four parts of water to one of orange juice. What volume of orange juice does she need?

11 How should the angle of 360° at the centre of the pie chart be divided to show areas of the circle in the ratio:
a 3:1
b 5:1
c 7:2
d 7:3
e 17:1?

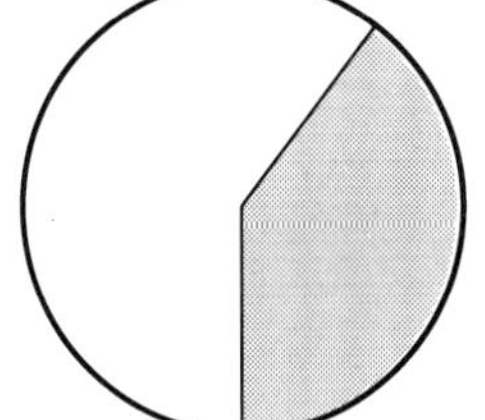

EXERCISE 2B

Sometimes it is useful to multiply the 'top' and 'bottom' of a ratio.

Examples

a $3:4 = \frac{3}{4} = \frac{3 \times 2}{4 \times 2} = \frac{6}{8}$, or 6:8

b Graham and Helen share a paper round.
They intend to share their wages in the ratio 3:2.
(i) How much will Helen get if Graham takes £15?

$\frac{\text{Graham's share}}{\text{Helen's share}} = \frac{3}{2} = \frac{3 \times 5}{2 \times 5} = \frac{15}{10}$. So Helen got £10.

(ii) How much did Graham take when Helen got £8?

$\frac{\text{Graham's share}}{\text{Helen's share}} = \frac{3}{2} = \frac{3 \times 4}{2 \times 4} = \frac{12}{8}$. So Graham got £12.

1 Find the missing numbers in these pairs of equal ratios.
a 3:2 = 9:□ **b** 1:4 = □:12
c 7:3 = 21:□ **d** 5:2 = 25:□
e 8:3 = □:21 **f** 3:5 = □:20

2 a Two numbers are in the ratio 3:4. If the smaller one is 9, what is the larger one?
b This time the ratio is 7:9, and the larger number is 54. What is the smaller number?

3

	Class	Number of boys:girls	Pupils in class	Question
a	2A	3:4	6 boys	How many girls?
b	2B	1:2	20 girls	How many boys?
c	2C	2:1	30 pupils	How many of each?

4 In Exercise 1C, question **3**, we told you that the ratio of the width to the height of the usual television screen is 4:3, and the ratio for the newer wide screen is 16:9.
The height of a screen of each type is 36 cm. Calculate:
a the width of each type of screen
b the ratio of the area of the larger screen to the area of the smaller one.

5

A basket of fruit contains apples, oranges and bananas in the ratio 3:2:5. (From this, the number of apples:oranges is 3:2, and oranges:bananas 2:5.)
a If there are 8 oranges, how many:
(i) apples (ii) bananas, are there?
b If there are 10 bananas, how many apples and oranges are there?
c If there are 30 pieces of fruit altogether, how many of each type are there?

EXERCISE 2C

1 Mrs Jones, Mr White, Miss Davis and Mr Baker have formed a pools syndicate. Each week they pay these amounts: Mrs Jones £3, Mr White £4, Miss Davis £2, Mr Baker £1.
On Saturday they hit the jack-pot, £1 million! How much will each person collect?

2 Gavin is making apple crumble, using butter, sugar and flour in the ratio 1:1:2.
a What weight of each would he need for 360 g of crumble?
b If he has 140 g butter, 130 g sugar and 275 g flour, what is the biggest weight of crumble that he can make?

3 The 'pitch' of a roof is the ratio height:span, usually given in the form $1:n$.
a Calculate the pitch of each of these roofs:
(i) height 4 m, span 28 m
(ii) height 3 m, span 42 m
(iii) height 5 m, span 12 m.

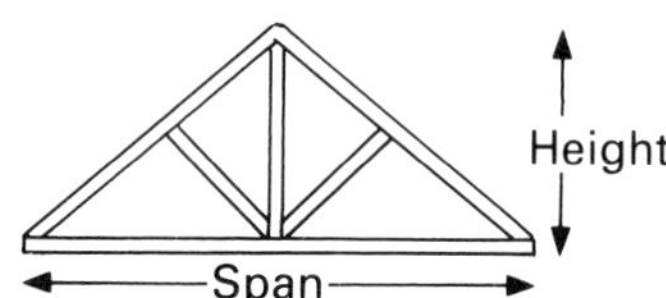

b A pitched roof has to be put on a flat roof to stop the rain leaks. The roof span is 36 m. Calculate the height of roof required for a pitch of:
(i) 1:6 (ii) 1:4 (iii) 1:4.5.

4 Rainbow D.I.Y. has a machine which lets you mix your own colour of paint.

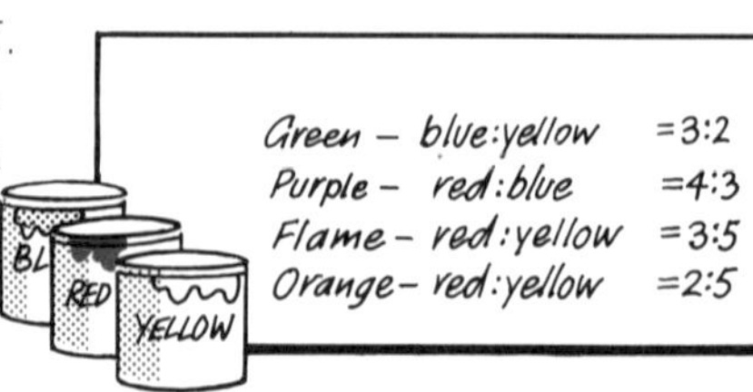

a Aileen mixed red and blue to get purple. She used 200 ml of red paint.
(i) How much blue did she need?
(ii) How much purple did she get?
b Leroy has 600 ml of red paint. How much yellow does he need to make:
(i) orange paint (ii) flame colour?
c Claire made 600 ml of green paint. How much blue and yellow paint did she need?
d What is the greatest volume of green paint that can be made if you have 100 ml blue and 100 ml yellow?

INVESTIGATION

a *The Fibonacci sequence is 1, 1, 2, 3, 5, 8, 13, . . . Calculate the following ratios:
2nd term:1st term, 3rd term:2nd term, . . . until you are able to estimate the 'final' value, as far as your calculator allows.*
b *This value also represents length:breadth of the sides of the 'golden rectangle' described in Chapter 13 on Rectangles and Squares in Book 1. Draw the rectangle as accurately as you can.*

SCALE—A SPECIAL RATIO

As we saw in Chapter 6, a scale is used for making drawings or models of distances or objects that are too large or too small to copy exactly.

The scale can be given by the ratio $\frac{\text{drawing length}}{\text{actual length}}$

Examples

a The classroom floor is 10 m by 8 m.
Scale 1 cm : 1 m. So 1 cm on the drawing represents 1 m on the floor.
Scale drawing 10 cm by 8 cm.

b A map of Britain
Scale 1 : 100 000
Map 1 cm on the map represents 100 000 cm, or 1000 m, or 1 km on the ground.

EXERCISE 3

1 This kitchen unit is drawn to the scale 1 : 20.
Measure each marked length, and calculate the actual lengths on the unit.

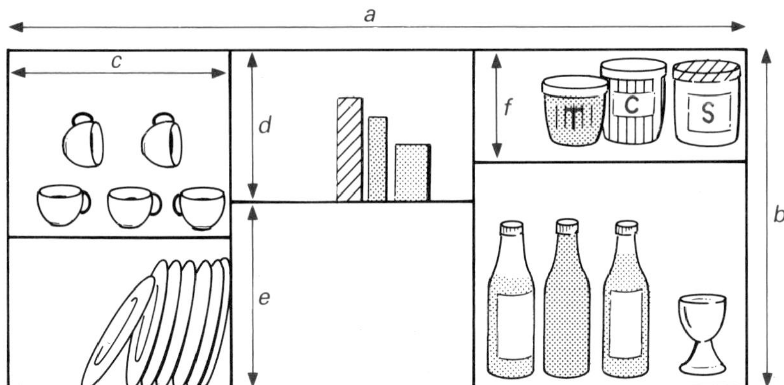

2 This plan of the Simpson's sitting room has a scale of 1 cm : 1 m.

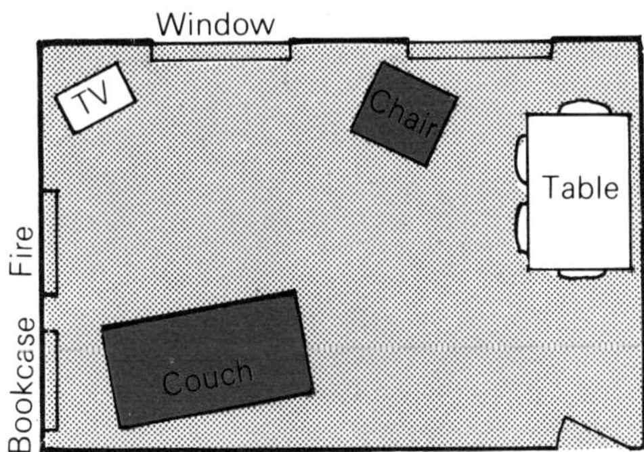

a Measure the length and breadth of the room in centimetres.
Then write down its actual size in metres.

b Repeat this for: (i) the table (ii) the couch.

3

Tony is making a model car to the scale 1 : 20.

a His model is 8 cm wide. How wide is the real car?

b The real car is 300 cm long. How long is his model?

4 The scale of a model of Concorde is 1 : 200.

a The model is 30 cm long. What is the actual length of Concorde?

b The wing span of Concorde is 26 m. What is the wing span of the model, in cm?

5 The scale of this road sketch is 1 cm : 1 km.

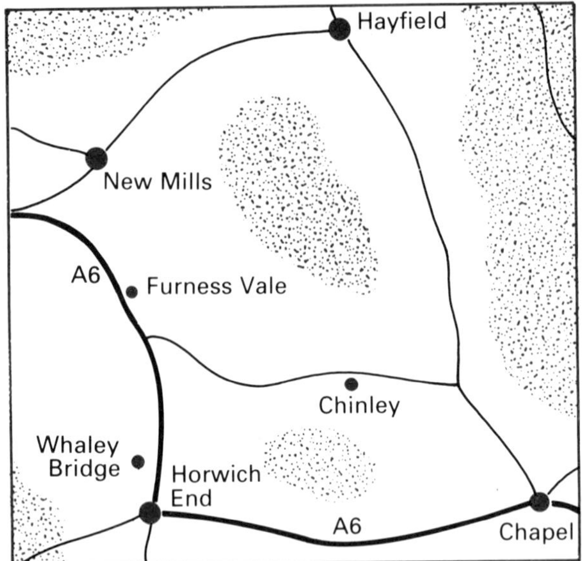

a How far is it roughly, by road, from:
(i) New Mills to Hayfield
(ii) Hayfield to Chapel
(iii) New Mills to Chapel along the A6?

b It is 14 km from New Mills to Stockport. What length would this distance be on the sketch?

6 Michelle's Dad is going to make her a doll's house in the same shape as their own house, which is 15 m long, 12 m broad and 9 m high.

a What scale should he use to make the doll's house 50 cm long?

b What would its breadth and height then be?

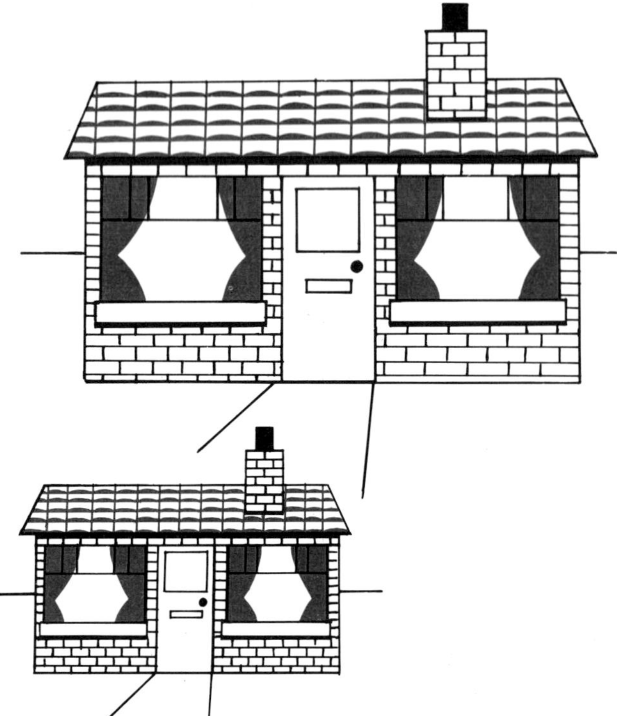

PRACTICAL PROJECTS

The drawing of the house has been enlarged by increasing the length of each side in the ratio 2 : 1. The scale of the enlargement is 2 : 1 (or × 2).

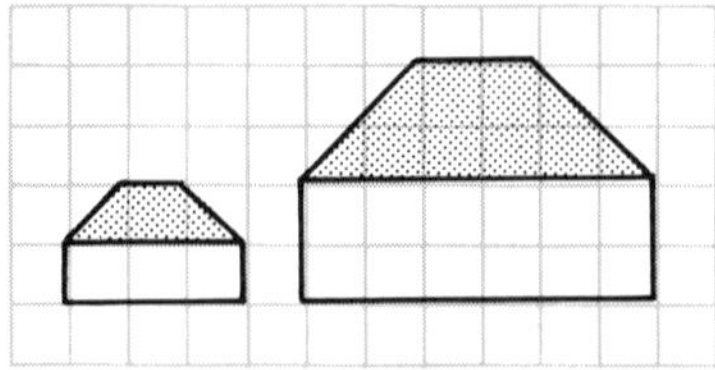

1 *On squared paper draw diagrams which:*

a *increase square A in the ratio 2 : 1 (scale × 2)*

b *enlarge rectangle B in the ratio 3 : 1 (scale × 3)*

c *reduce triangle C in the ratio 1 : 2 (scale × $\frac{1}{2}$).*

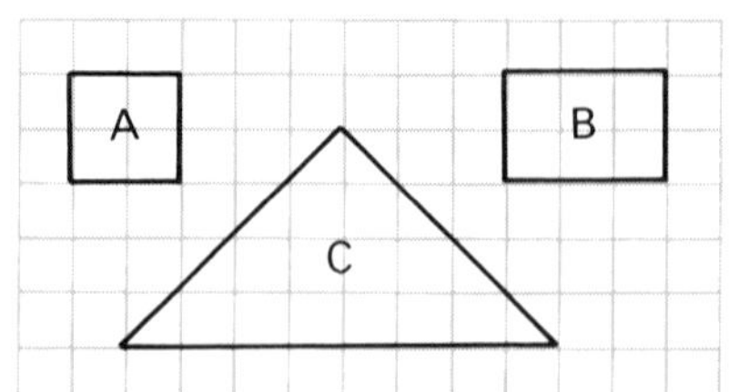

2 *Draw and colour the flag on squared paper, with all the lines twice as long as they are in this diagram (scale × 2).*

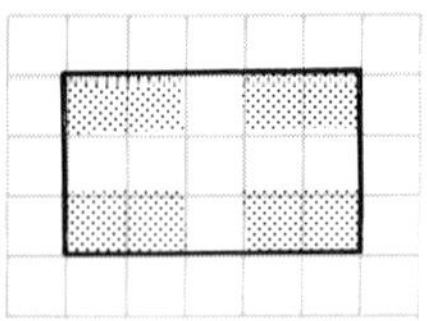

3 *Copy this guitar on cm squared paper to make a 3 : 1 enlargement of the diagram.*

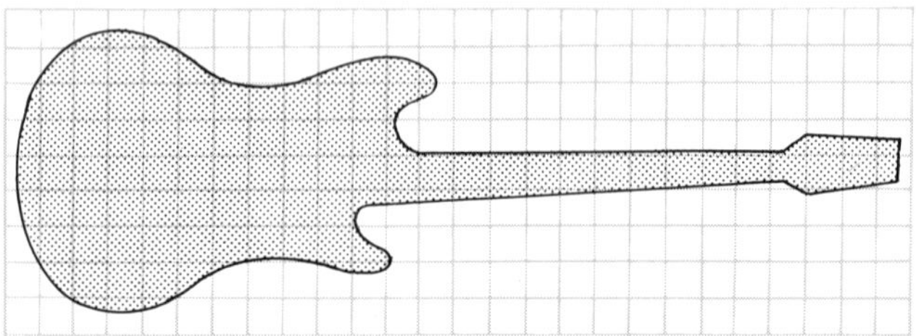

4 *Make a scale drawing of your classroom floor—remember to choose a good scale.*

DIRECT PROPORTION

Number of yoghurts	*Cost*
1 ⟷	30p (⟷ 'corresponds to')
2 ⟷	60p
3 ⟷	90p
4 ⟷	120p
5 ⟷	150p
6 ⟷	180p

Doubling the number of yoghurts **doubles** the cost. **Halving** the number of yoghurts **halves** the cost.
The number of yoghurts and their cost increase, or decrease, in the same ratio.
The cost is **directly proportional** to the number of yoghurts.

EXERCISE 4A

1 Copy and complete these tables. The cost is always directly proportional to the number of items of fruit.

a

Number of apples	*Cost*
1	15p
2	30p
3	45p
4	......
5	......

b

Number of bananas	*Cost*
1	20p
2	40p
3	......
4	......
5	......

2 The quantities in the tables below are in direct proportion. List the entries in the second row of each table.

a *The cost of diaries at 75p each*

Number of diaries	1	2	3	4	5
Cost (p)	75				

b *The number of calories in a helping of cereal*

Weight (grams)	10	20	30	40	50
Number of calories	30				

c *TV rental at £8 a month*

Number of months	3	6	9	12	15
Cost (£)					

d *Distance cycled at 15 km/h*

Number of hours	1	2	3	4	5
Distance (km)					

Assume that all these questions use direct proportion.

3 Dave walks 12 km in 3 hours. At the same speed, how far would he walk in:
a 1 hour **b** 2 hours **c** 5 hours?

4 Amy cycles 80 km in 4 hours. At the same speed, how far would she cycle in:
a 1 hour **b** 2 hours **c** 3 hours?

5 Phyllis knits 300 stitches in 10 minutes. How many would she knit in:
a 1 minute **b** 5 minutes **c** 15 minutes?

6 200 g of chips contain 600 calories. How many calories are there in:
a 1 g **b** 100 g **c** 800 g?

7 12 records need 3 cm of shelf space. How many records can be stored in:
a 1 cm **b** 8 cm **c** 15 cm?

8 50 bricks make a wall 0.5 m long. How many bricks are needed for a wall of length:
a 1 m **b** 4 m **c** 7.5 m?

9 6 m of rope cost £13.50. Find the cost of:
a 1 m **b** 5 m **c** 9 m.

10 Glynis pays £2.82 for 3 m of ribbon. How much would she pay for:
a 1 m **b** 2 m **c** 7 m?

11 Mr Amin earns £36 for 5 hours work.
a How much does he earn in:
(i) 1 hour (ii) 7 hours (iii) 12 hours?
b He works for 5 hours on Monday and 9 hours on Tuesday. How much more does he earn on Tuesday than on Monday?

EXERCISE 4B/C

For calculations with direct proportion, find the cost, weight, distance, etc., for 1 item.

Example
I travelled 18 km in 15 minutes. How far would I go in 20 minutes at the same speed?

$$15\text{ min} \longleftrightarrow 18\text{ km} \quad (15\text{ min corresponds to } 18\text{ km})$$
$$1\text{ min} \longleftrightarrow \tfrac{18}{15}\text{ km}$$
$$20\text{ min} \longleftrightarrow \tfrac{18}{15} \times 20 = 24\text{ km}$$

So I would travel 24 km in 20 minutes.
(*Common sense check:* more time, so greater distance.)

1 5 newspapers cost £2. Find the cost of 9 of them.

2 500 pencils cost £44. How much would 750 cost?

3 15 cassette tapes cost £27. How much would 10 cost?

4 Jenny can knit 300 stitches in 12 minutes. How many would she knit in 20 minutes at the same rate?

5 9 dollars can be bought for £5. How many dollars can be bought for £12 at the same rate?

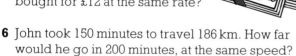

6 John took 150 minutes to travel 186 km. How far would he go in 200 minutes, at the same speed?

7 Mr Jones earns £112.68 in $7\frac{1}{2}$ hours. What would he earn in $37\frac{1}{2}$ hours at the same rate?

8 30 m of carpet cost £750. How much would 24 m cost?

9 Prices are not always directly proportional to the number of items bought. Which of these *are* in direct proportion?
a 2 chocolate biscuits for 36p, or 5 for 90p. (Calculate the cost of 1 in each case.)
b 2 comics for £1, or 4 for £2.
c 2 records for £6, or 4 for £10.
d 1 video tape for £6, or 10 for £50.
Why are prices not always in direct proportion?

10 The length of the shadow cast by an object is directly proportional to its height. At 10 am a man 6 feet tall casts a shadow 2 feet long.

a How long is the shadow of a house 30 feet high?
b What is the height of a tree which casts a shadow 17 feet long?
c At 11 am, the man's shadow is 1.5 feet long. Calculate the length now of the shadow of:
(i) the house (ii) the tree.

11 The cost of heating a room is directly proportional to its volume. A room which is 3 m by 4 m by 3 m costs 12p an hour to heat.
a Calculate the cost per hour of heating a room:
(i) 6 m by 5 m by 2 m (ii) 12 m by 6 m by 2 m.
b Find:
(i) the volume of a room which costs 32p per hour to heat
(ii) the height of a room with floor area 66 m^2 which costs 44p per hour to heat.
c A hall 13 m by 5 m by 2.4 m is hired for 6 hours. How much will the heating cost?

GRAPHS OF QUANTITIES IN DIRECT PROPORTION

This table shows the cost of yoghurts given on page 145.

Number of yoghurts	1	2	3	4	5	6
Cost (p)	30	60	90	120	150	180

Look at the graph.

The points lie on a straight line which passes through the origin.

This is always true for two quantities which are in **direct proportion**.

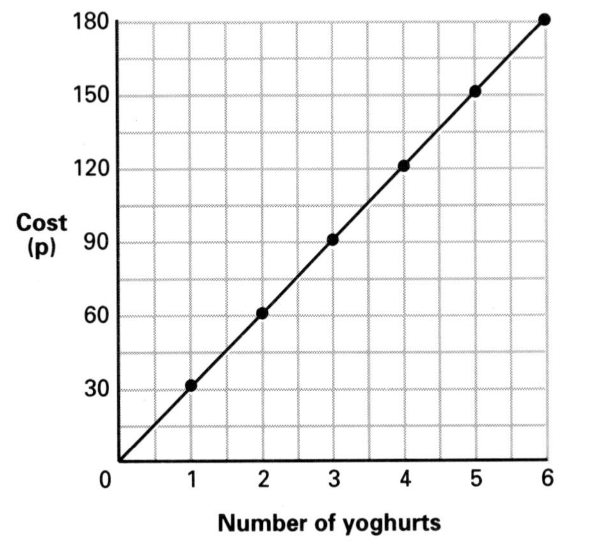

EXERCISE 5

1 a Copy and complete this table.

Number of oranges	1	2	3	4	5	6
Cost (p)	20	40				

b Using the same scales as in the graph above, plot the points.

c Draw a line through the points. Does the line pass through the origin?

2 Repeat question **1** for this table.

Number of Chews	1	2	3	4	5	6
Cost (p)	10	20				

3 a Copy and complete this table for a car travelling at 40 km/h.

Time (hours)	1	2	3	4
Distance (km)	40	80		

b Using scales of 2 squares to represent 1 hour on the horizontal axis, and 40 km/h on the vertical axis, plot these points and draw a line through them.

c Use your graph to find the distance travelled in 6 hours.

4 a Copy and complete this table for annual interest on money in a bank.

Money in bank (£)	0	100	200	300	400
Interest (£)	0	10	20		

b Using scales of 2 squares to represent £10 interest on the horizontal axis, and £100 in the bank on the vertical axis, plot these points and draw a line through them.

c Use your graph to find the interest on: (i) £500 (ii) £150.

5 a Using scales of 1 square to represent 14 cents on the horizontal axis, and 10p on the vertical axis, plot these two points.

Number of cents	0	180
Number of pence	0	100

b Join the points, to make a conversion graph between cents and pence.

c How many cents equal: (i) £1 (ii) 50p?

d How many pence equal 45 cents?

INVERSE PROPORTION

Sarah has become a teenager, and her parents have given her £60 to spend on a party for her friends.
The amount she can afford to spend on each friend depends on the number of people she invites to the party.

Number invited	1	2	3	4	5	6
Amount spent on each (£)	60	30	20	15	12	10

Can you see that:
(i) as the number of guests **increases**, the amount spent on each **decreases**?
Doubling one **halves** the other.
(ii) the number invited × the number of £s spent on each = 60 each time?
The amount spent per person is **inversely proportional** to the number invited.

EXERCISE 6A

1 The quantities in these tables are in inverse proportion. List the entries in the second row of each table.

a *Spending £50 on a party*

Number invited	1	2	5	10	25
Amount each (£)	50				

b *Sharing a box of 72 sweets*

Number of people	1	2	3	4	6
Number of sweets each	72				

c *Buying £480 worth of books*

Cost of a book (£)	5	6	8	10	12
Number bought	96				

d *A rectangle with area 36 cm²*

Length (cm)	2	4	6	9	12
Breadth (cm)	18				

e *Driving 150 km on holiday*

Speed (km/h)	75	50	30	25	20
Time (hours)	2				

f *Putting leaflets in envelopes*

Number of helpers	1	2	3	4	5
Time taken (hours)	6				

2

Lucy takes 1 hour to cycle to town at 10 km/h. How long would she take at 20 km/h. (Think! Faster speed, so more or less time? So . . .)

3 Alan and Roy can each paint doors in 4 hours. How long would it take if they painted a door together? (Think! More help, so more or less time? So . . .)

4 A video tape rewound in 4 minutes at 6 turns per second. How long would it take at 3 turns per second?

5 Hannah's parents gave her some money for her birthday. She calculated that she could buy 8 magazines costing £1.50 each. How many could she buy at:
a £3 each **b** 75p each **c** £1 each?

EXERCISE 6B/C

Again, for calculations with inverse proportion, find the time, cost, etc., for 1 item.

Example
25 pupils in a class take 24 minutes to put letters about the school concert into envelopes.
How long would 30 pupils take?

25 pupils ⟷ 24 minutes
1 pupil ⟷ $25 \times 24 = 600$ minutes
30 pupils ⟷ $600 \div 30 = 20$ minutes.

So 30 pupils would take 20 minutes.
(*Common-sense check*: more pupils would take less time.)

Good advice: use a common-sense check for your answers.

1 4 pupils take 20 minutes to set out chairs for Assembly. How long should:
a 1 pupil take **b** 5 pupils take?

2 At 6 km/h, Ian takes 10 minutes to walk to school. How long would he take at:
a 1 km/h **b** 5 km/h **c** 20 km/h?

3 Strathglen High School has a 6 period day, with each period 60 minutes long.
For the same length of school day, how long would each period be if the day had:
a 1 period **b** 5 periods **c** 8 periods?

4 A contractor estimates that 3 men could rewire the Sampson's house in 4 days. To please Mr Sampson he puts 4 men on the job.
How long should they take?

5 At 300 words a minute, John took 6 hours to read a detective story. How long would he take at 400 words a minute?

6 A car takes 3 hours for a journey at 80 km/h. What average speed would allow it to make the journey in $2\frac{1}{2}$ hours?

7 Clearview Double Glazing said that 2 of their men could fit new windows in a house in 3 days.

a How many men would be needed to fit the windows in 1 day?
b In the end 3 men did the work. How long did they take?

8 The electrical resistance in a circuit is inversely proportional to the current.
A fridge has a resistance of 120 ohms, and needs a current of 2 amps.
a What is the resistance of a toaster which needs a 4 amp current?
b What current does a kettle with resistance 30 ohms need?

9 A farmer reckons his field has enough grass to feed 48 cattle for 25 days.
a For how long would it feed 40 cattle?
b The field was let for 50 days. How many cattle could be grazed on it?

10 The population of the United Kingdom is about 56 million, giving a density of about 240 people to 1 km^2. Calculate the density if the population fell by 7 million.

EXERCISE 7 (MAKING SURE)

1 Which of these illustrate direct proportion (for example, 'double one quantity, double the other'), inverse proportion ('double one, halve the other'), or neither?

a The speed of travel, and the distance covered in a fixed time.
b The number of workers, and the time taken for a task.
c The age of a person, and his or her height.
d The speed of travel, and the time taken for a journey.
e The number of employees with the same wage, and the total wage bill.
f The length of the side of a square, and:
(i) the perimeter (ii) the area of the square.

Remember the common-sense check!

2 12 pairs of shoes cost £300. Find the cost of 28 pairs.

3 Isobel takes 24 minutes to cycle to a youth club at 15 km/h. How long would she take at 18 km/h?

4 A store has enough cattle feed for 10 cattle for 21 days. How long would the feed last for:
a 7 cattle **b** 35 cattle?

5 35 books cost £243.25. How much money would be saved by buying only 20 books?

6 The secretary can buy 144 pens at 5p each. How many *more* could she buy if they were 3p each?

7

5 men could complete a task in 18 days.
a How many *more* men would be needed to do it in 3 days?
b How long would 2 men take?

8 7 metres of fence cost £131.04. How much *more* would 9 metres cost?

9 A roll of tape can seal 400 panes of glass 20 cm wide. How many panes 50 cm wide could it seal?

10 Weights are placed on planks of different lengths to find the maximum load they can take before they break.

Length (m)	2	3	4	5
Breaking load (kg)	7.5	5	3.75	3

a Is the length directly, or inversely, proportional to the breaking load?
b What load could an 8 m plank take?

A PUZZLE

Only one of these questions has a sensible answer. Which question, and what answer?

1 *On sports day, 5 boys entered the 100 m race. How many entered the 500 m race?*

2 *One man can clean a telephone box in 4 minutes. How long would 8 men take?*

3 1 *costs 10p.* 12 *cost 20p.*

What is the cost of 127*?*

INVESTIGATION

Take an A4 sheet of paper, and fold it like this:

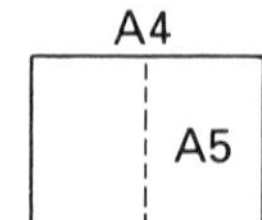

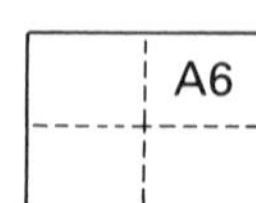

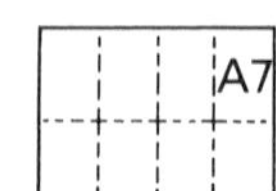

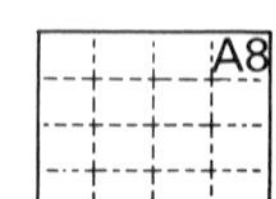

Copy and complete this table, giving the ratios in the form n:1, correct to 1 decimal place.

Paper size	A4	A5	A6	A7	A8
Length (cm)					
Breadth (cm)					
Length : Breadth					

What do you find? Compare your answer with the ratio of the diagonal to the side of a square. Investigate the length:breadth ratio for envelopes, newspapers, books, magazines, etc.

CHECK-UP ON RATIO AND PROPORTION

1 Ben has 5p, Tony has 15p and Kerry has 25p. Write these ratios of money in their simplest form.
a Ben's:Tony's **b** Tony's:Kerry's
c Kerry's:Ben's

2 Share 60p in the ratio:
a 2:1 **b** 3:2 **c** 1:5.

3 Two numbers are in the ratio 2:7. The larger one is 21. What is the other number?

4 Green paint is made by mixing blue paint and yellow paint in the ratio 3:1. How many litres of blue, and how many of yellow are needed for 12 litres of green paint?

5 By using the scale 1:5 make a scale drawing of the outside of this frame.

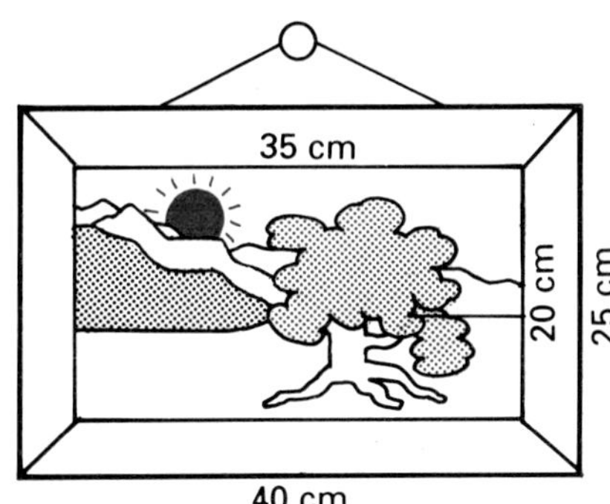

a Use your drawing to find the length of a diagonal of the frame.
b The inside frame is centrally placed. Add it to your drawing.

6 A model of a coach is made with a scale of 1:40.
a If the model is 6 cm wide, how wide is the coach?
b If the coach is 800 cm long, how long should the model be?

7 Which of these three tables have quantities in:
a direct proportion **b** inverse proportion?

(i)

x	1	2	3	4	5
y	240	120	80	60	48

(ii)

x	1	2	3	4	5
y	7	14	21	27	34

(iii)

x	1	2	3	4	5
y	1.5	3	4.5	6	7.5

Give reasons for your choices.

8 A 25 g portion of cheese spread contains 80 calories. How many calories does a 15 g portion contain?

9 A car journey takes 30 minutes at an average speed of 50 km/h.
If the speed is increased to 60 km/h, how long does the journey take?

10 An electric fire costs 15p per hour to run.
a Make a table of costs for 1, 2, 3, . . . 6 hours.
b Why would you expect the graph drawn by joining the points to be a straight line?
c Draw a graph of cost against time. Why does it pass through the origin?

MAKING SENSE OF STATISTICS 2

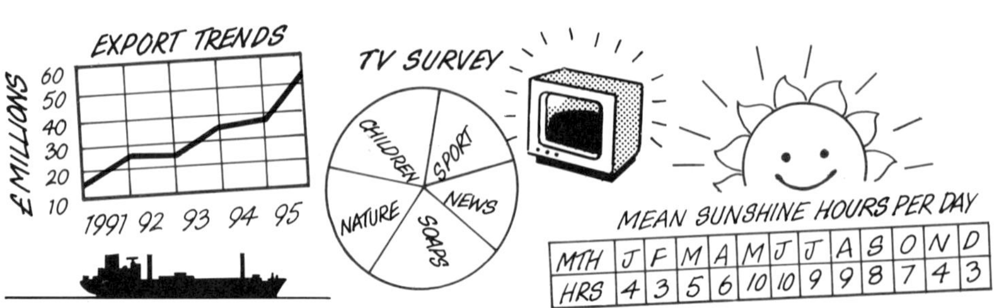

LOOKING BACK

1 Simon does a paper round and keeps a note of the magazines he delivers each day.

	Mon	Tue	Wed	Thu	Fri	Sat	Sun
Weekly	2	3	5	4	5	2	0
Fortnightly	2	1	3	3	1	2	1
Monthly	4	6	7	2	6	2	1

a On which day is his load:
(i) heaviest (ii) lightest?

b Illustrate the data by drawing three line graphs on the same diagram.

2 This table shows the arrival times of the 9A bus each morning in June.

8 43	8 51	8 56	8 53	8 55	9 00
9 04	8 55	9 03	8 57	8 48	8 54
8 54	8 49	8 58	8 46	8 53	8 56
8 57	9 03	8 50	8 58	8 59	8 52
8 51	8 58	9 02	8 46	8 57	8 56

a Organise the data in suitable groups of times, and draw a bar graph.

b Estimate the most likely time to catch the bus.

3 A company who make climbing rope test samples to find out its strength.
The table shows the breaking strengths, in kg, of 30 samples.

1840	1848	1943	1937	1963	1984
2041	1971	1867	1942	2035	2098
1992	2049	1957	1894	2006	1995
2010	1899	1990	1925	1879	2025
1997	2057	1919	1931	1984	2006

a Collect the data into groups:
1800 kg–1849 kg, 1850 kg–1899 kg, etc.

b Draw a bar graph, and write down the modal range of the breaking strengths.

4 Calculate the mean, mode, median and range of each list below.

a 1, 1, 1, 1, 2, 2, 3, 3, 4, 4, 4, 4, 4, 5, 5, 6, 6, 7, 8, 9

b 12, 12, 14, 14, 15, 16, 16, 16, 17, 18, 20, 20, 24, 38

c 1.5, 1.7, 1.9, 2.3, 2.7, 2.9, 4.7, 4.8, 4.9, 5.0

d £1.23, £1.45, £2.78, £2.79, £2.93, £3.58, £5.00, £10.00

5 How many marks above or below the mean has Fred scored in each subject?

																	Fred
Maths	12	15	16	27	28	35	46	57	58	58	59	60	72	75	90	94	59
English	22	26	29	45	47	62	72	79	89	90	90	91	93	94	97	99	62
French	22	23	23	26	35	37	38	48	49	49	56	58	59	63	65	72	26
Geography	23	24	46	53	54	57	57	58	58	61	63	69	70	73	74	76	69
Science	12	13	14	15	19	19	21	26	29	30	34	36	40	45	65	68	40

FREQUENCY TABLES

Gareth took a survey of the number of children in the families in his block of flats. He recorded his data in a **frequency table**, like this:

Number of children in family	Tally	Frequency	Number of children × frequency
0	~~IIII~~ I	6	0
1	~~IIII~~ IIII	9	9
2	IIII	4	8
3	II	2	6
4	II	2	8
5	I	1	5
	Total	24	36

His table helped him to calculate the mean number of children per family.

$$\textbf{Mean} = \frac{\text{total no. of children}}{\text{frequency}} = \frac{36}{24} = 1.5$$

Also: **mode** = 1 (the most common frequency)
median = 1 (the 12th or 13th number in the frequency column).

EXERCISE 1A

1 On Monday Miss Edgar asked her class of 30 pupils how many hours they had spent watching television at the weekend. Their answers were:

3 2 5 4 5
3 6 0 1 2
3 3 3 4 0
3 1 3 4 4
2 4 3 5 2
2 4 4 2 3

a Make a frequency table like the one above, with the headings:
Number of hours; Tally; Frequency; Number of hours × frequency.
b Calculate the mean number of hours.
c Write down the mode and median too.

2

The marks were:

5 8 2 6 5 9
7 6 7 7 8 7
5 4 5 7 3 7
8 9 10 5 5 10
9 5 7 9 3 7

a Make a frequency table, and calculate the mean mark, correct to 1 decimal place.
b Use the table to find the modal and median marks.

3 A consumer survey checked the price of small packs of Bran Crisps breakfast cereal in different shops. In a sample of 24 shops the prices, in pence, were:

74 74 77 75 78 73
74 69 74 69 76 76
74 72 76 74 77 70
73 74 77 73 78 74

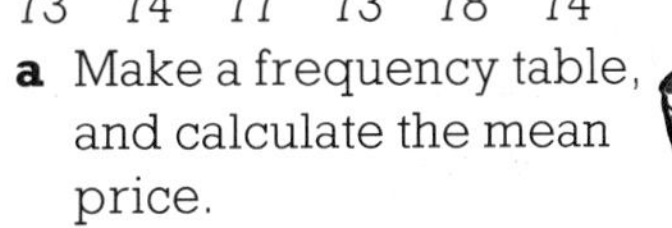

a Make a frequency table, and calculate the mean price.
b What is the range of prices?
c What are the modal and median prices?

4 In a sample of 20 boxes the number of matches per box was:

43 47 44 44 45 46 48 47 45 45
45 45 49 46 43 47 46 45 44 42

a Make a frequency table, and calculate the mean and range of the distribution.

b From this sample, would you say that the comment 'Average contents 45 matches' was reasonable?

5 a Make a frequency table for the number of hours of sunshine recorded at Action Academy's weather station in November:

5 3 2 4 0 1 0 3 6 5 2 0 0 2 3
4 1 0 1 2 0 1 0 1 3 0 5 0 1 0

b Calculate the mean, modal and median number of hours of sunshine.

c What fraction of the days had:

(i) no sunshine

(ii) more than 2 hours of sunshine?

EXERCISE 1B/C

1

The *Cola Cool* Company take samples of their product to check the volume of cola in a can.

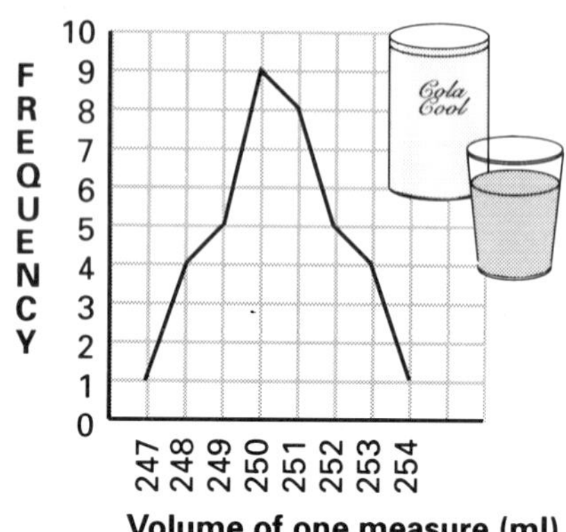

a Using the graph, copy and complete this table.

Volume	247	248						
Frequency	1	4						

b The company claim that the average contents of one of their cans is 250 ml. Is this claim correct?

2 The hours of sunshine at a holiday resort in June and July have been charted.

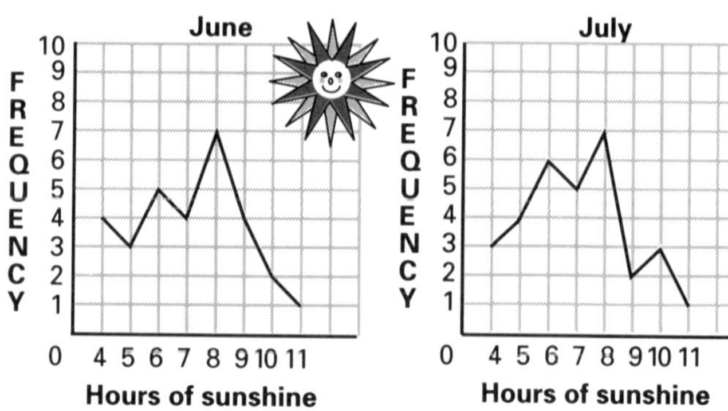

Which month had the:

a greater total amount of sun

b higher mean daily hours of sun?

3 Joe cast two dice 50 times. Each time he noted the sum of the two scores.

4	7	9	3	11	5	8	7	6	7
10	2	10	7	5	8	6	3	9	6
6	8	4	2	8	9	7	8	4	12
8	5	11	9	6	5	3	7	12	7
8	9	7	6	11	7	10	6	10	5

Organise the information in a table, and then calculate his mean score.

4 Try Joe's experiment with two dice. Is your mean score very different from his?

FREQUENCY DIAGRAMS

Twenty-four members entered the medal competition at the Whackit Golf Club.
Here are their scores in a frequency table.

Score	Frequency
67	1
68	1
69	2
70	4
71	7
72	5
73	3
74	0
75	1
Total	24

Using the table, the club secretary drew this **frequency diagram**. The frequencies of scores are represented by columns of equal width, drawn side by side.

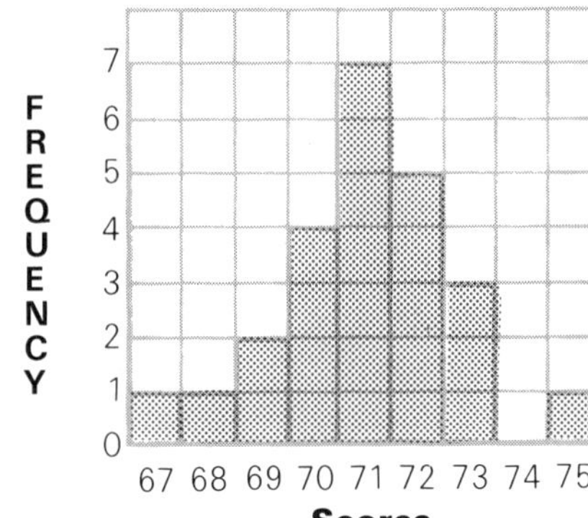

EXERCISE 2A

1 The number of goals scored by 26 Italian league teams one Saturday was:

0 1 0 3 4 2 2 1 2 3 2 5 2
3 2 1 0 4 1 1 2 2 3 0 0 1

a Make a frequency table, headed 'Score' and 'Frequency'.

b Draw a frequency diagram, using the axes and scales shown.

c What is the modal score?

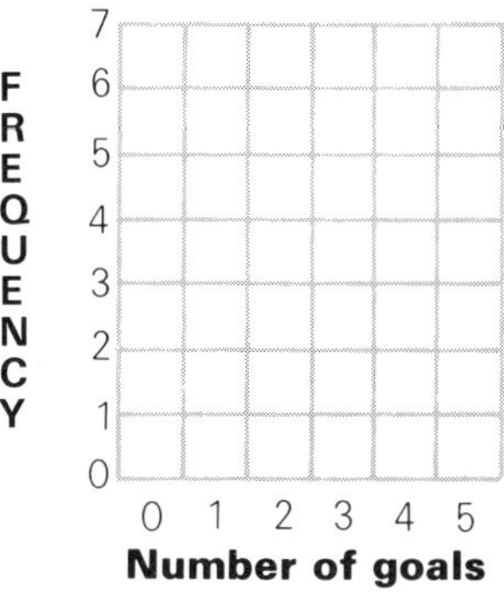

2 Draw a frequency diagram of the sample of matchboxes in question **4** of Exercise 1A. Use it to find the modal number of matches per box.

3 Draw a frequency diagram of the number of hours of sunshine in question **5** of Exercise 1A. Use it to write down the mode and median of the number of hours of sunshine.

4 Temperatures in °C recorded around the country on February 20th were:

5 3 4 2 1 3 1 0 2 1
−1 1 0 1 −2 0 4 −1 3

a Make a frequency table, and draw a frequency diagram with these figures.

b Write down the modal and median temperatures.

c Calculate the mean temperature.

5 To find out the demand for different collar sizes of shirt a shop kept records of sales of sizes 14–16½. The graph illustrates the result.

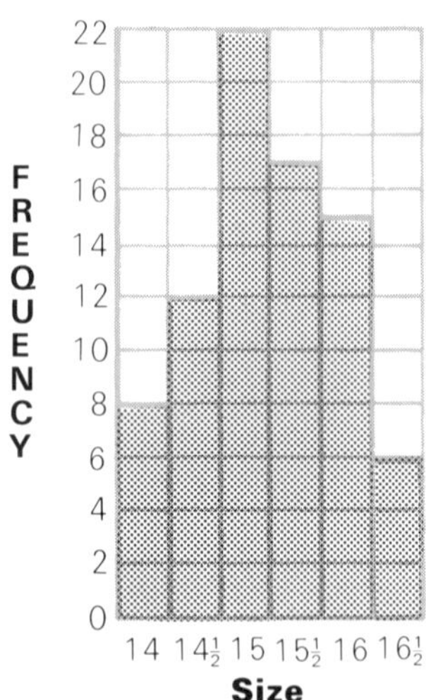

a What is the most popular size? Which type of average is this?
b What percentage of sales were: (i) size 14 (ii) size 16½?
c In a batch of 1000 shirts, how many size 16 should be ordered?

6 Use these frequency diagrams to find the mode, median and mean of each distribution.

a

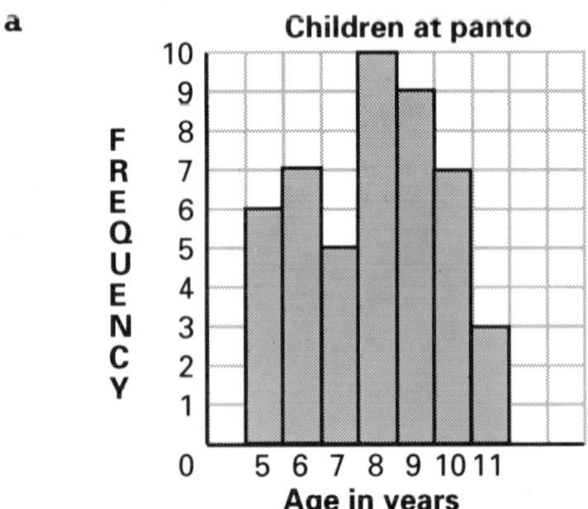

b

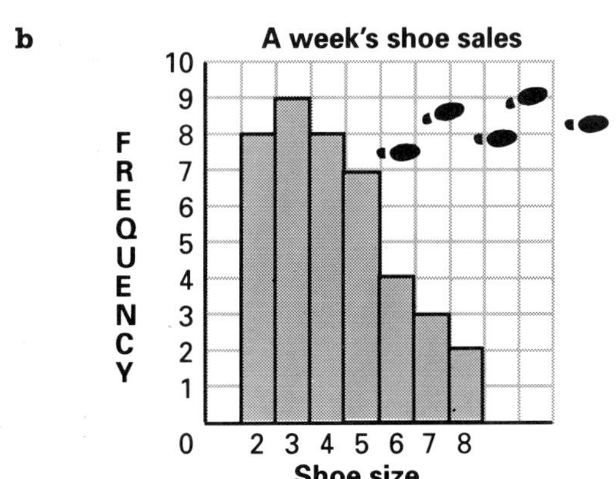

PRACTICAL PROJECTS

1

a *Throw a dice 100 times, and record the scores in a frequency table.*
b *Draw a frequency diagram, and calculate the mean score.*
c *What do you think the mean score would be if you threw the dice 1000 times?*

2 *Note the year registration letters on a large sample of cars.*

a *Make a frequency table, with the letters in order, oldest first.*
b *Draw a frequency diagram, and calculate the mean age of the cars.*
c *Write down the modal and median ages.*

FREQUENCY POLYGONS

Using a frequency diagram, another graph called a **frequency polygon** can be constructed by joining the midpoints of the tops of the columns. The first and last midpoints are joined to zero frequencies, as shown.

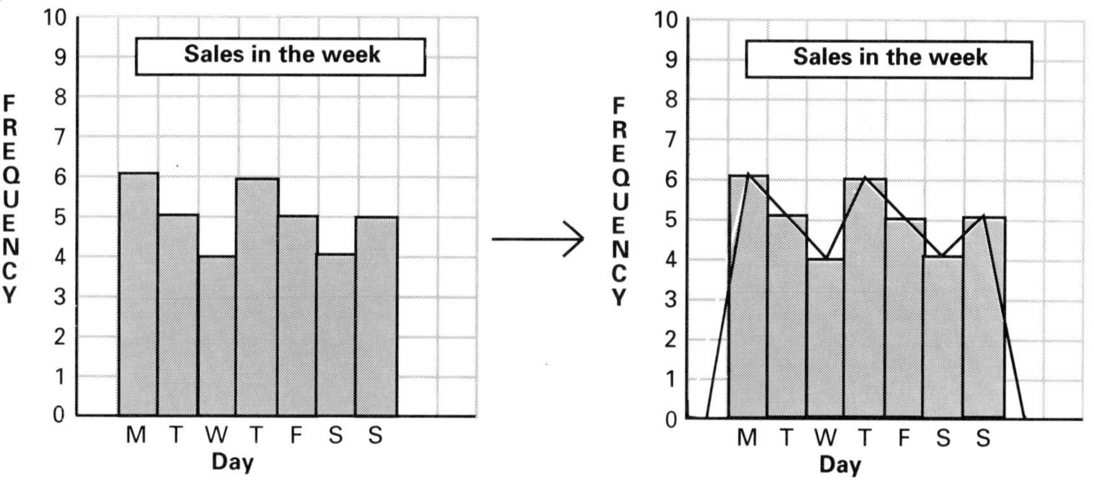

EXERCISE 2B/C

1 a Copy this frequency diagram of a batch of letter weights.

b Draw the frequency polygon on this diagram.

c Which are the most common and the least common weights of letters in the batch?

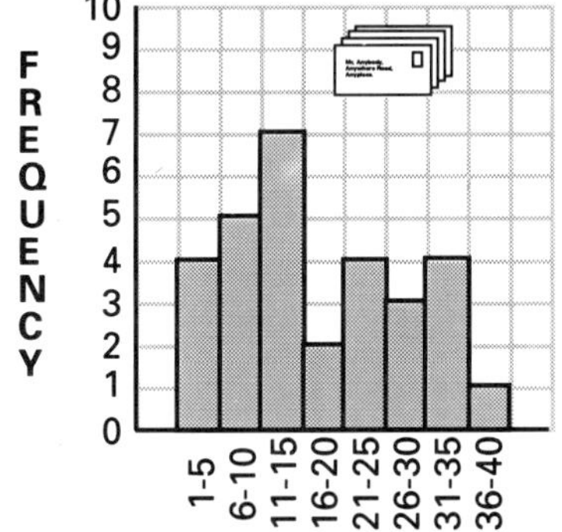

2 A shopkeeper keeps records through the week of his three best-selling flavours of ice lollies. His figures are shown in the table below.

Flavour	Mon	Tue	Wed	Thu	Fri	Sat	Sun
Lime	7	9	7	4	1	8	10
Lemon	2	4	3	5	2	9	8
Strawberry	1	2	5	7	1	3	7

a Draw a frequency diagram and a frequency polygon for his total daily sales.

b What was probably:
(i) the sunniest day (ii) a rainy day?

3

The Maths Department sets three exams, one each term, for Second Year pupils.
Here are all the marks.

Marks		1–20	21–40	41–60	61–80	81–100
Frequency	Term 1	4	21	23	48	4
	Term 2	6	22	51	19	2
	Term 3	12	55	18	14	1

a Draw a frequency diagram and a frequency polygon for each exam.

b Which exam do you think was:
(i) hardest (ii) easiest?

CLASS INTERVALS FOR LOTS OF DATA

EXERCISE 3

1

The 150 pupils in the third year at Action Academy were tested on the number of sit-ups they could do in $1\frac{1}{2}$ minutes. The number ranged from 40 to 74, a range of 34 sit-ups. The PE staff decided to group them in class intervals of 5.

Number of sit-ups	Frequency
40–44	11
45–49	20
50–54	28
55–59	36
60–64	24
65–69	19
70–74	12
Total	150

a Copy and complete this frequency diagram.

b In which class interval is:
(i) the mode
(ii) the median?

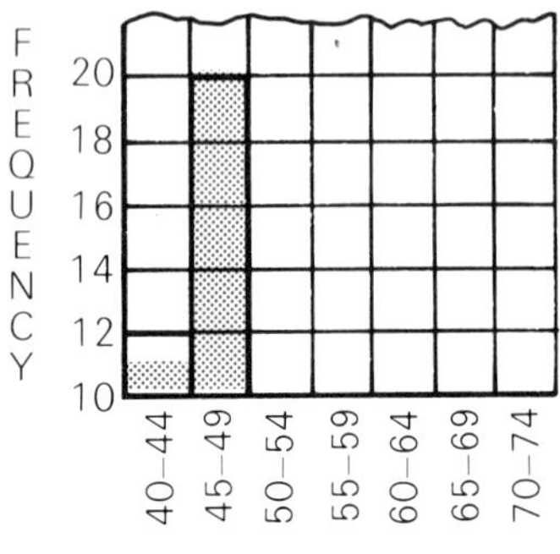

2 The best throws in the javelin contest at a school sports day are given below, to the nearest metre:

28 26 31 37 33 41 28 37 48 52 39 47
35 31 25 29 38 40 49 35 34 37 42 44
22 48 32 34 45 43 38 37 42 42 40 36

a Make frequency tables, using class intervals of (i) 20–29, 30–39, . . . (ii) 20–24, 25–29, . . .

b Show the results in a frequency diagram.

c Within which class intervals did most throws fall?

d What fraction of the throws were 45 m or more?

3 This frequency diagram shows the number of learners who needed 0–2, 3–5, . . . hours' tuition before taking their driving test.

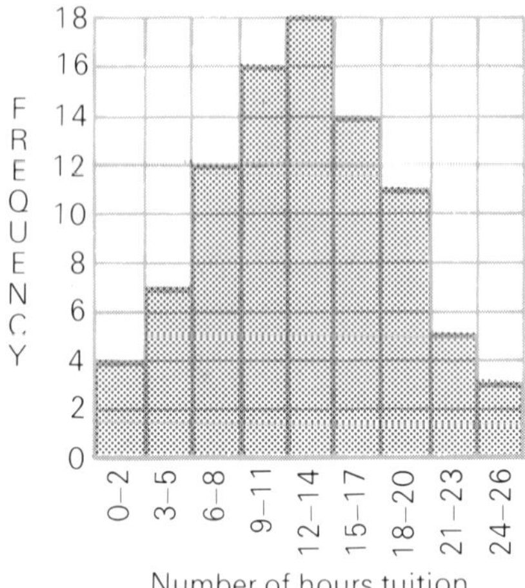

a In which class interval is:
(i) the mode (ii) the median?

b How many needed more than 20 hours?

4 The noise level at an open-air concert site has been measured at various places in the neighbourhood.

Noise (decibels)	81–90	91–100	101–110	111–120	121–130
Frequency	8	18	16	6	2

a Illustrate the data in a frequency diagram.

b Organisers expect complaints from places where the noise level goes above 110 decibels. How many complaints do they expect?

SCATTER DIAGRAMS

When collecting and graphing data it is often useful to spot connections between the items or variables. Here are the Maths, Physics and Art marks for class 2A.

Candidate / Subject	1	2	3	4	5	6	7	8	9	10	11	12	13	14	15	16	17	18	19	20
Maths	82	60	48	41	33	39	19	40	52	55	10	60	71	69	90	71	80	96	61	70
Physics	71	61	51	53	37	21	23	30	35	55	7	40	55	43	67	67	56	76	66	60
Art	23	89	42	11	88	60	72	77	66	23	15	30	62	12	53	30	90	36	66	80

Are each pupil's three marks related? If you are good at Maths, are you likely to be good at Physics? Pairs of Maths–Physics results, (82, 71), (60, 61), . . . are plotted to make a **scatter diagram** of points. Maths–Art pairs are shown in the second diagram.

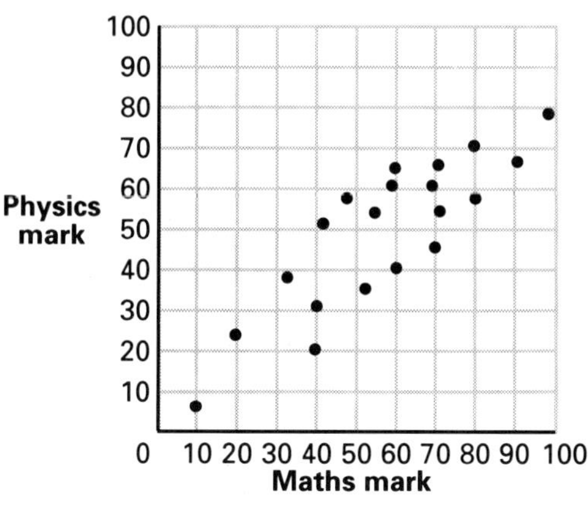

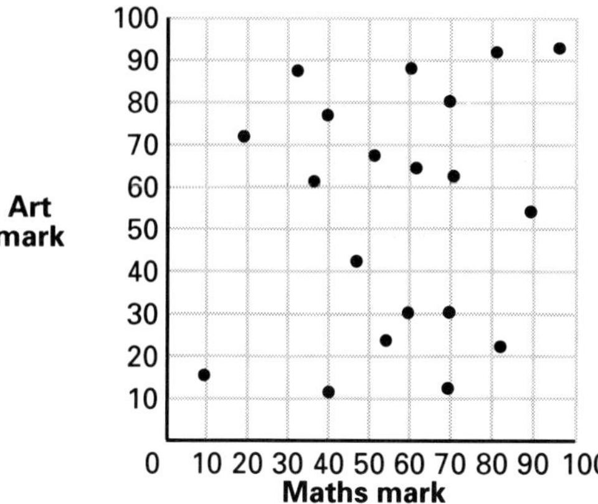

There is a general connection between the Maths and Physics marks—better Maths marks go with better Physics marks. But there is no obvious connection between the Maths and Art marks.

EXERCISE 4A

1 What does each of these scatter diagrams tell you about the connection between the two variables involved?

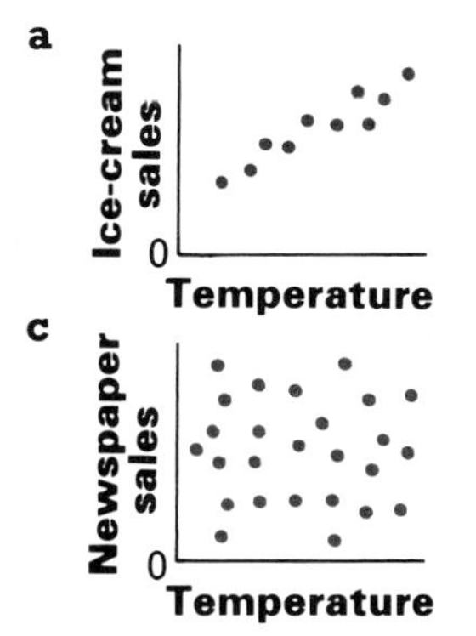

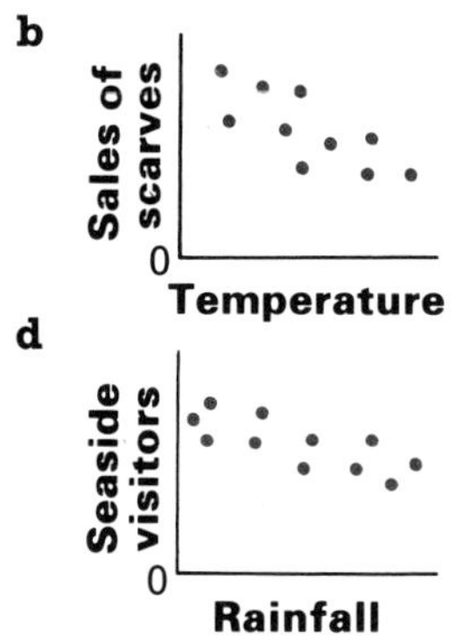

2 Copy these axes and titles. Draw scatter diagrams of dots which you think represent the situations.

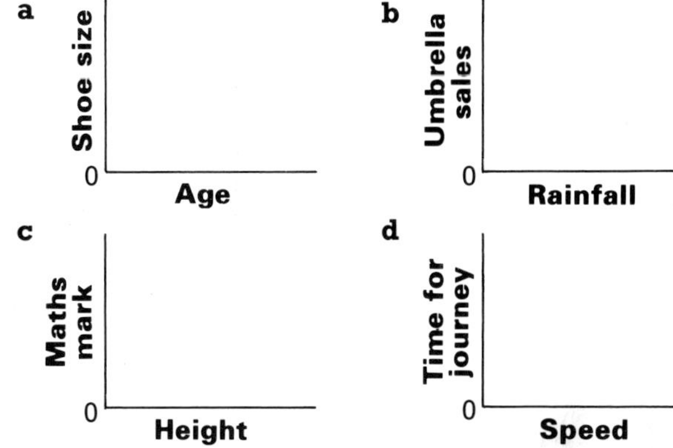

3

Question \ Pupil	1	2	3	4	5	6	7	8	9	10	11	12	13	14	15	16	17	18	19	20
Age	13	15	16	13	12	11	15	14	17	18	17	17	14	16	15	12	11	12	11	13
Height (cm)	172	181	180	169	139	140	201	181	198	190	189	180	170	191	180	160	159	171	149	150
Weight (kg)	56	64	71	53	45	50	70	55	66	75	77	72	60	76	67	60	54	55	41	45

A survey of 20 pupils chosen at random produced the above figures.

a Draw scatter diagrams to find if there is a relation between:
(i) age and height (ii) age and weight
(iii) height and weight.
(Keep these diagrams—you will need them in Exercise 4B/C.)

b Can you suggest a reasonable range of heights and weights for pupils who are:
(i) 15 years old (ii) 12 years old?

4 Experiments were carried out on the length of tyre mark left as a car braked. The table below shows the results of 20 trials.

a Draw a scatter diagram to find any relation between the speed of the car and the length of tyre mark on the road.

b After an accident the skid length was 35 m. Estimate the car's speed.

c What length of tyre mark might be left by a car braking hard at 80 km/h?

Speed (km/h)	10	52	19	31	60	58	41	12	70	49	23	40	62	68	50	11	31	39	32	20
Length of mark (m)	4	51	12	23	59	64	30	5	81	40	7	27	53	79	43	2	12	23	16	6

PRACTICAL PROJECT

Collect data to construct a scatter diagram connecting one or more of these:

a *Pupils': (i) shoe sizes and heights*
(ii) arm spans and heights
(iii) home distances from school and travel times.

b *Weights of chocolate bars and their prices.*

LINE OF BEST FIT

By looking at the pattern of dots on the Maths–Physics scatter diagram, we can draw a **'best-fitting' straight line** through the cluster of dots.

This can be used, for example, to estimate the Physics mark of an absent pupil who scored 60 in Maths; in this case it would be about 50.

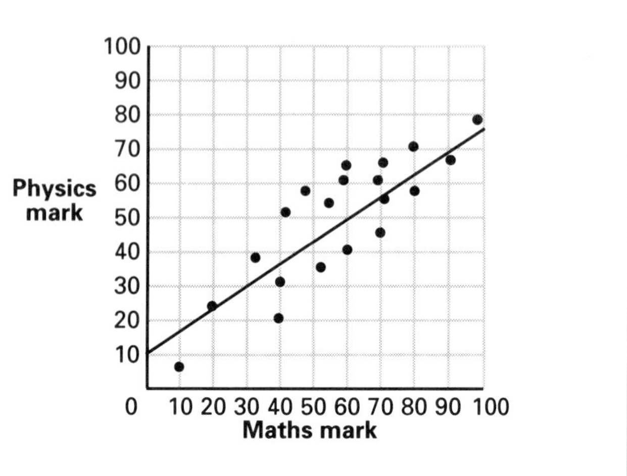

EXERCISE 4B/C

1 Draw lines of best fit on the diagrams you drew in Exercise 4A, question **3**.
Use them to estimate, for that group of pupils:

a the age of someone whose height is:
(i) 175 cm (ii) 150 cm

b the weight of someone aged: (i) 12 (ii) 16

c the height of someone whose weight is:
(i) 65 kg (ii) 50 kg.

Place your ruler on its edge in the diagrams in questions **2** and **3** below to represent the line of best fit.

2

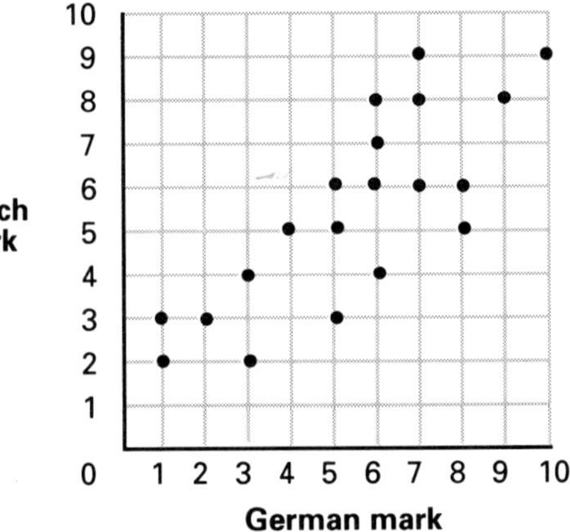

Estimate:

a the French mark that goes with a German mark of: (i) 6 (ii) 2

b the German mark for a French mark of:
(i) 4 (ii) 1.

3

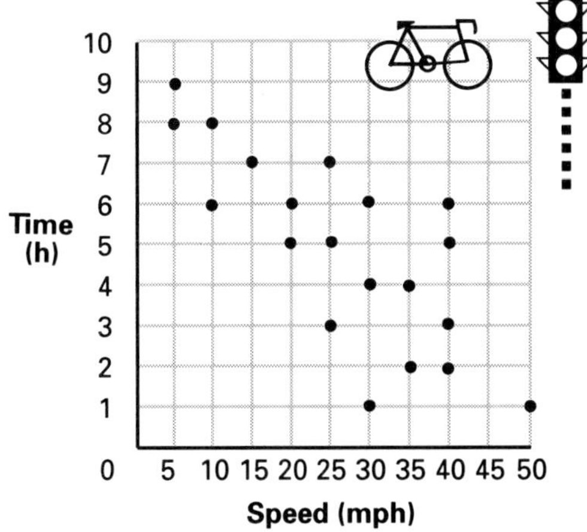

Estimate:

a the speed for a journey of:
(i) 3 hours (ii) 7 hours

b the time taken when the speed is:
(i) 30 mph (ii) 40 mph.

4 Some samples of springs are tested for their compression under various weights.

The results are given in the table below.

a Construct a scatter diagram, and draw a line of best fit.

b Estimate the compression for weights of:
(i) 3 kg (ii) 9 kg.

c Estimate weights which give compressions of:
(i) 70 mm (ii) 180 mm.

Sample number	1	2	3	4	5	6	7	8	9	10	11	12	13	14	15	16	17	18	19	20
Weight (kg)	4	3	4	2	1	6	7	7	6	3	2	1	8	3	1	5	6	5	8	7
Compression (mm)	101	93	62	54	40	105	128	150	133	39	23	32	131	69	10	102	82	80	163	109

CHECK-UP ON MAKING SENSE OF STATISTICS 2

1 a Make a frequency table for the thicknesses of the books in the pile.

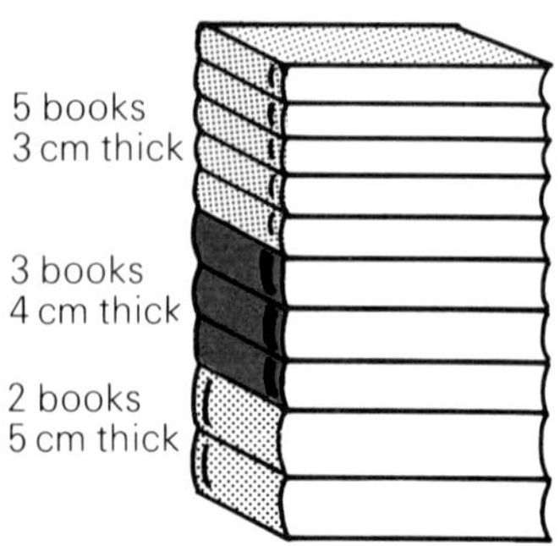

b What is the height of the pile?
c Calculate the mean thickness of the books, to 1 decimal place.
d Write down the modal and median thicknesses.

2 A note was taken of the ages of pupils at the school disco.

Age	11	12	13	14	15	16	17
Frequency	15	22	31	41	16	12	13

a What percentage of the dancers are:
(i) under 13 (ii) over 14 (to nearest percent)?
b Draw a frequency diagram and a frequency polygon of the data.

3

Samantha noted the length of her telephone calls, to the nearest minute.

2 4 3 15 5 5 7 8 17 35 3 5 6 9 11
34 2 4 5 12 4 9 6 13 17 5 9 4 3 33

Calculate the mean length of one of her calls.

4 A TV programme claimed that the custom of sending Christmas cards was on the way out. A survey of pupils showed the number of cards they received, to the nearest 5.

Last year

Number of cards	5	10	15	20	25	30	35	40	45
Number of pupils	2	18	27	20	25	21	12	10	15

This year

Number of cards	5	10	15	20	25	30	35	40	45
Number of pupils	4	18	25	22	21	23	10	11	14

a Find the mode, median and mean of the number of cards received each year.
b Is the TV claim justified?

5 Make a frequency table for the telephone calls in question **3**, and use it to calculate the mean length of a call.

Length of call (min)	1–5	6–10
Frequency		

6 A pop magazine invites readers to award artists marks out of 10 for their looks and their voices. Hold your ruler along the line of best fit on the scatter diagram, and predict:
a a voice score for a looks score of 9
b a looks score for a voice score of 2.

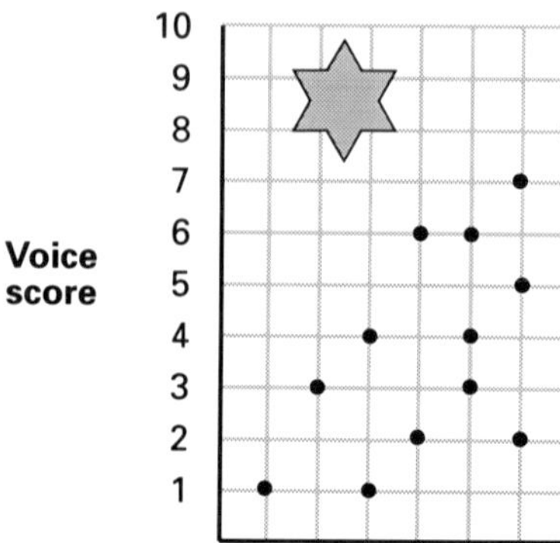

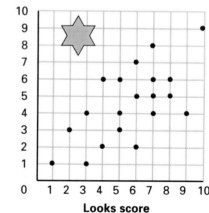

14 KINDS OF QUADRILATERAL

LOOKING BACK

1

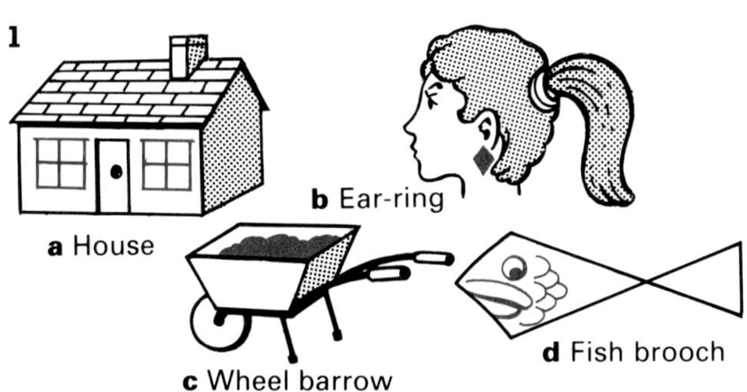

List the names of the two-dimensional shapes as they appear in the pictures above.

2 a Calculate the sizes of the unmarked angles in these three triangles, $\angle ACB = \ldots$, etc.

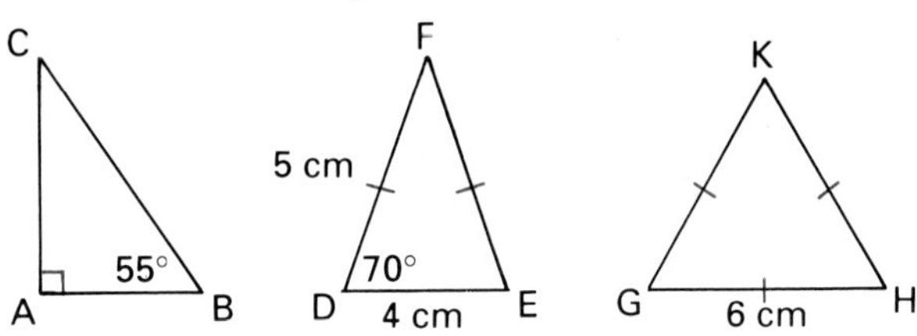

b Write down the lengths of as many sides as you can.

c Sketch the triangles which have axes of symmetry, and draw the axes as dotted lines.

3 a Copy the square and the rectangle below and mark:

(i) the sizes of the angles
(ii) the lengths of all the sides
(iii) the axes of symmetry.

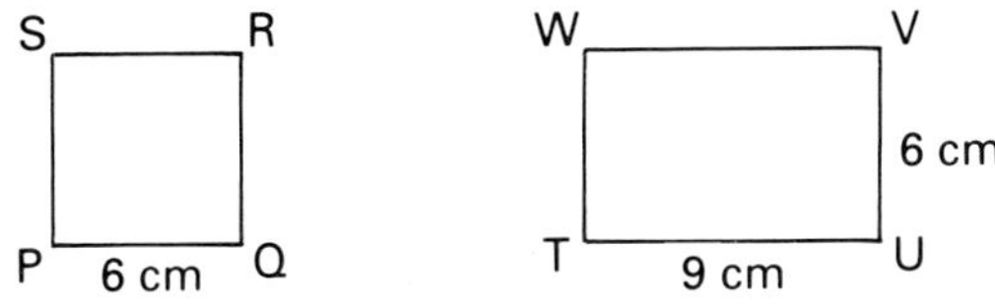

b Calculate the area of each shape.

4 Calculate the areas of these triangles.

a

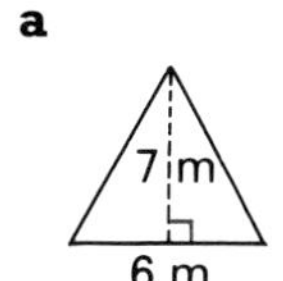

b

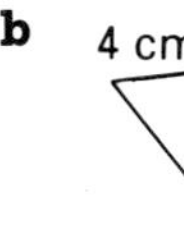

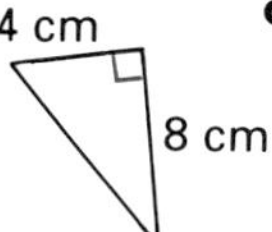

c

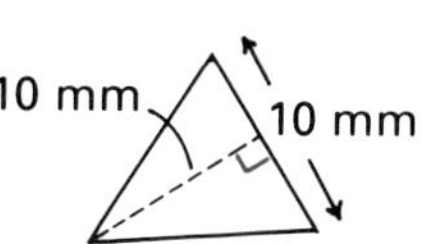

5 Copy the diagrams below, and fill in all the angles.

a 60° **b** 140° **c**

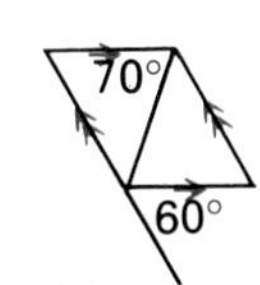

6 a A(2, 1) and B(2, 5) are corners of square ABCD. Find possible coordinates for C and D.

b R(1, 3) and S(7, 3) are corners of rectangle RSTU, which has length 6 units and breadth 5 units. Find possible coordinates for T and U.

INVESTIGATION

*Four-sided shapes are called **quadrilaterals**. Draw some pairs of crossing lines for diagonals, and join the ends to make a number of different quadrilaterals.*

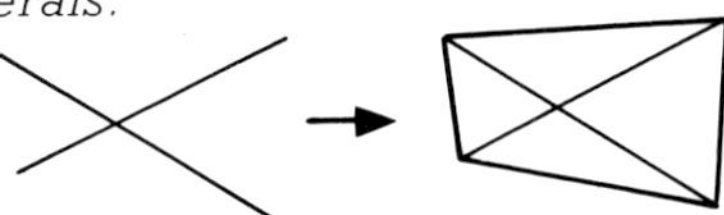

Investigate all the different shapes of quadrilateral you can draw by making the diagonals equal, or at right angles, or bisecting each other, . . .
Have you found any of these shapes yet? Which ones can you name?

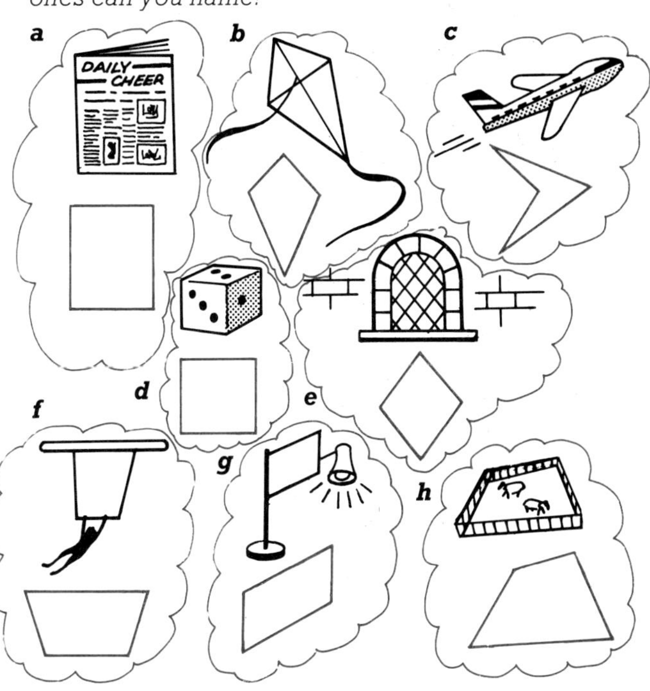

THE KITE

A kite is a quadrilateral with one axis of symmetry.

Draw a kite on squared paper, cut it out and fold it about its axis of symmetry.

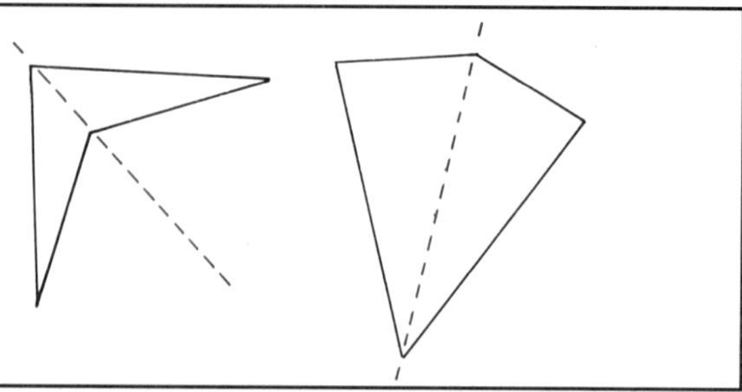

EXERCISE 1A

1

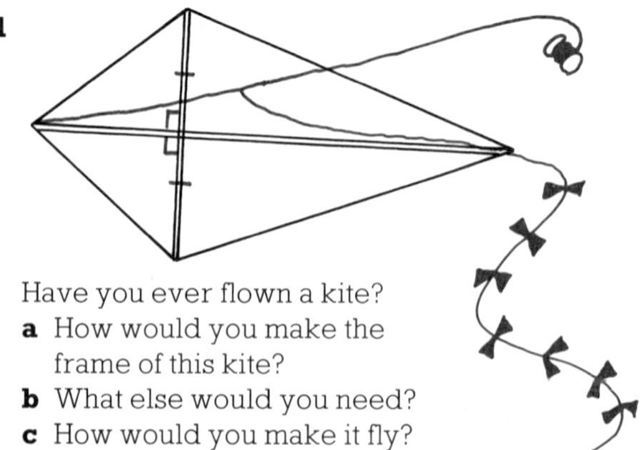

Have you ever flown a kite?

a How would you make the frame of this kite?
b What else would you need?
c How would you make it fly?

2 If this kite is to balance, diagonal AC has to be an axis of symmetry. So AC bisects BD at right angles.

a Copy the kite and mark equal lines and angles.
b Name two isosceles triangles.
c Name a triangle congruent to △ABC.

3 Sketch these kites, and mark as many lengths and angles as you can. Remember that each kite has an axis of symmetry.

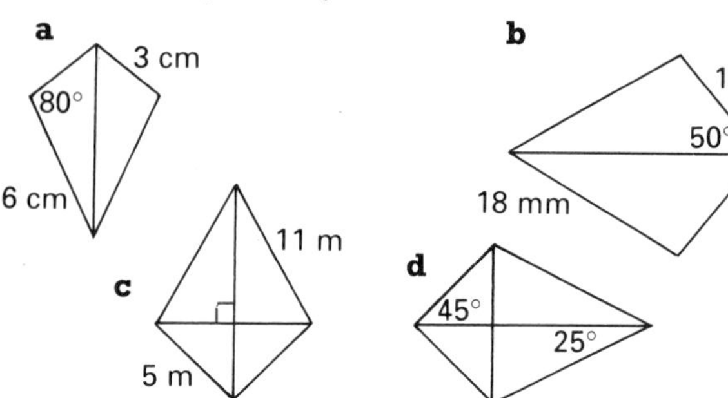

4 a Copy these pairs of diagonals on squared paper, and draw the five kites.

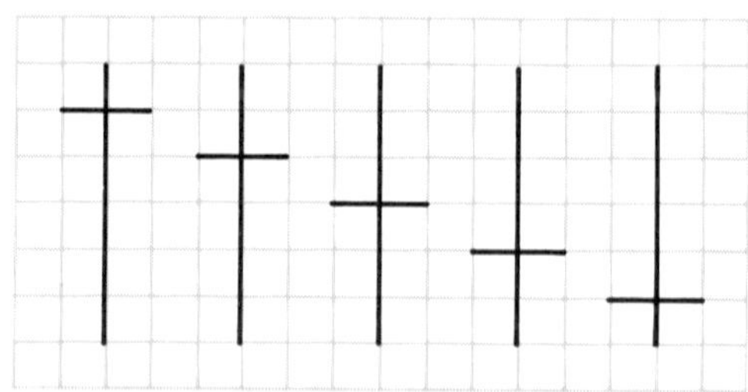

b Which pairs are congruent?
c One kite is special. Which one, and why?

5 A(4, 8), B(1, 5) and C(4, 1) are corners of kite ABCD. Find the coordinates of:
a D **b** the point where the diagonals cross.

6 a Copy the kite in question **2**, and mark AB = 25 mm, BC = 40 mm, ∠ABD = 40° and ∠CBD = 60°.
b Fill in the sizes of all the angles and the lengths of all the sides.
c What is the sum of the four angles of the kite?
d Why is the sum of the angles of every quadrilateral 360°?

7 a Copy the two kites below and mark the sizes of all the angles.
b Check that the sum of the angles in each kite is 360°.

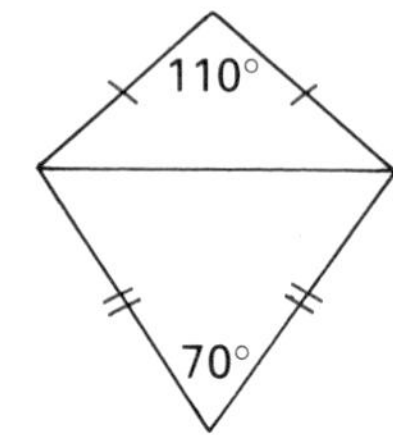

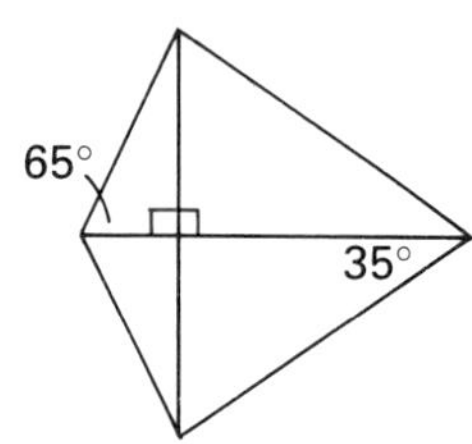

EXERCISE 1B/C

1 Copy these kites, and mark the sizes of all the angles.

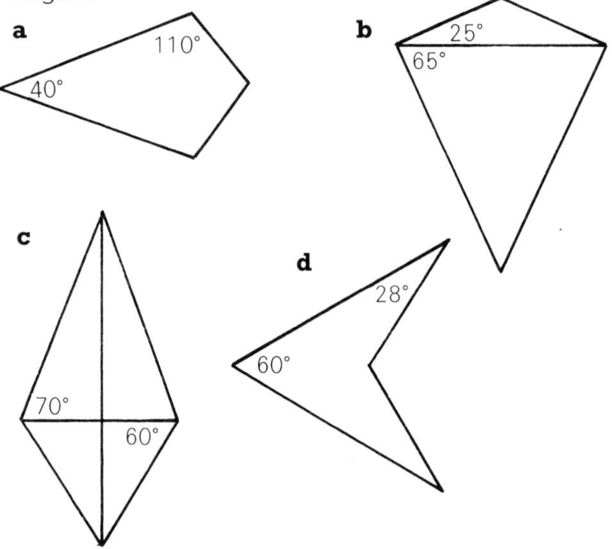

2 A square piece of paper is folded along its diagonal AC, and opened out again. AB is folded onto AC, then AD is folded onto AC. These are opened out, and the paper is creased along EF.

a Can you name three kites?

b Why is $x^\circ = 22\frac{1}{2}^\circ$? Copy the diagram, and mark the sizes of all the angles.

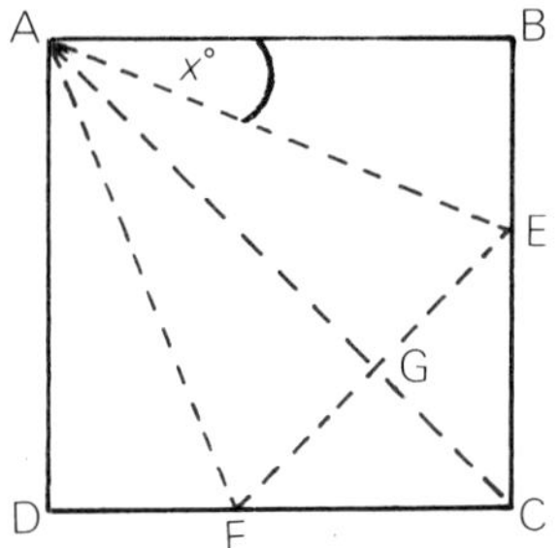

3 The main frame of this bicycle is a kite. $\triangle$ABD is equilateral.

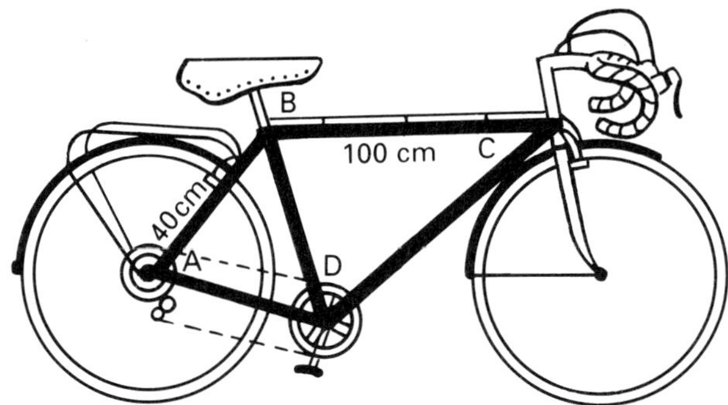

a Sketch the kite and its diagonal BD, and calculate the length of tubing needed for the frame.

b Use a scale drawing to find the distance between A and C.

4 An ice cream container is made from a semi-circular net of kites, as shown below.

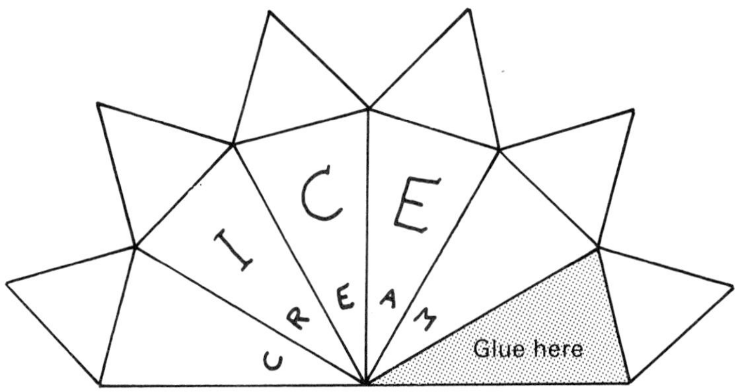

a Calculate each kite angle at:
(i) X (ii) Y.

b Now calculate these two angles for:
(i) a semi-circle with seven kites (making a six-sided container)
(ii) an n-sided pyramid container.

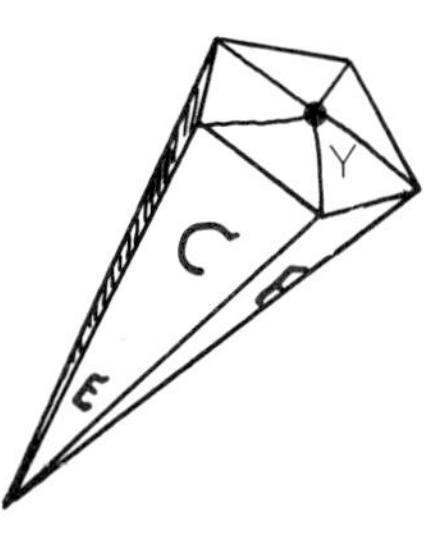

PRACTICAL PROJECT

Draw the net, and make the ice cream container in question **4**.

INVESTIGATION

a

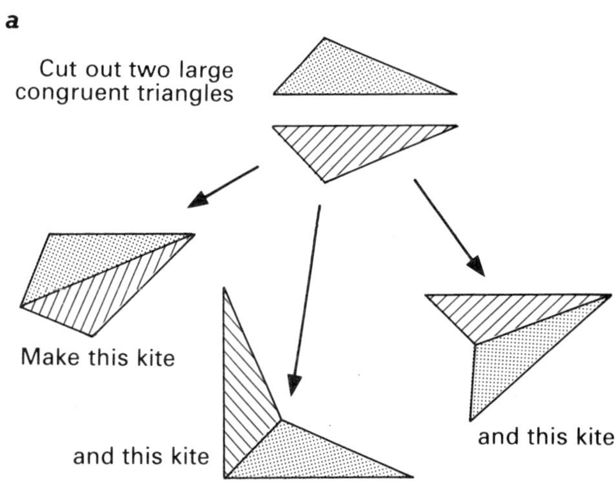

b *Investigate the shape of triangle you start with if you can only make:*
(i) one shape of kite (ii) two shapes of kite.

THE RHOMBUS

A rhombus is a special kite with two axes of symmetry.

Draw a rhombus on squared paper, cut it out and fold it about each axis of symmetry.

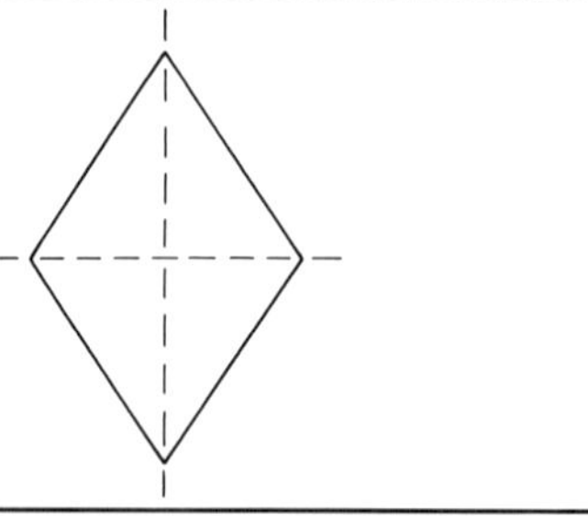

EXERCISE 2A

1 When the rhombus ABCD is folded about its axis of symmetry AC, where do these go?
a AB **b** BC **c** ∠ABC.

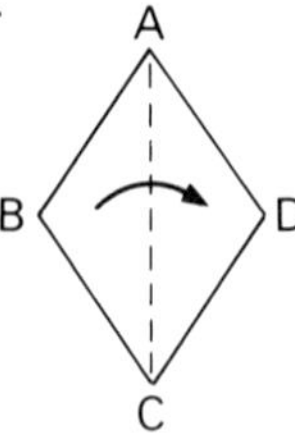

2 When the rhombus is folded about its axis of symmetry BD, where do these go?
a AB **b** AD **c** ∠BAD.

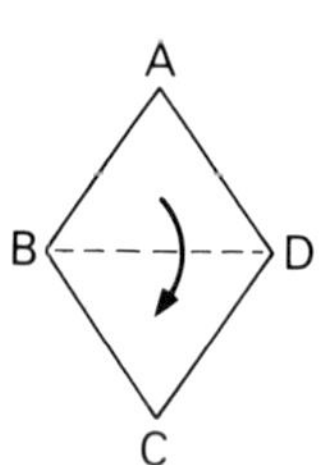

All the sides are equal, and opposite angles are equal.

3 Four equal strips are hinged at each corner. The shape swings round about the bottom strip AB, to the left and right.

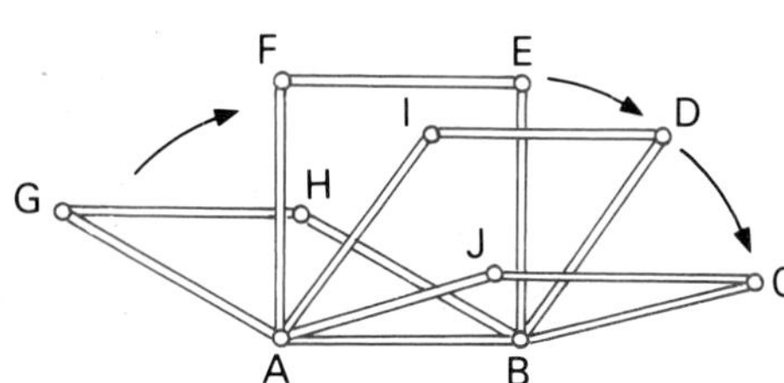

a How many rhombuses can you see?
b Which one is special? Why? What is it called?

4 **a** Use axis of symmetry AC to copy and complete:
(i) BO = . . .
(ii) ∠AOB = ∠AOD = . . .°
b Use axis of symmetry BD to copy and complete:
(i) AO = . . .
(ii) ∠BOA = ∠. . . = . . .°

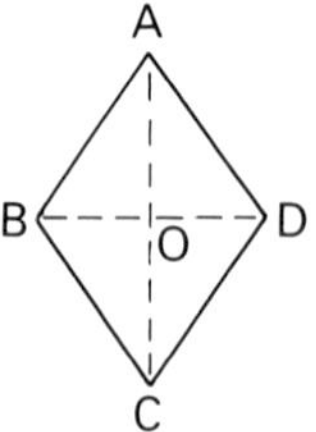

The diagonals of a rhombus bisect each other at right angles.

5 Sketch this rhombus and mark all the equal lines and angles.

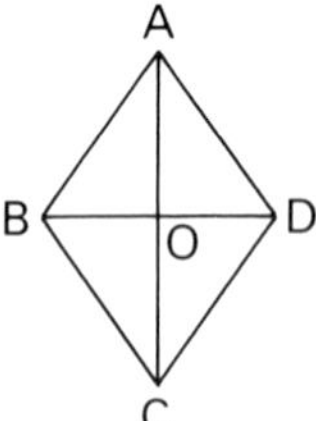

6 Copy each rhombus below, and fill in the sizes of all the angles.

a

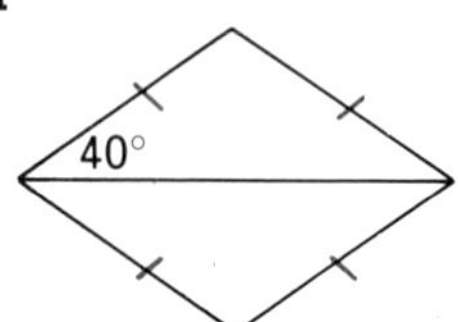

b

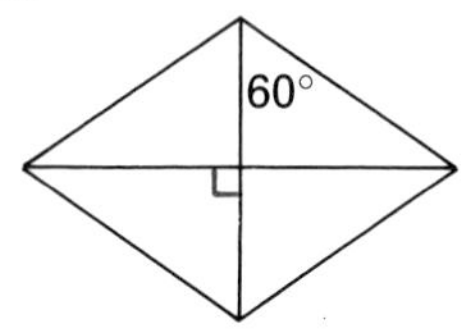

7 **a** Draw a rhombus on squared paper with diagonals 8 cm and 6 cm long.
b Measure the length of each side, and the size of each angle of the rhombus.

8 Sketch each rhombus below, and mark as many lengths as you can.

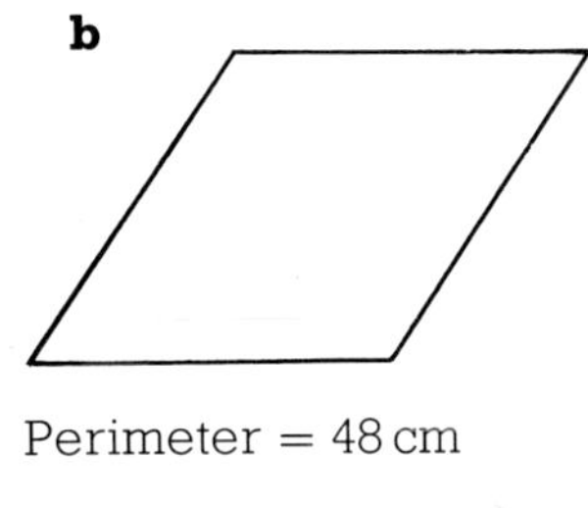

Perimeter = 48 cm

9 R is (3, 5), S(8, 6) and T(7, 1). Find the coordinates of the fourth vertex of the rhombus:
a RSTU **b** RSVT.

10 K(5, 6), L(1, 4) and N(9, 4) are corners of a rhombus KLMN. Find:
a the coordinates of M
b the lengths of diagonals KM and LN
c the point where KM and LN cross.

11 DRAW YOUR OWN RHOMBUS

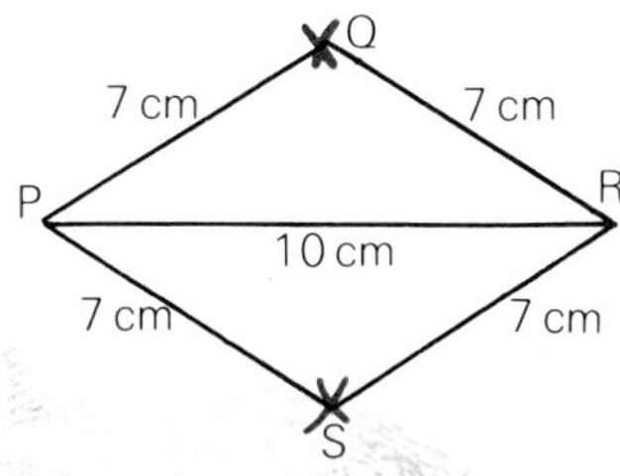

Make PR 10 cm long. Use compasses to mark the points Q and S.

Measure the four angles of the rhombus.

12 a Trace the rhombus you drew in question **11**.
b Check by folding that it has two axes of symmetry.
c Find out by turning it about its centre whether it has:
(i) half-turn symmetry
(ii) quarter-turn symmetry.

EXERCISE 2B/C

1 Farms and gardens often have fences like this.

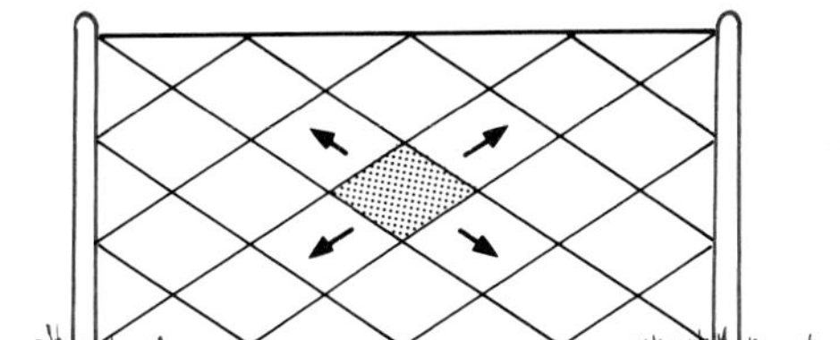

a How many small rhombuses can you see in the picture?
b Trace one, and check that it can slide along the fence in the directions shown. It can do this because its opposite sides are parallel.

The opposite sides of a rhombus are parallel.

2 In question **1** the side of each small rhombus is 10 cm long. What is the length of the perimeter of the largest wire rhombus you can see in the picture?

3 a Draw a tiling of four rhombuses which make one larger rhombus.
b Draw the axes of symmetry of the large rhombus.

4 The Allenbys made a drawing of a star-shaped flower bed for their garden. They drew five congruent rhombuses, of side 4 cm, round a point.

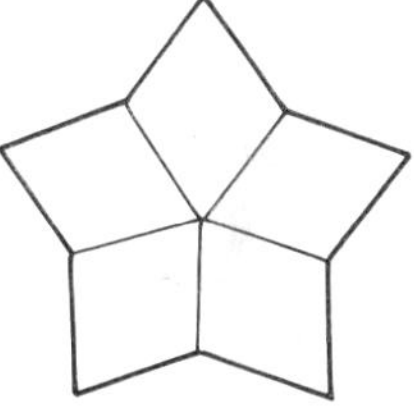

a Calculate the size of each angle at the centre.
b Draw the five angles at the centre, using a protractor.
c Draw the pattern with a ruler and compasses.

5 ABCD is a mirror in the shape of a rhombus, set in a rectangular frame.
CE = 30 cm,
DE = 40 cm and
CD = 50 cm.

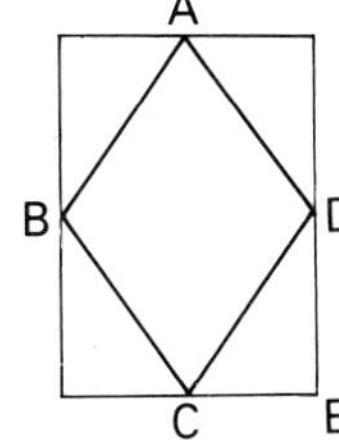

Calculate the perimeter of:
a the mirror **b** the window frame.

6 The diagonals of rhombus ABCD intersect at O. Copy and complete:
Under a half-turn about O, A → . . . , B → . . . , so AB → . . . .
It follows that AB is equal and parallel to . . . .

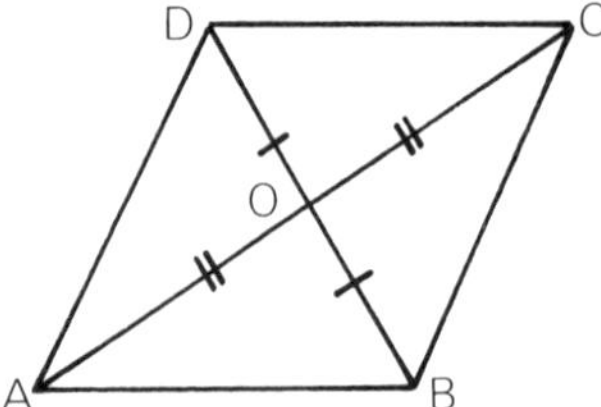

In the same way, AD is . . . . . . and . . . . . . to . . . .
The opposite sides of a rhombus are . . . . . . and . . .

PRACTICAL PROJECT

Construct a net of four rhombuses, as shown, to make a party decoration. Colour them, then cut out and fold the net to make the 'octahedron decoration'. Put a flap on a side of each rhombus if you want to glue them together.

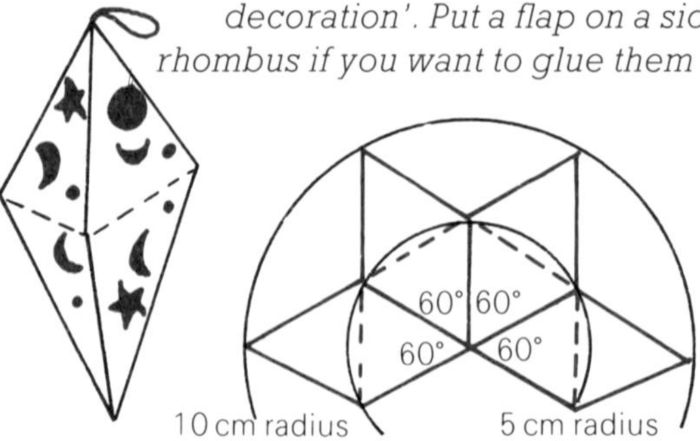

SOME SPECIAL CONSTRUCTIONS USING RULER AND COMPASSES

EXERCISE 3

1 To bisect a line

a Draw AB 10 cm long. Follow the pictures, keeping your compasses at the same setting.

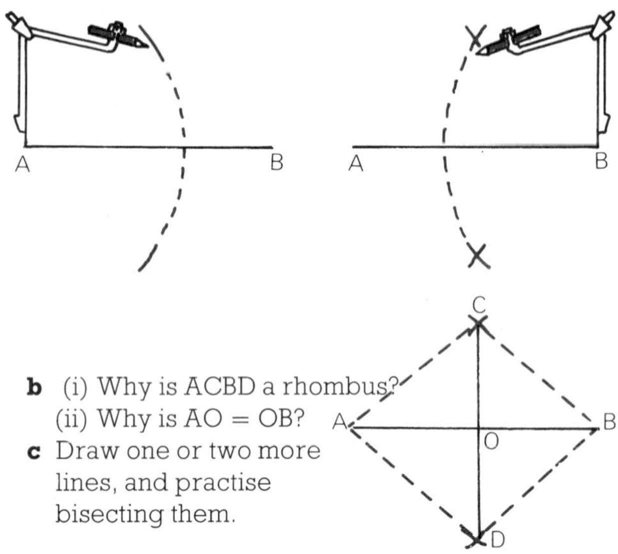

b (i) Why is ACBD a rhombus?
(ii) Why is AO = OB?

c Draw one or two more lines, and practise bisecting them.

2 To bisect an angle

a Draw an acute angle ABC. Follow the pictures, keeping your compasses at the same setting.

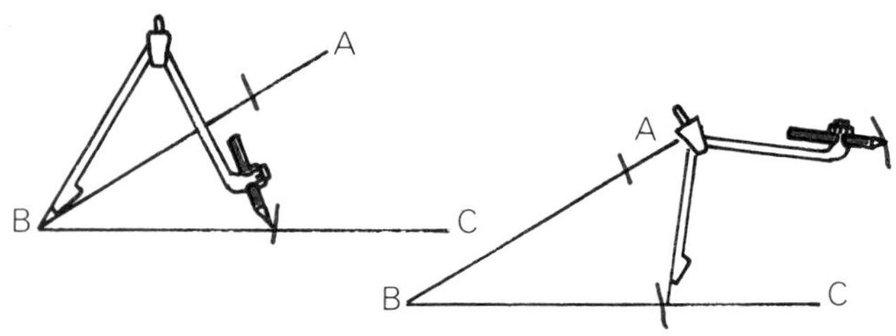

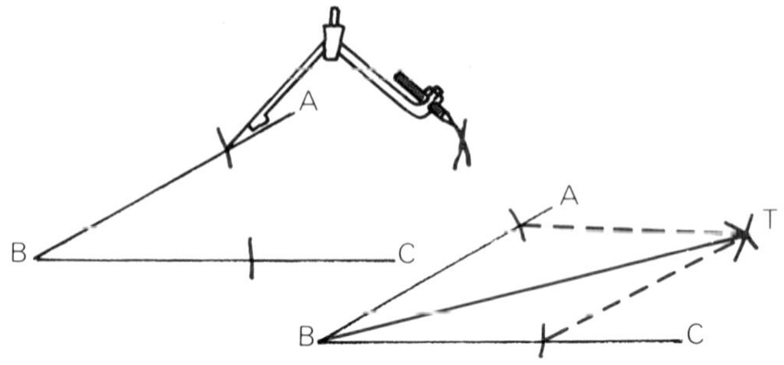

b Why does BT bisect ∠ABC?

c Draw some more angles, including a straight angle and an obtuse angle, and practise bisecting them.

3 Draw a large triangle, and bisect each angle. The three lines bisecting the angles should meet at the same point. Did you find this?

4 Construct an angle of 60°, using ruler and compasses only. In this angle construct angles of 30°, 15°, ... How far can you go?

CHALLENGE

Draw a line XY, and mark a point O somewhere above it.

Using compasses and ruler only, construct the shortest possible line from O to XY.

THE PARALLELOGRAM

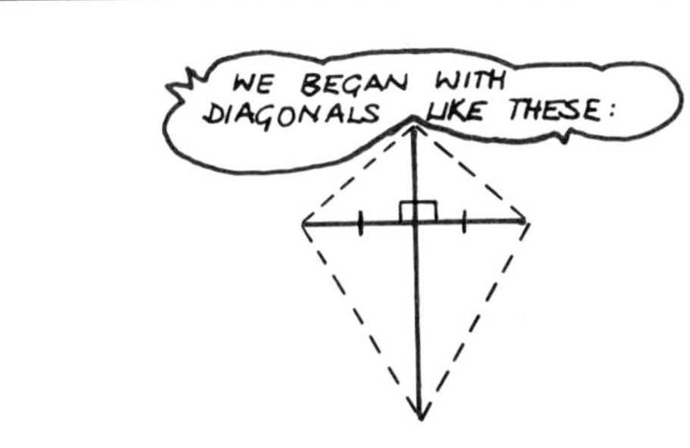

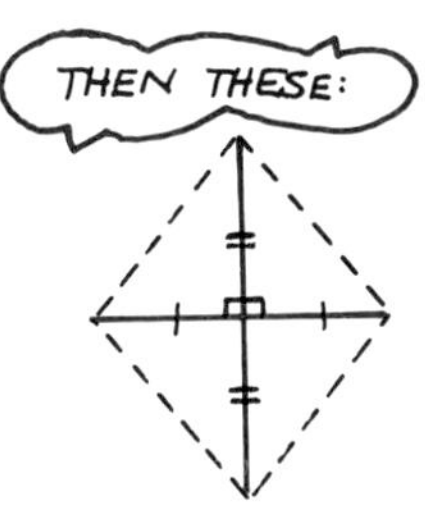

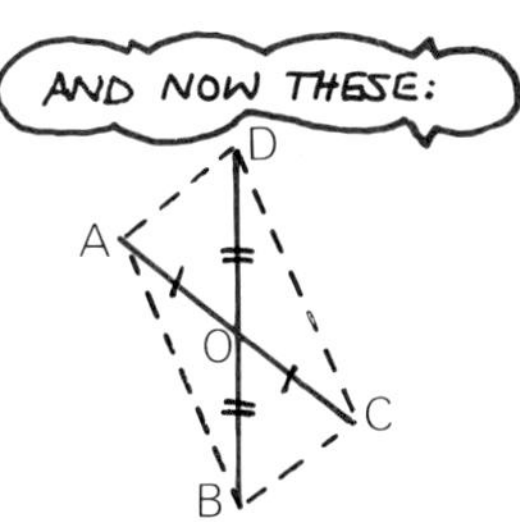

Trace the parallelogram ABCD, and give it a half turn about its centre of symmetry O.

EXERCISE 4A

1 Under a half turn about O, where do these go?
a (i) A (ii) B (iii) AB
b (i) A (ii) D (iii) AD.
So AB = DC and AB || DC;
AD = BC and AD || BC.

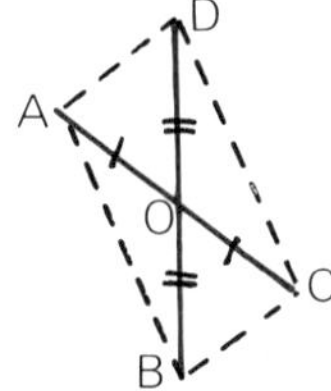

The opposite sides of a parallelogram are equal and parallel.

2 Two pairs of equal strips are hinged at each corner. The shape swings about AB, to the left and right.

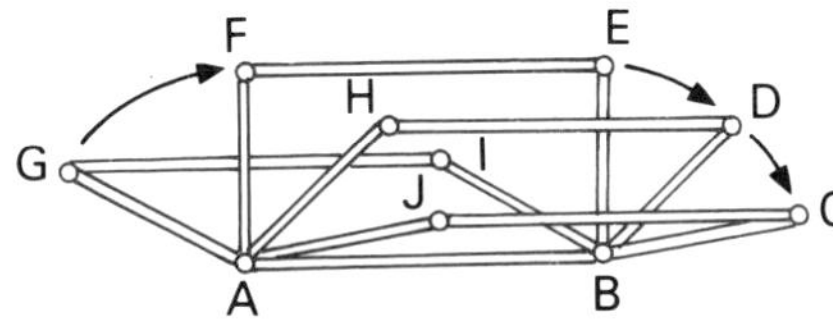

a How many parallelograms can you see?
b Which one is special? Why? What is it called?

3 a Trace, or copy and cut out, parallelogram ABCD.

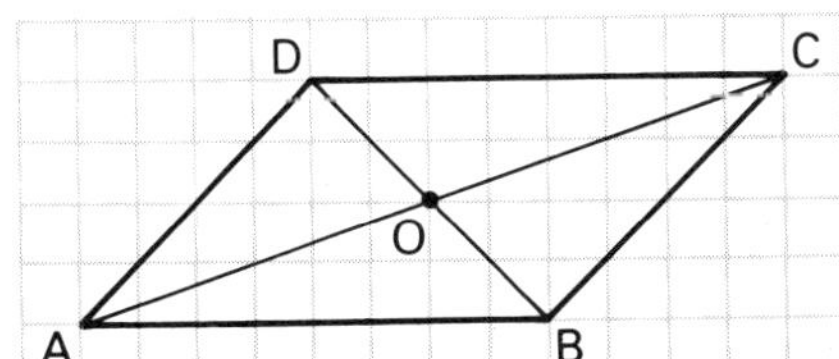

b Check, by folding, that ABCD has **no axes of symmetry**.
c Check again, by turning, that O is a centre of symmetry under a half turn.

4 In the diagram for question **3**, name a line equal to: **a** AB **b** AD **c** AO **d** DO.

5 Copy these parallelograms, and mark as many lengths as you can.

a

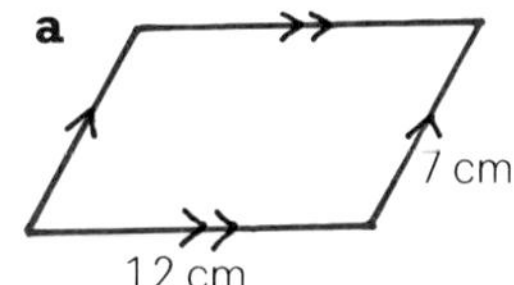

b

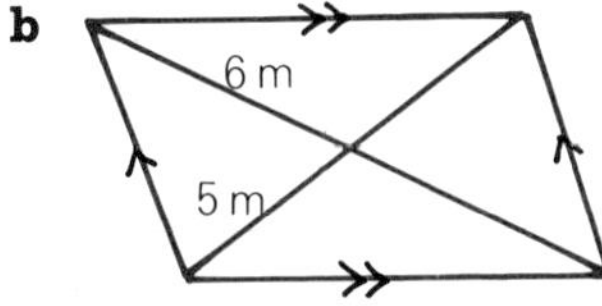

6 Which angles go in the spaces below?
Under a half turn about O:
a ∠ABC → ∠ . . . ,
so ∠ . . . = ∠ . . .
b ∠BAD → ∠ . . . ,
so ∠ . . . = ∠ . . .

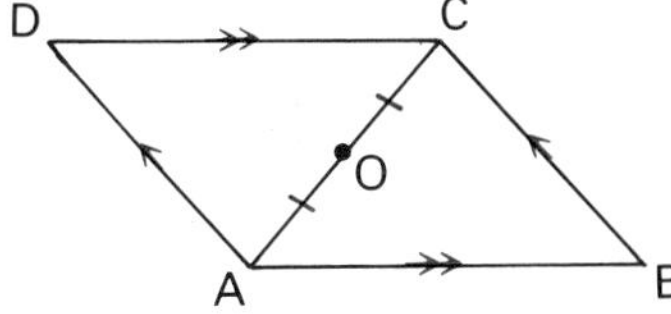

The opposite angles of a parallelogram are equal.

7 a Copy and continue this tiling of parallelograms on squared paper.

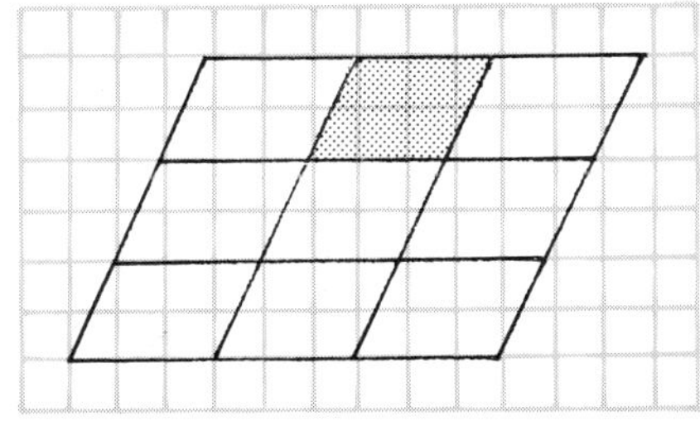

b Why can the shaded tile slide along or down the 'tracks'?
c The angle at the top right-hand corner of the tiling is 65°. Fill in all the angles in the parallelograms in your tiling.

8 Under a half turn about T, which angle is equal to:

a ∠PQS
b ∠SQR
c ∠PTS?

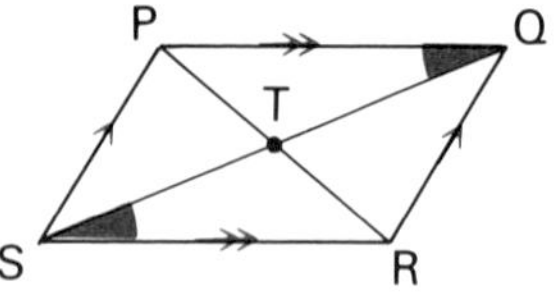

Reminders **For parallel lines:**

(i) **alternate angles are equal.**

(ii) **corresponding angles are equal.**

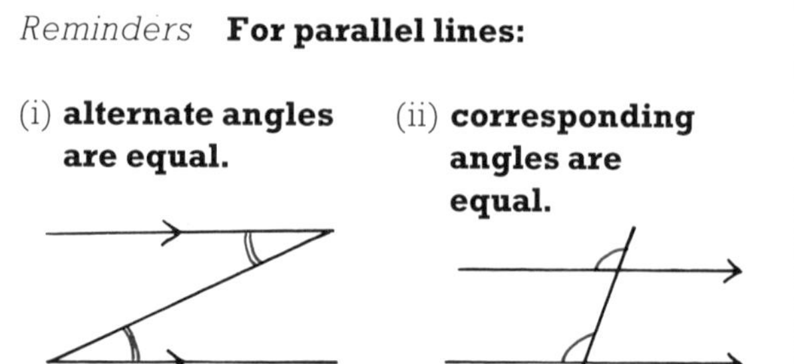

9 Sketch these parallelograms, and fill in all the angles.

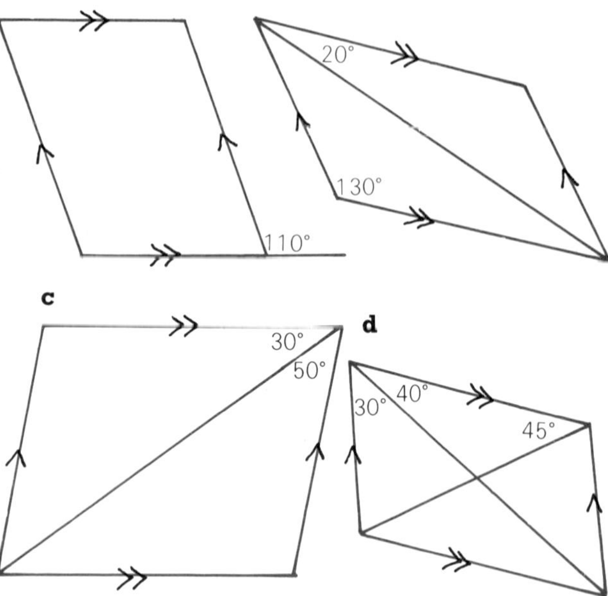

10 a Calculate the length of the route from A to B on this tiling.

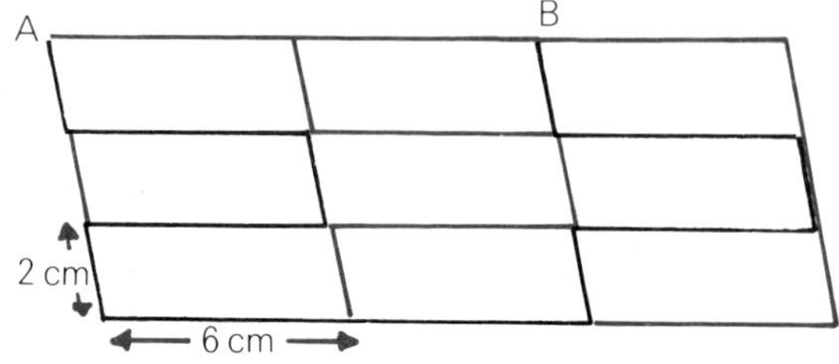

b How many horizontal parallel lines are there in the diagram?

c If the angle at A is 77°, what sizes are the angles in the tiling at B?

11 The pattern round the drum looks like this if it is opened out:

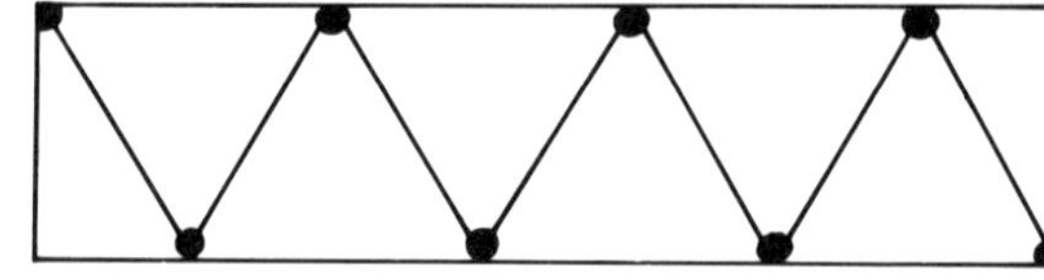

a How many small parallelograms can you see?

b How many larger ones?

12 Draw a parallelogram ABCD, and join A to C. Explain why the sum of the angles of the parallelogram is 360°.

A QUICK PUZZLE

Which diagonal is longer? Measure to make sure!

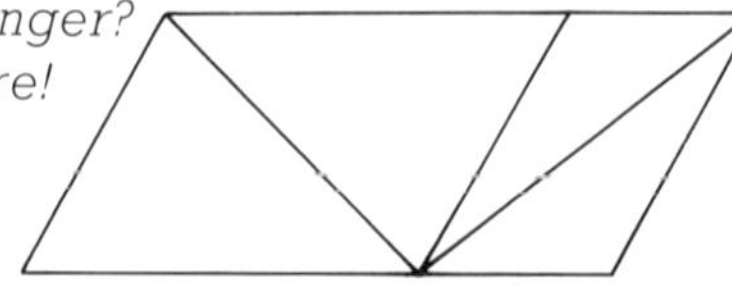

CHALLENGE

Explain how each of these things works, and the way in which the parallelogram helps.

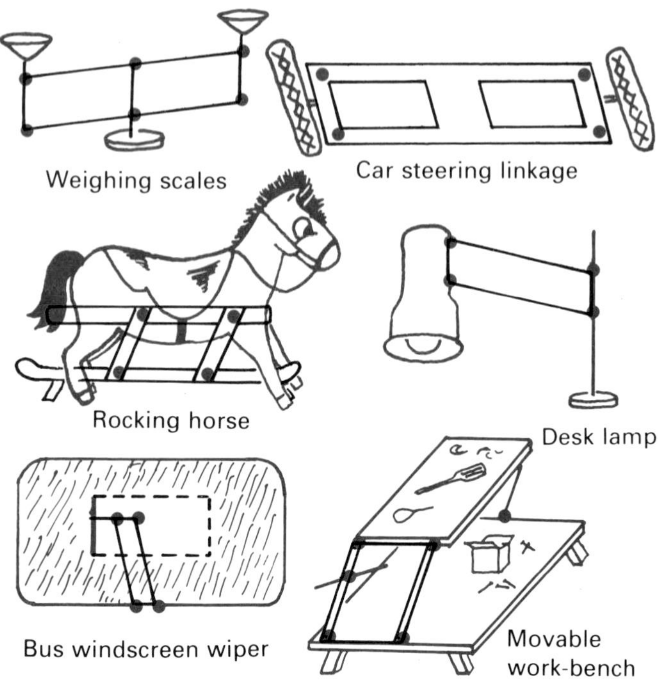

EXERCISE 4B/C

1 What properties does a rectangle have that a parallelogram doesn't have?

2

Can you draw a shape to show that Aisha is right?

3 Elizabeth is climbing the stairs.

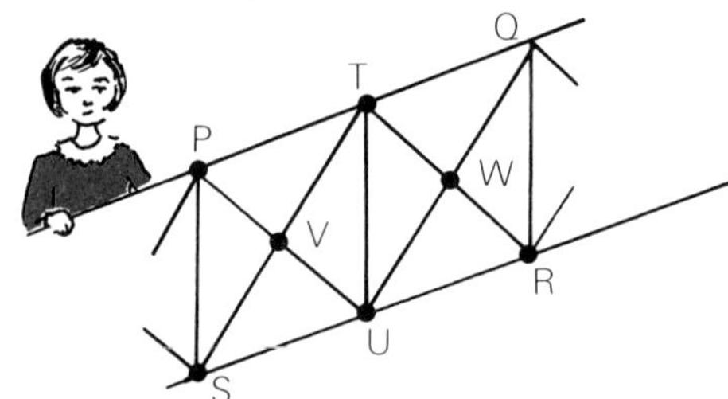

a Why is PQRS a parallelogram?
b Name six parallelograms in the drawing.

4

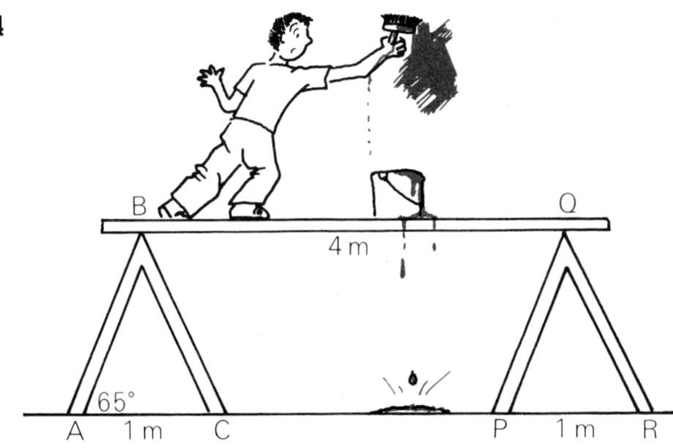

a Name two parallelograms in the picture.
b AC = PR = 1 m, BQ = 4 m. Calculate CP.
c Copy the part of the diagram between BQ and AR, and fill in as many angles as you can.

5 Crystals of calcite are parallelepipeds. Their opposite faces are congruent parallelograms. How many:
a parallelograms
b sets of parallel lines are there?

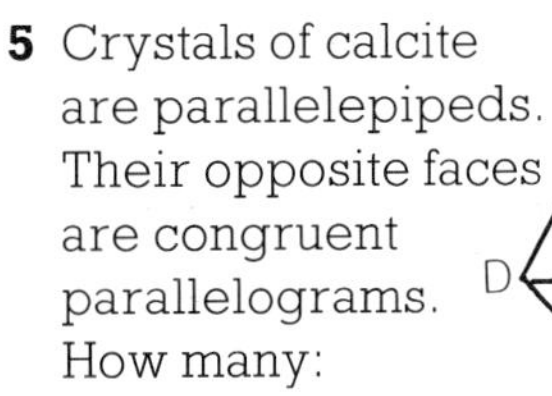

6 PQRS is a parallelogram, and $\angle PQT = \angle UQR = 55°$.

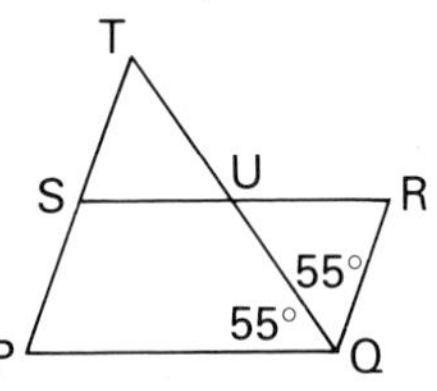

a Name an angle:
(i) corresponding to $\angle PQT$
(ii) alternate to $\angle UQR$.
b Copy the diagram, and fill in the sizes of all the angles.
c What type of triangle is $\triangle STU$?

INVESTIGATION

Draw these lines on a square piece of card, or thick paper. What shapes can you see? Cut them out, and make the goose.

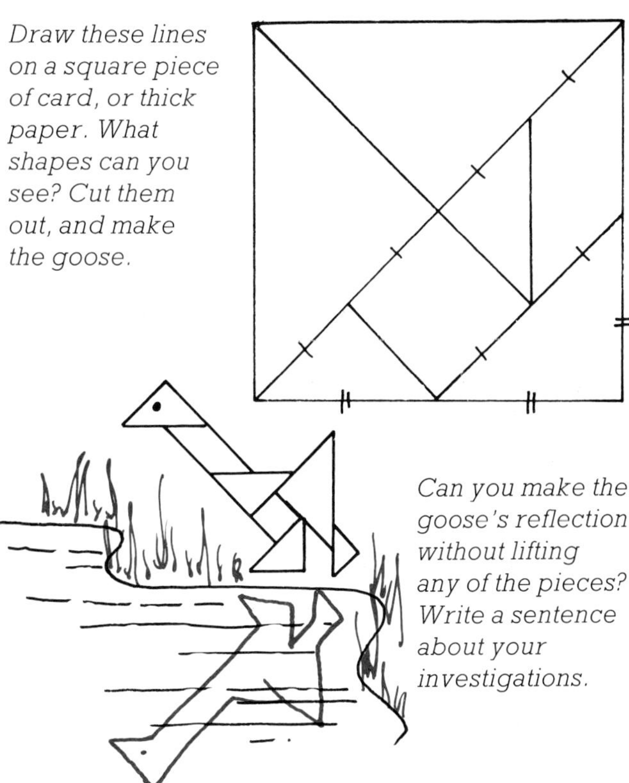

Can you make the goose's reflection without lifting any of the pieces? Write a sentence about your investigations.

BRAINSTORMER

In a competition, Sanjay has to find the distance across the river from A. He walks 20 yards along the bank to B, and puts a stick in the ground. He then walks 20 yards further to C. Explain the rest yourself.

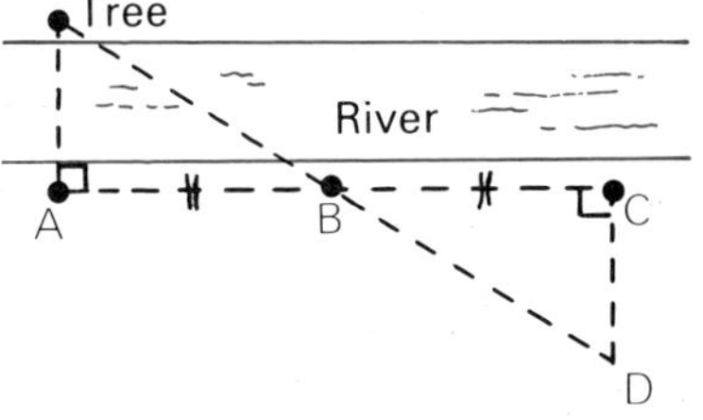

THE TRAPEZIUM

A trapezium is a quadrilateral with one pair of parallel sides.

EXERCISE 5

1 Trace the shapes drawn in black in the diagrams below. Each one is a trapezium.

2 a Draw the quadrilateral with vertices (3, 2), (6, 13), (6, 1), (3, 7).
  b Explain why it is a trapezium.

3 Draw some different trapezia on squared paper. Include one which has an axis of symmetry.

4 a Sketch this special trapezium which has an axis of symmetry through E and F.

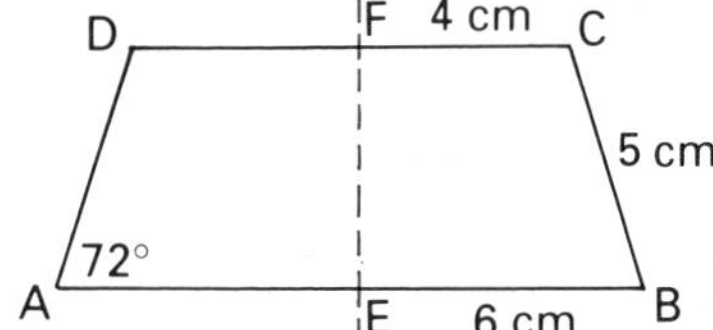

  b Fill in:
   (i) the lengths of AE, AD and DF
   (ii) the sizes of all the angles.

5 A(1, 4), B(3, 1) and C(8, 1) are corners of a trapezium ABCD. Find the coordinates of:
  a D, so that AD is parallel to BC and the trapezium has an axis of symmetry
  b the image of the trapezium under reflection in the x-axis.

6 Sketch each diagram below, and name the trapezium in each. Fill in all the angles, and check that the angles of each trapezium add up to 360°.

a A, 60°, D, 40°, E, B, C

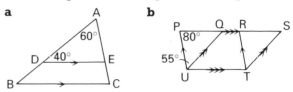

b P, Q, R, S, 80°, 55°, U, T

7 The shape between the frets on a guitar is a trapezium. The heights of the trapezia are h_1, h_2, h_3, . . . and $h_2 = \frac{17}{18}h_1$, $h_3 = \frac{17}{18}h_2$, and so on.

88°; h_1; h_2; h_3; Trapezium; Frets

  a Copy the enlarged diagram.
  b Mark the sizes of all the angles.
  c How many trapezia can you see in it?
  d Calculate h_2 and h_3, correct to 0.1 mm, given that $h_1 = 35$ mm.

INVESTIGATION

Investigate all the different (non-congruent) trapezia you can make on sets of 10 dots like these.

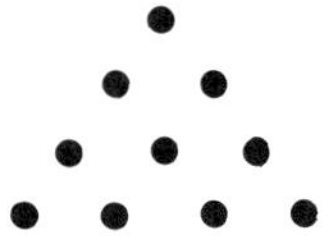

MAKING SURE

The **kite** has one axis of symmetry.

The **rhombus** has two axes of symmetry (and a centre of symmetry).

The **parallelogram** has a centre of symmetry (and opposite sides parallel).

The **trapezium** has one pair of parallel sides.

The **rectangle** has two axes of symmetry, and all angles 90°.

The **square** has four axes of symmetry, and all angles 90°.

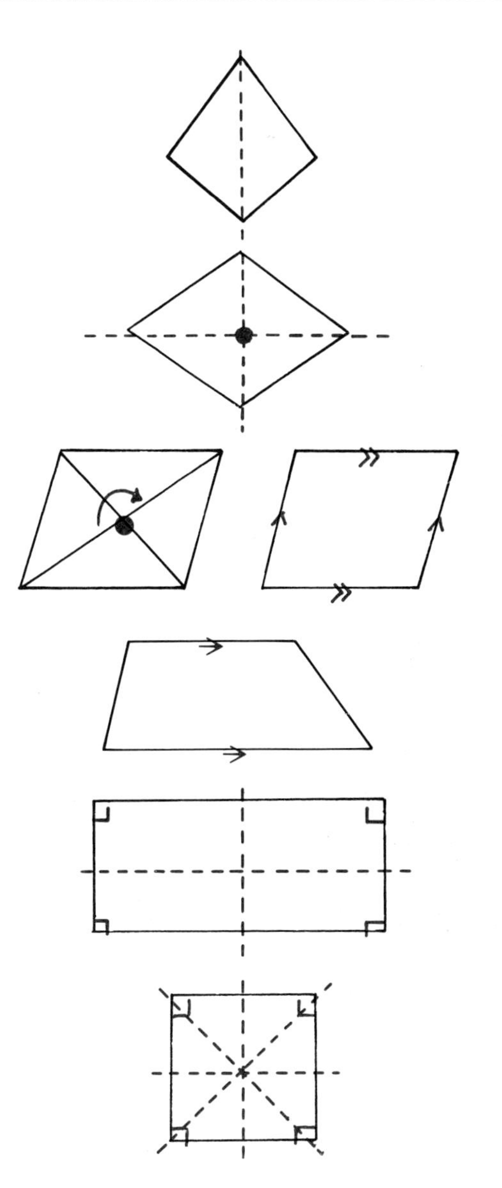

EXERCISE 6

1 Which types of quadrilateral have:

a all their sides equal
b all their sides and angles equal
c two pairs of parallel sides
d no axes of symmetry
e just one axis of symmetry
f diagonals which bisect each other at right angles
g half-turn symmetry
h quarter-turn symmetry?

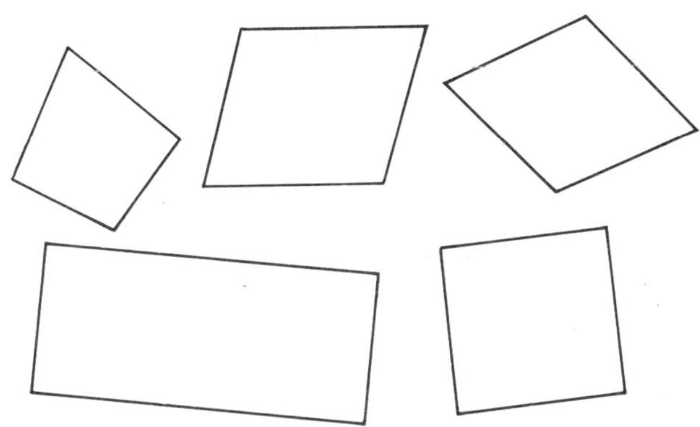

2 Number the axes 0–26 and 0–16, as shown, on a sheet of squared paper.

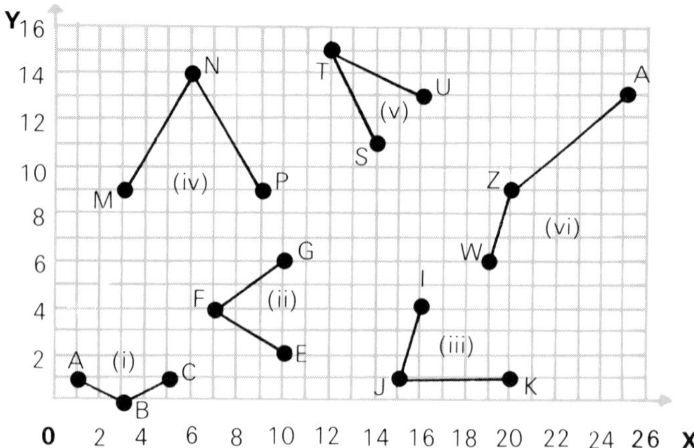

a	**Copy:**	**Complete:**	**Give coordinates of:**
(i)	ABC	Kite ABCD, with diagonal BD 5 units long	D
(ii)	EFG	Rhombus EFGH	H
(iii)	IJK	Parallelogram IJKL	L
(iv)	MNP	Kite MNPR, with diagonal NR 7 units long	R
(v)	STU	Rhombus STUV	V
(vi)	WZA	Parallelogram WZAB	B

b Add some shapes of your own, and write down the coordinates of their vertices.

CHALLENGE

Draw five patterns of equally spaced dots and join them up to make:

a *a rhombus*
b *a kite, which is different from the rhombus*
c *two different parallelograms*
d *three different sizes of square*
e *two different trapezia.*

BRAINSTORMER

Find which type of quadrilateral you get when you join the midpoints of the sides of:

a *a rectangle* ***b*** *a square* ***c*** *a rhombus*
d *a kite* ***e*** *a parallelogram*
f *a trapezium with an axis of symmetry.*

INVESTIGATION

Choose a quadrilateral. Follow the arrows on first one, then the other, flowchart until you find which box it goes in.
Repeat for other types of quadrilateral until you have found entries for all the boxes.

a

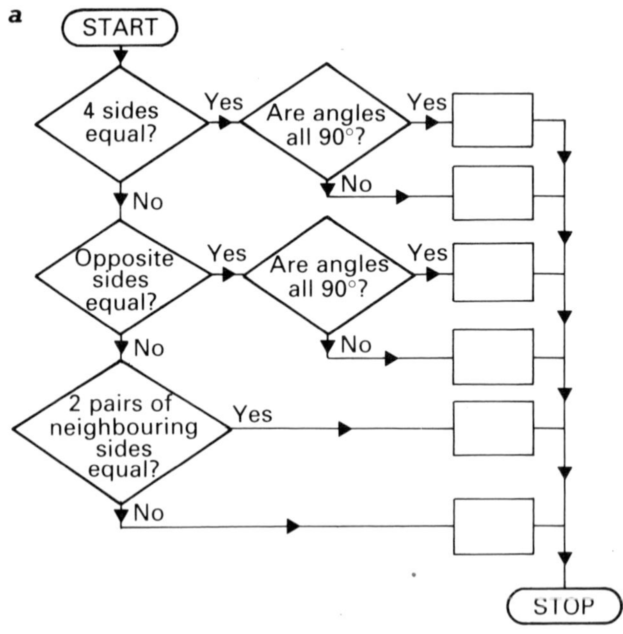

b

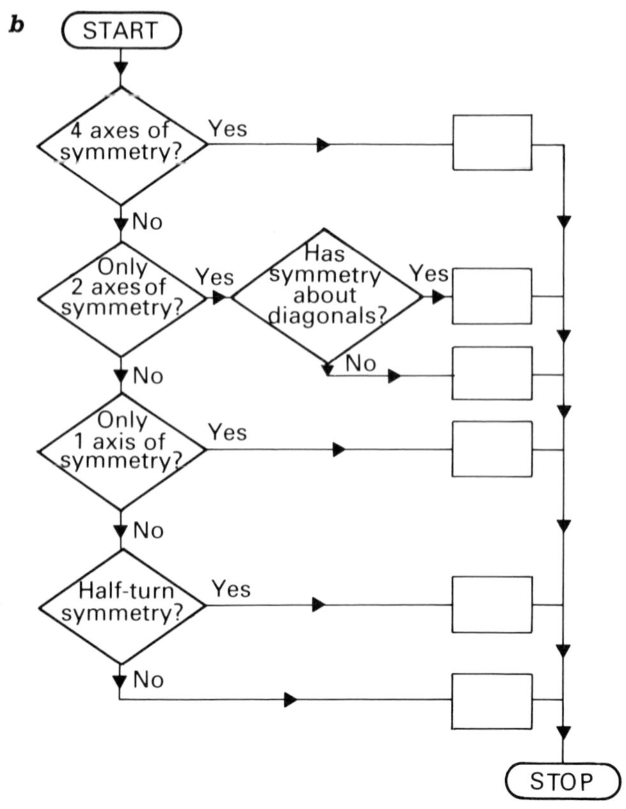

CHECK-UP ON KINDS OF QUADRILATERAL

1 Which types of quadrilateral are these?

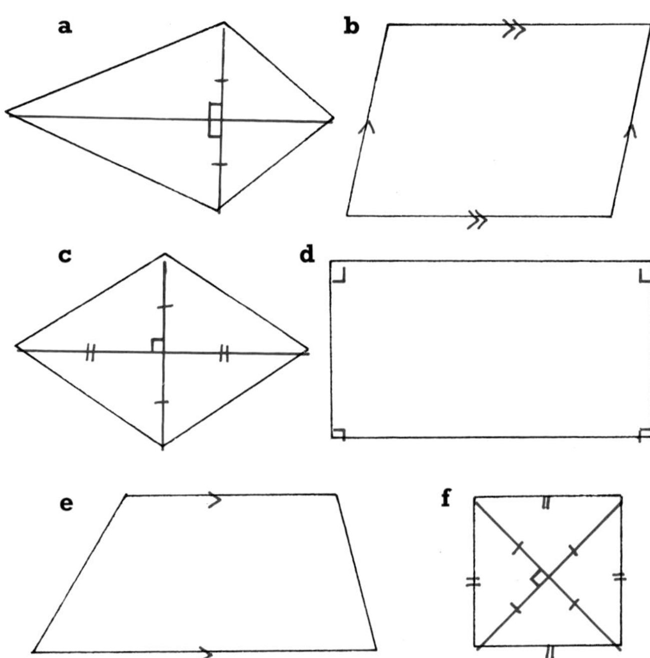

2 Which of the shapes above have:
a just two axes of symmetry
b no axis of symmetry
c a centre of symmetry?

3 Copy the kite ABCD and the rhombus PQRS, and fill in as many lengths and angles as you can.

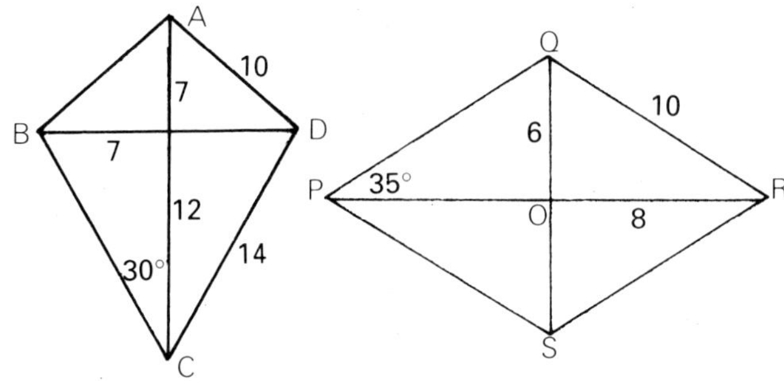

4 DEFG is a parallelogram.

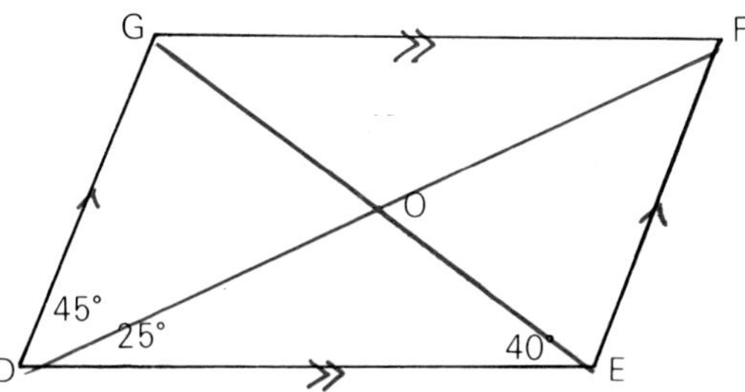

a Under a half turn about O, where do these go?
(i) DO (ii) DE (iii) ∠EFG (iv) ∠ODE
(v) △DOG
b Copy the parallelogram, and fill in all the angles.

5 a A is the point (4, 3), B is (8, 5), C(12, 3) and D(8, 1). Explain why ABCD is a rhombus.
b Find the coordinates of P, Q, R and S, the midpoints of AB, BC, CD and DA. Explain why PQRS is a rectangle.

6 Calculate:
a the size of each angle in the parallelogram
b the perimeter of the parallelogram.

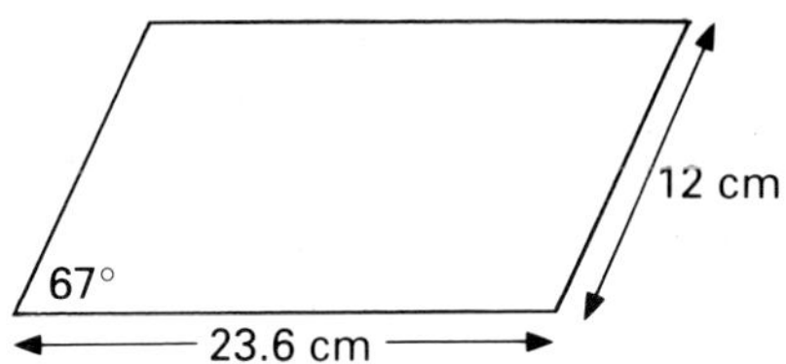

7 a In the diagram, name:
(i) a parallelogram (ii) a trapezium.
b Copy the diagram, and mark four sets of equal angles.

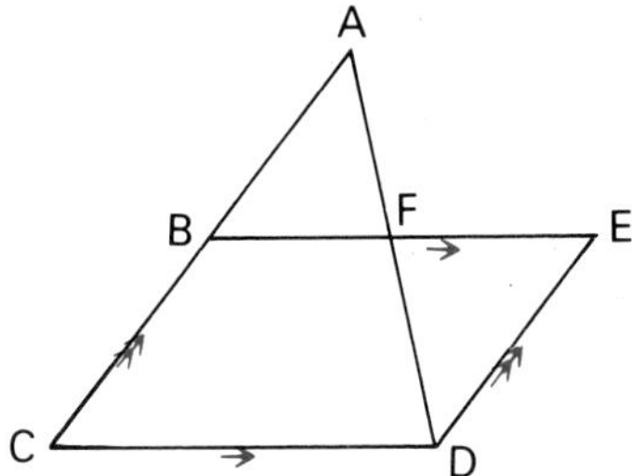

15 SOME SPECIAL NUMBERS

LOOKING BACK

1 Find two more numbers in each sequence.
a 2, 4, 6, . . . **b** 3, 6, 9, . . . **c** 5, 10, 15, . . .
d 9, 18, 27, . . .

2

The traffic lights turn red every 2 minutes. Will they turn red after:
a 4 minutes **b** 10 minutes **c** 15 minutes
d 1 hour?

3 Which gives the greater number in each pair?
a $2 \times 2 \times 2$, or 3×3 **b** 5×5, or $3 \times 3 \times 3$
c 10×10, or $1 \times 1 \times 1 \times 1$

4 Find the missing numbers.

a

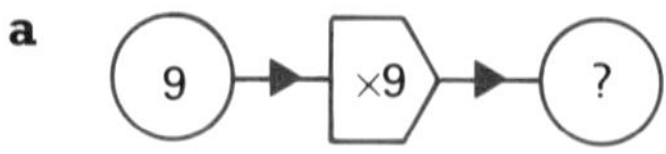

b

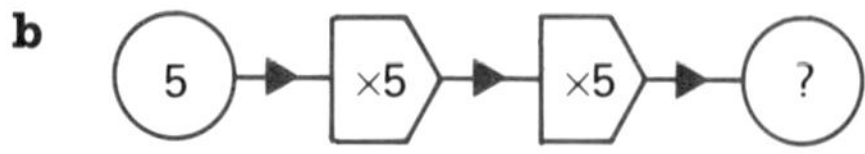

c

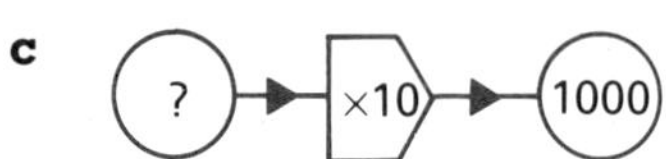

d

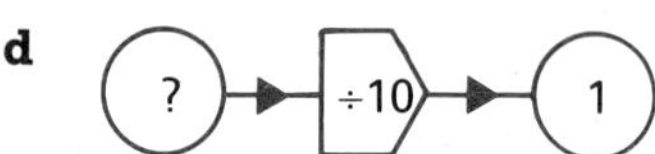

5 True or false? 'The first number divides exactly into the second number.'
a 2, 10 **b** 3, 18 **c** 4, 14 **d** 5, 100
e 6, 26 **f** 7, 50 **g** 8, 64 **h** 9, 72

6 Find two more numbers for each sequence.
a 1, 5, 9, 13, . . . **b** 4, 8, 12, 16, . . .
c 4, 8, 16, 32, . . .

7 Can you find any numbers, apart from 1 and the number itself, which divide exactly into each of the numbers in the picture?

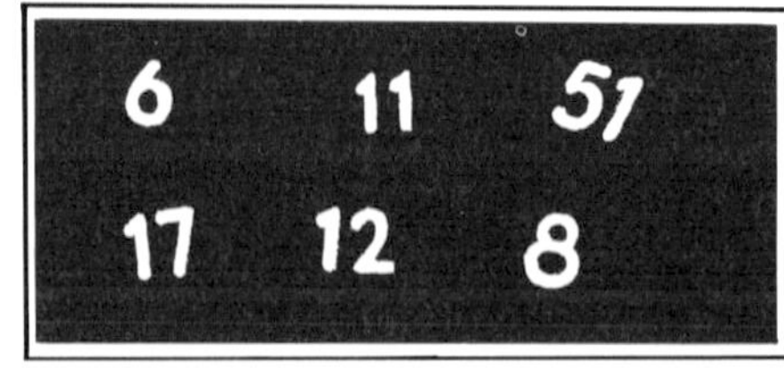

8 $15 = 15 \times 1$ or 5×3. Copy and complete in the same way:
a $10 = 10 \times \ldots$, or $5 \times \ldots$
b $16 = 16 \times \ldots$, or $8 \times \ldots$, or $4 \times \ldots$

9 Along the road from the traffic lights in question **2**, another set of lights turn red every 3 minutes. Both sets turn red together. After how many minutes will they both turn red together again? Can you find more than one time?

POWERS AND INDICES

A colony of bacteria doubles in number every hour.

Time	Number of bacteria	Shorter form
Start	1	—
After: 1 hour	2	—
2 hours	2×2	2^2
3 hours	$2 \times 2 \times 2$	2^3
4 hours	$2 \times 2 \times 2 \times 2$	2^4
...	...	...
8 hours	...	2^8

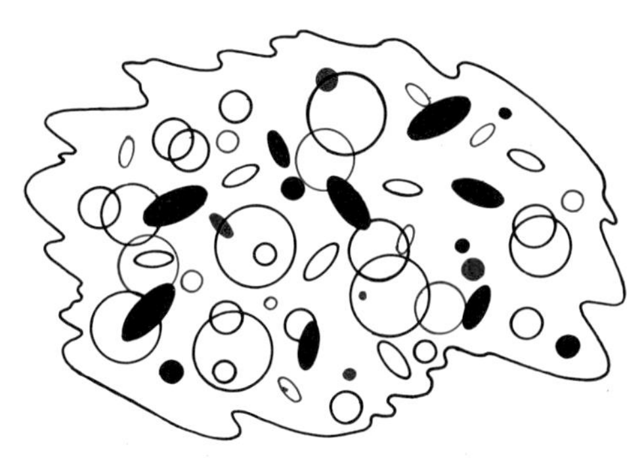

2^4, or 2 **to the power** 4, is a short way of writing $2 \times 2 \times 2 \times 2$.
In 2^4, 4 is called the **index** of the power (plural, 'indices').

EXERCISE 1A

1 Write each of these in a shorter form. For example, $5 \times 5 \times 5 = 5^3$.
a $2 \times 2 \times 2$ **b** 7×7 **c** $3 \times 3 \times 3 \times 3$
d $8 \times 8 \times 8 \times 8 \times 8$

2 Copy and complete:
a $1 \times 1 = 1^2$
$2 \times 2 = \dots$
$3 \times 3 = \dots$
b $1 \times 1 \times 1 = 1^3$
$2 \times 2 \times 2 = \dots$
$3 \times 3 \times 3 = \dots$
c $10 \times 10 = 10^2$
$7 \times 7 \times 7 = \dots$
$4 \times 4 \times 4 \times 4 \times 4 = \dots$

3

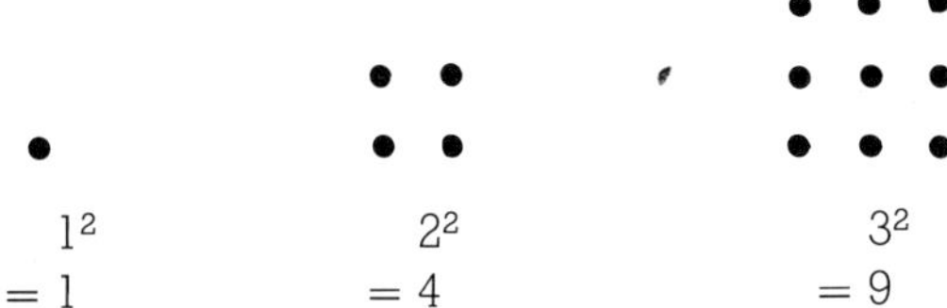

$1^2 = 1$ $2^2 = 4$ $3^2 = 9$

a Draw the next three dot patterns, and count the number of dots in each.
b Why is 3^2 sometimes called '3 squared'?

4 Calculate 12^2:
a as 12×12
b using the x^2 key on your calculator.

5 Check that:
a $3^2 + 4^2 = 5^2$ **b** $5^2 + 12^2 = 13^2$
c $10^2 + 11^2 + 12^2 = 13^2 + 14^2$

6 a Calculate the values of 10^2, 10^3, 10^4, 10^5 and 10^6.
b What do we call 10^3 and 10^6?

7 Calculate the values of these numbers, and list them in order, smallest first.

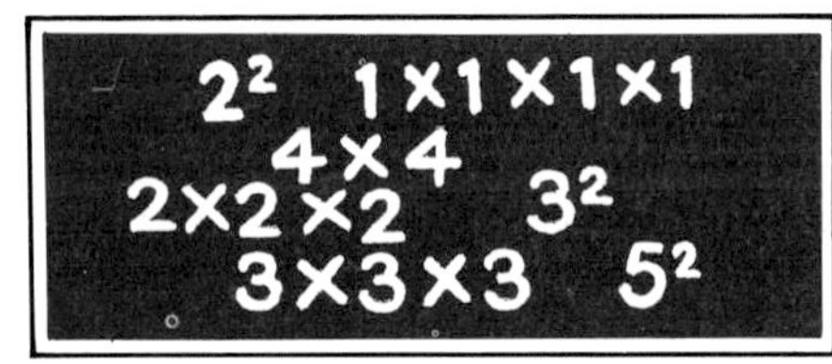
2^2 $1 \times 1 \times 1 \times 1$
4×4
$2 \times 2 \times 2$ 3^2
$3 \times 3 \times 3$ 5^2

8 Find the number of bacteria in the colony at the top of this page after 2, 3, 4, ..., 8 hours by calculating $2^2, 2^3, 2^4, \dots, 2^8$.

9 Wasik puts £10 into a video games company that aims to treble his money each year. How much should he have after:
a 1 year **b** 2 years **c** 3 years
d 4 years **e** 5 years?

10 Write these numbers as powers of 2 or 3. For example, $32 = 2^5$, $81 = 3^4$.
a 4 **b** 9 **c** 8 **d** 27 **e** 16

BRAINSTORMER

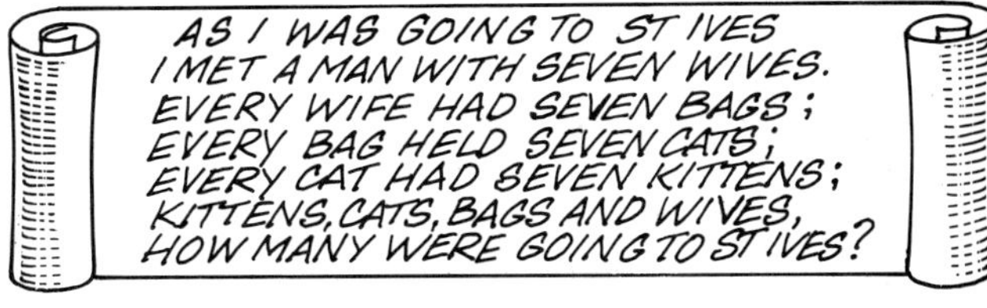
AS I WAS GOING TO ST IVES
I MET A MAN WITH SEVEN WIVES.
EVERY WIFE HAD SEVEN BAGS;
EVERY BAG HELD SEVEN CATS;
EVERY CAT HAD SEVEN KITTENS;
KITTENS, CATS, BAGS AND WIVES,
HOW MANY WERE GOING TO ST IVES?

a *Calculate the number of:*
(i) wives (ii) bags (iii) cats (iv) kittens.
b *Answer the riddle.*

EXERCISE 1B

1 Write in a shorter form as powers of 4, 5, etc:
a $4 \times 4 \times 4$ **b** $5 \times 5 \times 5 \times 5$ **c** $6 \times 6 \times 6 \times 6 \times 6$

2 Calculate the value of:
a 5^4 **b** 8^3 **c** 10^5 **d** 2^6 **e** 3^3

3 Which is greater in each pair?
a 3^2 or 2^3 **b** 4^3 or 3^4 **c** 2^4 or 4^2 **d** 5^3 or 11^2

4 The index in the power 10^5 is 5. Write down the index in:
a 2^7 **b** 3^4 **c** 7^1 **d** 10^9 **e** 1^5

5 These powers are all less than 100.
Arrange them in order, smallest first.

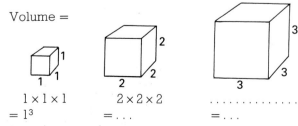

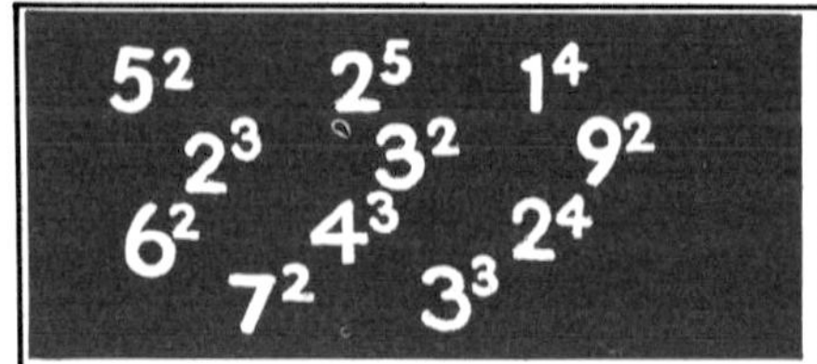

6

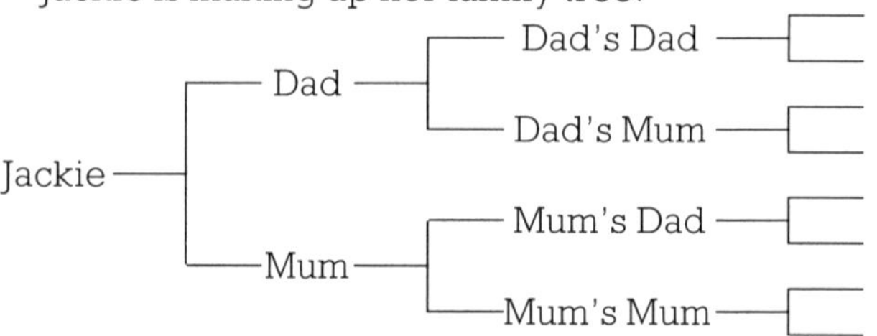

Jackie is making up her family tree.

Jackie — Dad — Dad's Dad
Jackie — Dad — Dad's Mum
Jackie — Mum — Mum's Dad
Jackie — Mum — Mum's Mum

a In the next generation the first entry will be Dad's Dad's Dad.
How many people belong to that generation?

b If we use D^3 to stand for the number of people in the generation headed by Dad's Dad's Dad, for example, what is the value of:
(i) D^3 (ii) D^4 (iii) D^5?

EXERCISE 1C

1 Find the numbers which go in the circles below.
For example,

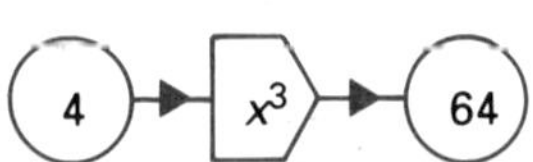

a
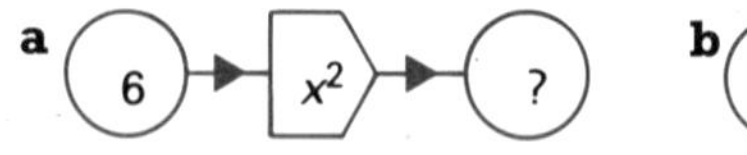

b
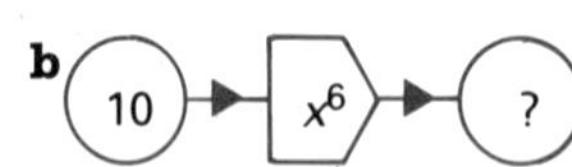

c
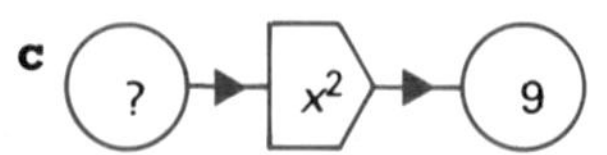

d
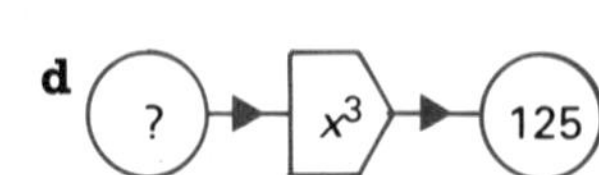

e
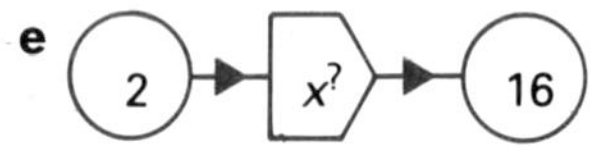

f
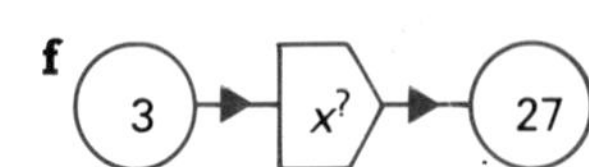

2 a Copy and complete the calculations for these cubes, and the next two in the sequence.

Volume =

1, 1, 1	2, 2, 2	3, 3, 3
$1 \times 1 \times 1$	$2 \times 2 \times 2$	. . . .
$= 1^3$	= . . .	= . . .

b Why is 2^3 sometimes called '2 cubed'?

3

Wasik's £10 doubles in value each year. How long will it be before its value is:
a £320 **b** £1280?

4 If $3^n = 9$, then the value of the index n must be 2, because $3^2 = 9$.
Find the value of the index in each of these:
a $2^n = 8$ **b** $3^n = 27$ **c** $10^n = 10\,000$ **d** $5^n = 625$

5 a Calculate 2^4, and check your answer using the y^x key on your calculator.

b Use your calculator to find the value of:
(i) 2^{10} (ii) 3^8 (iii) 5^6 (iv) 10^6 (v) 1^{100}

6 Find x, given that:
a $6^2 + 8^2 = x^2$ **b** $9^2 + 12^2 = x^2$ **c** $7^2 + 24^2 = x^2$

CHALLENGES

1 *One billion is 'one thousand million'. Write this number:* **a** *in figures* **b** *as a product of 10s* **c** *as a power of 10.*

2 **a** *Copy and extend this pattern:* $2^2 = 1 \times 3 + 1$
$3^2 = 2 \times 4 + 1$
$4^2 = 3 \times 5 + 1$
. . . .

b *Use it to calculate: (i)* 9^2 *(ii)* 99^2 *(iii)* 999^2

3 *The sides of these squares are 20 cm, 30 cm, 40 cm, 50 cm and 60 cm long. The sum of the areas of two of the squares is equal to the area of a third square. Which three squares?*

SQUARES AND SQUARE ROOTS

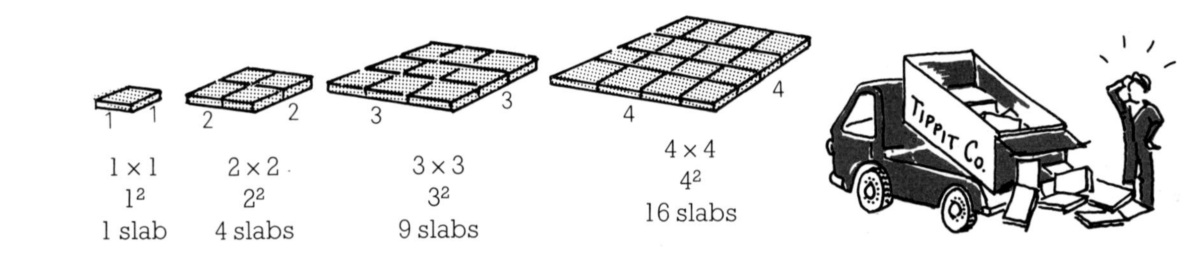

1×1	2×2	3×3	4×4
1^2	2^2	3^2	4^2
1 slab	4 slabs	9 slabs	16 slabs

How many slabs are needed to make a square with 13 slabs along each side?

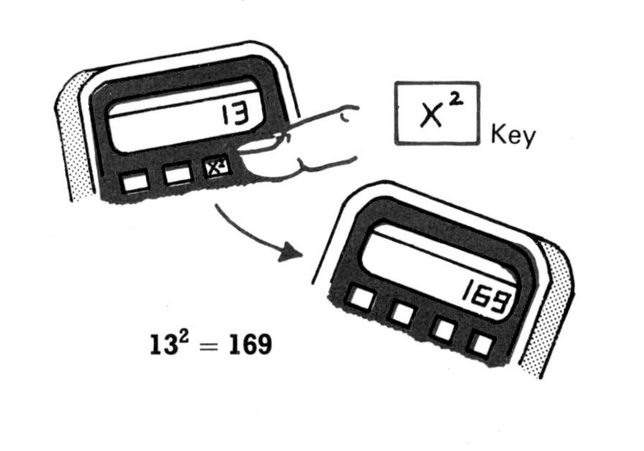

$$13^2 = 169$$

With 256 slabs in a square, how many slabs are there along each side?

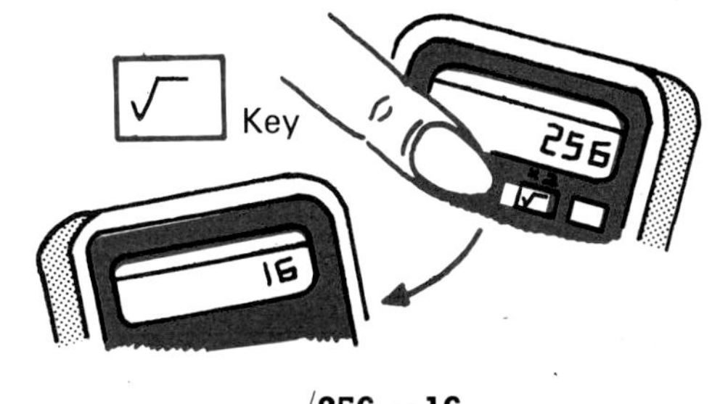

$$\sqrt{256} = 16$$

The square root of 256 is 16, because $16^2 = 256$.

EXERCISE 2A

1 How many slabs are needed to make these square paved areas?

a **b**

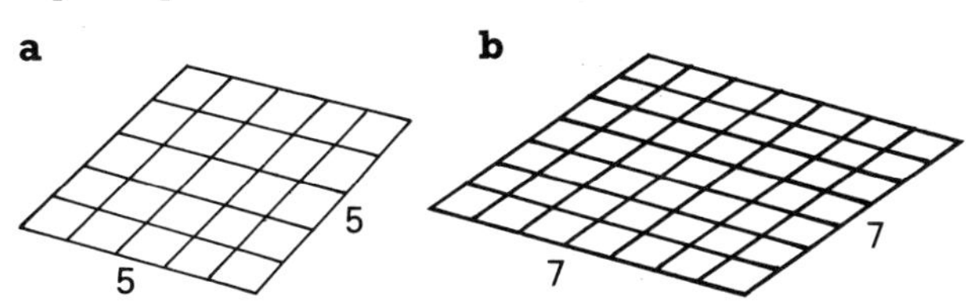

2 How many slabs are there in square paved areas made up of:
a 8 rows of 8 slabs **b** 10 rows of 10 slabs?

3 How many rows of slabs with 1 metre sides would be needed for these paved areas?

a

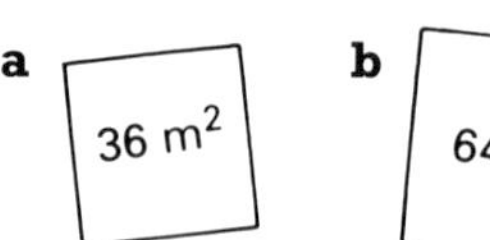

b

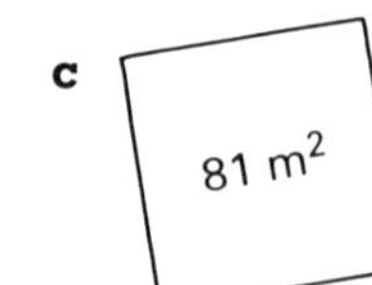

c

81 m²

4 How many rows of slabs with 1 metre sides are needed for a square paving of area:
a $16\,m^2$ **b** $144\,m^2$ **c** $4\,m^2$?

5 Use the x^2 and √ keys on your calculator to find the missing numbers in these tables.

a

Number of slabs in each side	8	11	15	20	39	77
Number of slabs in square						

b

Number of slabs in square	25	81	100	169	256	625
Number of slabs in each side						

6 a Rearrange the slabs in each diagram below to make a square paved area.

(i) (ii) (iii)

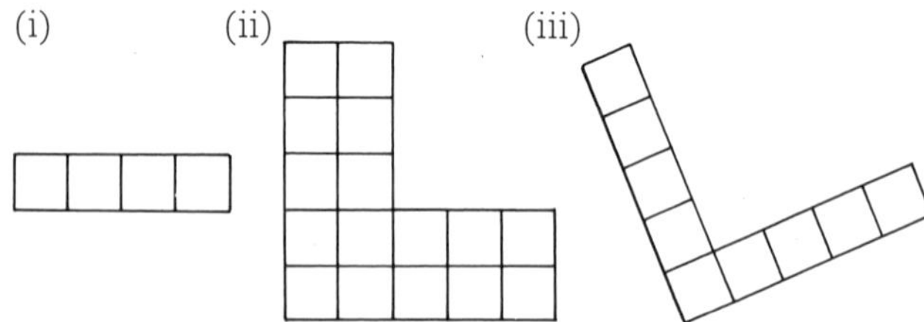

b In your diagrams, how many slabs are there along each side, and in each square?

7 a *No calculator here!*
Which of the following could be 18^2?
Hint $8 \times 8 = 6\mathbf{4}$.
(i) 36 (ii) 180
(iii) 226 (iv) 324

b Check with your calculator.

8 a Which of the following could be $\sqrt{361}$?
(i) 17 (ii) 18 (iii) 19 (iv) 20

b Check with your calculator.

9 Use your calculator to find:
a 16^2 **b** 125^2 **c** 3.4^2 **d** 0.9^2 **e** 0.1^2
f $\sqrt{484}$ **g** $\sqrt{81}$ **h** $\sqrt{7.84}$ **i** $\sqrt{0.64}$ **j** $\sqrt{5625}$

10 a Press these keys on your calculator, and explain what happens.

(i) 8 → x^2 → √ = ?

(ii) 1 0 0 → x^2 → √ = ?

b Without using your calculator, what is the answer to:

2 7 → x^2 → √ = ?

11 a Between which two whole numbers does the value of each square root lie?
(i) $\sqrt{10}$ (ii) $\sqrt{30}$ (iii) $\sqrt{8}$ (iv) $\sqrt{70}$?

b Find the values of the square roots, correct to 1 decimal place.

12

a The floor of Fiona's bedroom is square. Each edge is 2.8 m long. Calculate its area.

b The kitchen in her house is also square. The floor is covered with 169 square tiles. How many tiles are along each edge?

13 The perimeter of a square school playground is 240 m. Calculate:

a the length of one side

b the area of the playground.

EXERCISE 2B/C

1 Rearrange the slabs in each pair to make a square paved area, if you can.

a

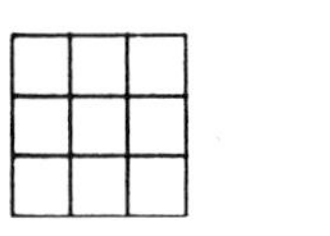

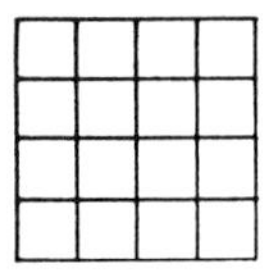

b

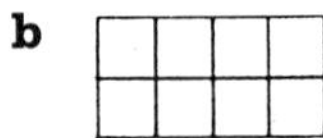

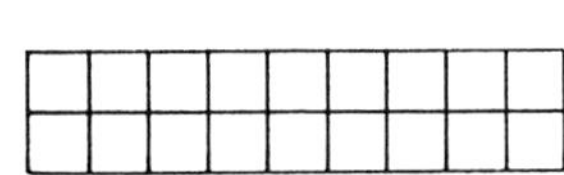

2 Calculate, correct to 2 decimal places:
a 1.35^2 **b** 0.19^2 **c** $\sqrt{2}$ **d** $\sqrt{1004}$ **e** $\sqrt{0.023}$

3 Find the missing IN and OUT numbers.

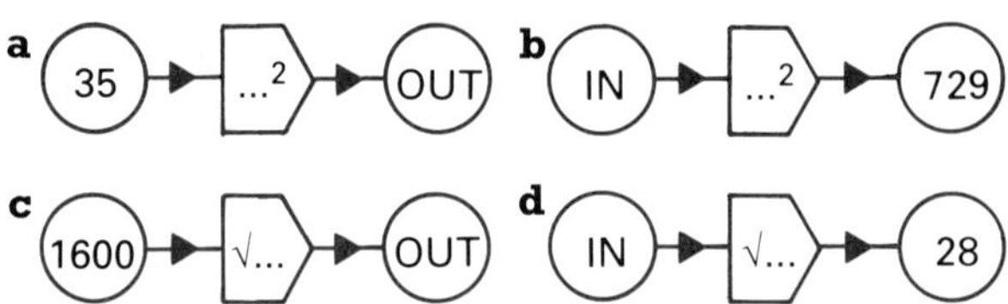

4 a Write down an *estimate* for each of these, to 1 decimal place.
(i) 1.9^2 (ii) 3.5^2 (iii) $\sqrt{20}$ (iv) $\sqrt{50}$
b Check, using your calculator.

5 Find the largest square you could make with 500 slabs. How many slabs are along each edge? How many slabs are left over?

6 Try question **5** again for:
a 88 slabs **b** 1000 slabs.

7

The distance, d metres, fallen by a bungee jumper after t seconds is given by the formula $d = 5t^2$. Calculate:
a d, when: (i) $t = 4$ (ii) $t = 6$.
b t, when:
(i) $d = 45$
(ii) $d = 60$ (to the nearest tenth of a second).

8

Hanif's calculator does not have a square root key. So he has to find $\sqrt{70}$ by a 'trial and improvement' method:

His guess	*The square*	
8^2	64	What number should
8.5^2	72.25	he try next?

Continue his calculation to find $\sqrt{70}$ correct to:
a 1 decimal place **b** 2 decimal places
c 3 decimal places.

9 Repeat Hanif's method to find $\sqrt{40}$, $\sqrt{95}$ and $\sqrt{420}$, each correct to 1, 2 and 3 decimal places.

10 Use a 'trial and improvement' method (no $\sqrt{\ }$ key allowed) to solve:
a $x^2 = 30$
b $x^2 + x = 100$, each correct to 3 decimal places.

BRAINSTORMERS

1 *Which two consecutive numbers add up to 481 when they are squared?*

2 *Which three consecutive numbers when squared add up to 149?*

3 *Find two consecutive numbers which, when squared, have the same digits, but in a different order.*

4 a *Which year was the last exact square?*
b *Which year will be the next exact square?*

MULTIPLES

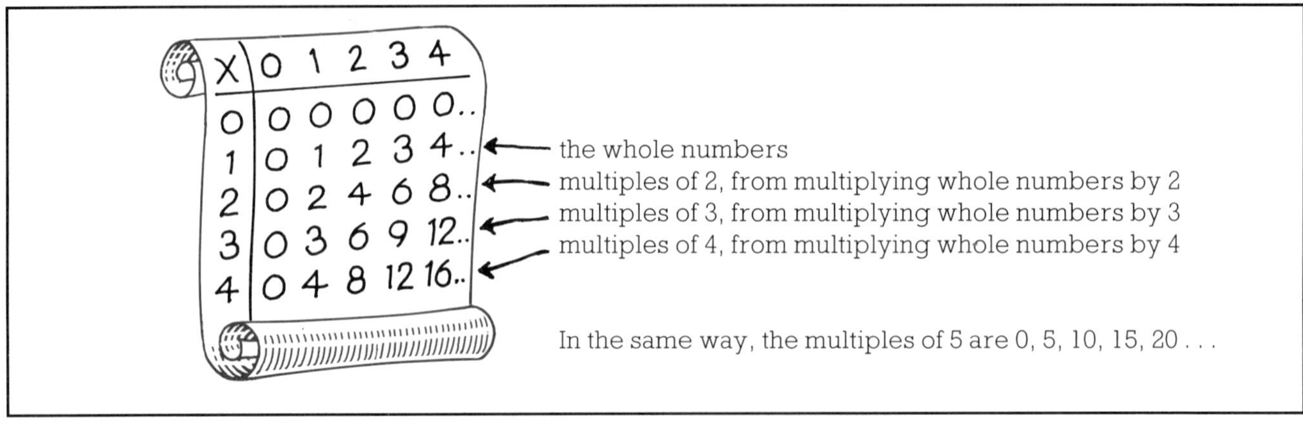

EXERCISE 3A

1 List all the multiples of:
a 2, up to 20 (0, 2, 4, . . .)
b 3, up to 15 **c** 4, up to 24.

2 Say whether or not each number can go into the envelope.

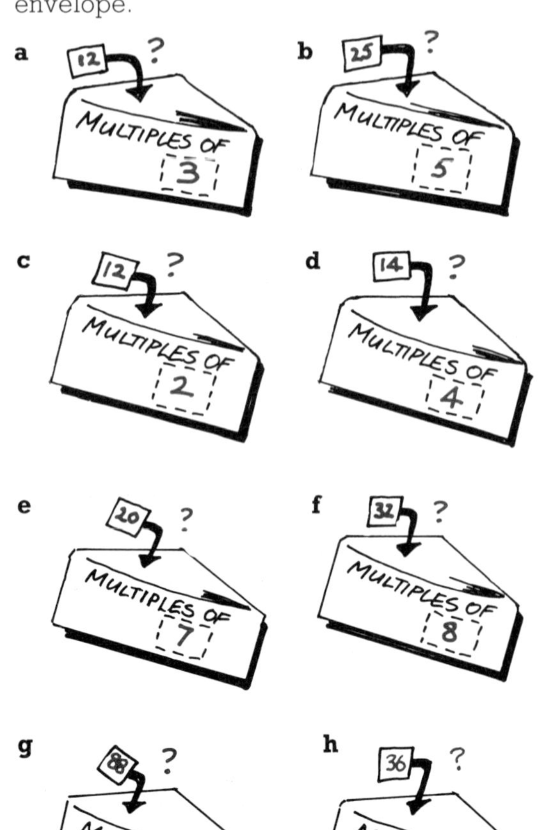

3 List the given numbers that can go into these envelopes.

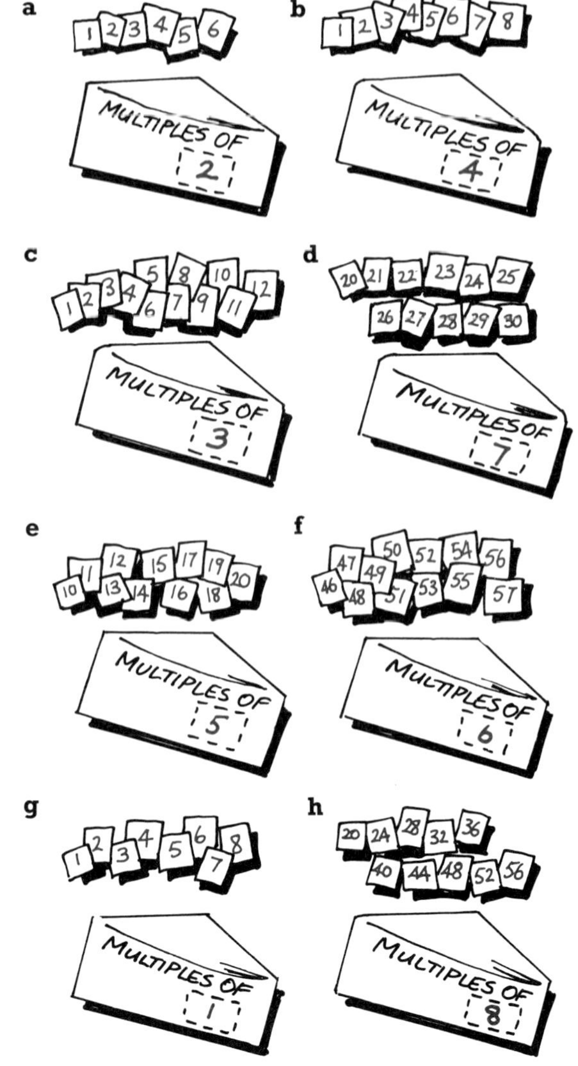

4 List the first five multiples of: **a** 5 **b** 10.

5 a List the whole numbers from 10 to 20.
  b (i) Underline the multiples of 5.
   (ii) Circle the multiples of 10.
  c Which numbers are multiples of 5 *and* 10?

6 a List the whole numbers from 1 to 12.
  b (i) Underline the multiples of 2.
   (ii) Circle the multiples of 3.
  c Which numbers are multiples of 2 *and* 3?

7 

are all multiples of 7.

Which numbers are these sets multiples of?

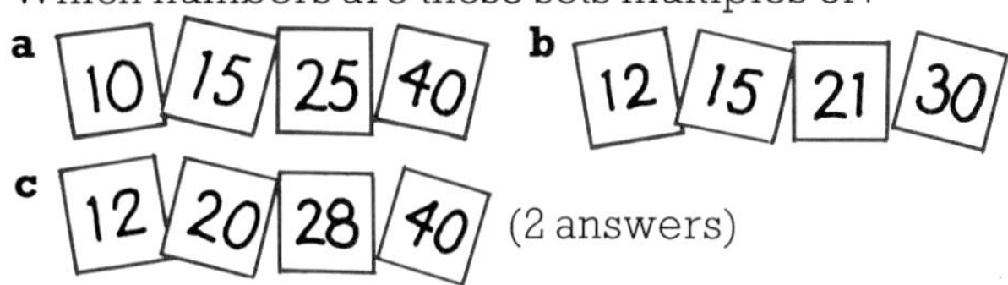

(2 answers)

8 a List all the numbers that have 30 as one of their multiples.
  b Repeat **a** for 36 instead of 30.

EXERCISE 3B/C

> Multiples of 2: **0**, 2, 4, **6**, 8, 10, **12**, 14, 16, **18**, . . .
> Multiples of 3: **0**, 3, **6**, 9, **12**, 15, **18**, 21, 24, 27, . . .
> The **common multiples** of 2 and 3 are 0, 6, 12, 18, . . .
> The **least common multiple (lcm)** of 2 and 3 is 6 (zero is excluded).

1 Make lists of the multiples of the numbers on the envelopes until you have at least two common multiples. Say which is the least common multiple (lcm).

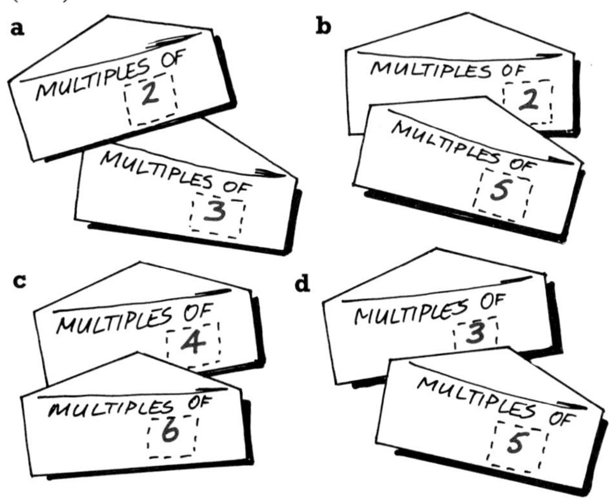

2 Find the lcm of:
a 3 and 4 **b** 6 and 9 **c** 5 and 10 **d** 4 and 10

3 The smallest number of rectangular tiles 4 cm by 3 cm that can make a square is 12. Find the smallest number of the following tiles that can make a square:
a 2 cm by 1 cm **b** 3 cm by 2 cm
c 5 cm by 4 cm **d** 12 cm by 9 cm.
Sketch the squares for **a** and **b**.

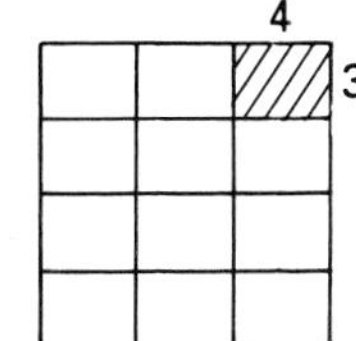

4 Ellie sells eggs from her farm. With 30 eggs she could fill five boxes of 6 eggs, or three boxes of 10 eggs. What is the least number of eggs she needs to fill sets of boxes of:
a 5 eggs or 9 eggs **b** 8 eggs or 12 eggs?

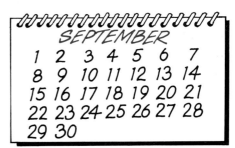

5 a Chris and Sheena join a Youth Club. Chris goes first on September 3rd, and then drops in every 3 days.
Sheena joins on September 4th, and then attends every 4 days.

SEPTEMBER

1	2	3	4	5	6	7
8	9	10	11	12	13	14
15	16	17	18	19	20	21
22	23	24	25	26	27	28
29	30					

(i) Write down the dates in September when Chris goes to the Youth Club, and the dates when Sheena goes.
(ii) On which dates will they both be there?
(iii) On which date could they first meet there?

b Naima goes to the Youth Club on September 6th, and then every 6 days.
Terri goes on the 8th, and then every 8 days.
Repeat part **a** for Naima and Terri.

c Is there any date on which all four girls might meet at the Youth Club?

6 Find the lcm of:
a 2, 17 **b** 6, 18 **c** 8, 10 **d** 10, 14 **e** 4, 6, 9

CHALLENGES

1 a *Choose a number from 10 to 99, reverse its digits, and add the two numbers. For example, $62+26=88$.*

b *Repeat part **a** for other numbers. What can you say about the sum of the two numbers every time?*

c *Subtract the smaller number from the larger this time. What can you say about the difference between the numbers each time?*

d *Try **a–c** for numbers from 100 to 999.*

2

a *Which of these numbers are multiples of:*
(i) 2 (ii) 5 (iii) 10 (iv) 100?

b *How could you tell?*

BRAINSTORMER

May, Ben, and Kris deliver leaflets to the houses along their road.
May delivers at 2, 4, 6, . . .
Ben at 3, 6, 9, . . .
Kris at 5, 10, 15, . . .

a *Which houses receive three leaflets?*

b *Next time they deliver at multiples of 3, 5 and 6. Which houses receive three leaflets now?*

FACTORS

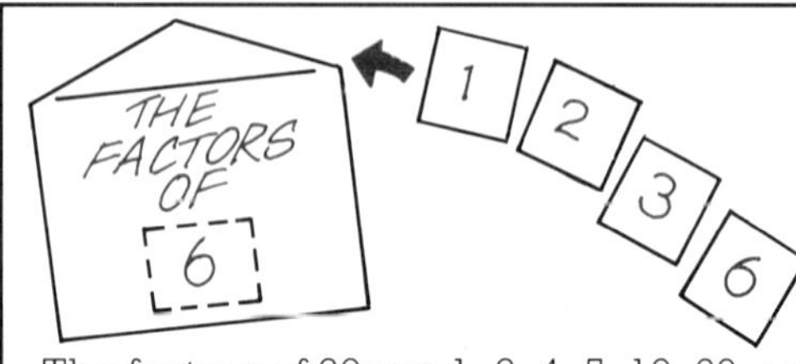

All these cards can go into the envelope, as the numbers are **factors** of 6.
They divide into 6 exactly.

The factors of 20 are 1, 2, 4, 5, 10, 20, as these are the only numbers that divide into 20 exactly.

EXERCISE 4A

1 Can the number go into the envelope in each diagram below?

a

b

c

d

e

? 7 THE FACTORS OF 7

2 Each of these envelopes holds *all* the factors of the number on it. List the factors in each envelope.

a

b

THE FACTORS OF 21

c

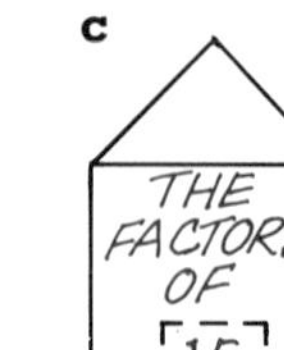

d

THE FACTORS OF 27

e

f

g

h

3 Each factor has a partner.
For example, $28 = 1 \times 28$, 2×14 or 4×7.

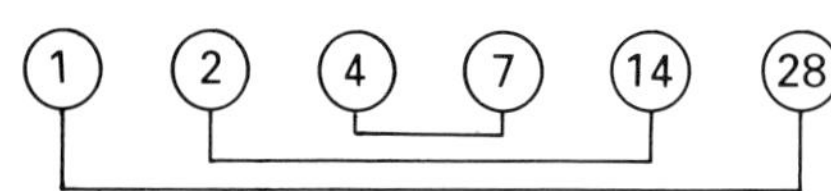

Write down all the pairs of factors of:
a 26 **b** 33 **c** 40 **d** 45 **e** 60

4 Chocolate bars are made in rectangles. Two possible 4-piece bars are shown. They illustrate $4 = 4 \times 1$ or 2×2.

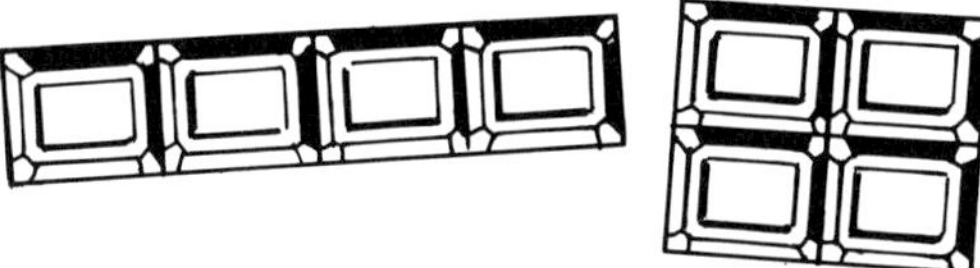

Draw all possible bars that can be made from these, and list pairs of factors of the numbers.
a 8 pieces **b** 5 pieces **c** 9 pieces
d 12 pieces **e** 18 pieces

5

The Puffin Pipe Band always marches in a single line or in a rectangular formation.
a List all possible formations for these numbers of pipers:
(i) 42 (ii) 32 (iii) 50 (iv) 55 (v) 25
(vi) 100
b Which of your answers are square formations?

EXERCISE 4B

1 List all the factors in these envelopes.

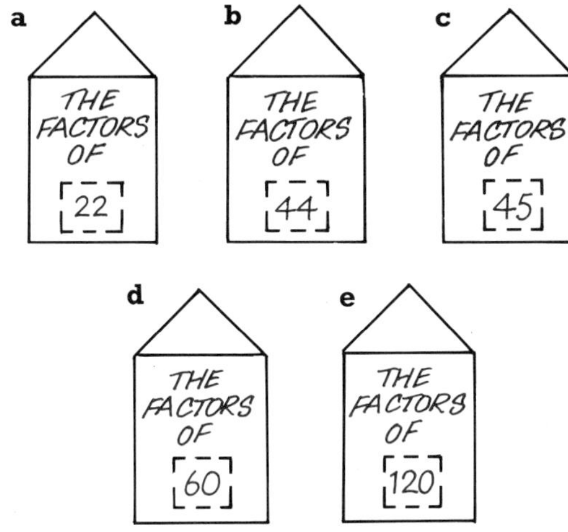

2 a List all the factors of the square numbers 1, 4, 9, . . . , 36.
b Is there always an odd or even number of factors?

3 List all the factors of the cubic numbers 1, 8, 27, . . . , 125.

4

The factors of 20 are 1, **2**, **4**, 5, 10, 20; and the factors of 28 are 1, **2**, **4**, 7, 14, 28. Apart from 1, the **common factors** of 20 and 28 are 2 and 4.
Find all the common factors of these pairs of numbers.
a 6, 9 **b** 8, 12 **c** 10, 20 **d** 12, 18
e 24, 36 **f** 30, 40 **g** 35, 49 **h** 60, 80
i 48, 72 **j** 120, 144

EXERCISE 4C

The highest common factor of 20 and 28 is 4. Find the highest common factor of each of these sets of numbers.
1 42, 60 **2** 48, 64 **3** 36, 72 **4** 51, 68
5 48, 60 **6** 25, 35, 45 **7** 32, 36, 40 **8** 42, 63, 98
9 36, 54, 72

CHALLENGE

Find all twenty-four factors of

PRIME NUMBERS

These are examples of **prime numbers**.

A prime number is divisible only by itself and by 1, but 1 is not a prime number.

EXERCISE 5

1 a Copy and complete this list:

Number	Factors
2	1, 2
3	1, 3
4	1, 2, 4
5	.
.	.
.	.
20	.

b Use your list to write down the numbers which have only two factors—the prime numbers from 1 to 20.

2 Use this 'program' to find all the prime numbers between 1 and 100.

1. Start a list with the numbers 2, 3, 5.
2. Add the next odd number to the list as a target to test.
3. Try to divide the target by each number already in the list.
4. If the target is divisible by any of the numbers, score it out and go to line 2.
5. Otherwise leave the target number in the list.
6. Go to line 2.

Check that you have all 25 prime numbers between 1 and 100.

3 a Why is no even number, apart from 2, a prime number?

b Why are these not prime numbers:
(i) 9 (ii) 25 (iii) 33?

4 How can you tell that each of these numbers is not prime?
a 120 **b** 155 **c** 802 **d** 777 **e** 363

5 The Puffin Pipers still march in rectangular formation or in a single line. Which of these numbers of pipers can only march in single file (like 7 × 1, for example)?
11, 12, 15, 23, 39, 43, 51, 68, 81, 97, 100, 101

6

a List all the prime numbers up to 100 given by these instructions.

b Whose method gives most prime numbers?

7 Evan said he had found a way to get all the prime numbers from 5 to 97. 'Take all the multiples of 6, add 1 to each and subtract 1 from each. Score out any that are not prime.' Is he correct?

8 The largest known prime number in 1992 was $2^{756839}-1$, a number with 227 832 digits. Calculate 2^2-1, 2^3-1, 2^4-1, . . . and find as many prime numbers as you can in this way.

CHALLENGE

a *Make a number spiral like this from 1 to 100, starting* ***4 3*** *.*
1 2

17	*16*	*15*	*14*	*13*
18	*5*	***4***	***3***	*12*
19	*6*	***1***	***2***	*11*
20	*7*	*8*	*9*	*10*
21	*22*	*23*	*24*	*. . .*

b *Circle all the prime numbers.*

c *Join up diagonal rows of prime numbers. Find the longest row you can.*

d *Look for other patterns of prime numbers.*

BRAINSTORMER

Sunnyside Avenue has odd-numbered houses on one side from 1 to 63. Janine's house has a prime number, but the two houses on either side of hers do not. What is the number of her house?

PRIME FACTORS

After all her work on factors and prime numbers, Molly noticed that every number had a set of prime factors. Can you see the set of prime factors of 60?

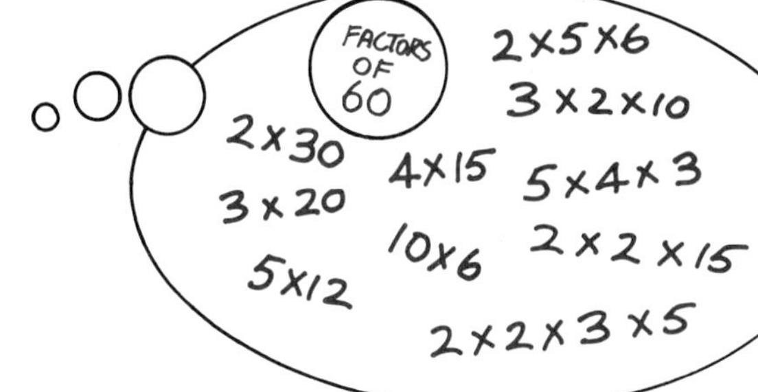

Molly tried to think of a good system for finding these prime factors. 'Like peeling an onion,' she thought. 'One prime factor at a time.'

For 60, peel off 2: 2×30
Peel off 2 again: $2 \times 2 \times 15$
Now peel off 3: $2 \times 2 \times 3 \times 5$
$= 2^2 \times 3 \times 5$

or like branches of a tree:

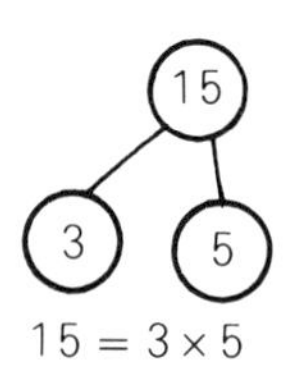

$15 = 3 \times 5$

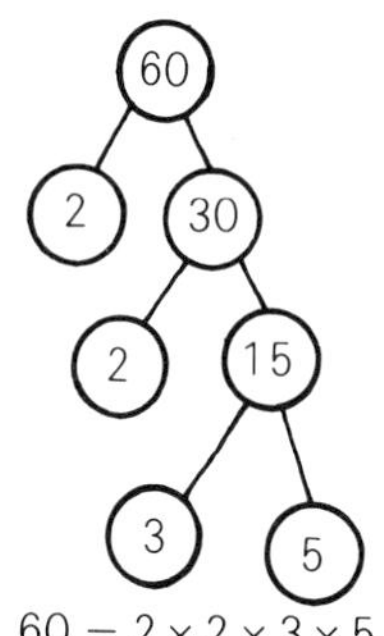

$60 = 2 \times 2 \times 3 \times 5$

EXERCISE 6A

Copy and complete these 'trees'. Then write each number as a product of prime factors, as in the example, $60 = 2 \times 2 \times 3 \times 5$.

1 a 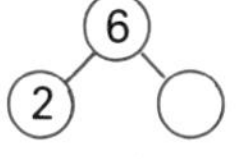

$6 = 2 \times \ldots$

b 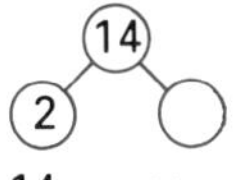

$14 = \ldots \times \ldots$

c 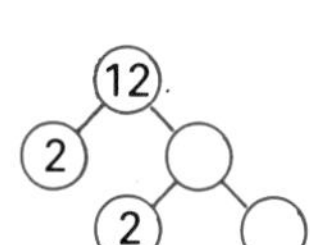

$12 = 2 \times 2 \times \ldots$

d 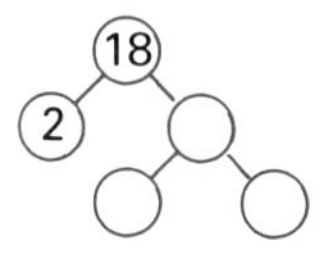

$18 = \ldots \times \ldots \times \ldots$

2 a 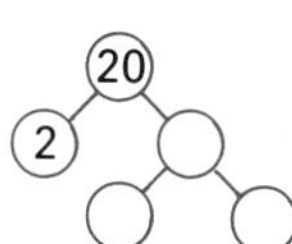

b 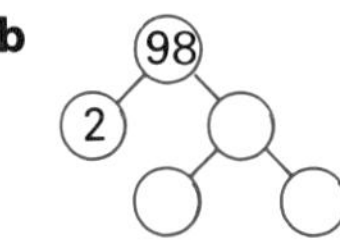

c 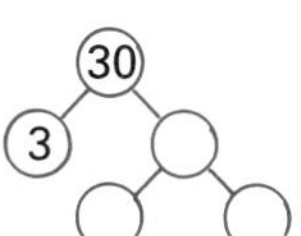

d 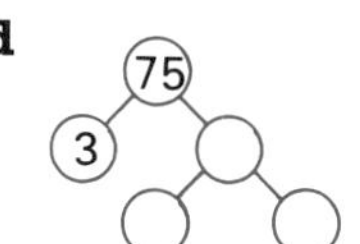

3 a 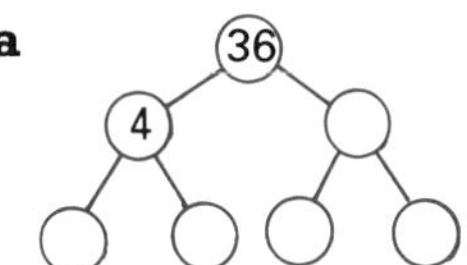

b 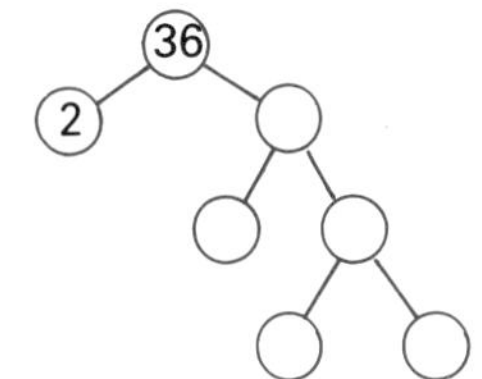

Write each number in the two questions which follow as a product of its prime factors.

4 a 21 **b** 35 **c** 27 **d** 45 **e** 50

5 a 100 **b** 120 **c** 125 **d** 144 **e** 180

EXERCISE 6B/C

Write each number in questions **1** and **2** as a product of prime factors in index form.
For example, $72 = 8 \times 9 = 2 \times 4 \times 3 \times 3 = 2 \times 2 \times 2 \times 3 \times 3 = 2^3 \times 3^2$.

1 a 32 **b** 36 **c** 90 **d** 88 **e** 24

2 a 48 **b** 100 **c** 81 **d** 192 **e** 400

Prime factors are useful for finding lowest common multiples.
$24 = 2 \times 2 \times 2 \times 3$, and $30 = 2 \times 3 \times 5$. So the lcm of 24 and 30 must contain each of these products of factors.
The lcm of 24 and 30 $= 2 \times 2 \times 2 \times 3 \times 5$.

3 Use this method to find the lcm of:
a 8 and 12 **b** 20 and 35 **c** 18 and 15
d 32 and 48

4

One lighthouse shines out every 30 seconds, the other every 100 seconds. At one instant they shine together. How long will it be until this next happens?

5 A planet has three moons. Alphos goes round it in 12 days, Betos in 4 days and Gammos in 8 days.

a The moons were all in a straight line on 1st July. On what date will they next be in a straight line?
b A comet knocks Gammos into a different orbit. It now goes round the planet in 7 days. How does this affect your answer to **a**?

BRAINSTORMER

To the nearest 10 years, the times taken by Jupiter, Saturn, Uranus, Neptune and Pluto to revolve round the sun are 10, 30, 80, 160 and 250 years. How long must it be between occasions when they are all in a straight line out from the sun?

INVESTIGATIONS

1 *Make a list of factors of numbers from 1 to 50, like this:*

Number	Factors	Number of factors
1	1	1
2	1, 2	2
3	...	...
...	...	...

Investigate numbers in your list that:
a *have only one factor*
b *have exactly two factors. What are these called?*
c *have an odd number of factors. What do you call these numbers? List the first ten*
d *have exactly three factors*
e *have most factors*
f *are the smallest ones with 1, 2, . . . , 6 factors.*

2 a *The factors of 6 are 1, 2, 3, 6. The 'factor sum' of 6 is 1 + 2 + 3 = 6. (Don't count the number itself.) Investigate numbers which have factor sums one less than the numbers themselves. What do you call these numbers?*
b *A 'perfect number' has a factor sum equal to itself. Find two perfect numbers in the list you made in part **1**.*

CHECK-UP ON SOME SPECIAL NUMBERS

1 a Write $2 \times 2 \times 2 \times 2$, $5 \times 5 \times 5$ and 10×10 in index form.
b Calculate the value of each.

2 Which is greater in each pair below?
a 2^4 or 3^2 **b** 5^2 or 3^3 **c** 4^3 or 8^2

3 Which of these numbers of square slabs can be arranged as a square paved area?
a 4 **b** 14 **c** 24 **d** 36 **e** 49 **f** 80

4 Calculate:
a 14^2 **b** 7.7^2 **c** $\sqrt{900}$ **d** $\sqrt{289}$ **e** $\sqrt{4.84}$

5

A checked tablecloth covers the top of a square table exactly. It has 1225 centimetre squares on it. What is the length of each edge of the table?

6 a Write these terms of the sequence 1, 4, 9, 16, . . . as squares.
b What is the nth term?
c Which term is one million?

7 List the first four multiples of:
a 2 **b** 6 **c** 9 **d** 10

8 Of which numbers (apart from 1) are these all multiples?
a 3, 6, 9 **b** 22, 10, 6 **c** 15, 5, 10 **d** 21, 35, 14

9 List all the factors of:
a 7 **b** 15 **c** 12 **d** 19 **e** 20

10 Which of these are prime numbers?
2, 3, 27, 29, 30, 31, 32, 33

11 Write these as products of prime factors.
a 12 **b** 18 **c** 16 **d** 48 **e** 135

12 Write each answer in question **11** in index form.

13 a Write down the first six multiples of 3 and of 5.
b What is their least common multiple (apart from 0)?

14 Find the least common multiple (lcm) of 8 and 20.

15 At the school disco a green light flashes every 3 seconds, and a red light every 4 seconds. If they start together, when will they next flash at the same time? A blue light starts at the same time and flashes every 5 seconds. When will all three lights next flash together?

16 FORMULAE AND SEQUENCES

LOOKING BACK

1 Find three more numbers for each sequence:
a 1, 3, 5, 7, . . . **b** 4, 8, 12, 16, . . .
c 20, 18, 16, 14, . . . **d** 30, 25, 20, 15, . . .
e 1, 2, 4, 8, . . . **f** 2, 3, 5, 8, . . .

2 Solve these equations:
a $x+2=8$ **b** $y+4=17$ **c** $n-8=10$
d $5t+1=11$ **e** $4u+2=22$ **f** $8v-3=21$

3 Use the patterns in the rows and columns to find the numbers the letters stand for.

	A	B	C	D
1	3	5	x	y
2	6	8	10	b
3	9	11	13	15
4	a	k	16	m

4 $t=2$. Find the value of:
a $5t$ **b** $t+3$ **c** $3t+1$ **d** $t-2$
e t^2 **f** $3t^2$ **g** $(3t)^2$ **h** $(t-1)^2$

5 Find the value of $2n-4$ when $n=$
a 1 **b** 2 **c** 3 **d** 4 **e** 5

6 a Count the number of small triangles in each picture.

  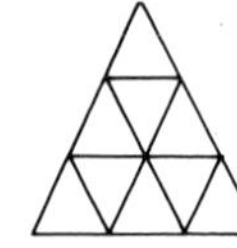 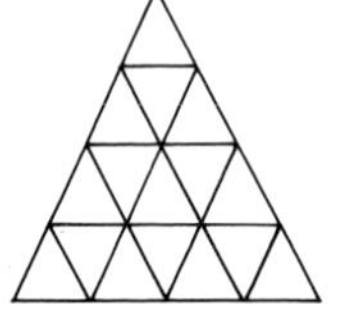 . . . .

b Write down three more numbers in the sequence.

7 Write down a formula for the perimeter P of each shape below.

a
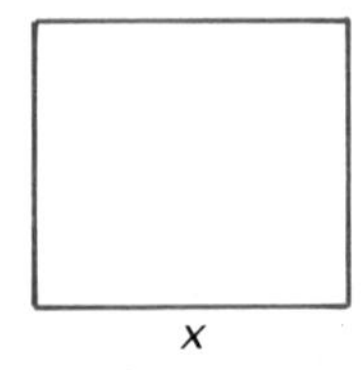

Square

b

Rectangle

c
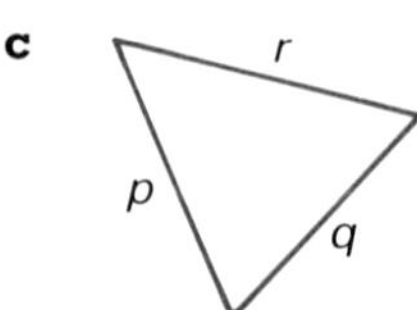

Triangle

d
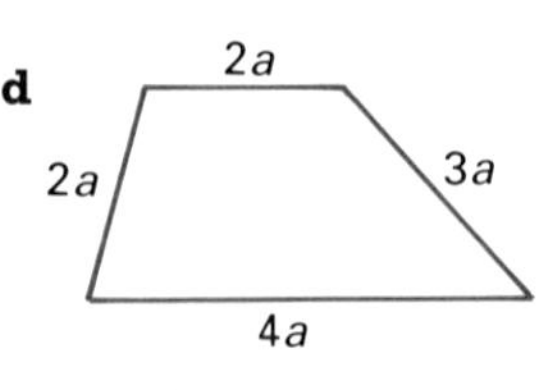

Trapezium

8 The volume of the cylinder is given by $\pi r^2 h$. Calculate its volume when $r=2.5$ and $h=5.2$. Use the π key on your calculator, and give your answer to the nearest cm³.

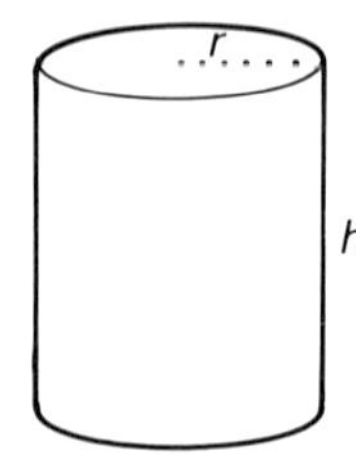

9 a Copy and continue this pyramid of numbers.

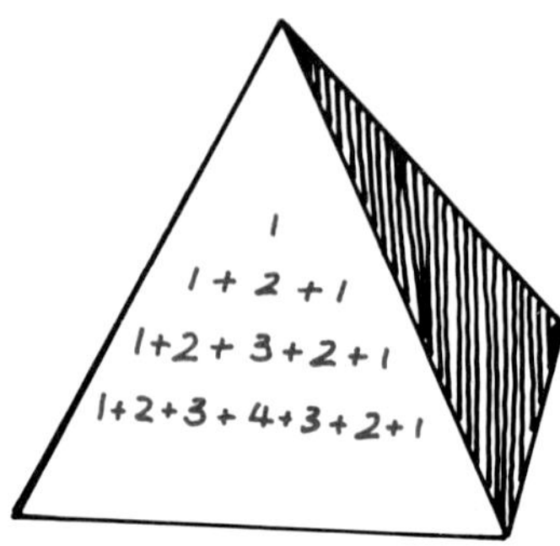

b (i) Calculate the sum of each of the given rows.
(ii) What is the sum of:
$1+2+3+\ldots+99+100+99+\ldots+3+2+1$?

MAKING AND USING FORMULAE

Life is full of rules!

Throw a 6 to start

SCHOOL HOURS 9am–4pm

Entrance £2.50

COOK FOR 30 MINUTES PLUS 20 MINUTES PER KG

HORROR OF HORRORS NOW SHOWING 18

30

Be home by 11pm!

CLOSED 1pm–2pm

Can you think of any others?

Mathematics has its own rules.

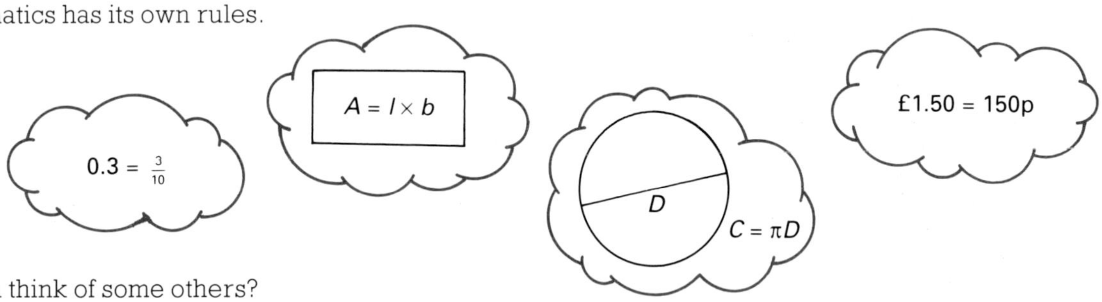

$A = l \times b$

$0.3 = \frac{3}{10}$

$C = \pi D$

£1.50 = 150p

Can you think of some others?

Who needs mathematics? Everyone! Employed or unemployed, in the home and outside it; joiners, teachers, scientists, sales personnel, hairdressers, shopkeepers, plumbers, dressmakers, engineers, cooks, . . . All of them make use of rules, or formulae, at some time.

EXERCISE 1A

1

'How much tea?'
'One spoonful per person and one for the pot.'
How many spoonfuls for:
a 3 people **b** 6 people **c** x people?

2 'What's the selling price?'
'Double the cost price.'
Calculate the selling price if the cost price is:
a £5 **b** £9 **c** £y.

3 'What is the perimeter of a square window?'
'Four times the length of the window.'
Calculate the perimeter for a length of:
a 2 m **b** 50 cm **c** w m.

4 'How far is it round the circumference of the pipe?'
'About three times its diameter.'

How far is it for a diameter of:
a 30 mm **b** 15 cm **c** d m?

5

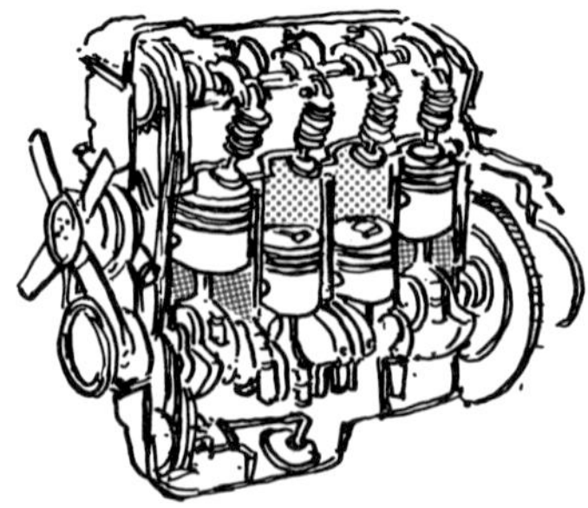

'What is the area of the end of each piston?'
*'About three **times** the radius, **times** the radius.'*
Calculate the area if the radius is:
a 4 cm **b** 10 cm **c** r cm.

6 The weather forecaster's rough rule for changing a Celsius temperature to Fahrenheit is: 'Double it, and add 30.'
Change these to Fahrenheit:
a 16°C **b** 25°C **c** x°C.

7

a In four games, Rovers scored 2, 1, 5 and 3 goals. How many goals did they score altogether?
b In four more games they scored p, q, r and s goals. Make a formula for the total T goals in these four games, $T = p + \ldots + \ldots + \ldots$.

8 **a** Toby saved £2, £5, £3, £1 and £4. He then spent £8. How much had he left?

b He saves £a, £b, £c, £d and £e, and then spends £f. Make a formula for the amount £A he has left.
$A = a + \ldots + \ldots + \ldots + \ldots - \ldots$.

9 Mr Jones runs a music shop.

a Calculate his profit on CDs which:
(i) he buys for £8 and sells for £9.50
(ii) he buys for £9.50 and sells for £11.25.
b If he buys for £c and sells for £s his profit is £p. Which is greater, £c or £s? Make a formula for the profit, $p = \ldots - \ldots$.

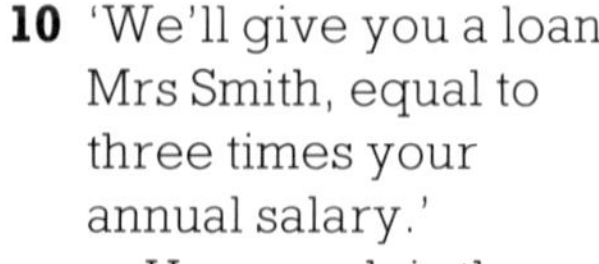

10 'We'll give you a loan, Mrs Smith, equal to three times your annual salary.'

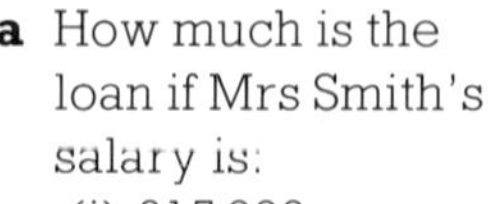

a How much is the loan if Mrs Smith's salary is:
(i) £15 000
(ii) £18 500?
b The loan is £L and Mrs Smith's salary is £A.
Make a formula, $L = \ldots$.

11 Jayne has a starting salary of £9000, and annual increases of £800.
a What is her salary after:

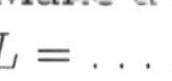

(i) 1 year (ii) 2 years (iii) 3 years?
b Her salary after n years is £s. Make a formula, $s = 9000 + \ldots$.

12 **a** Jim, Frank and Ian weigh themselves. Their weights are 60 kg, 70 kg and 74 kg. Calculate:
(i) their average weight
(ii) the average weight of three friends who weigh 55 kg, 65 kg and 72 kg.
b W kg is the average of x kg, y kg and z kg.

Make a formula,

$W = \dfrac{\ldots}{\ldots}$.

EXERCISE 1B/C

1 'How much material is in this triangular scarf?'

'Multiply the lengths of the two sides about the right angle, and divide by 2.'
Calculate the area for lengths of:
a 10 cm and 12 cm **b** 20 cm and 30 cm
c p cm and q cm.

2 'How many metres deep is the well?'

'Drop a stone down the well. Count the number of seconds for it to fall. Multiply this number by itself, then by 5.'
Calculate the depth of the well for a stone taking:
a 2 seconds **b** 5 seconds **c** t seconds to fall.

3 'How could you find the area of the path around the pool?'

'Add the radii of the two circles, then subtract them. Multiply the two answers together, then multiply by 3.'
Calculate the area of the path if the radii are:
a 4 m and 5 m **b** 10 m and 12 m
c u m and v m **d** r m and $2r$ m.

4 Kim buys a personal stereo on hire purchase.
a How much does she pay altogether if the shop asks for:
(i) a deposit of £23 and 12 monthly payments of £1.50
(ii) a deposit of £25 and 10 monthly payments of £1.40?
b For a deposit of £D there are n payments of £p each. Make a formula for the total sum £S to pay, $S = D + \ldots.$

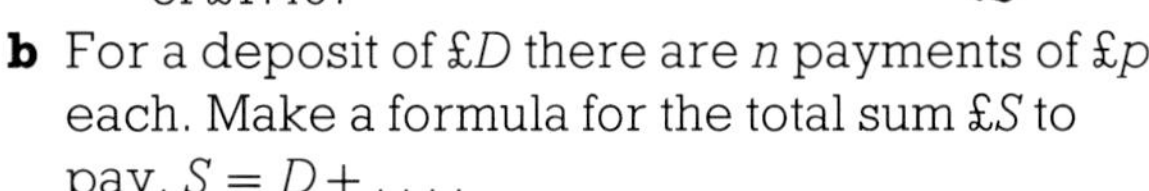

5 

'Cooking time,' said Ali 'is 60 minutes plus 40 minutes per kg of meat.'
a How many minutes would be needed for:
(i) 2 kg
(ii) 4 kg
(iii) 6 kg?
b (i) t minutes are needed for x kg. Make a formula $t = 60 + \ldots.$
(ii) If $t = 160$, find x.

6 For each table below, find a simple formula for y in terms of x, $y = \ldots$

a

x	1	2	3	4
y	2	4	6	8

b

x	10	20	30	40
y	5	10	15	20

c

x	2	4	6	8
y	1	3	5	7

d

x	1	2	3	4
y	1	4	9	16

FORMULAE IN ALL SHAPES

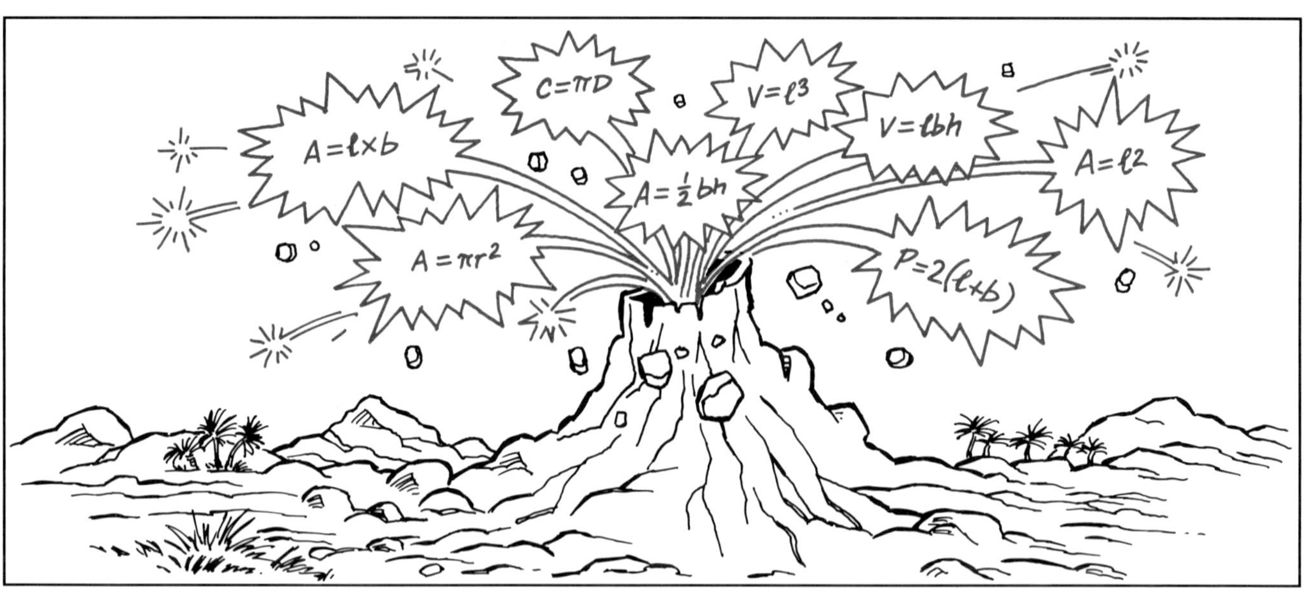

EXERCISE 2A

Choose the formulae you need from the volcano to calculate these lengths, areas and volumes. Use the π key on your calculator (or take $\pi = 3.14$), and give answers correct to 3 significant figures.

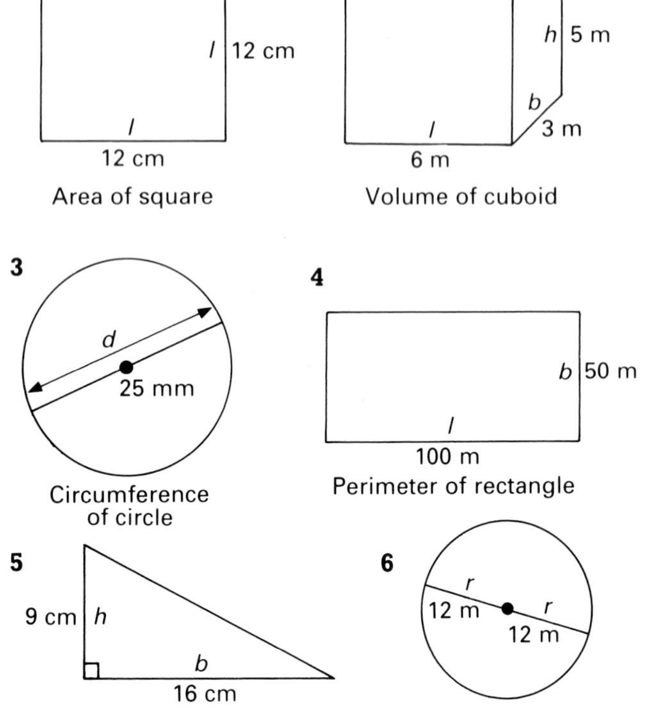

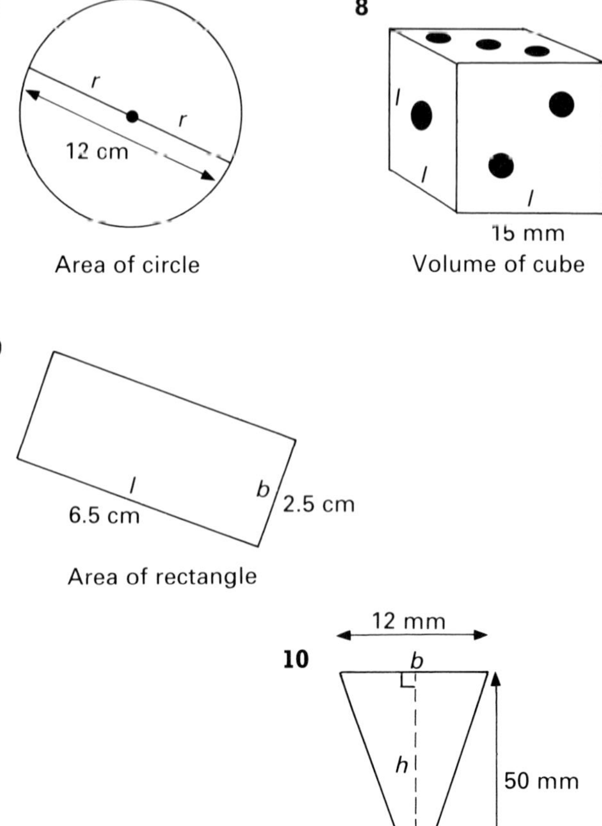

EXERCISE 2B/C

All lengths are in centimetres. Give answers correct to 3 significant figures.

1 Find the area of the trapezium below when $a = 2$, $b = 6$ and $h = 3$.

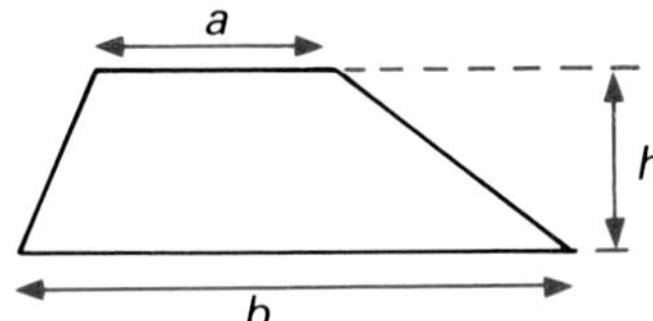

Area $A = \frac{1}{2}h(a+b)$

2 Find the area of the ring in the diagram below when $R = 10$ and $r = 8$.

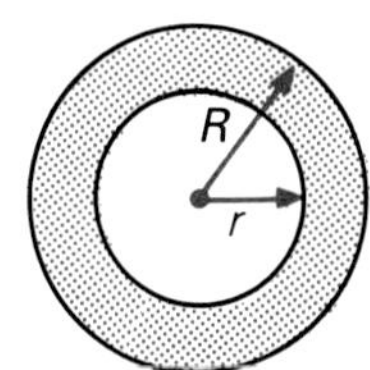

Area $A = \pi(R^2 - r^2)$

3 Calculate the surface area and volume of a football 28 cm in diameter.

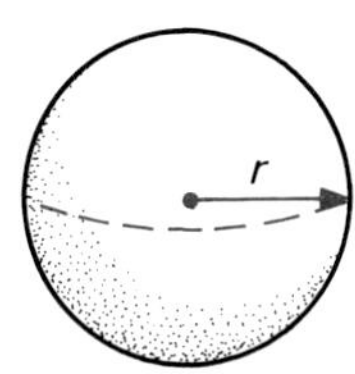

Surface area $A = 4\pi r^2$
Volume $V = \frac{4}{3}\pi r^3$

4 On this cone, when $r = 5$, $s = 13$ and $h = 12$, calculate:

a the curved area
b the total area
c the volume.

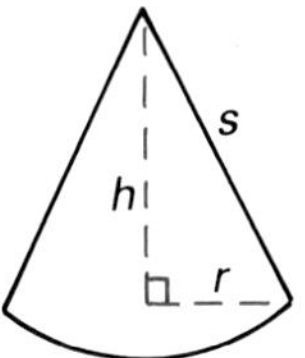

Curved area $= \pi rs$
Total area $= \pi r(r+s)$
Volume $= \frac{1}{3}\pi r^2 h$

5 Calculate side c in right-angled $\triangle$ABC, when:

a $a = 12, b = 9$
b $a = 7.5, b = 4$.

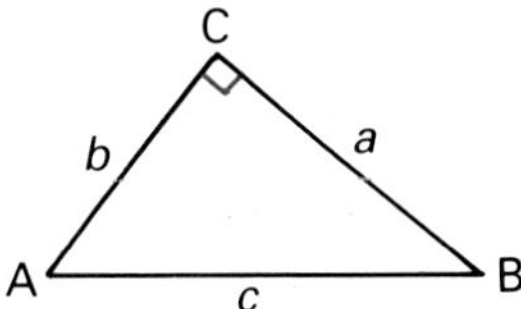

Side $c = \sqrt{(a^2 + b^2)}$

6 For this ellipse calculate:

a the area when $a = 12.2$ and $b = 8.5$
b the circumference when $a = 8$ and $b = 6$.

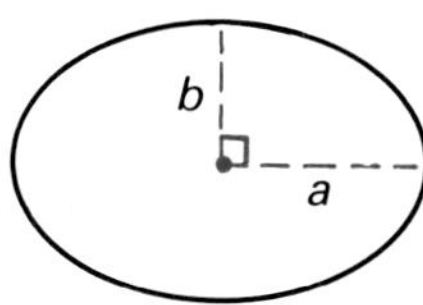

Area $A = \pi ab$
Circumference $C = 2\pi\sqrt{\left(\frac{a^2 + b^2}{2}\right)}$

SEQUENCES EVERYWHERE

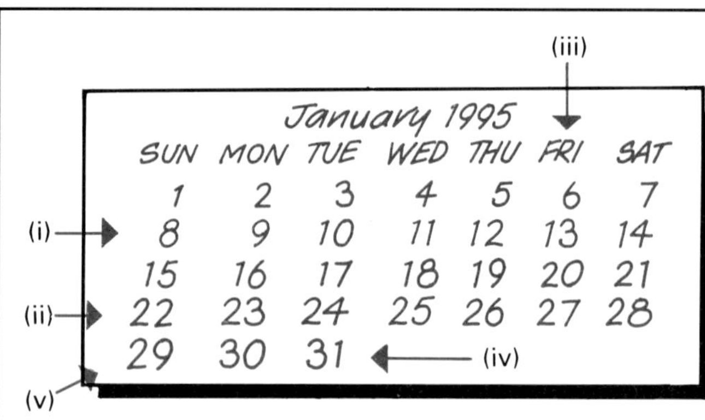

Each arrow points to a sequence of numbers.
The direction of the arrow is important.
A rule for sequence (i) is 'Add 1'.

EXERCISE 3A

1 Give possible rules for sequences (ii)–(v) above.

2 Give a rule for each sequence of numbers in the examples below:

a Wedding Anniversaries

15 Crystal
20 China
25 Silver
30 Pearl
35 Coral

b Shoe Sizes

3
$3\frac{1}{2}$
4
$4\frac{1}{2}$
5

c

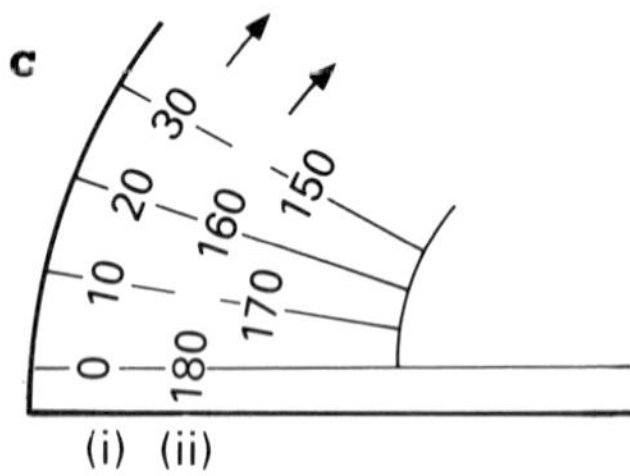

d

(i)	Number of hours	1	4	7	10	13
(ii)	Number of bacteria	1	2	4	8	16

e

(i)	Number of weeks	34	49	64	79	94
(ii)	Number of trees	729	243	81	27	9

3 Write down the first five terms in each sequence.

a **Date**
May 1st
⋮
Every 3 days

b

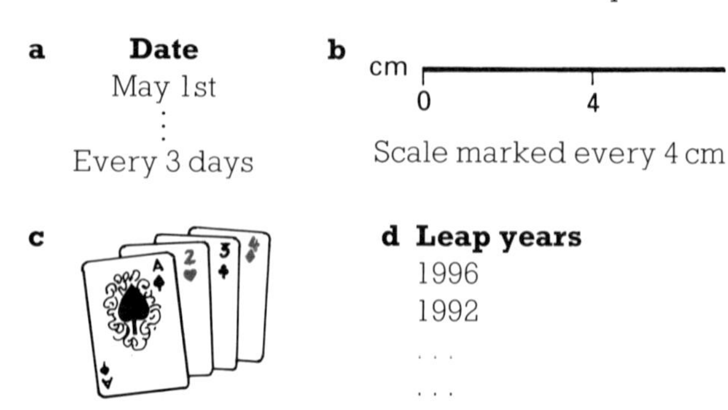

Scale marked every 4 cm

c The ace and every second card

d Leap years
1996
1992
. . .
. . .
. . .

e January

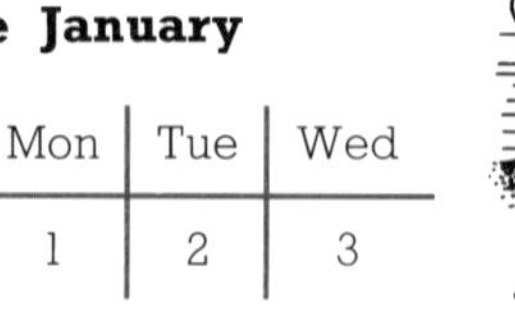

Mon	Tue	Wed
1	2	3

The dates of Tuesdays

f

650
625
600

25 feet contours

4 List the first four numbers in each sequence.

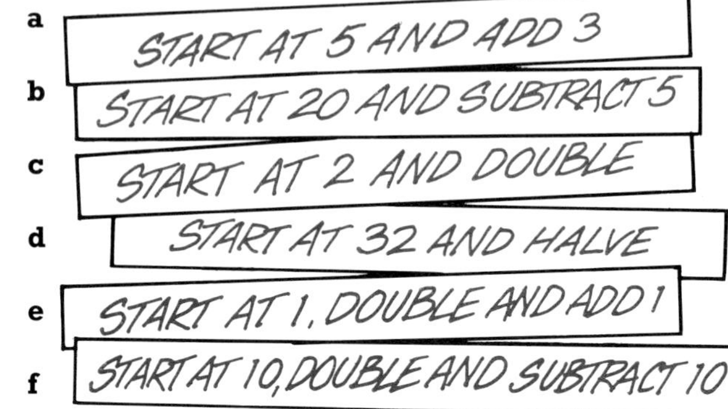

a START AT 5 AND ADD 3
b START AT 20 AND SUBTRACT 5
c START AT 2 AND DOUBLE
d START AT 32 AND HALVE
e START AT 1, DOUBLE AND ADD 1
f START AT 10, DOUBLE AND SUBTRACT 10

EXERCISE 3B/C

1 Find the missing numbers in each sequence below. Each number (apart from the first and second) is the sum of the two previous numbers.
a 1, 1, 2, 3, __, __ **b** 1, −1, __, −1, __, __
c __, 3, __, 8, __, 21 **d** 4, −3, __, −2, __, __

2 Here are descriptions of some rules:

A	Add on the same number	B	Multiply by the same number
C	Subtract the same number	D	Divide by the same number
E	Add the two previous numbers	F	Add on regularly increasing numbers

Which description above best fits a rule for each of these sequences?

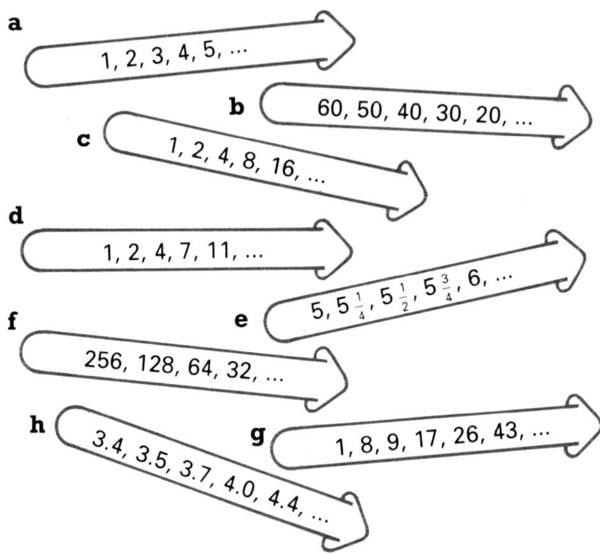

a 1, 2, 3, 4, 5, ...
b 60, 50, 40, 30, 20, ...
c 1, 2, 4, 8, 16, ...
d 1, 2, 4, 7, 11, ...
e 5, 5¼, 5½, 5¾, 6, ...
f 256, 128, 64, 32, ...
g 1, 8, 9, 17, 26, 43, ...
h 3.4, 3.5, 3.7, 4.0, 4.4, ...

3 This diagram shows a rule for making the sequence 7, 12, 17, 22, . . .
Check this.

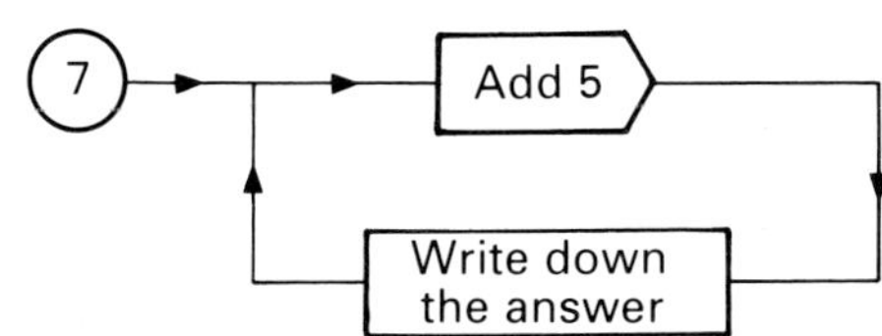

Draw similar diagrams which give possible rules for these sequences:
a 3, 7, 11, . . . **b** 30, 25, 20, . . .
c 1, 10, 19, . . . **d** 1, 10, 100, . . .
e 19, 17, 15, . . . **f** 81, 27, 9, . . .

4 Copy and complete these sequences, using the given rules.
a __, 2, __, __, __, . . . Double the previous number.
b __, __, 5, __, __, . . . Square the previous number and add 1.
c __, __, __, −1, 3, 2, __, __, . . . Add the two previous numbers.
d 1, 4, __, __, __, __, . . . If the previous number is odd, multiply it by 3 and add 1, otherwise halve it.

5 In each of these sequences:
(i) Use the patterns in the differences between one term and the next to write down the next five numbers.
(ii) Sketch the diagram for the fourth number, and check it with your answer in (i).

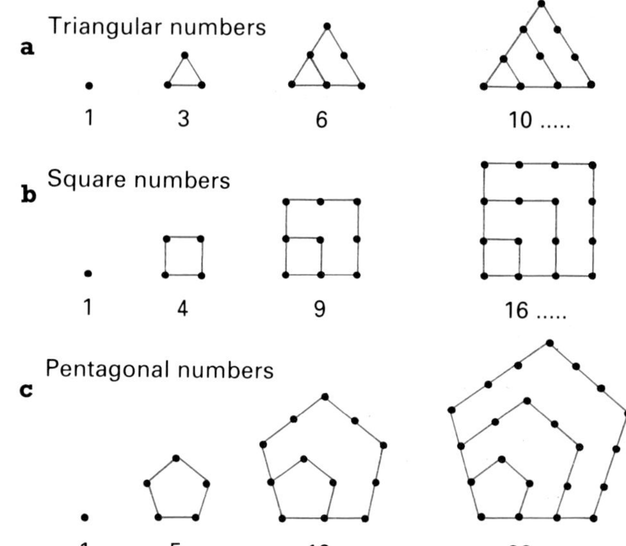

BRAINSTORMERS

1 *In each sequence below the difference between each term and the one that follows it is constant. Find the missing numbers.*
a *4, __, __, 13* ***b*** *19, __, 9* ***c*** *6, __, __, __, 22*

2 ***a*** *Can you find rules which give different sets of numbers to follow these in a sequence?*
1, 2, 4, . . .
b *Did any of your rules give 7 or 8 as the fourth term? If not, try to find the rules which do this.*

3 ***a*** *Repeat **2a** for a sequence starting 2, 3, 5, . . .*
b *Did your rules give 7, 8 or 9 as the fourth term? If not, try to find these rules.*

CHALLENGE

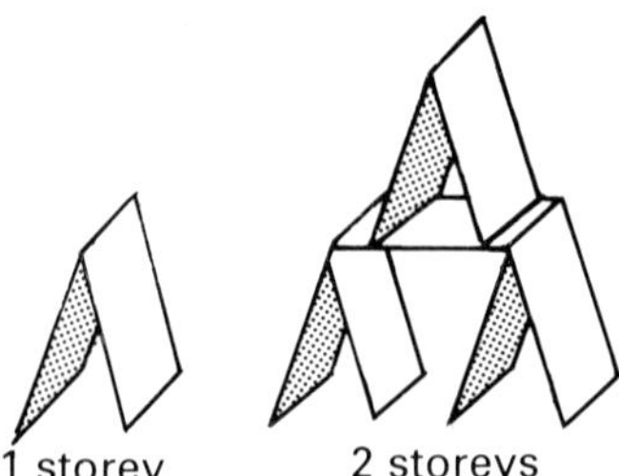

a *Count the cards in each 'house of cards'.*
b *Draw the third house in the sequence, and count the cards in it.*
c *Are there enough cards in two packs, each with 52 cards, to build an eight-storey house?*

INVESTIGATION

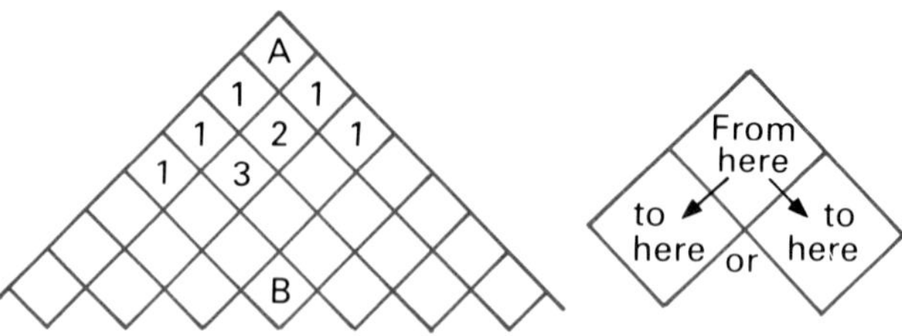

Colin was investigating the number of different routes from A to B. Allowable routes from one square to another are shown on the right.
a *Copy his board, and fill in all the possible number of routes. How many routes are there from A to B?*
b *Investigate the sequence of numbers formed from the sums of the numbers in horizontal rows.*

MAKING AND USING FORMULAE FOR SEQUENCES

Using a formula for the *n*th term, $2n-1$.

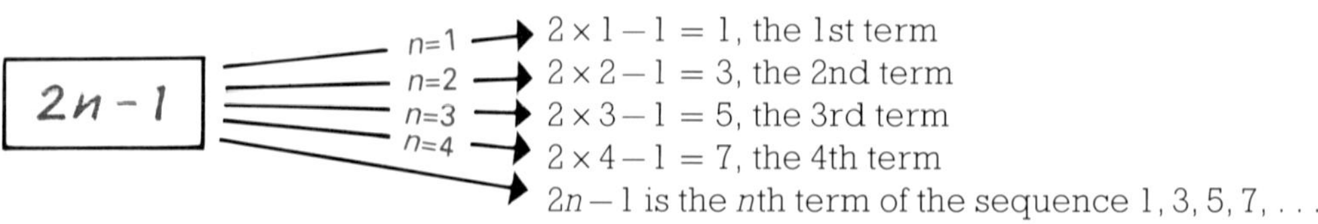

2n - 1
n=1
n=2
n=3
n=4
2 × 1 − 1 = 1, the 1st term
2 × 2 − 1 = 3, the 2nd term
2 × 3 − 1 = 5, the 3rd term
2 × 4 − 1 = 7, the 4th term
2n − 1 is the nth term of the sequence 1, 3, 5, 7, . . .

Using the nth term formula, the 100th term is $2 \times 100 - 1 = 199$.

EXERCISE 4A

1 Use the nth term formula to find the first four terms of each sequence below, by putting $n = 1, 2, 3$ and 4.
a $3n$ **b** $3n+1$ **c** $3n-1$ **d** $3n+5$

2 Use the nth term formula to find the terms asked for in each of these sequences.

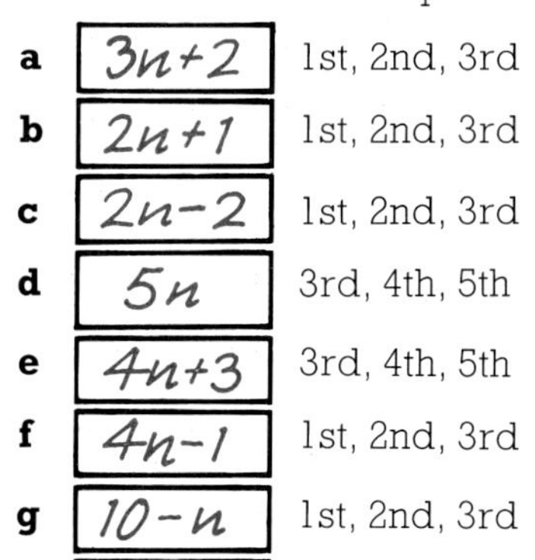

3 Match each nth term formula with the correct sequence, by putting $n = 1, 2$ and 3.

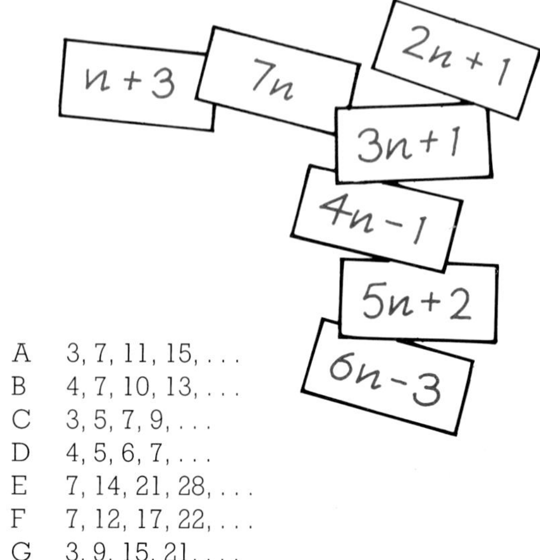

A 3, 7, 11, 15, . . .
B 4, 7, 10, 13, . . .
C 3, 5, 7, 9, . . .
D 4, 5, 6, 7, . . .
E 7, 14, 21, 28, . . .
F 7, 12, 17, 22, . . .
G 3, 9, 15, 21, . . .

The nth term formula $2n-1$ gives the sequence

The difference between pairs of terms is constant.

4 Use each nth term formula to find the first four terms of the sequence. List the difference between terms as shown in the box above.

a $n+4$ **b** $2n+5$ **c** $3n-2$
d $4n$ **e** $4n+1$ **f** $5n$
g $5n-1$ **h** $10n$ **i** $10n+5$
j $6n+2$ **k** $3n+9$ **l** $8n+7$

Can you see the connection between each nth term formula and the differences? Try to describe it.

5 Use each nth term formula to find the given terms.

a $\frac{1}{2}n$ first three terms

b n^3 first three terms

c $10-n$ 5th, 6th, 7th terms

d $2(n+1)$ first three terms

FINDING AN *n*th TERM FORMULA

Here is a sequence: 7 10 13 16 . . .
(differences: 3 3 3)

The differences are all 3, so the nth term is $3n+\ldots$

The first term is $3\times 1+\ldots = 7$, so the missing number is 4.

The nth term formula is $3n+4$.

Check:

n	1	2	3	4
$3n+4$	7	10	13	16

EXERCISE 4B/C

1 Find a formula for the nth term of each sequence. Check it for $n = 1, 2, 3$ and 4 each time.

a 2, 4, 6, 8, . . . **b** 4, 8, 12, 16, . . .
c 6, 12, 18, 24, . . . **d** 3, 4, 5, 6, . . .
e 9, 11, 13, 15, . . . **f** 4, 7, 10, 13, . . .
g 4, 10, 16, 22, . . . **h** 10, 16, 22, 28, . . .
i 0, 1, 2, 3, . . .

2 Find a formula for the nth term, and check it for the first few terms of these sequences. Use it to find the given term.

a 8, 11, 14, 17, . . . 20th term
b 2, 7, 12, 17, . . . 12th term
c 5, 9, 13, 17, . . . 25th term
d 2, 9, 16, 23, . . . 100th term

3 a Find a formula for the cost of a party of n people going to the pantomime.
b Use it to find the cost for 20 people.

Number of people	1	2	3	4
Total cost (£)	7	10	13	16

4 This table shows Century Car Hire's charges.

Miles travelled	1	2	3	4
Total hire charge (£)	26	27	28	29

a Find a formula for the charge for n miles.
b Use it to find the cost for 100 miles.

5 In chemistry, carbon atoms combine with hydrogen atoms in special ways to create the hydrocarbons.

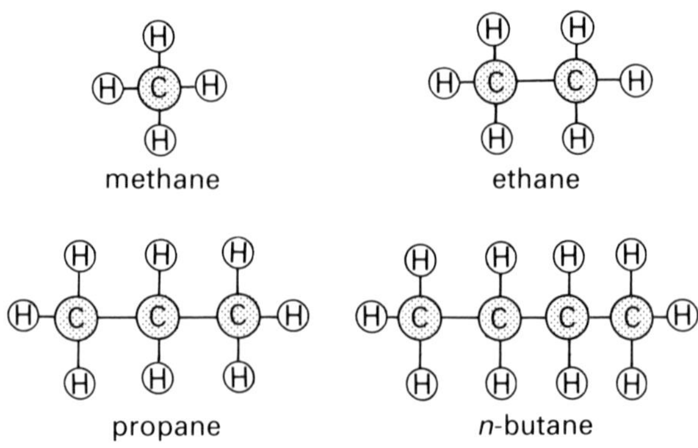

a Copy and complete the table.

Number of C atoms	1	2	3	4
Number of H atoms	4			
Total number of atoms	5			

b A hydrocarbon has n carbon atoms.
Find a formula for the number of:
(i) hydrogen atoms
(ii) hydrogen and carbon atoms.

c Use these formulae for a hydrocarbon with 12 carbon atoms, saying what you have found.

6

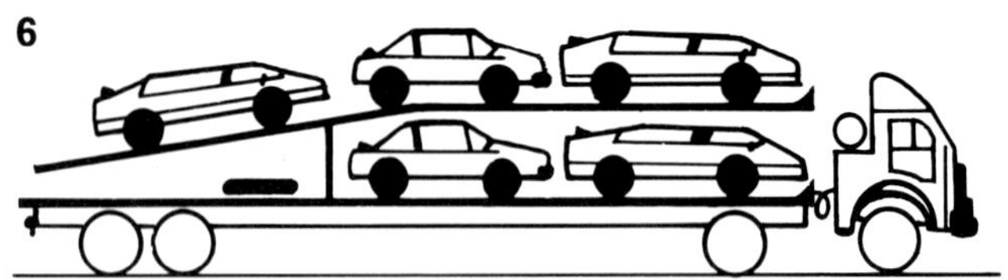

The total weight of the car transporter depends on the number of cars it is carrying.

Number of cars	1	2	3	4
Total weight (t)	27	29	31	33

a Find a formula for its weight when carrying n cars.

b Calculate its weight with eight cars.

7 The thickness of a book depends on the thickness of its covers and its leaves, as shown in the table. Find the thickness of a book with 200 leaves.

Number of leaves	1	2	3	4
Thickness (mm)	2.2	2.4	2.6	2.8

DIFFERENCE TABLES POINT THE WAY AHEAD

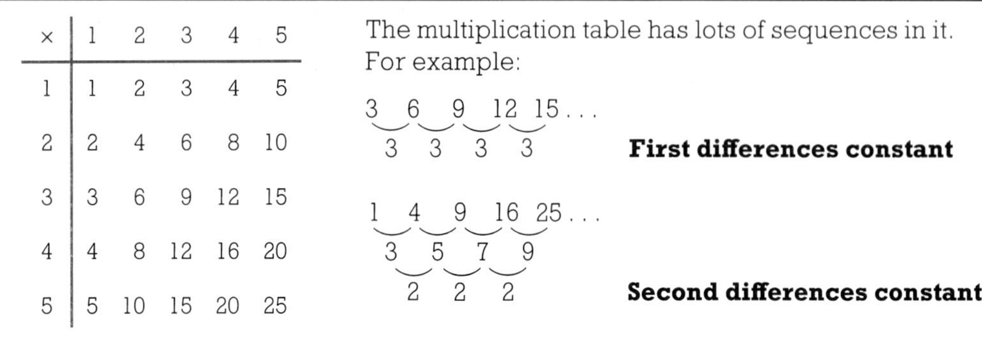

×	1	2	3	4	5
1	1	2	3	4	5
2	2	4	6	8	10
3	3	6	9	12	15
4	4	8	12	16	20
5	5	10	15	20	25

The multiplication table has lots of sequences in it. For example:

First differences constant

Second differences constant

Assuming the differences stay constant, we can predict more numbers in the sequence.

EXERCISE 5A

1 Make 'difference tables' for these sequences. Say which difference (first or second) is constant.

a 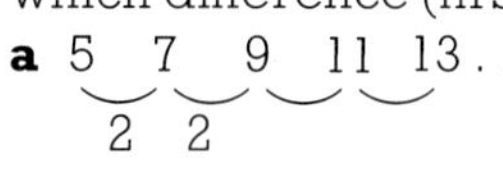

b 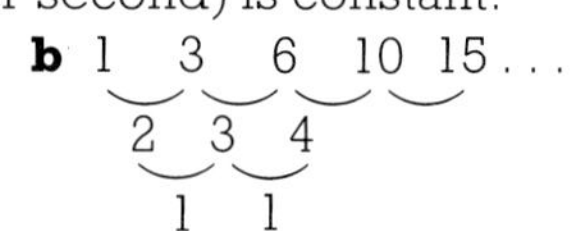

c 13, 17, 21, 25, 29, . . . **d** 3, 6, 11, 18, 27, . . .
e 1, 2, 4, 7, 11, . . . **f** 5, 10, 18, 29, 43, . . .

2 Copy and complete these difference tables, and so extend the sequences as shown.

a 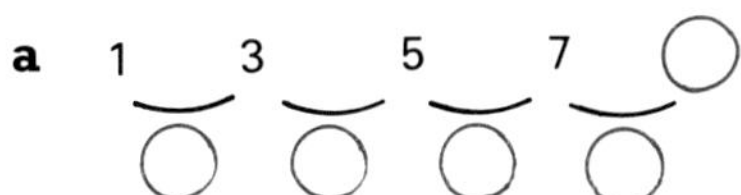

b *First differences constant*

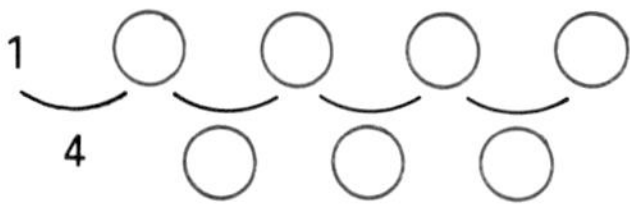

c *First differences constant*

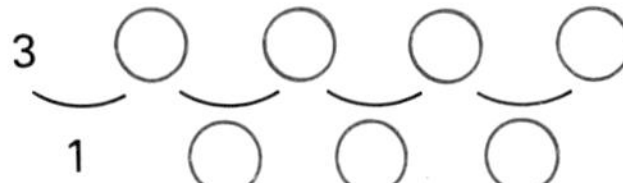

d *First differences constant*

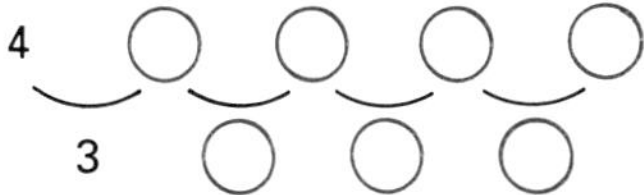

e *Second differences constant*

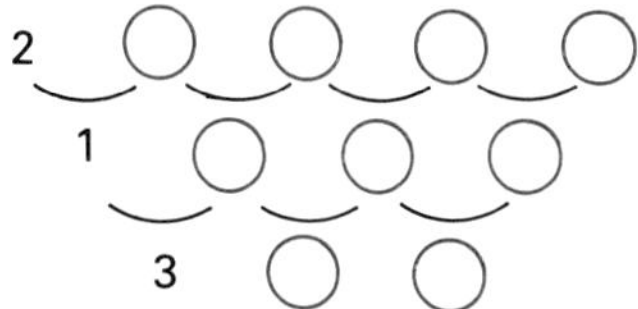

f *Second differences constant*

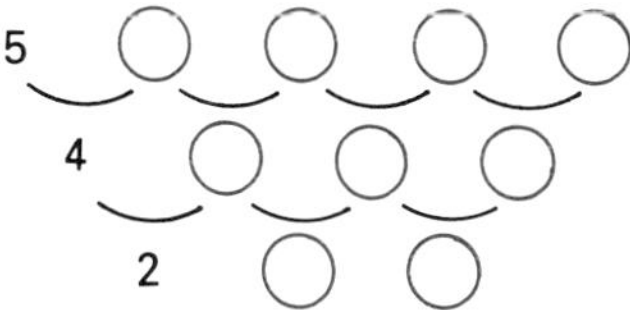

3 Construct difference tables for these sequences, then use them to find the next three numbers. A calculator might help you with this.

a 5, 18, 31, 44, . . . **b** 7, 24, 41, 58, . . .
c 3, 14, 27, 42, . . . **d** 11, 25, 42, 62, . . .
e 14, 41, 68, 95, . . . **f** 4, 47, 91, 136, . . .

4 Use difference tables to find the number of matchsticks in the fifth picture of each sequence of matchsticks shown below.

a

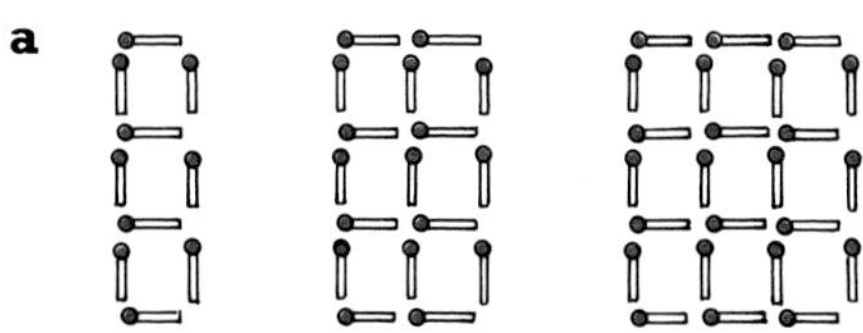

b

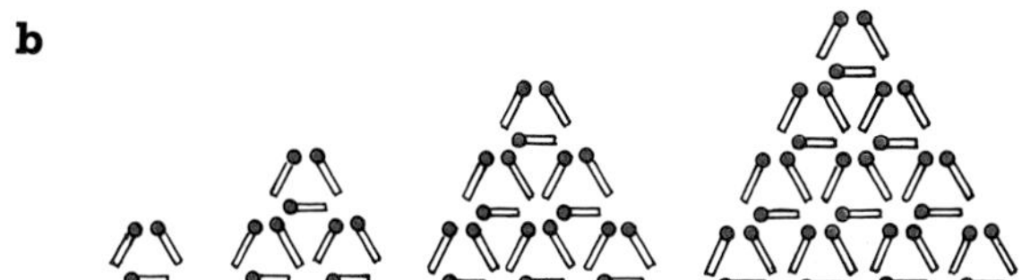

EXERCISE 5B

1 This number pattern is based on a number spiral which starts

5←4←3
↓ ↑
6 1→2

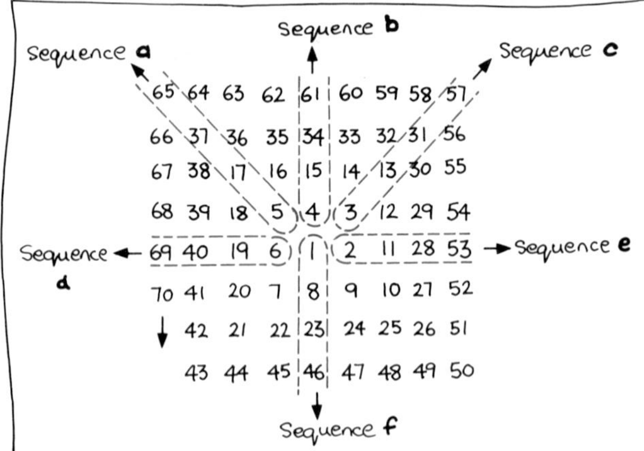

(i) Make a difference table for each sequence (**a–f**), and use it to find the next term.
(ii) Check your answers by extending the spiral.

2 Use difference tables to find the next term in each of these sequences.

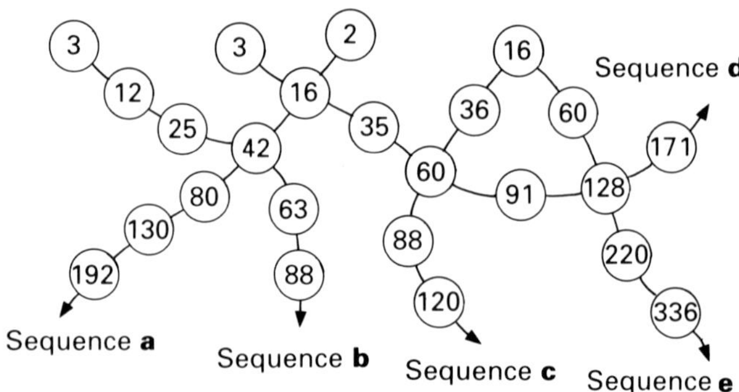

3 Negative numbers may appear in difference tables. Construct difference tables for these sequences, and find the next three terms in each sequence.
a 18, 13, 8, 3, . . .
b 10, 6, 4, 4, 6, . . .
c −10, −7, −5, −4, −4, . . .
d 10, 8, 3, −5, −16, . . .

EXERCISE 5C

1 a How many different pairs of vertically opposite angles can you see in these diagrams?

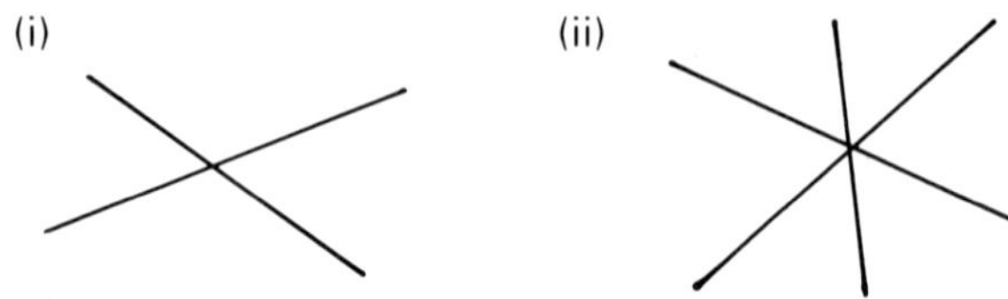

b Copy and complete the table.

Number of crossing lines	2	3	4	5
Pairs of vertically opposite angles			12	20

c Can you predict how many pairs there would be for: (i) 6 (ii) 8 crossing lines?

2 Douglas is investigating the number of moves needed in this leapfrog game for the red and white counters to change places. Check his results for one and two counters on each side.

A counter can:
(i) slide one square
(ii) jump over an opponent.

Number of counters each side	1	2	3	4	5
Least number of moves	3	8	15	24	35

Assuming that the pattern continues, how many moves would be needed in a game with nine counters on each side?

3 a Check that the magic sum in this 3 by 3 square is 15.

2	7	6
9	5	1
4	3	8

b Assuming that the table is correct, can you predict the magic sum for a 10 by 10 square?

Magic square	3 × 3	4 × 4	5 × 5	6 × 6	7 × 7
Magic sum	15	34	65	111	175

c It is thought that the magic sum formula for an n by n square is $\frac{1}{2}n(n^2+1)$. Does this work for the 3 × 3, . . . , 7 × 7, and the 10 × 10 squares?

4 Celia was counting the total number of squares *of all sizes* on different playing boards.

a Copy and complete her table.

Size of board	1 × 1	2 × 2	3 × 3	4 × 4	5 × 5
Total number of squares				30	55

b Use the sequence to predict the total number of squares on a chessboard (8 × 8).

CHALLENGE

1	2	3	4	5	6	7	8	9	10
11	12	13	14	15	16	17	18	19	20
21	22	23	24	25	26	27	28	29	30

a *If 3 is replaced by n in the T-shaped box, what would replace 2, 4, 13 and 23?*

b *If 7 is replaced by n in the rectangular box, write down replacements for the other numbers in the box.*

BRAINSTORMER

An astronomer was studying some of the moons of the planet Uranus. Using four of the moons, he guessed 'third differences constant' for their sequence of distances from the planet.

Name of moon	Distance from Uranus (1000s km)
Cressida	62
Desdemona	63
Juliet	64
Portia	66

a *From his table he predicted the distances of:*
(i) the next moon outwards
(ii) the next two nearer the planet. What were his predictions?

b *Using the extra data in the table below, check his predictions.*

Name of moon	Distance (1000s km)
Cordelia	49
Ophelia	54
Bianca	59
Rosalind	70
Belinda	75

c *Check his predictions for Cordelia and Belinda also.*

CHECK-UP ON FORMULAE AND SEQUENCES

1 a Write down formulae for the perimeters P of these shapes.

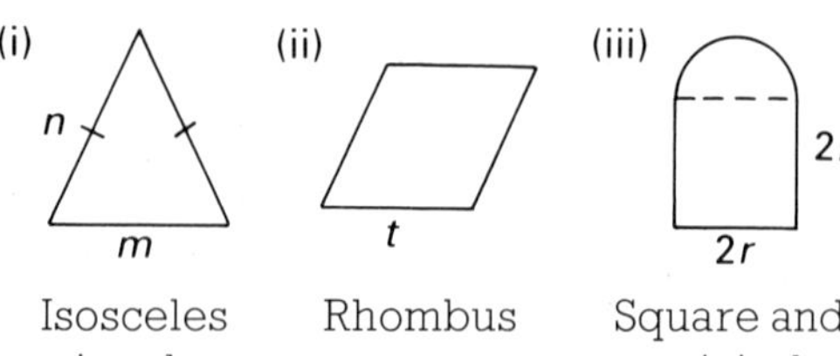

b Calculate the perimeters when $m = 12$, $n = 15$, $t = 15$, $r = 5$ (all in cm).

2 Write down a formula for the area A of the shape in question **1a**(iii).

3 a How far can you cycle at 12 km/h in:
(i) 2 hours (ii) 3 hours (iii) H hours?

b Make a formula for the distance D km in H hours at S km/h, $D = \ldots$

4

A queue of people is p metres long. More people make it q metres longer, but some leave, reducing it by r metres.

a If the queue is now n metres long, make a formula connecting n, p, q and r.

b Calculate n if $p = 6$, $q = 15$ and $r = 9$.

5 Describe a possible rule used to form each sequence below:

a 2, 7, 12, 17, ... **b** 2, 6, 18, 54, ...
c 50, 49, 47, 44, ...

6 Use the nth term formula to find the given terms:

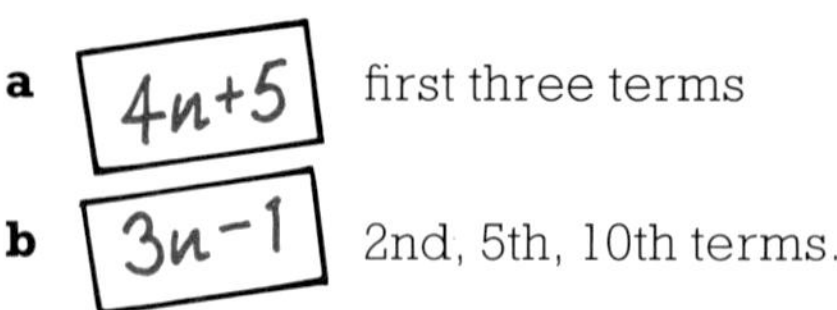

7 Find the nth term formula for each sequence, and check it for $n = 1$, 2 and 3.

a 1, 6, 11, 16, ... **b** 9, 13, 17, 21, ...
c 7, 13, 19, 25, ...

8 Which term will the given number be in each sequence?

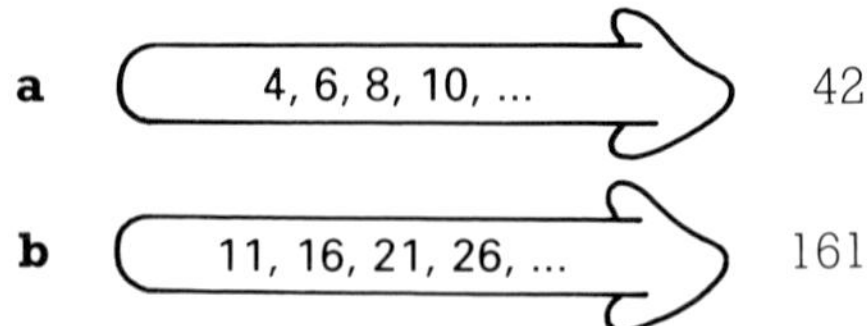

9 Construct difference tables for these sequences, and use them to calculate the next three terms in each.

a 7, 10, 13, 16, ... **b** 17, 20, 27, 38, ...
c 1, 2, 6, 15, 31, ...

10 Jane made up this table of the number of diagonals in shapes with various numbers of sides. Check it by drawing the diagonals in a quadrilateral and a pentagon.

Number of sides	3	4	5	6
Number of diagonals	0	2	5	9

Assuming that the pattern of differences continues, how many diagonals has a nonagon (9 sides)?

11 a Use a systematic method to count the number of upward-pointing triangles of all sizes in each picture below.

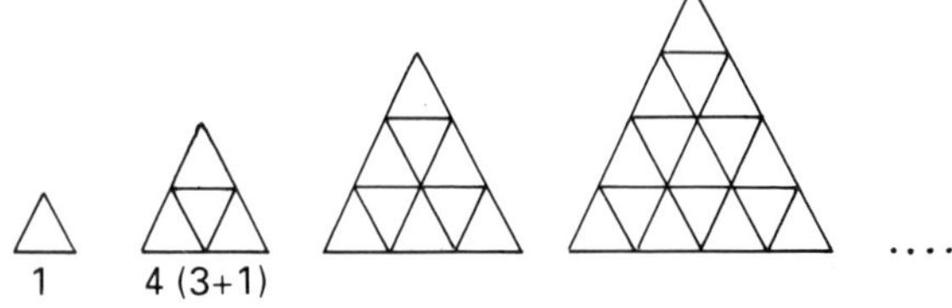

b Explain how to continue the sequence.

c How many upward triangles are there in the sixth picture?

17 PROBABILITY

LOOKING BACK

1 Which of these statements are true?

a

'There is an even chance of Heads.'

b

'A six is the most likely score.'

c

'The square is certain to have four sides.'

d

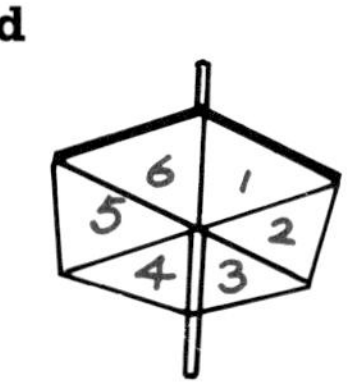

'The probability of an even score is $\frac{1}{3}$.'

2 Which is more likely in each of these?

a Scoring 6 with a dice, or getting Heads when tossing a coin.

b Scoring a 4 or scoring a 6 when rolling a dice.

3 **a** What fraction of these sweets is red?

b One is selected at random. What is the probability that it is not red?

4 A letter is chosen at random from the word PROBABILITY. Calculate:

a P(Y) **b** P(B) **c** P(vowel).

5 Fifty letters are posted first class, and forty-five arrive the following day. What is the probability that one of the letters, chosen at random, arrived the following day?

6 One item of cutlery is taken at random from the drawer. Calculate:

a P(spoon)

b P(knife)

c P(fork).

7 A class vote produced this pie chart of the popularity of TV sport.

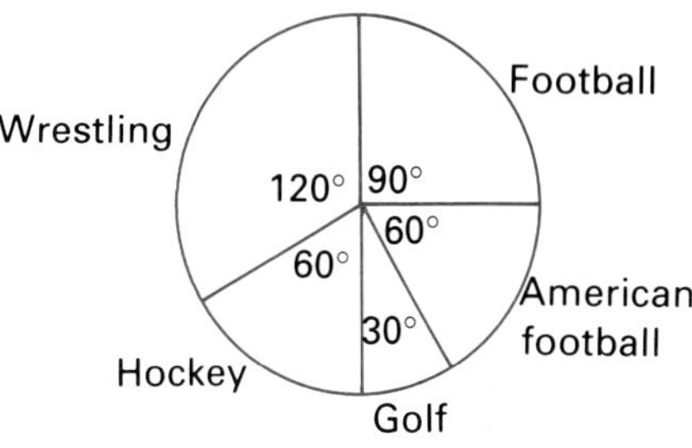

If one pupil's vote is chosen at random, what is:

a P(golf) **b** P(wrestling) **c** P(hockey)

d P(football—both kinds)?

8 How could you:

a check whether or not a dice is loaded (unfair)

b find which pop group is most popular in your school

c calculate the probability that a child born in Britain will be a boy?

PROBABILITY AND TREE DIAGRAMS

Reminder
Where there are several equally likely outcomes of an event, the **probability** of a favourable outcome

$$= \frac{\text{number of favourable outcomes}}{\text{number of possible outcomes}}$$

Example Probability of winning, $P(W) = \frac{2}{8} = \frac{1}{4}$
$P(L) = \frac{6}{8} = \frac{3}{4}$

In a tree diagram, 

Notice that $P(W)+P(L) = 1$

EXERCISE 1A

1 A 10p coin is tossed.
a What are the possible results?
b Calculate P(H) and P(T).
c Copy and complete the tree diagram.

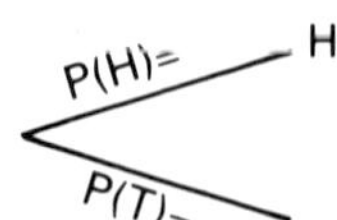

2 **a** On how many edges can the spinner land?
b Calculate P(W) and P(L) for the spinner.
c Copy and complete the tree diagram.

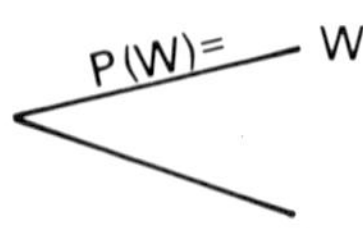

3 **a** List all possible outcomes when the dice is tossed.
b Calculate the probability of getting:
(i) an even number (ii) an odd number.
c Copy and complete the tree diagram.

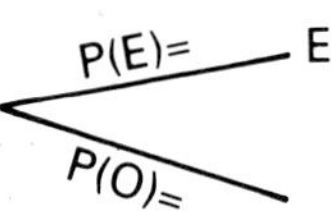

4 A bag contains 6 black counters and 3 green ones. One is taken out at random.
a Calculate: (i) P(black) (ii) P(green).
b Draw a tree diagram for this.

5 A box holds 6 red buttons and 4 white ones. One is taken out at random.
a Calculate: (i) P(red) (ii) P(white).
b Draw a tree diagram.

6 There are 15 girls and 5 boys in a class.
a What is the probability that a pupil, chosen at random, is:
(i) a girl (ii) a boy?
b Show the probabilities in a tree diagram.

7 Dave has to pick a card to find which day he is on duty as Prefect.

a Calculate the probability that he picks:
(i) Friday (ii) another day.
b Draw a tree diagram.

8 Mike chooses a number at random from the hat.

a What is the probability that he takes:
(i) an odd number (ii) an even number?
b Show the results in a tree diagram.

EXERCISE 1B/C

1 A bag holds 12 marbles. Look at the tree diagram and decide how many are:
a yellow **b** green.

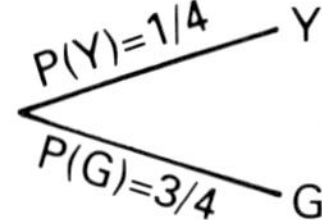

2 Draw a tree diagram to show P(6) and P(not 6) when a dice is rolled.

3 Kay throws a dart which sticks in the whirling disc. Each circle is marked off in congruent parts.
(i) Calculate P(R) and P(W) for each disc, as decimal fractions, correct to 1 decimal place.
(ii) Draw a tree diagram for each pair of probabilities.

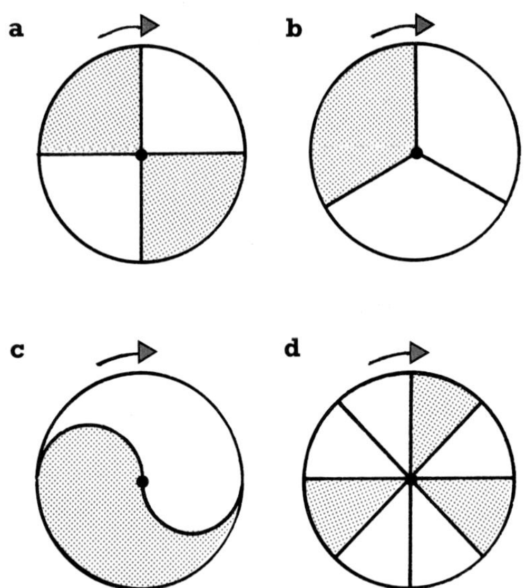

4 Fruit gums are made in four colours, as shown in the tree diagram.

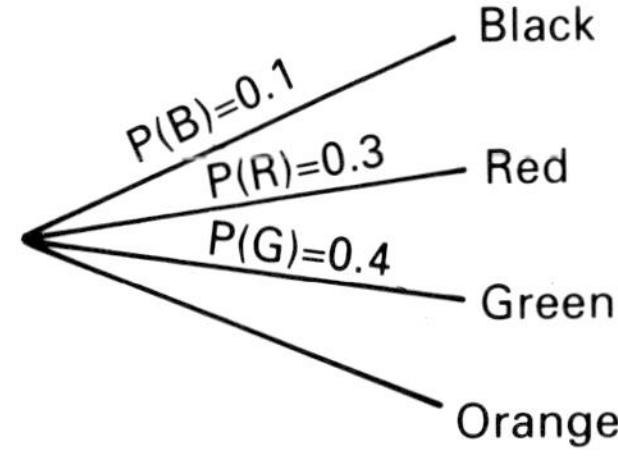

a Calculate:
(i) P(orange) (ii) P(B)+P(R)+P(G)+P(O).
b How many of each colour would you expect in a packet of 40 gums?

5 The bar graph tells you about the number of pupils in each year who are members of the Film Society.

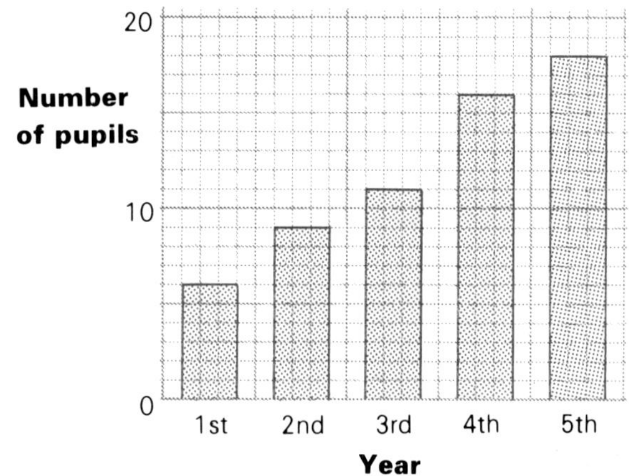

a How many members are there?
b Calculate the probability that a member chosen at random is:
(i) a Fifth Year pupil
(ii) not a Fifth Year pupil.
c Show the probabilities in a tree diagram.

6

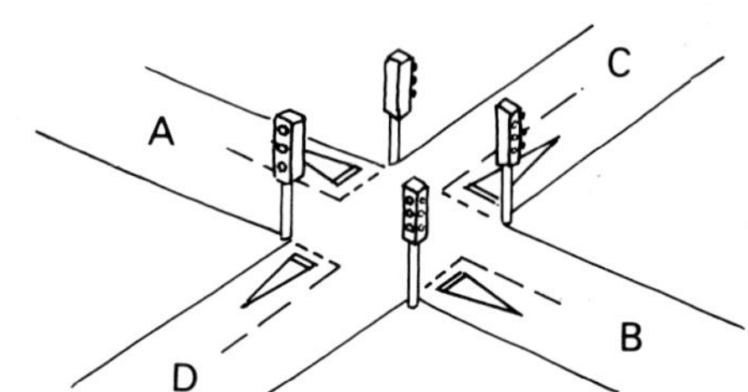

The traffic lights show green as follows: A and B together for 90 seconds, then C for 60 seconds, then D for 30 seconds. Calculate the probability that when a car arrives at the junction the lights are green at: **a** A **b** C **c** D.

CERTAIN SUCCESS AND CERTAIN FAILURE

Sandra has a two-headed coin. She tossed it 5, 10, 20 times, and wasn't surprised when she got Heads every time, never Tails.
So P(H) = 1, and P(T) = 0.

P(certain success) = 1, and P(certain failure) = 0

This can be shown on a line, with all the other probabilities in between.

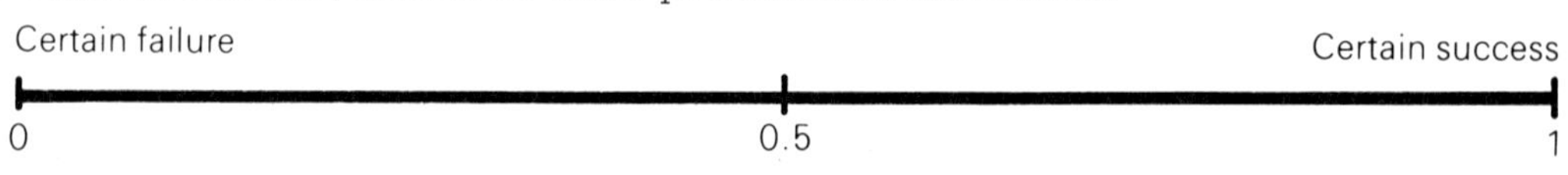

EXERCISE 2A

1 Draw a scale like the one above. Put arrows on it to show the probabilities that:
a a team captain wins the toss of a coin
b you will live forever
c your maths lesson will finish today
d you will score 7 with a dice
e a card chosen at random from a full pack will be a club
f a slice of toast will fall with the buttered side up.

2

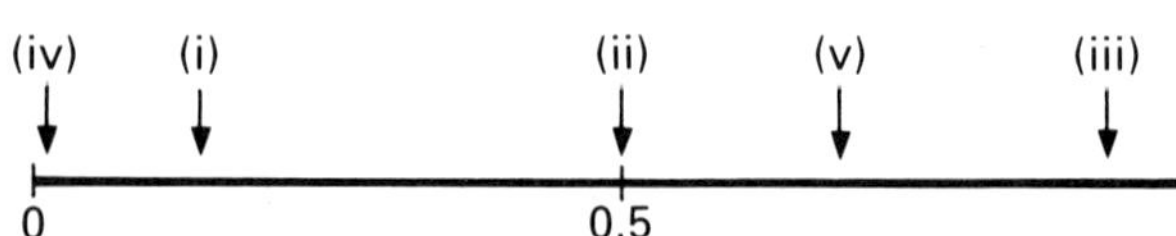

Pair the arrows and the probability that:
a you choose a red card at random from a pack
b a person chosen at random will be right-handed
c a person entering a lottery will win a prize
d you will get 6 when you roll a dice once
e you will score at least 3 on rolling a dice.

3

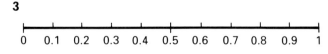

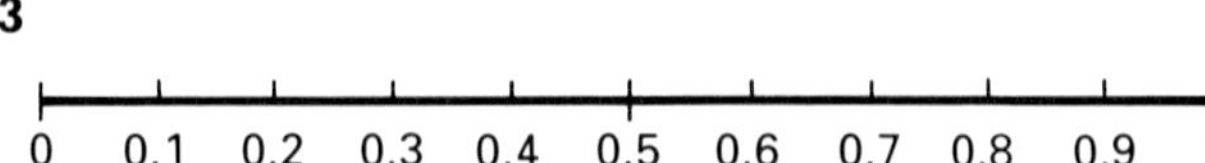

Copy the scale. A letter is chosen at random from the word ASSESSMENT.
Calculate these probabilities below, and mark them on the scale.
a P(A) **b** P(vowel) **c** P(consonant) **d** P(E)
e P(S)

4 This table shows the results of a survey of items owned in 1000 households.

Computer	Colour TV	Video	Microwave	Telephone
200	900	550	400	850

Copy the scale in question **3**. Calculate and mark the probabilities that a household, chosen at random, would possess each one of the items.

EXERCISE 2B/C

1 A coin is tossed.
a Calculate: (i) P(Head) (ii) P(not a Head).
b Add your answers in **a**. You should get 1. Why?

2 A dice is rolled.
a Calculate: (i) P(6) (ii) P(5).
b Add your answers in **a**. Why do you not get 1?

3 In a game, the arrow spins freely. Calculate:
a P(1) **b** P(2) **c** P(3) **d** P(4)
e P(1)+P(2)+P(3)+P(4).

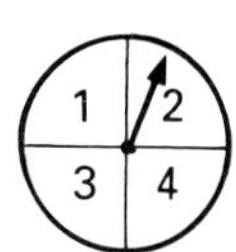

4

0 — 1

Copy this scale, and mark the probability that:

a you will score 5 when you roll a dice
b you will not score 5 when you roll the dice
c the spinner will land on WIN
d the spinner will land on LOSE
e a letter chosen at random from the word TRAPEZIUM will be:
(i) a vowel (ii) a consonant.

5 The probability of a seed germinating in a trial of new flowers is 0.9. What is the probability that it will *not* germinate?

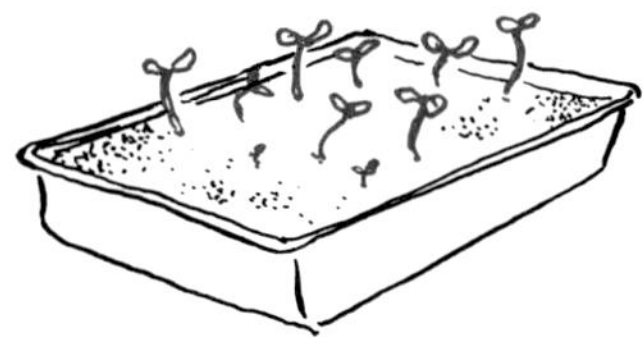

6 The High Risk Insurance Company calculates that the probability of a client having a car accident in a particular year is 0.15. What is the probability that a client will *not* have an accident in that year?

7 a Patients given a certain treatment have a 72% probability of recovery. Calculate the probability of the treatment failing.
b Another treatment has a 50–50 chance of success. Which treatment is more successful, **a** or **b**?

8 Rovers have a dismal record at away games. Their manager estimates P(defeat) = 0.8 and P(draw) = 0.15. Calculate P(away win).

9 Here are the nets of some special dice.

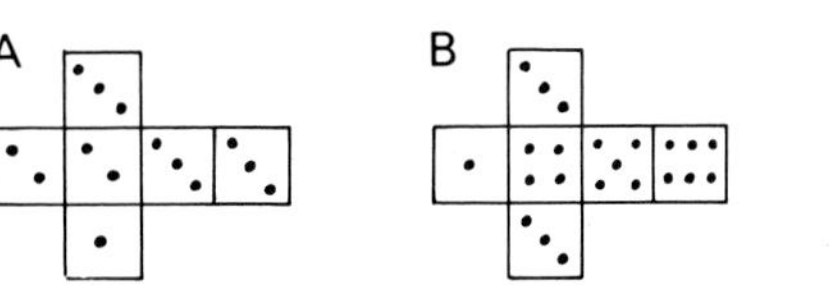

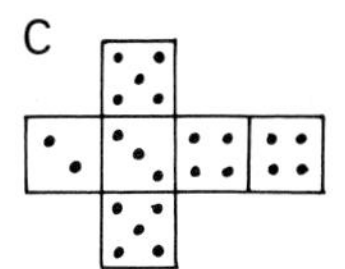

For which of the dice is:

a $P(2) = \frac{1}{6}$ **b** $P(2) = \frac{1}{3}$ **c** $P(2) = 0$
d $P(6) = \frac{1}{6}$ **e** $P(6) = 0$ **f** $P(\text{not } 2) = \frac{5}{6}$
g $P(\text{not } 3) = \frac{2}{3}$?

BRAINSTORMERS

1 *Robin is on the first floor. He presses one of the buttons at random.*

a *Calculate:*
(i) P(going up) (ii) P(going down).
b *Add your answers in* **a**. *Why is their sum not 1?*
c *What is the probability that the lift does not move when he presses a button?*

2 *Lindsay is trying to raise money for charity. Each label has a football team on it. She sells them at 20p each to friends who write their names on their chosen labels. When the card is full she takes off all the labels and finds where the £5 is hidden.*

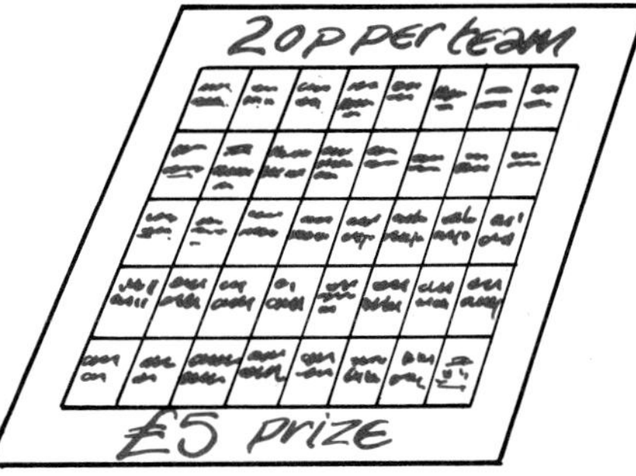

a *Help Lindsay's friend Cameron to work out the probability of winning the £5 prize if he buys:*
(i) one label (ii) a column of labels
(iii) a row of labels.
b *How much does a full card raise for charity? (Remember the prize!)*

ESTIMATING AND CALCULATING PROBABILITIES

EXERCISE 3

Which of the above methods would you use to estimate or calculate the probability that:

1. a pupil in your school, chosen at random, has red hair
2. there will be more than 10 cm of rain in London in July
3. a box of matches will contain more than the number stated on the box
4. a 10p coin will land Heads when tossed
5. a birth will produce twins
6. an ace will be selected from a pack of cards
7. a dice with a corner chipped off will give a score of 6
8. a pupil in your class will be absent ill for at least a week in one year
9. a first class letter will arrive next day
10. a bungee jumper will have an injury on his or her first jump
11. a pencil sharpener will land blade-up when tossed
12. two sixes will be obtained on tossing two dice
13. all the digits of a telephone number are different
14. a pupil chosen at random has more than £5 pocket money a week
15. an earthquake will occur in San Francisco this year?

PRACTICAL PROJECTS

1

a *Predict the number of red cards you would get if you cut a complete pack of cards 10, 20, 50 times.*

b *Compare your predictions with results by experiment. Show all your data in a table.*

*2 Repeat **1** for cutting an ace in a complete pack of cards.*

3 Choose a spinner and compare the number of 1s, evens (or something else of your choice) that you get, by experiment and by calculation, for increasing numbers of spins. Again show the results in a table. Write a sentence about any conclusions you can make.

FORETELLING THE FUTURE—EXPECTATION

CLASS DISCUSSION

Where are probabilities needed for these?

Motor car insurance

Total claims in a year?

Company must make a profit

Max Games Company

How many games will be sold in a year?

How much stock?

Drugs Company

Is a new drug better than older ones?

Is it worth marketing?

Example

Dr Jones is trying to decide how much time he can give to each patient. He has 4000 people on his list.
The probability that one of them has to see him in the course of a year is 0.23.
How many patients should he expect to visit him each year?
The number expected would be $0.23 \times 4000 = 920$.

EXERCISE 4A

1 a What is P(Heads) on tossing a coin?
b Copy and complete the table:

Number of tosses	Number of heads expected
10	
50	
100	

2 a What is P(6) on rolling a dice?
b Copy and complete the table:

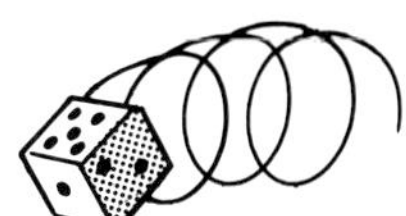

Number of throws	Number of 6s expected
6	
60	
120	

3 a What is P(5) for the spinner?
b Copy and complete the table:

Number of spins	Number of 5s	Number of even numbers
10		
100		
1000		

4 a What is P(dishwasher in house)?
b Copy and complete the table:

1 in 10 homes has a dish-washer

Number of homes	Number of dishwashers
20	
100	
300	

5 The probability that a person chosen at random is male is 0.48. How many males would you expect to live in a village of 4000 people?

6 At the maternity unit the probability of a male birth was 0.52. After 1200 births how many boys and how many girls could the doctor expect?

7 The probability of winning the jackpot on a fruit machine is 0.01.
a How many jackpots would you expect to win in 300 spins?
b A spin costs 10p, and the jackpot is £3. Who makes a profit, and how much, in 300 spins?

8 In a sample of houses, three-quarters had central heating.
How many houses would you expect to have central heating in an area with:
a 8000 houses **b** 60 000 houses?

9 If the claim on the tin is true, how many dogs would choose Woof in a street where 60 dogs live?

EXERCISE 4B

1

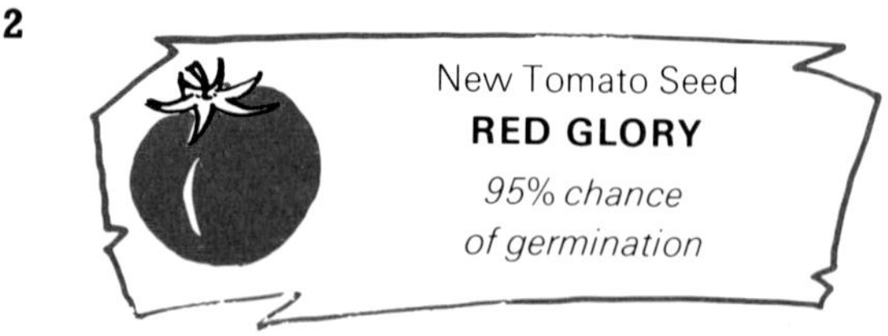

This is a memo to salespeople in the High Risk Insurance Company. Calculate the expected number of claims.

2

The owners of a nursery garden sow 3000 seeds. How many plants can they expect eventually?

3 **a** What is the probability that a person chosen at random is right-handed?

b Copy and complete the table:

Number of people	Number expected to be right-handed
30	
200	
800	

c A sports shop orders 100 sets of golf clubs. How many right-handed sets, and how many left-handed sets should they buy?

4 Using this spinner, how many wins, and how many losses would you expect in these numbers of spins?
a 12 **b** 30 **c** 45 **d** 120

5 Ann is on the 'Wheel of Fortune' TV show. The wheel is equally likely to stop at any point.

a Calculate the probability that Ann will win £100.
b Out of 20 contestants, how many might win £100?

6 A double glazing firm sends out special offer leaflets ('60% off') to homes. It expects about 5% of them to reply. How many replies should they expect from: **a** 200 **b** 2000 **c** 5000 leaflets?

7 A fountain pen manufacturer carries out some research, and finds that about 15% of his sample of people use italic handwriting. How many italic pens should be included in batches of:
a 200 **b** 1000?

8 How many 6s should Paul expect if he rolls a dice:
a 60 times
b 100 times (to the nearest whole number)?

9 In 180 rolls of a dice a 5 or 6 is thrown 120 times. Do you think that the dice is fair? Explain your answer in a sentence.

EXERCISE 4C

1 In trials, 170 out of 200 patients responded to a new drug.
a Calculate the probability that the drug is successful.
b How many patients could hope to benefit if the use of the drug is extended to:
(i) 500 people (ii) 10 000 people?

2

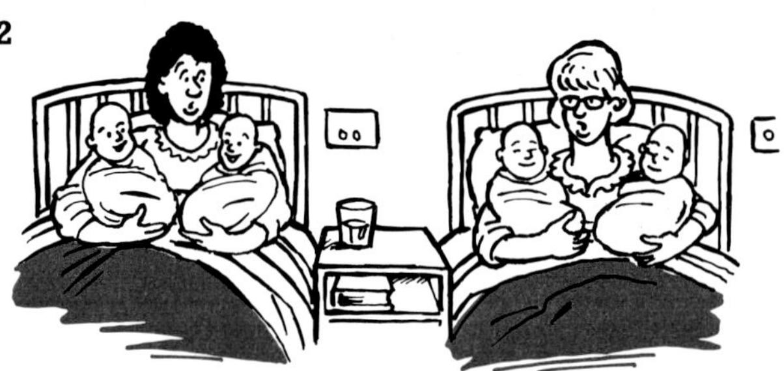

At a maternity unit the probability of twins is 0.0125. About how many sets of twins would be expected in:
a 320 births **b** 1000 births?

3 More maternity unit statistics are given in the table.

	Probability
Triplets	$\frac{1}{2500}$, or 0.0004
Quadruplets	$\frac{1}{64\,000}$, or 0.000 015 6

Of 1 000 000 births, estimate the number of sets of:
a triplets **b** quadruplets.

4 Starlight Electric Light Bulb Company tested 200 bulbs.

Life of bulb (hours)	up to 100	101–150	151–200
Number of bulbs	5	20	60
Life of bulb (hours)	201–250	251–300	over 300
Number of bulbs	75	35	5

a Calculate the probability (as a percentage) that a bulb chosen at random will have a life of:
(i) up to 100 hours (ii) 201–250 hours.
b The company replaces bulbs that fail within 100 hours. If bulbs cost £1.20 each how much will they pay for failures in a batch of 1000 bulbs?

5 A sample of 3000 people is chosen at random. How many would you predict were born:
a in a leap year **b** on 29th February?

6

P (accident to driver under 25) = 0.2
P (accident to driver 25 or over) = 0.12
Number of clients
(i) Under 25—1200
(ii) 25 or over—3500

This is a memo to salespeople in the Stoutheart Insurance Company. Estimate for the year:
a the total number of accident claims expected
b the number of clients aged 25 or over who are not expected to make a claim.

BRAINSTORMER

Anita has a board game that 'plays' golf. If the spinner stops at 4 she counts 4 strokes for that hole.

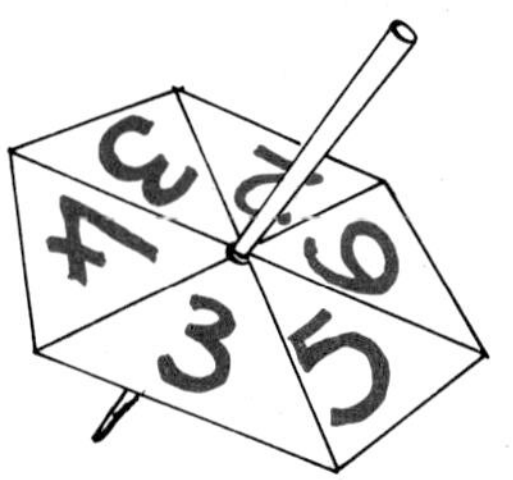

a *Calculate the probability of taking:*
(i) 2 strokes (ii) 3 strokes (iii) 6 strokes.
b *What score could she expect for a round of 18 holes?*

INVESTIGATION

Using the statistics in question **2** *of Exercise 4C, how many sets of twins would you expect there to be in your school? Find out how many there actually are. How did the theory compare? . . . Are there any triplets?*

PRACTICAL PROJECT

a *Calculate the probability that a domino, chosen at random from a complete set of 28, is a double.*
b *(i) Place a set face down, and ask a friend to shuffle the pieces. Choose a domino, note what it is, replace it, and have the set shuffled again. Do this 50 times.*
(ii) How many doubles did you get? Compare with a calculation of the expected number and comment on the results.

CHECK-UP ON PROBABILITY

1 The bag contains two red sweets and four white ones.

a When a sweet is chosen at random, calculate:
(i) P(R) (ii) P(W).

b Show the probabilities in a tree diagram.

2

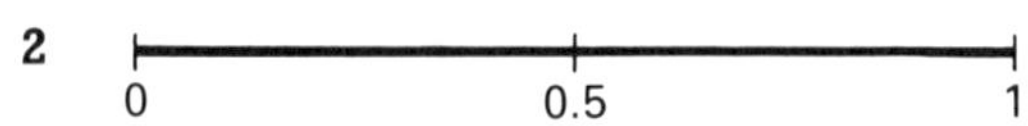

Copy the scale, and use arrows to mark the probabilities that:

a a badminton racket lands smooth side up when spun

b a stone will fall when dropped in mid-air

c you'll have an odd number when you add two even numbers

d lightning will strike twice in the same place

e a pupil in your class chosen at random has brown hair.

3 The weatherman says that there is a 5% chance of snow tomorrow. What is the probability that it will not snow?

4 How could you estimate the probability that:

a a pupil, chosen at random, owns a bicycle

b August will be the driest month next year?

5 From past experience, the probability that a car will fail to complete the Moor and Mountain Rally is $\frac{2}{5}$. Calculate the likely number of failures and finishers in an entry of 150 cars.

6 In a survey, 350 out of 500 adults said they found it difficult to pay all their household bills. If one person is chosen at random, what is the probability, as a percentage, that he or she is having this difficulty?

7 There are 1000 straws. 50 of them are £1 winners.

a Calculate the probability of picking a winner.

b Mr Barlow picks 20 straws. How much can he expect to win?

8 Greg's spinner is equally likely to land on any sector when it stops.

Calculate:

a P(even number) **b** P(prime number)

c P(multiple of 3) **d** P(factor of 8)

e P(n, where $n > 3$).

9 The Longlife Car Company examines 500 cars of each age in the table for rust.

Age of car (years)	1	2	3	4	5	6	7 or more
Number with rust	0	0	2	15	26	142	315

Calculate the probability (as a percentage) of rust in a car:

a 2 years old **b** 4 years old

c 7 or more years old.

18 INFORMATION TECHNOLOGY

The need to handle data means that computers are very useful in many areas of life. A computer stores, displays and processes information, and its operation is greatly assisted by databases and spreadsheets.

DATABASES

A database is a computer program which organises files of information on any given subject.
Using it, the computer can: **search**, **sort** and **calculate**.
Horizontal rows of information are called *records*, and vertical columns of information are called *fields*.

Example
Jeff collects British stamps, and has entered some information about his collection into a computer database. His database has 15 records (rows) and 6 fields (columns).

Year	Watermark	Perforation	Face Value	Colour	Value (£)
1840	small crown	none	1d	black	140
1840	small crown	none	2d	blue	300
1841	small crown	none	1d	red	3.50
1841	small crown	none	2d	blue	35
1847	VR	none	6d	lilac	400
1847	none	none	10d	brown	550
1847	none	none	1/-	green	350
1854	small crown	16	1d	red	3.50
1854	small crown	16	2d	blue	35
1857	large crown	14	1d	red	1
1857	large crown	14	2d	blue	25
1858	half penny	14	$\frac{1}{2}$d	red	5
1858	large crown	14	1d	red	0.50
1858	large crown	14	$1\frac{1}{2}$d	red	18
1858	large crown	14	2d	blue	4

The actual layout on the screen depends on the software; data in the first record might appear like this:

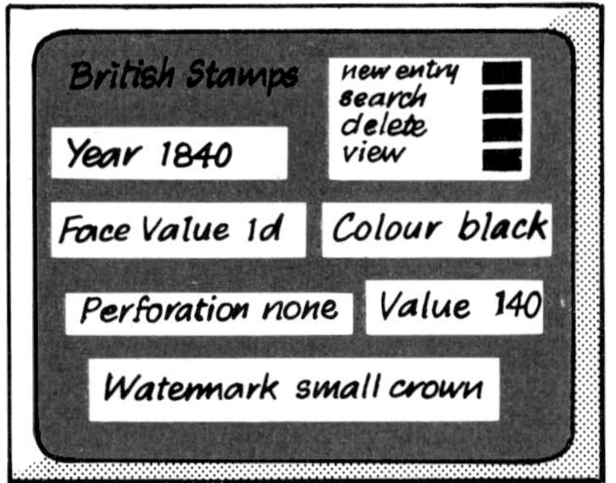

EXERCISE 1

Here you have to act the part of a computer. Use Jeff's data to answer questions **1–5**. Its records, or rows, are numbered 1 to 15.

1 Searching Which records hold data on stamps with:
a colour black **b** face value 6d
c year 1857 **d** watermark large crown
e perforation 14?

2 Sorting If records are sorted:
a alphabetically by colour, which colour is first and which is last?
b by increasing value (£), which value is first and which is last?

3 Calculating In the database:
a how many small crowns are there?
b what is the total value (£) of all the records?

4 Which records hold these stamps?
a 1857 and blue **b** Face value $1\frac{1}{2}$d and red
c Small crown and perforation 16 and blue.

5 Create the following lists:
a the year and colour of stamps which are not red
b the year and face value of red stamps with a large crown
c the watermark and face value of stamps with perforation 16 *or* value over £300.

6 Vicky entered this information about the planets in the solar system in a database.

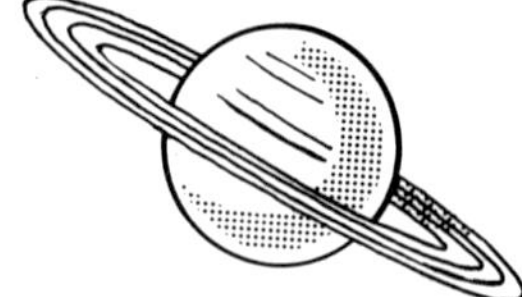

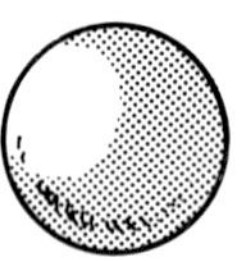

Planet	Millions of miles from sun	'Year' length		Diameter (miles)	Mass compared to Earth	'Day'	
		Years	Days			Hours	Mins
Mercury	36	0	88	3008	0.05	25	42
Venus	67	0	225	7600	0.82	23	21
Earth	93	1	0	7913	1.00	23	56
Mars	141	1	322	4200	0.11	24	37
Jupiter	483	11	314	90 254	318	9	50
Saturn	886	29	167	75 000	95	10	14
Uranus	1783	84	7	31 900	15	10	49
Neptune	2793	164	280	32 900	17	15	48
Pluto	3700	248	0	3600	0.1	?	

a List the planets which:

(i) are more than 500 million miles from the sun

(ii) have a 'day' which is longer than the Earth's day

(iii) have a 'day' which is longer than the Earth's *and* a 'year' that is shorter than the Earth's

(iv) have a mass greater than the Earth's *or* a diameter that is less than the Earth's.

b Compare the list of planets which are further from the sun than Earth, with those which have a longer 'year' than the Earth has.

c Make a scatter diagram to compare 'distance from sun' and 'year length'.

SPREADSHEETS

A spreadsheet is a computer program which enables you to use tables of data. It consists of rows, columns and cells, like this:

Columns

		A	B	C	D	E
Rows	**1**		5			15 ← Cell E1
	2					

A cell can hold a label (usually a row or column heading), a number or a formula.
If 5 is in cell B1, and the formula 3 × B1 is in cell E1, the computer prints 15 there.

EXERCISE 2

1 Jenny starts a spreadsheet to work out the takings at the school shop. She types in this information:

	A	B	C	D	E
1	Item	Number	Cost per	Total	Takings
2		sold	item (p)	cost	
3	Crisps	30	18	B3 × C3	
4	Choc bars	12	25	B4 × C4	
5	Biscuits	26	12	B5 × C5	
6	Juice	40	12	B6 × C6	
7					Sum (D3, D6)
8					

This is what appears on the screen:

	A	B	C	D	E
1	Item	Number	Cost per	Total	Takings
2		sold	item (p)	cost	
3	Crisps	30	18	540	
4	Choc bars	12	25	300	
5	Biscuits	26	12	312	
6	Juice	40	12	480	
7					1632
8					

What appears on the screen in the cells listed below? Give the entry, and say whether it is a label, a number or a formula.

a A1 **b** B2 **c** B4 **d** C2
e C5 **f** D1 **g** D6 **h** E7

2 The price for a chocolate bar should be 26p. Describe the changes required in the cells in the spreadsheet to correct this.

3 The next week, sales figures change and prices increase.
Columns B and C become:

B	C
Number	Cost per
sold	item (p)
26	19
15	26
31	12
38	13

Redraw columns D and E.

4 Steven has a spreadsheet for costing the job of carpeting and wallpapering a house. Lengths are in metres.

a (i) What formula should go into E2?
(ii) Copy and complete column E.

b (i) Show that the formula for F2 is $2 \times D2 \times (B2 + C2)$.
(ii) Copy and complete column F.

c What other columns are needed in the spreadsheet in order to do the costing?

	A	B	C	D	E	F
1	Room	Length	Breadth	Height	Floor area	Wall area
2	Kitchen	5	3	3	15	48
3	Dining	8	3	3		
4	Lounge	7	5	3		
5	Bathroom	6	5	3		
6	Bedroom	9	5	3		
7						
8						

5 Mr Khawaja, the Maths teacher, uses a spreadsheet to keep a record of the class marks.

a E3 comes from the formula $C3 + D3$.
Copy and complete column E.

b C9 and D9 both contain the totals for their column.
Use 'Sum' to write formulae for these cells.

c F3 is E3 expressed as a percentage.
(i) Write down a formula for F3.
(ii) Copy and complete column F.

	A	B	C	D	E	F
1	First	Second	Test 1	Test 2	Total	Percentage
2	Name	Name	(100)	(80)	(180)	%
3	Adam	Smith	74	52	126	70
4	Margaret	Lee	67	68		
5	Helen	Long	62	46		
6	Harry	Reid	49	41		
7	John	Kenneth	22	23		
8	Anne	Jones	44	37		
9						

CHALLENGE

Make up a database in the form of a table with at least ten ***records*** *(rows) and four* ***fields*** *(columns) for one or more of these:*

a *statistics on friends*
b *the Top 10 records*
c *quadrilaterals*
d *trees.*

REVISION EXERCISES

REVISION EXERCISE ON CHAPTER 1: NUMBERS IN ACTION

1 Mrs Cooper pays her council tax in ten monthly instalments. Calculate each payment if her annual charge is:
a £845 **b** £442.75 (to the nearest penny).

2 Write each mileometer reading below to the nearest:
(i) 100 km (ii) 10 km (iii) km (iv) 0.1 km.

a
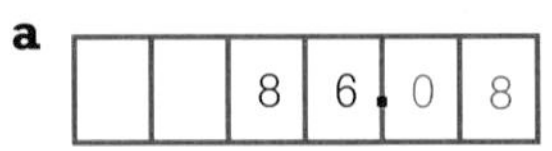

b
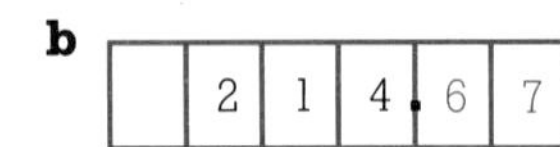

c
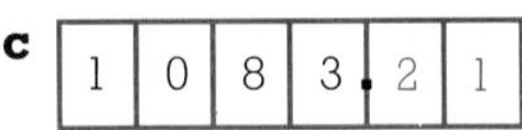

3 Estimate, then calculate correct to 1 decimal place:
a $6.95 + 8.37 - 2.19$ **b** $0.38 \times 7.7 \times 19.56$

4 a Use 1 significant figure in each number to *estimate* the area of each rectangle.

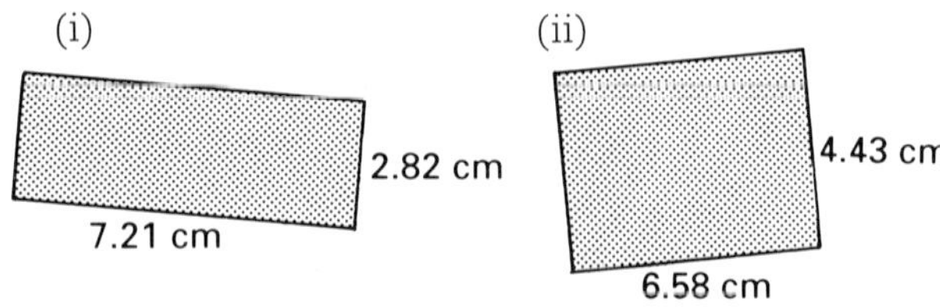

b Calculate the area of each rectangle, correct to 1 decimal place.

5 Calculate $100 \div 7$, correct to:
a 2 significant figures **b** 2 decimal places.

ORDER
( )
× ÷
+ −

6 a *Without using a calculator*, calculate:
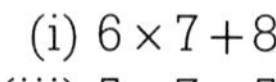
(i) $6 \times 7 + 8$ (ii) $2 + 6 \times 8$
(iii) $5 \times 7 - 5$ (iv) $(12 - 2) \div 5$
(v) $3 \times (6 - 6)$ (vi) $20 \div (8 - 4)$
b Check your answers with a calculator.

7 a Calculate the unit price of the coffee in each jar (the cost of 1 gram of coffee).

b Which is the best buy? How many decimal places do you have to look at?

8 Nicola is writing a 2000 word essay. So far she has written 1250 words.

a How many words has she still to write?
b She writes 250 words per page. How many pages does her essay take?

9 a Calculate the thickness of each page of the book in millimetres, correct to 2 decimal places.

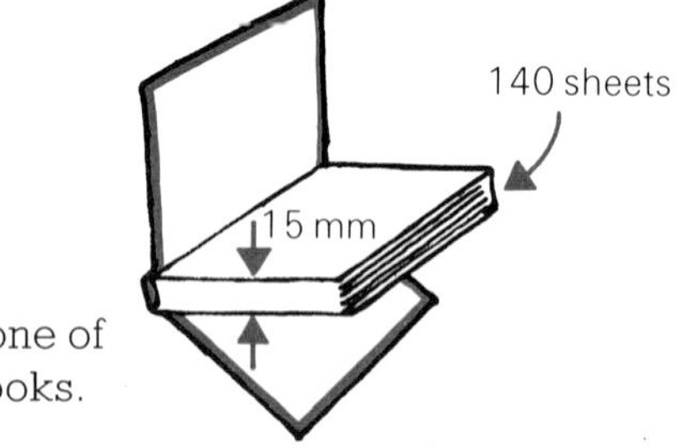

b Try this for one of your own books.

10 Calculate to 3 decimal places using:
(i) brackets (ii) the calculator memory.
a $\dfrac{1795 + 1325}{85.7 + 334.3}$ **b** $\dfrac{21.6 \div 8.4}{4.5 \times 6.5}$

REVISION EXERCISE ON CHAPTER 2: ALL ABOUT ANGLES

1 ∠ABC = 75°. What is the size of its:
a complement **b** supplement
c vertically opposite angle?

2 ABC is the end of a tent on horizontal ground XY. AD is a vertical pole.

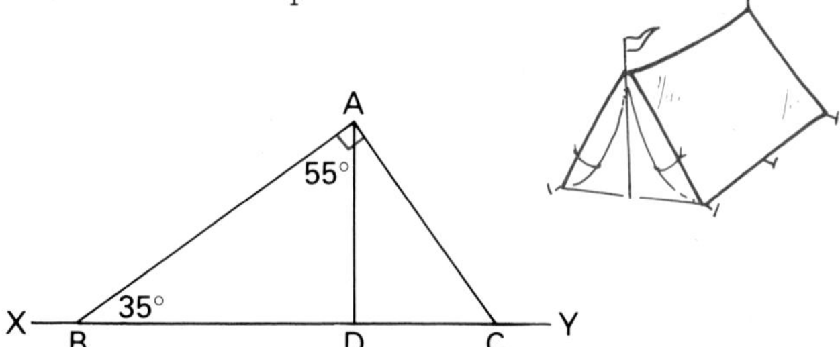

a Calculate angles ADB, ABX and DAC.
b How many pairs of:
(i) complementary (ii) supplementary angles are there?

3 Calculate x in each diagram below.

a

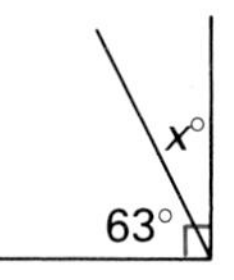

b

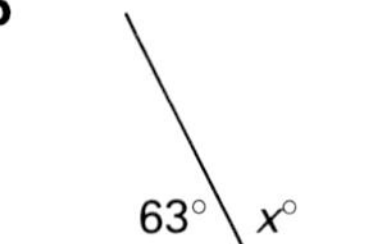

c

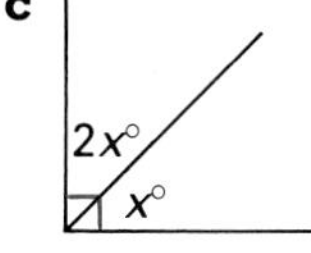

d

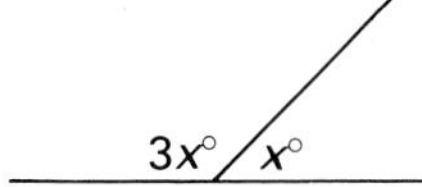

4 Copy these diagrams, and fill in as many angles as you can.

a

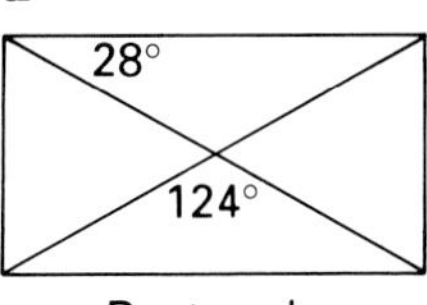

Rectangle

b

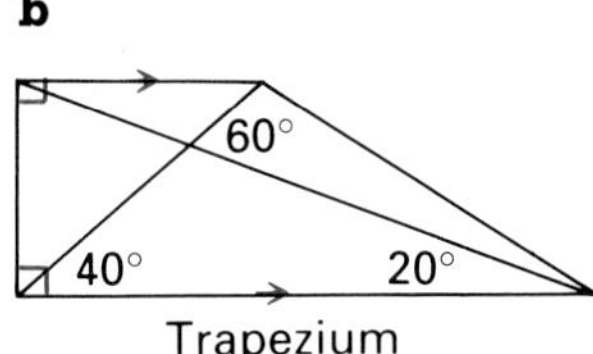

Trapezium

5 Howard is drawing plans for a new glider. Calculate $a°$, $b°$, $c°$, $d°$ and $e°$, giving a reason for each answer.

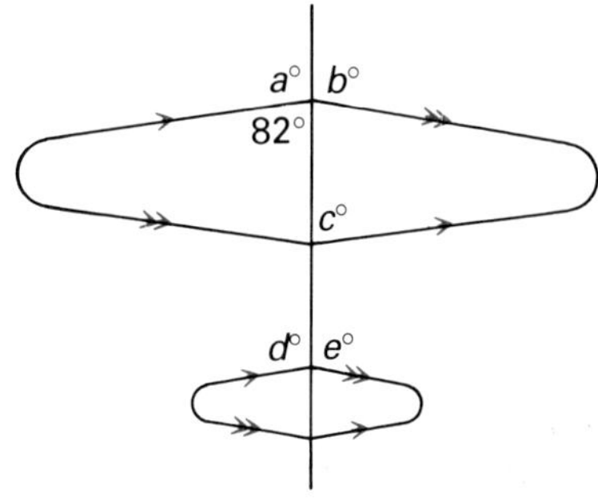

6 Find the values of a, b, c, . . . in these diagrams.

a

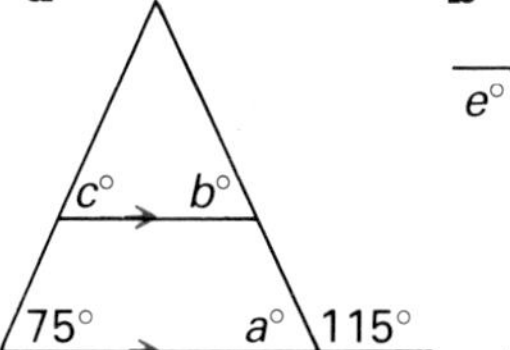

b

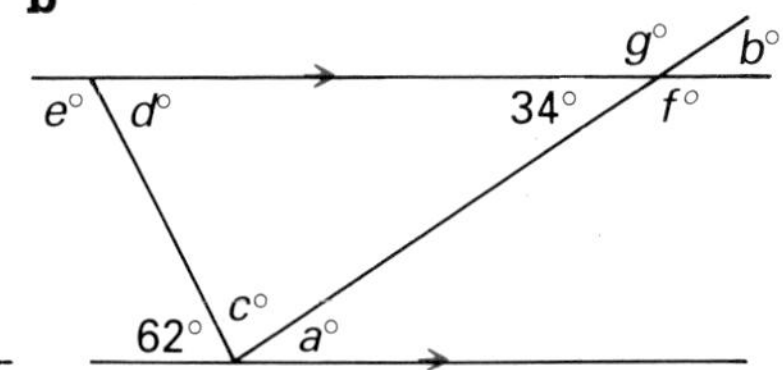

7 Calculate x.

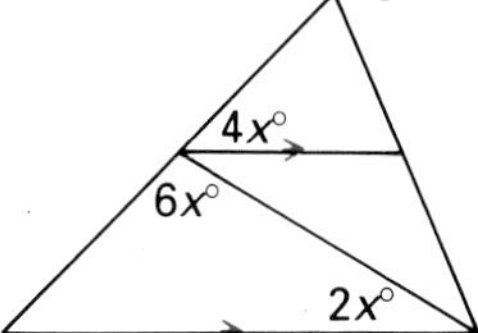

8 Prove that $a° + b° + c° = 180°$.

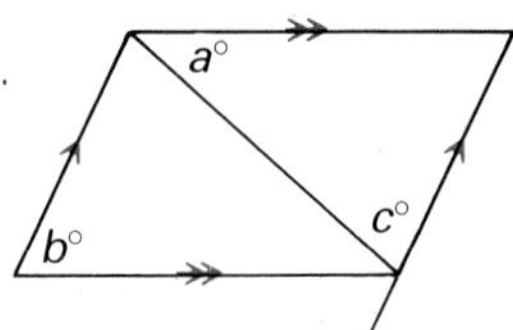

9 Find an angle whose supplement is three times its complement.

REVISION EXERCISE ON CHAPTER 3: LETTERS AND NUMBERS

1 Write in a shorter form:
a $3n+2n$ **b** $t+t+t+t$ **c** $3k-k$
d $m \times m$ **e** $5 \times 5 \times 5 \times 5$ **f** $x \times x$

2 Simplify:
a $3x+x+2$ **b** $5y-y-2$ **c** $2a \times 2a \times 2a$

3 $u = 3$, $v = 2$, $w = 1$. Find the value of:
a $u+v+w$ **b** uvw **c** u^2 **d** v^2
e w^2 **f** $2u+v$ **g** $3v-w$ **h** $(uvw)^2$

4 Find the output from each machine for each of the three inputs shown.

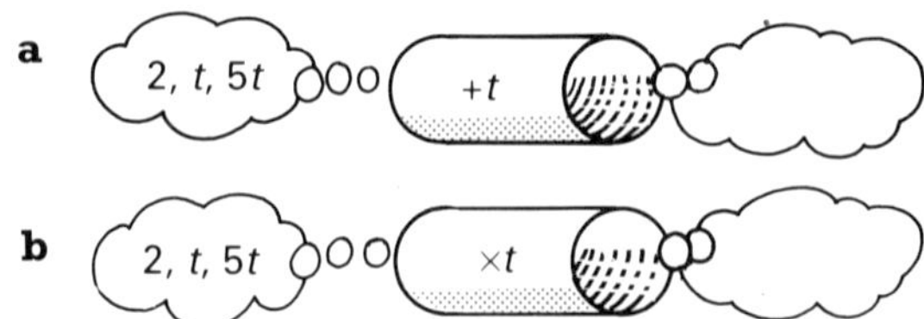

5 Find expressions for the unmarked snakes and ladders.

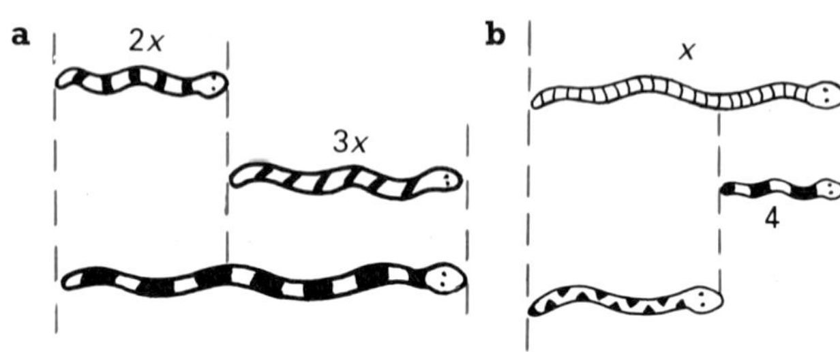

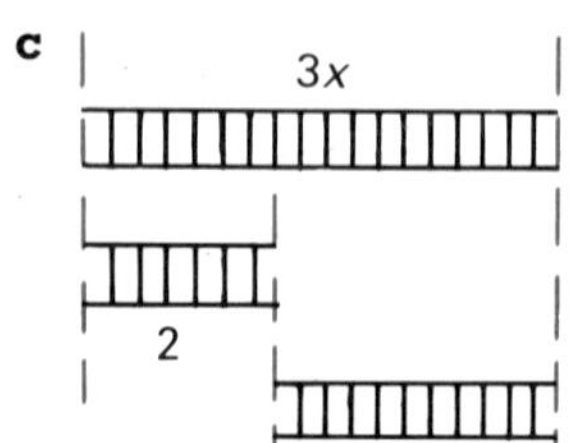

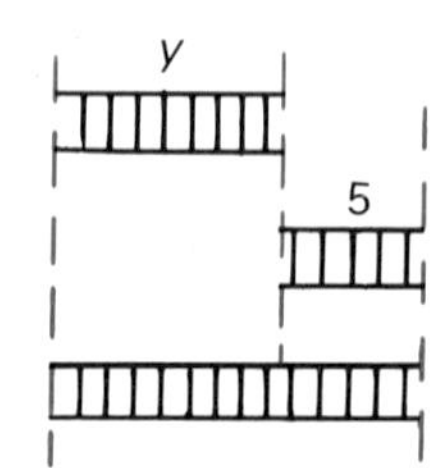

6 Find expressions for the perimeters and areas of these shapes, which are based on rectangles. The lengths are in cm.

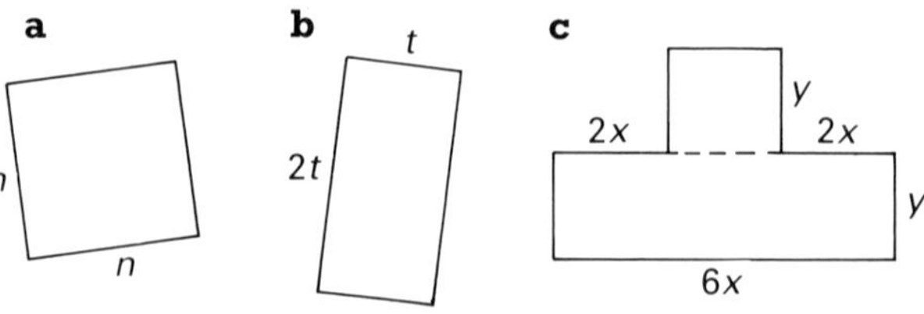

7 Find the areas of these rectangular gardens. Lengths are in metres.

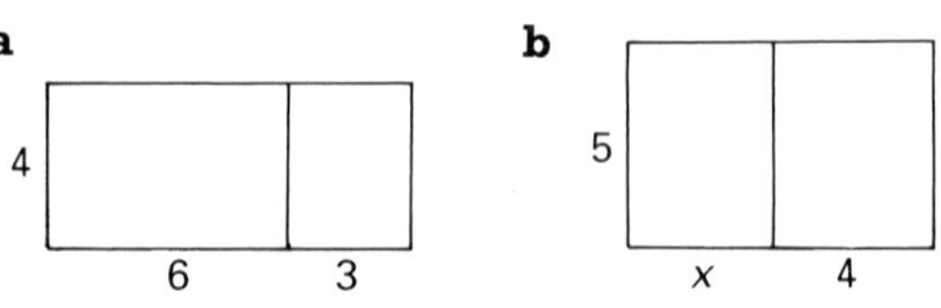

8 Write these without brackets:
a $5(x+2)$ **b** $6(y+1)$ **c** $3(z-4)$
d $2(p-1)$ **e** $8(1+t)$ **f** $7(2+n)$
g $4(3-v)$ **h** $9(4-w)$ **i** $2(2a+1)$
j $3(4b+2)$ **k** $5(2c-1)$ **l** $7(6-9d)$

9 Find an expression for the volume of these steps. Lengths are in cm.

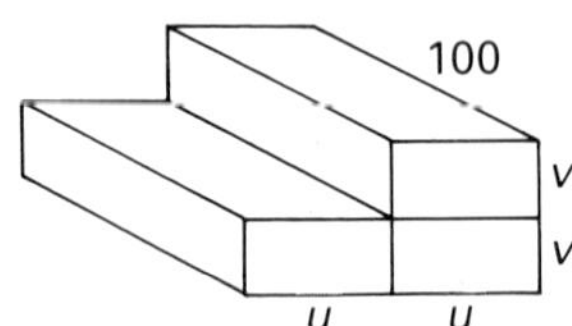

10 $x = 4$, $y = 2$, $z = 0$. Find the value of:
a x^2 **b** $2x^2$ **c** $(2x)^2$ **d** xyz
e y^2 **f** $(2y)^2$ **g** $2y^2$ **h** $(x+y+z)^2$

11 Simplify:
a $3(x+1)-3$ **b** $2(y-2)+4$ **c** $x(x+1)+x$
d $x(x+y)+xy$ **e** $x(1-x)-x^2$ **f** $x^2(x+1)+x$

REVISION EXERCISE ON CHAPTER 4: MAKING SENSE OF STATISTICS 1

1 This graph shows a shop's newspaper sales over a period of 15 days.

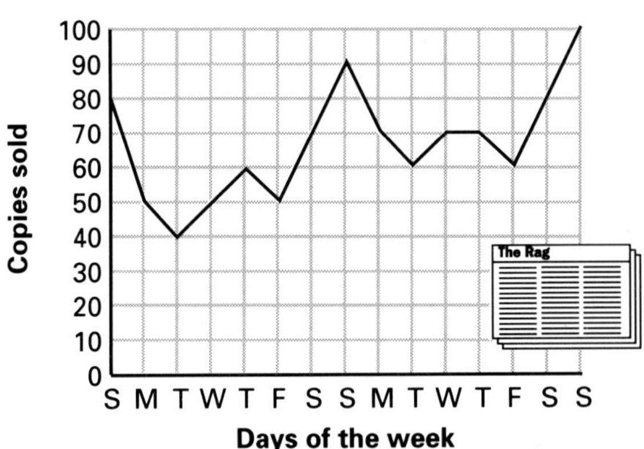

a Calculate the average Monday sales.
b Which is the best weekday (M–F) for sales?
c Describe two patterns in the graph.
d Is there an overall trend in the sales?

2 During some weather experiments the temperature was measured daily at noon.

Day (Week 1)	1	2	3	4	5	6	7
Temperature (°C)	15	12	13	16	17	13	13

Day (Week 2)	1	2	3	4	5	6	7
Temperature (°C)	14	17	16	17	15	13	12

Find the mean, median, mode and range of temperatures for:
a the first week **b** the second week
c the whole fortnight.

3 a Draw a line graph of the temperatures each week in question **2**.
b Describe any patterns of temperature which the graphs show.

4 Rovers have a new manager.

a Calculate the total number of goals scored each season.

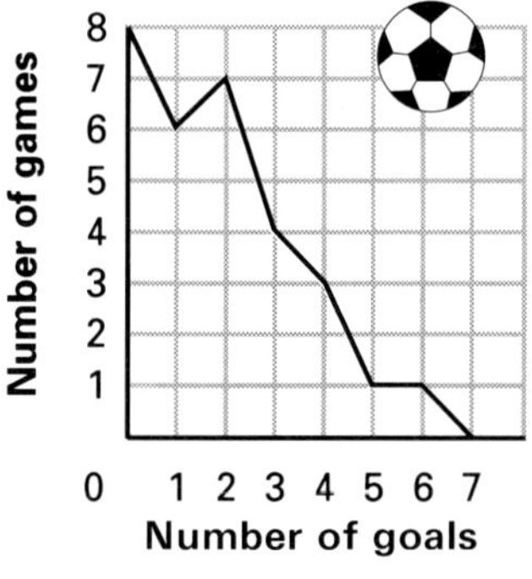

(30 matches this season)

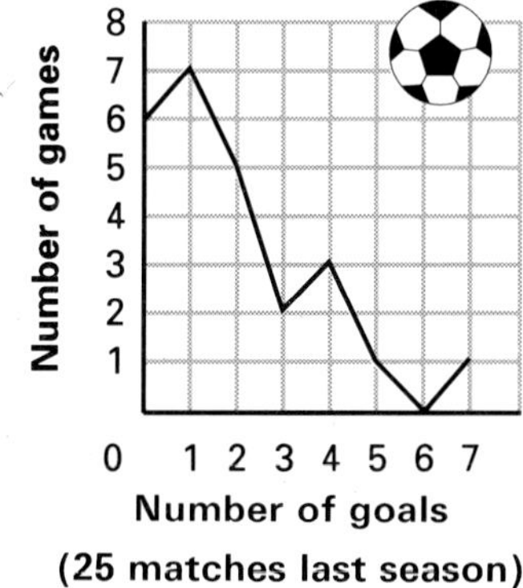

(25 matches last season)

b Find the mean number of goals each season, correct to 1 decimal place.
c The new manager claims that the team has played much better this season. Is his claim justified?

5 When metal is heated it expands. A number of metal rods 1 metre long were heated by the same amount and their expansions, in millionths of a metre, were measured and recorded in the table below.
a Calculate the mean expansion of the metal rods.
b Summarise the table by quoting the median and range of the expansions.
c Draw a bar graph of the data in the table.

Metal	Lead	Zinc	Aluminium	Tin	Brass	Copper	Nickel	Steel	Iron
Expansion	27.6	26.0	25.5	21.4	18.9	16.7	12.8	11.9	10.2

REVISION EXERCISE ON CHAPTER 5: FRACTIONS, DECIMALS AND PERCENTAGES

1

What fractions, decimal fractions and percentages of these shapes are:
a rectangles **b** triangles **c** circles?

2 Calculate:
a 25% of £18 **b** 40% of £650 **c** 3% of £1.

3 A rectangular garden is 35.5 m long and 15.6 m wide. Calculate:
a its area
b the length of fence needed to go round the sides.

4 In a sale a £60 watch is reduced by 15%. How much is:
a the reduction in price
b the actual price you would have to pay?

5 The Safe as Houses Building Society offers interest at 7%. Calculate the interest after one year, on:
a £300 **b** £1000 **c** £75.

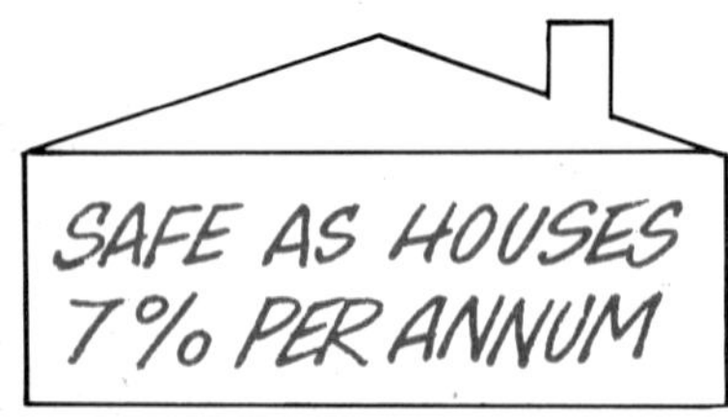

6 A car dealer buys a used car for £2500, and sells it for £2800.
a Calculate:
(i) his profit
(ii) his percentage profit, based on the cost price.
b What would the selling price be if he wants to make a profit of 14%?

7 Change these fractions to decimal form in order to find which is larger in each pair.
a $\frac{3}{5}$ or $\frac{11}{20}$ **b** $\frac{1}{3}$ or $\frac{4}{11}$ **c** $\frac{5}{9}$ or $\frac{7}{11}$ **d** $\frac{5}{7}$ or $\frac{8}{11}$

8 Jane sorted TV programmes into five types: children, family, adult, news and sport. She compared BBC and ITV one Saturday and drew these pie charts.

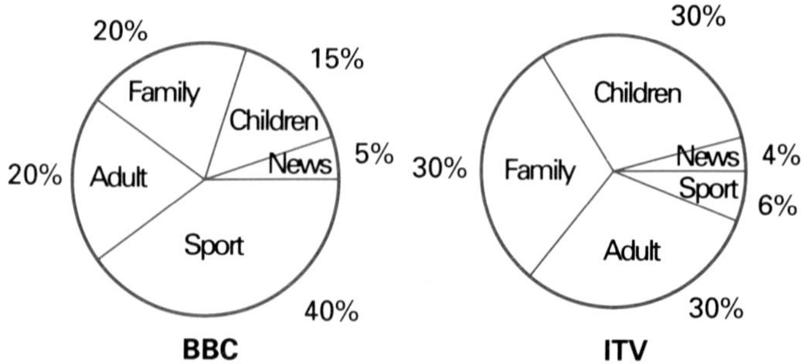

a Which channel gave a higher percentage of its time to:
(i) sport
(ii) family viewing
(iii) news?
b BBC gave 3 hours of time to family viewing. How much time did it give to:
(i) sport (ii) news (iii) children's programmes?

9 80 000 tickets are sold for the football cup final. Each of the two clubs playing in the final receives 16 000 tickets. What percentage of the tickets does each club get?

10

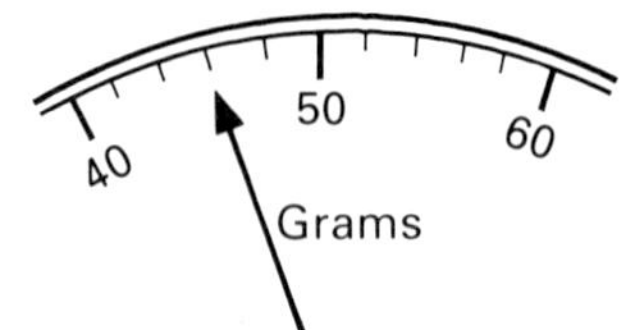

The true weight of the letter is 40 g.
a Calculate the percentage error in the reading on the scales.
b What would the reading be for a true weight of 60 g?

11 A supermarket buys video cassettes for £5 each. It sells them at £8 each, or 3 for £20.
a Calculate the profit on:
(i) 1 tape (ii) 2 tapes (iii) 3 tapes.
b Calculate the percentage profit on 1, 2 and 3 tapes.
c Investigate the percentage profit for 4, 5, 6, . . . tapes.

REVISION EXERCISE ON CHAPTER 6: DISTANCES AND DIRECTIONS

1 A rectangular piece of scaffolding has a diagonal strut AC to strengthen it. AB = 150 cm and BC = 300 cm.

a Use a scale of 1 cm to 50 cm to make a scale drawing.

b Measure the length of AC, and the size of ∠BAC.

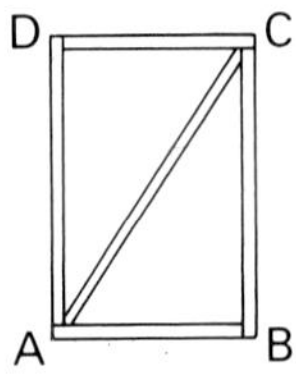

2 Mark's kite is caught at the top of the tree. Use a scale of 1 cm to 5 m to make a scale drawing of △TJK, and find the height of the kite above the ground.

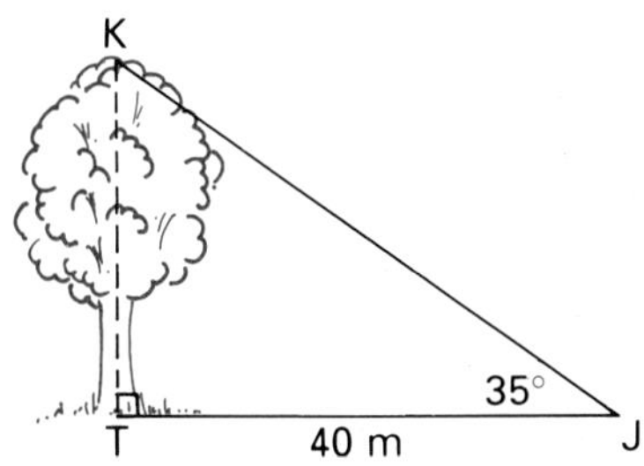

3 The school sports field is 85 m long, and slopes up from one end to the other. The angle of elevation of one end from the other is 8°. Use a scale drawing to find the difference in height between the two ends.

4 You are facing east. On your right there is a tree. On your left there is a house. Behind you there is a large rock.

a Is the house north or south of the tree?

b In which direction must you face to see the rock?

c What instructions would take you to the house?

5 M, P, T, K, B, R are the first letters of the names of villages on Coral Island.

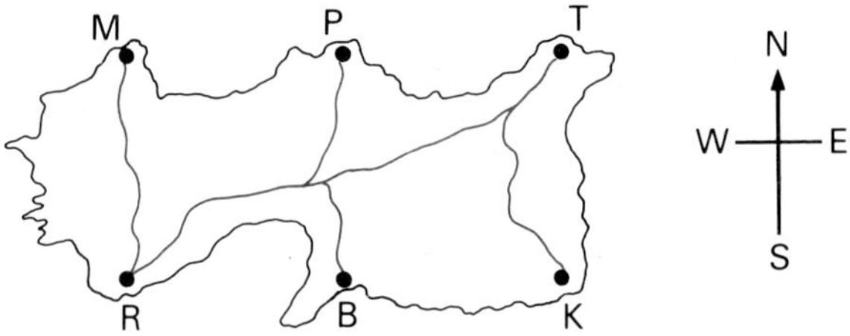

a Which village is:
(i) SE of P (ii) NE of R (iii) NW of B
(iv) SW of T?

b The straight-line distance from M to R is 12 km. Find the straight-line distances from M to:
(i) T (ii) K.

6 Two boats leave the same harbour. One sails 12 km on a bearing of 035°, and the other sails 9 km on a bearing of 165°.

a How far apart are they?

b What bearing should the first boat take if it wishes to join the other one?

7 A TV detector van at a point A picks up a signal on a bearing of 047°. It moves 100 m east along a road to B, and picks up the signal on a bearing of 305°. Use a scale drawing to find how far the signal is:

a north of the road **b** east of A.

REVISION EXERCISE ON CHAPTER 7: POSITIVE AND NEGATIVE NUMBERS

1 Here are some mid-January night temperatures: London, -6°C; Edinburgh, -8°C; Newcastle, -7°C; Inverness, -10°C; Aberdeen, -9°C.

a Which place is:
(i) warmest (ii) coldest?

b List the temperatures in order, coldest first.

2 a Which number is:
(i) 2 greater than -2
(ii) 3 less than -2?

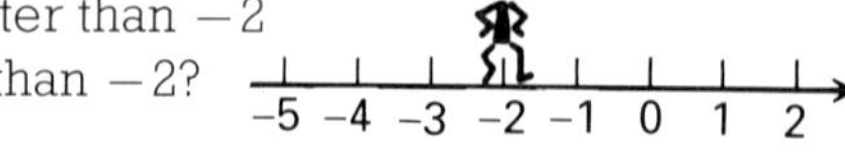

b Which numbers shown on the number line are:
(i) less than -2 (ii) greater than -2?

3 List the numbers in each set for x.

a $-4, 2, -2, 0, 1$; x is less than -1

b $-3, 1, -2, -1, 0$; x is greater than -2.

4 Write down calculations for these walks.

a

b

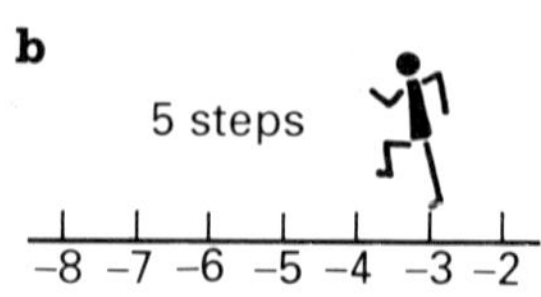

c

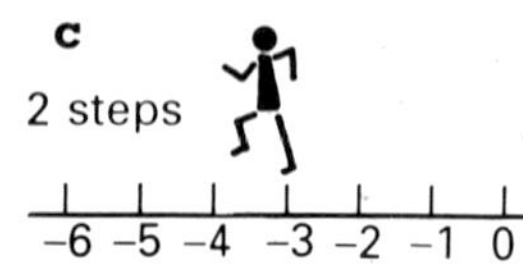

d

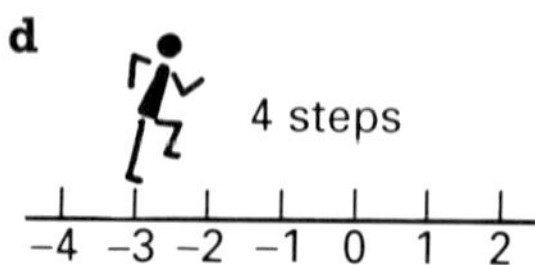

5 Copy and complete:

a
$2+1 = \ldots$
$1+1 = \ldots$
$0+1 = \ldots$
$-1+1 = \ldots$
$-2+1 = \ldots$
. . . .
$-9+1 = \ldots$

b
$2+1 = \ldots$
$2+0 = \ldots$
$2+(-1) = \ldots$
$2+(-2) = \ldots$
$2+(-3) = \ldots$
. . . .
$2+(-9) = \ldots$

c
$3-2 = \ldots$
$3-1 = \ldots$
$3-0 = \ldots$
$3-(-1) = \ldots$
$3-(-2) = \ldots$
$3-(-3) = \ldots$
. . . .
$3-(-9) = \ldots$

6 Calculate:

a $3-7$ **b** $3+(-7)$ **c** $3-(-7)$
d $-3-7$ **e** $-3+(-7)$ **f** $-3-(-7)$
g $-3+7$ **h** $-2+(-5)$ **i** $1-9$
j $9+(-1)$ **k** $9-(-1)$ **l** $-9-1$

7 Use the 'cover up' method to solve these equations:

a $x+2=1$ **b** $4-x=-2$ **c** $5-x=7$

8 a Copy the diagram, and mark C, the image of A under reflection in the x-axis, and D, the image of B under reflection in the y-axis.

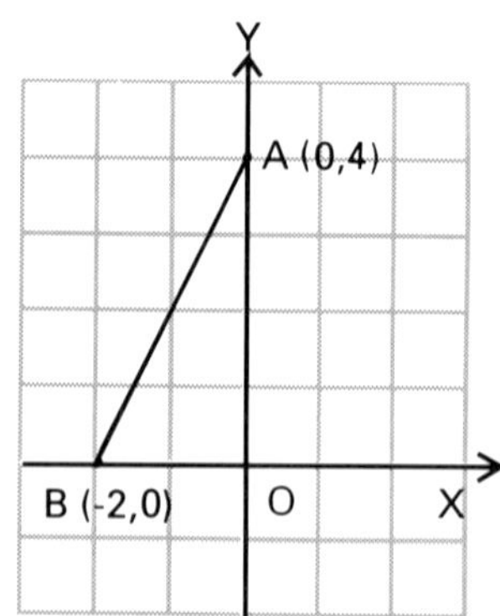

b Write down the lengths of AC and BD.

9 a Add across and down.

0	-1	
-2	3	

b Subtract across and down.

0	-1	
-2	3	

10 $p=-3$, $q=-5$, $r=-7$. Calculate the values of:

a $p+q$ **b** $p-q$ **c** $r+p$ **d** $r-p$

REVISION EXERCISE ON CHAPTER 8: ROUND IN CIRCLES

1 Each wheel of a bicycle has a circumference of 2 m. How far will the bicycle go in:
a 1 turn of a wheel **b** 10 turns
c 100 turns **d** $\frac{1}{2}$ turn?

2 Calculate:
a the diameter of the top of the plant pot
b the radius of the bottom.

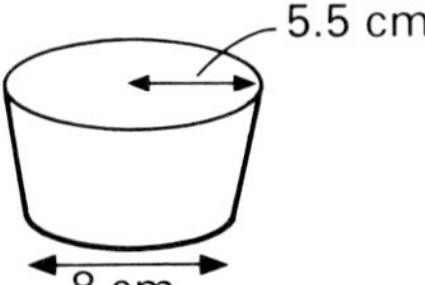

3 Using $\pi = 3$, calculate the circumferences of the top and bottom of the plant pot in question **2**.

4 Use $\pi = 3$ to calculate these circumferences.

a
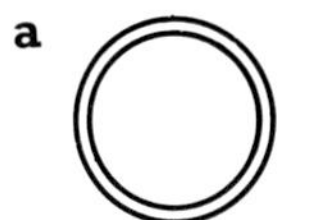
Child's hoop (diameter 60 cm)

b
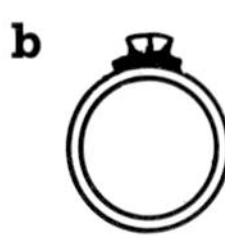
Ring (diameter 1.5 cm)

c

Steering wheel (radius 18 cm)

5 Calculate the circumference of each object to 3 significant figures:

a the inlay round the hole in the guitar

b the red trim round each piece of this china.

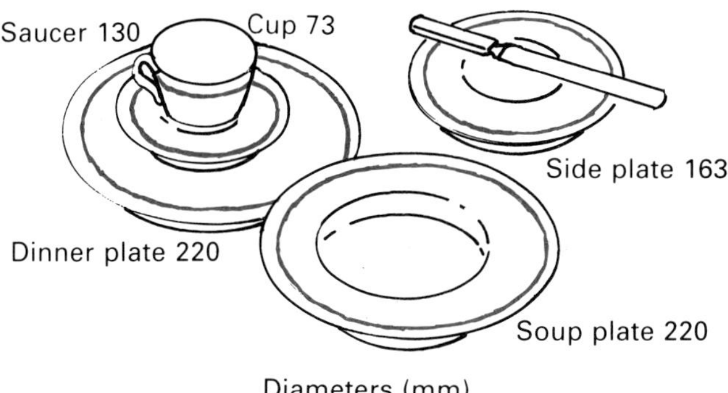

Diameters (mm)

6 The opening at the top of the lampshade is 6 cm in diameter.

The opening at the bottom is 20 cm across. Use $\pi = 3$ to estimate the areas of the two openings.

7 The diameter of the camera lens is 80 mm. Calculate its area, correct to 3 significant figures.

8 At the Royal Mint 2p coins are stamped out of strips of metal.

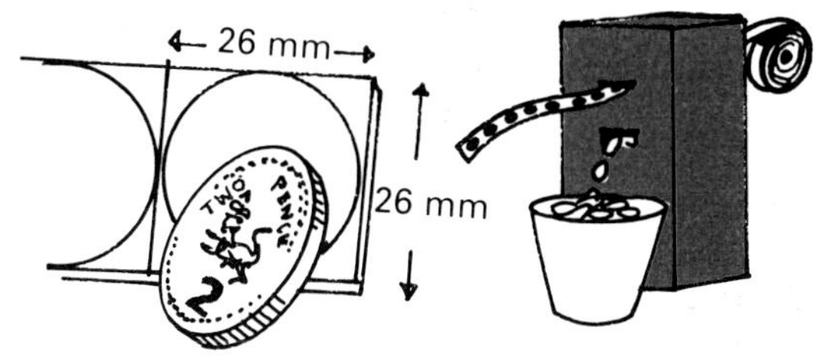

a Calculate the area of waste for each coin, to the nearest mm^2.
b How much waste will there be when £5 worth of coins have been stamped out?

REVISION EXERCISE ON CHAPTER 9: TYPES OF TRIANGLE

1 Calculate $a, b, c, \ldots$

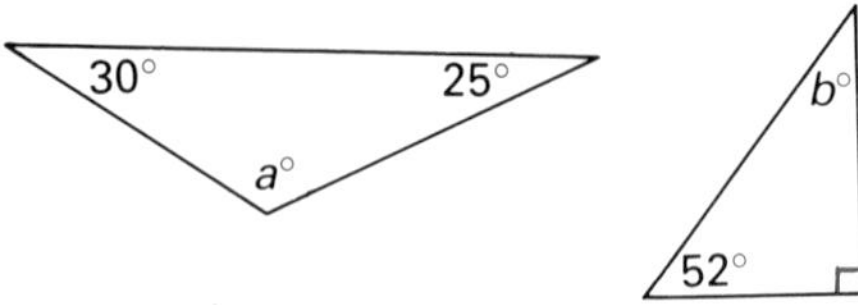

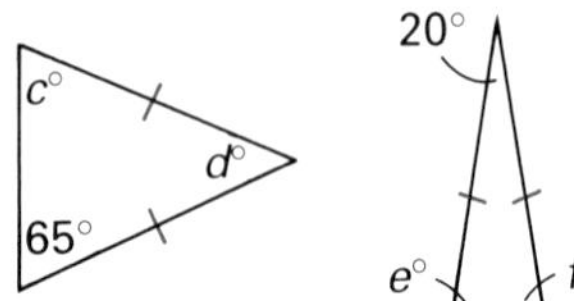

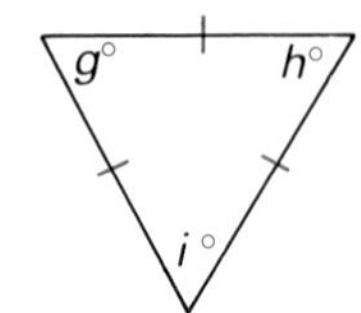

2 Calculate the areas of these isosceles triangles.

a

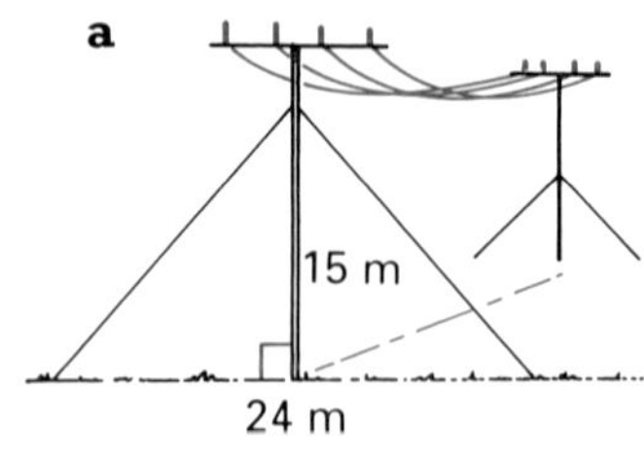

b

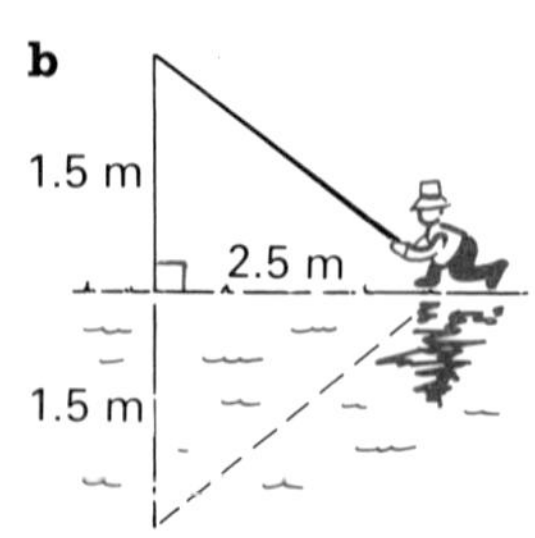

c

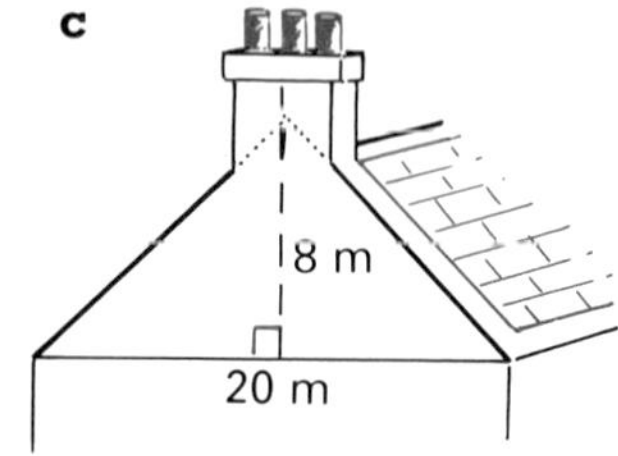

d

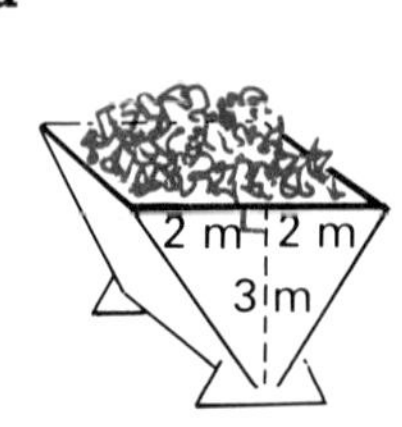

3 Describe each triangle below: acute-angled, obtuse-angled, right-angled, isosceles.

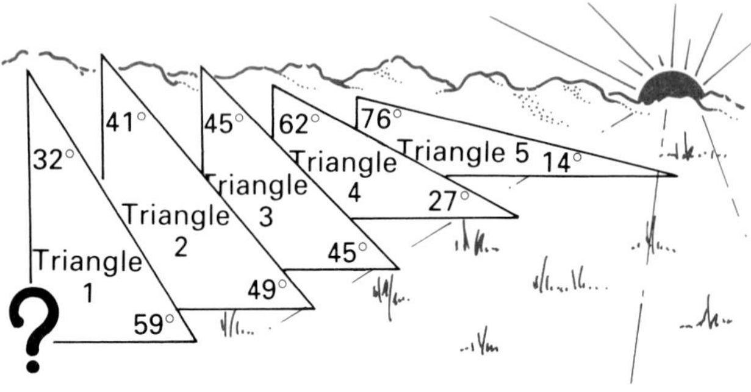

4 Copy these diagrams, and fill in all the angles. Describe each type of triangle.

a **b**

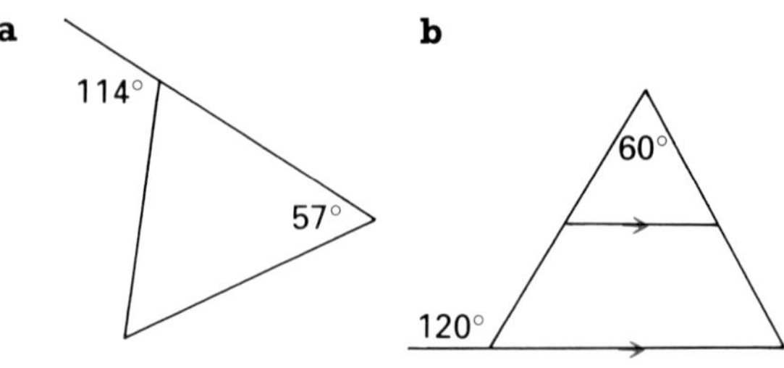

c **d**

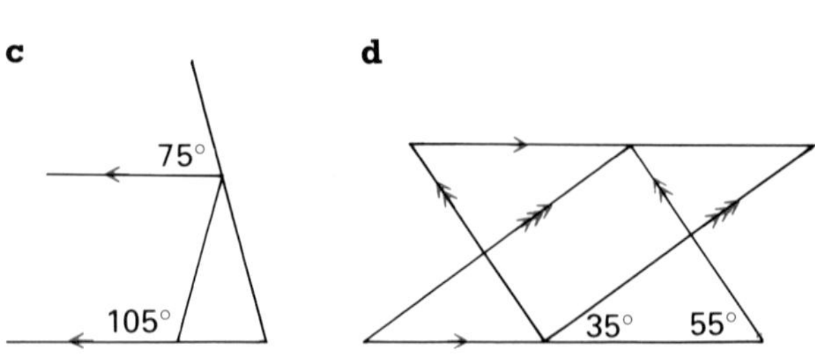

5 A machine is programmed to cut out metal triangles which are used to make paint strippers.

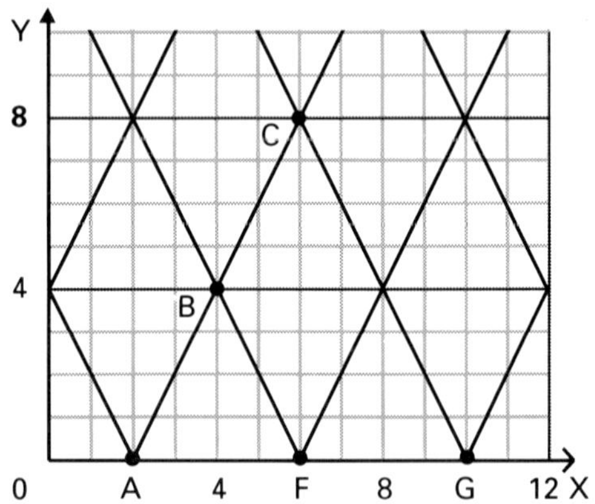

a Write down the coordinates of the points A, B and C.

b D and E are the next two points on the line ABC. What are their coordinates?

c Calculate the area of each metal triangle.

d Notice that there is wastage along the sides. What is the total area of metal wasted in each horizontal strip?

6 A ladder 6 m long leans against the wall of a house. It makes an angle of 72° with the horizontal ground.

a Make a sketch, and fill in all the angles.

b Make a scale drawing, and find how far up the wall the ladder goes.

REVISION EXERCISE ON CHAPTER 10: METRIC MEASURE

1 Change each of these to the units given in brackets:
a 4 km (m) **b** 15 mm (cm) **c** 1.2 kg (g)
d 5 litres (ml)

2 On squared paper draw a rectangle ABCD with AB = 60 mm and BC = 45 mm.
a Measure the length of its diagonal AC in mm.
b Calculate its perimeter in: (i) mm (ii) cm.

3 Calculate these shaded areas, correct to 2 significant figures.

a

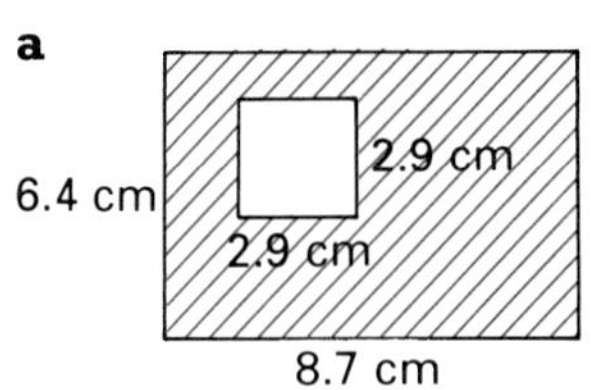

b

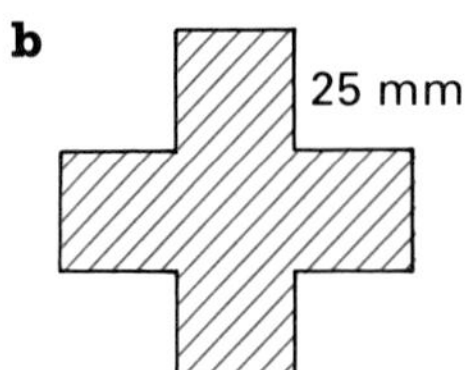

4 **a** Each small square has side 1 cm long. Calculate the area and perimeter of this shape.

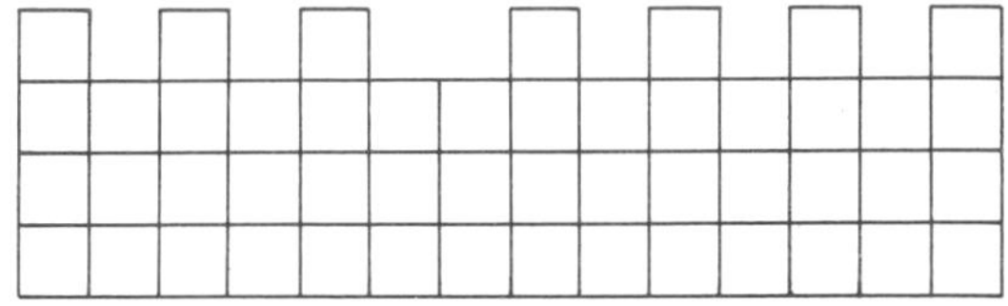

b The shape can be remade as a square. Find its area and perimeter.

5 The carpet is 2.5 m wide. A hotel orders 1 hectare (10 000 m^2) of carpet.

a What length of carpet is this?
b What is the cost, at £18.99 per m^2?
c What is the volume of the carpet in m^3, if it is 2 cm thick?

6 What are the capacities of **a** and **b**, in millilitres, and the volume of liquid in the jug in **c**?

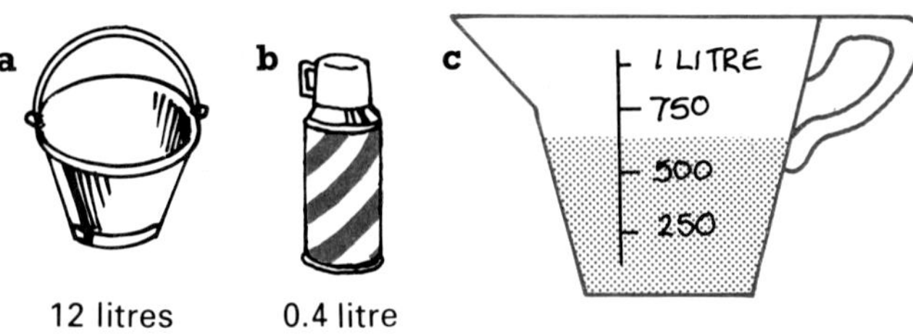

7 A chemical was dried out by heating.
a What was its weight before and after being dried?

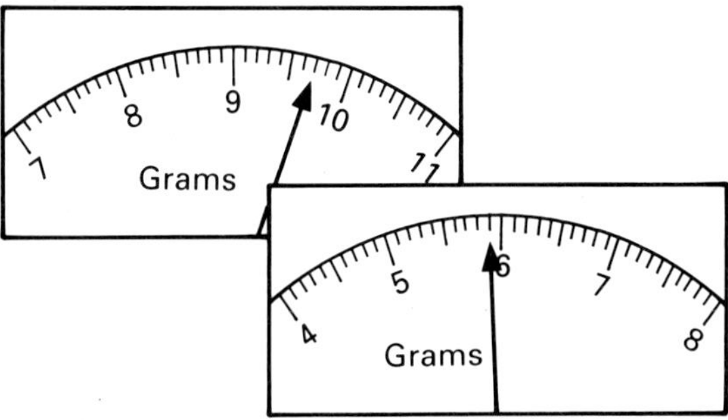

b (i) What weight of water was removed?
(ii) What was the volume of this water?

8 Calculate the weight of metal in this casting, in kg, given that 1 cm^3 weighs 9.5 g.

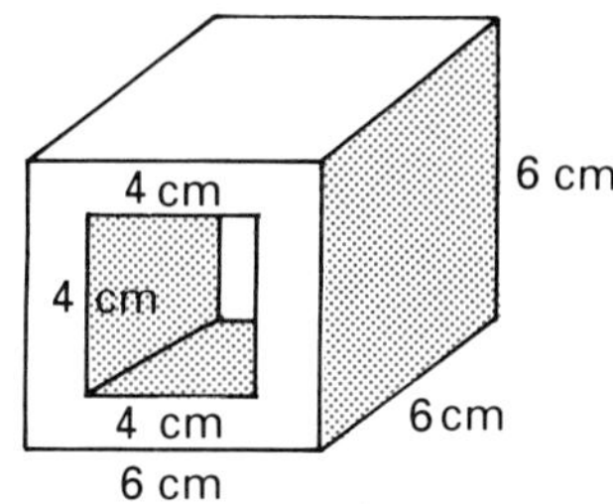

9 A cuboid of modelling clay is 9 cm by 6 cm by 4 cm. If it is remade as a cube, what length is each side of the cube?

REVISION EXERCISE ON CHAPTER 11: EQUATIONS AND INEQUATIONS

1 Solve these equations:
a $2x+1=9$ **b** $3x-1=14$ **c** $4x+7=19$
d $3y=y+8$ **e** $8y=3y+50$ **f** $y=4-y$

2 (i) Make an equation for each picture, and solve it.
(ii) Write down the cost of a box in **a**, and the weight (in kg) of a bag in **b**.

a

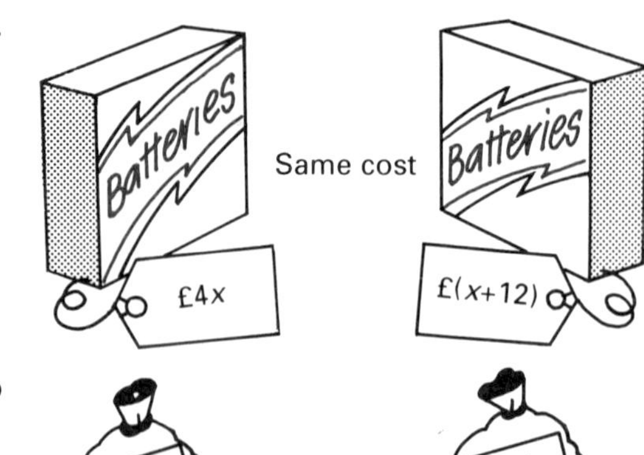

b

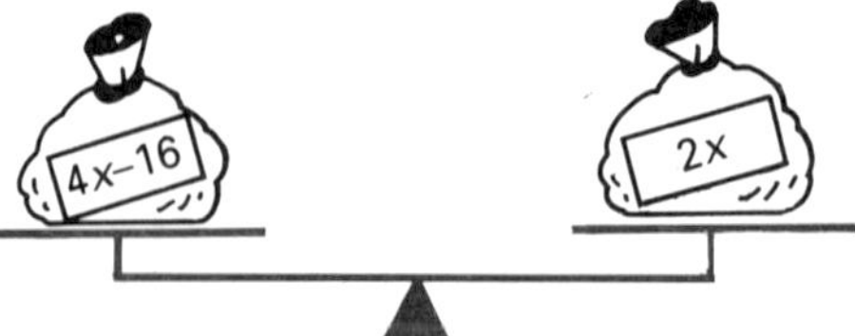

3 Solve:
a $2(x+3)=12$ **b** $7(x+2)=21$
c $3(x-2)+1 - 13$

4 Write down an inequation for the weights in each picture below.

a

b

5 Solve these equations:
a $2x+5=x+10$ **b** $3y-4=y+8$

6 Make equations for these pictures, and solve them. Weights are in kg.

a

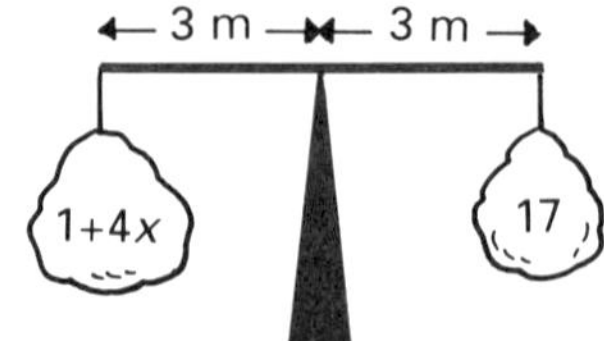

b

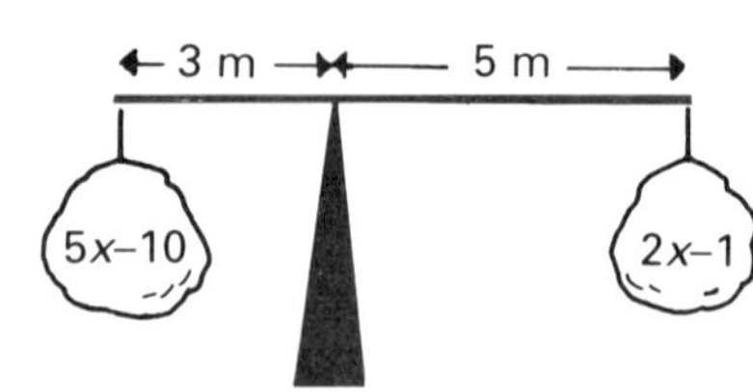

7 Solve these equations:
a $x-5=-1$ **b** $3=-5+2x$ **c** $4t=t+9$

8 a True or false?
(i) $-2<-1$ (ii) $-2<0$ (iii) $-2>-5$
(iv) $-2>1$
b List the numbers in each set for x.
(i) $-3, 1, 2, -1; x>-2$
(ii) $0, -1, 1, 2, -2; x<0$

9 The area of the first rectangle is double the area of the second one. Make an equation and solve it to find the length and breadth of each rectangle.

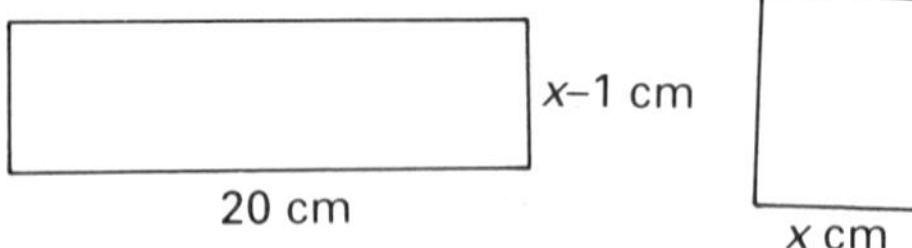

10 If the perimeter of the first rectangle in question **9** is double the perimeter of the second one, find the dimensions of the two rectangles.

11 Choosing replacements from $\{-2, -1, 0, 1, 2\}$, solve:
a $x<0$ **b** $4>x+3$ **c** $-2+x \leqslant -3$

REVISION EXERCISE ON CHAPTER 12: RATIO AND PROPORTION

1 Write down, in simplest form, the ratio of the number of:

a faces : vertices

b vertices : edges

c faces : edges

d dots you can see: dots you cannot see on the dice.

2 Share £1 in the ratio:

a 1 : 1 **b** 3 : 1 **c** 3 : 2.

3 A rich sweet pastry is made by mixing flour, icing sugar and butter in the ratio 4 : 1 : 2.

a What weight of each is needed for 350 g of pastry?

b George has 100 g flour.

(i) What weights of icing sugar and butter does he need?

(ii) What weight of pastry does this make?

4 Photograph A is enlarged to B in the ratio 3 : 2.

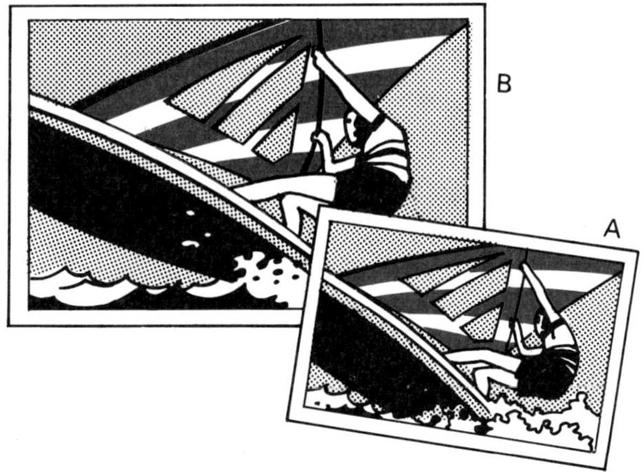

a A is 120 mm wide. What is the width of B?

b The mast in A is 36 mm tall. How tall is it in B?

c B is 114 mm long. What is the length of A?

5 Marie earns £33 for six hours as a waitress in the Peartree restaurant. How much does she earn in:

a 1 hour **b** 3 hours **c** 5 hours?

6 List the entries in the second row of each table.

a Mortgage interest at 10% of mortgage

Mortgage (£)	1000	2000	3000	5000	10 000
Interest (£)					

b Sharing out £5 equally

Number of shares	1	2	5	10	20
Value of each (£)					

7 A garden path can be covered with 56 paving slabs, each 1.5 m long. (Assume that the slabs are all the same width.)

a How many are needed if each is 1.2 m long?

b What is the length of a paving slab if 168 are needed?

8 Pam travels around Britain selling perfume.

Use her travel records:

a to show that the number of miles she travels seems to be directly proportional to the number of gallons of petrol she uses

b to find the values of (i), (ii), (iii) and (iv) in her records.

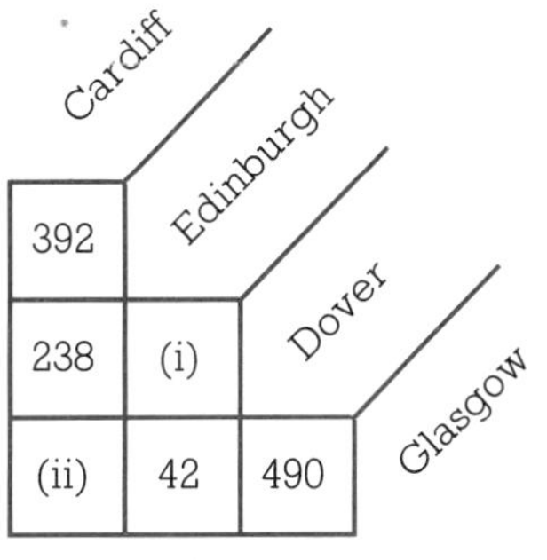

	Cardiff	Edinburgh	Dover
Edinburgh	392		
Dover	238	(i)	
Glasgow	(ii)	42	490

Distance (miles)

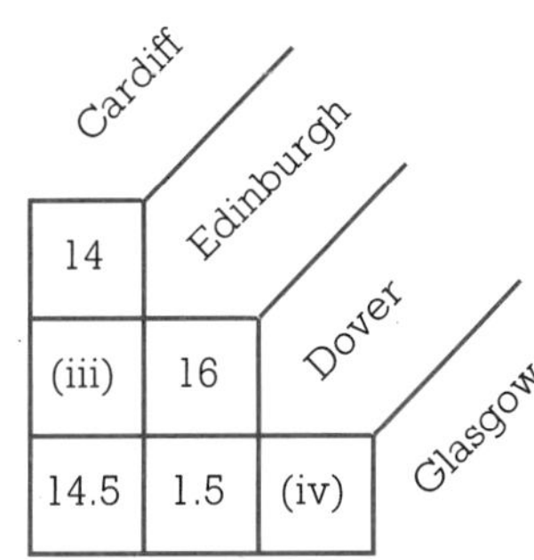

	Cardiff	Edinburgh	Dover
Edinburgh	14		
Dover	(iii)	16	
Glasgow	14.5	1.5	(iv)

Petrol used (gallons)

REVISION EXERCISE ON CHAPTER 13: MAKING SENSE OF STATISTICS 2

1

Peter checked how much memory was left on each of his floppy discs, to the nearest 1 k. Calculate the mean memory left on his discs.

Memory left	1	2	3	4	5	6	7	8	9
Number of discs	6	8	4	10	2	13	1	2	4

2

June was testing a spinner for fairness. She numbered its sectors 1–8, and got these results in 42 spins.

1	6	5	5	7	7	5
4	3	5	2	4	7	2
8	8	1	3	5	1	5
2	7	4	7	6	4	8
7	1	8	2	1	8	5
8	4	5	7	4	2	8

a What would the mean, median and mode be for a fair spinner?
b Calculate the mean, median and mode for June's spinner and comment.

3 A survey of the values of flats in a city gave these results.

Value (£1000's)	20	30	40	50	60	70	80
Frequency	7	5	8	2	4	2	2

a Draw a frequency diagram and a frequency polygon to illustrate the values.
b Find the mean, median and modal values.

4 a Arrange these test marks in a frequency table in classes 1–5, 6–10, . . .

11	25	36	31	38	10	20	27	30	28
32	25	31	25	14	29	20	18	16	28
25	19	29	22	17	35	24	22	36	32

b Calculate the mean mark.
c What is the modal class interval?

5 A small company makes luxury cars. This diagram shows a year's output.

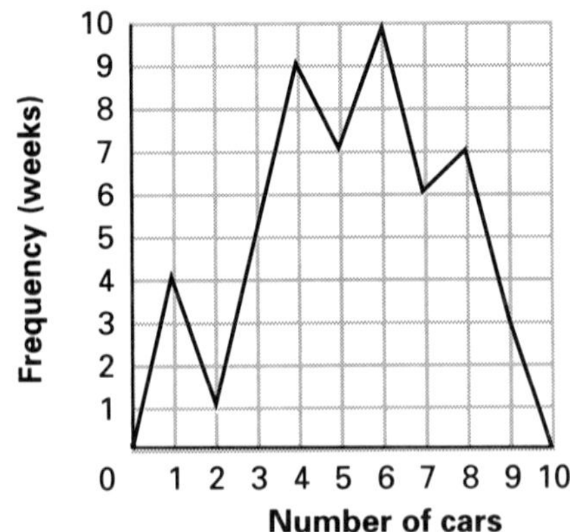

a Make a table:

Number of cars	
Number of weeks	

b Calculate the mean number of cars made each week.

6 A cotton mill is making a new fibre. The fibre's strength is tested by finding how much weight it takes to break threads of different diameters.

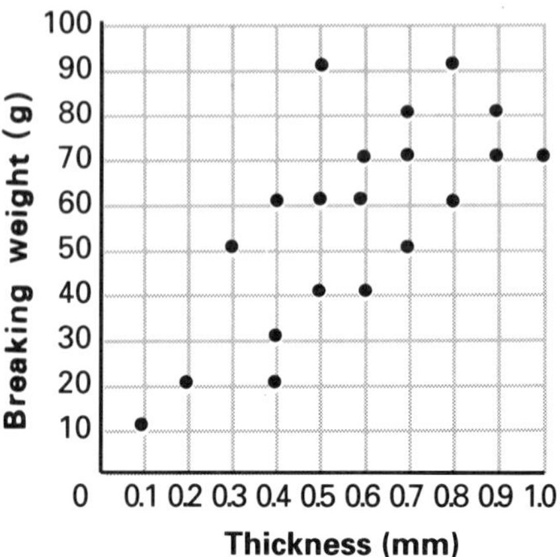

a Find the mean thickness of the sample.
b Find the mean breaking weight of the sample.
c Hold your ruler along the line of best fit, and estimate the diameter for a breaking weight of 45 g.

REVISION EXERCISE ON CHAPTER 14: KINDS OF QUADRILATERAL

1 Draw a kite ABCD, and its axis of symmetry. Mark pairs of equal sides and angles.

2 Draw a parallelogram PQRS, and its centre of symmetry. Mark pairs of equal sides and angles, and parallel sides.

3 Which quadrilaterals have:
- **a** exactly two axes of symmetry
- **b** opposite sides parallel
- **c** angles which add up to 360°
- **d** half-turn symmetry?

4 Copy these quadrilaterals, and fill in the sizes of as many angles and the lengths of as many lines as you can. The lengths are in cm.

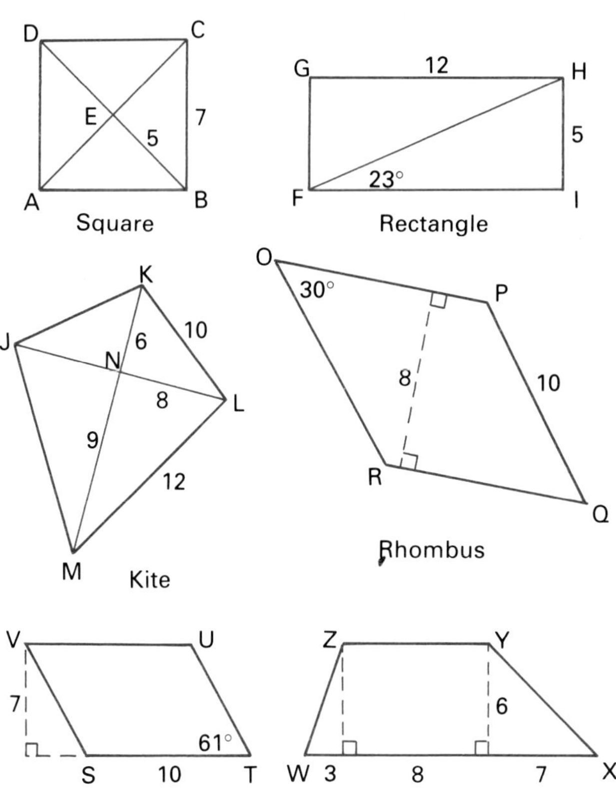

5 O(0, 0), A(6, 4) and B(6, 6) are corners of a parallelogram OABC.
- **a** Write down the coordinates of:
  (i) C (ii) the point where OB and AC cross.
- **b** Find the coordinates of the images of O, A, B and C under:
  (i) reflection in the x-axis
  (ii) reflection in the y-axis
  (iii) a half turn about O.

6 In the diagram below, R = rhombus, P = parallelogram and T = trapezium. Copy the diagram and fill in all the angles.

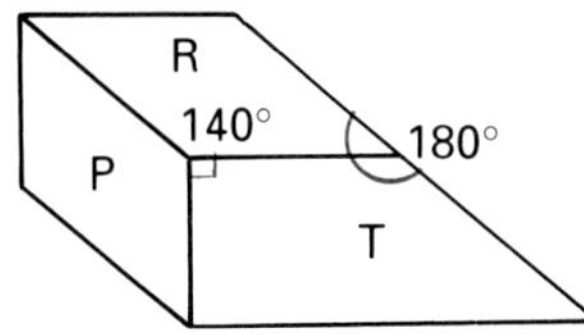

7 This star is made up of six congruent kites.

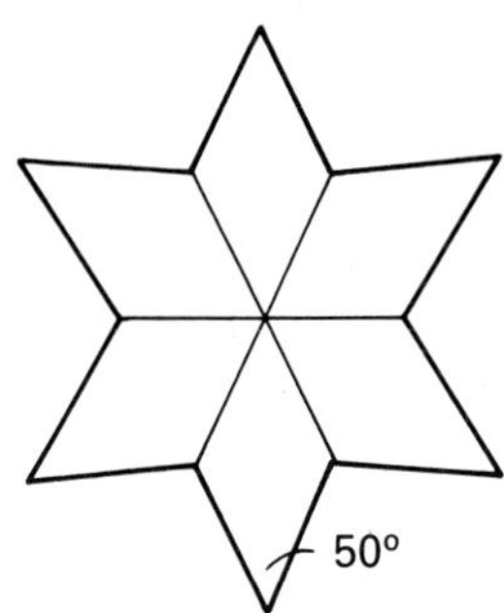

- **a** How many axes of symmetry has the star? What is its order of symmetry?
- **b** Make a sketch of the star, and fill in all the angles.

8 A gymnasium box has a square top and trapezium sides that slope outwards. Sketch the box, and a net you could use to make a model of it.

REVISION EXERCISE ON CHAPTER 15: SOME SPECIAL NUMBERS

1 Calculate the value of:
a 5^2 **b** 3^4 **c** 2^5 **d** 10^6

2 Write in index form:
a $7 \times 7 \times 7$ **b** $2 \times 2 \times 2 \times 2 \times 2 \times 2$ **c** 25×25

3 Write down all the numbers between 20 and 90 which are squares of whole numbers.

4 What is the length of the edge of each square?

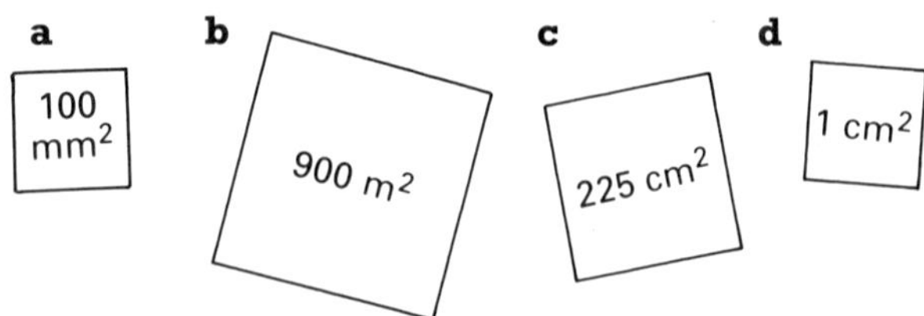

5 Between which two consecutive whole numbers does each square root lie?
a $\sqrt{5}$ **b** $\sqrt{31}$ **c** $\sqrt{96}$ **d** $\sqrt{70}$

6 *Estimate*, then check, the value of each square root in question **5**, correct to 1 decimal place.

7 Which numbers go in the envelopes?

a

b

8 Which numbers go in the envelopes?

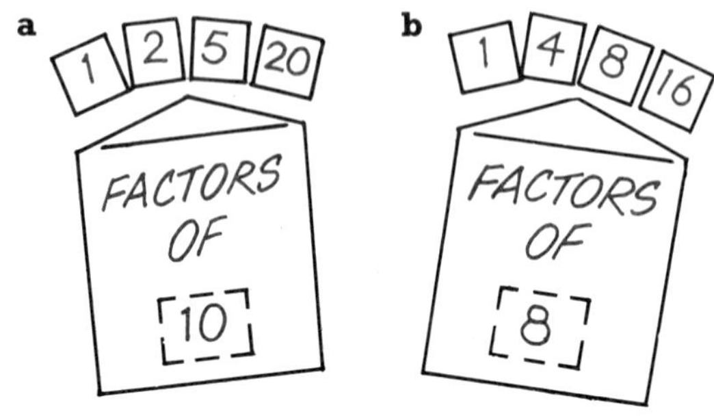

9 From the list 9, 10, 11, 12, 13, 14, 15, 16, 17, 18, 19, 20, 21, 22, write down the:
a square numbers
b numbers with exact square roots
c multiples of 7
d prime numbers
e factors of 44.

10 Find two consecutive prime numbers that add up to 60.

11 Copy and complete these prime factor trees:

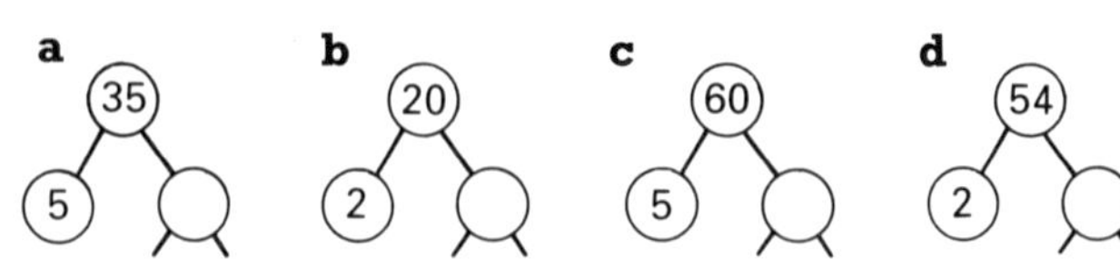

12 Find all the factors of:
a 28 **b** 150

13 Find the least common multiple (lcm) of:
a 8 and 11 **b** 6 and 9 **c** 12 and 14

14 Write each number below as a product of prime factors:
a 9 **b** 36 **c** 110 **d** 400

15 Write the answers to question **14** in index form.

16 Three lights flash at the same time.
The first one flashes every 10 seconds, the next one every 12 seconds and the third one every 15 seconds. After how long will they all flash together again?

REVISION EXERCISE ON CHAPTER 16: FORMULAE AND SEQUENCES

1 a Make a formula for the distance d from S to F along the route shown (distances are in km).

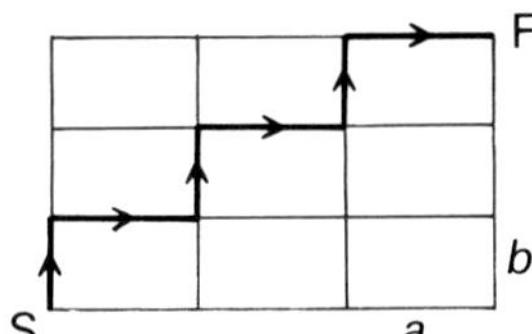

b Use the formula to calculate d when $a = 125$ and $b = 75$.

2 a Find formulae for the perimeter P and area A of the sports arena with semi-circular ends in diagram (i), and the garden shaped like a kite in diagram (ii).

(i)

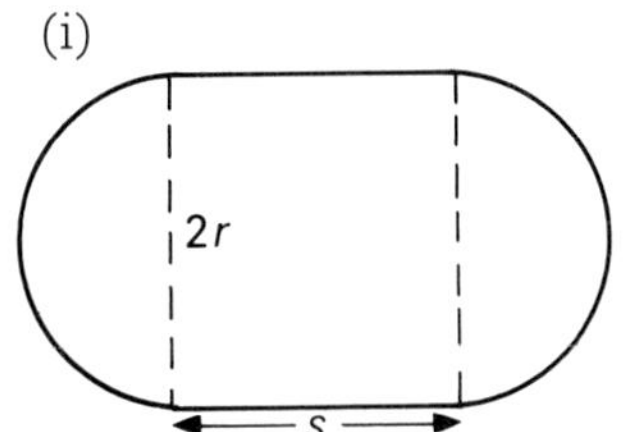

(ii)

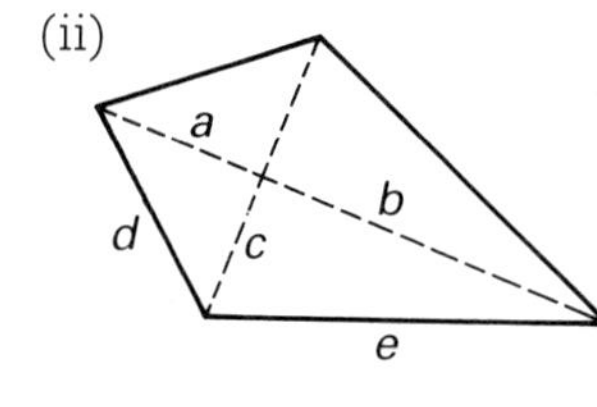

b Use your formulae to calculate the perimeters and areas when $r = 50$, $s = 80$, $a = 9$, $b = 16$, $c = 12$, $d = 15$ and $e = 20$.
(Lengths are in metres. Give answers correct to 3 significant figures.)

3 Write down the first four terms of sequences given by the rules:
a start at 8, and add 8
b start at 8, and subtract 8
c start at 5, double and subtract 6.

4 Find the missing numbers in these sequences:
a __, 13, 15, __, 19 **b** __, 15, __, 23, 27

5 Given the nth terms, find the terms listed:

a 2n+4 first three terms

b 7n−3 4th, 5th, 6th terms.

6 Find the nth terms of these sequences:
a 6, 11, 16, 21, . . . **b** 9, 18, 27, 36, . . .
c 7, 15, 23, 31, . . .

7 Find which term the given number is in these sequences:

a

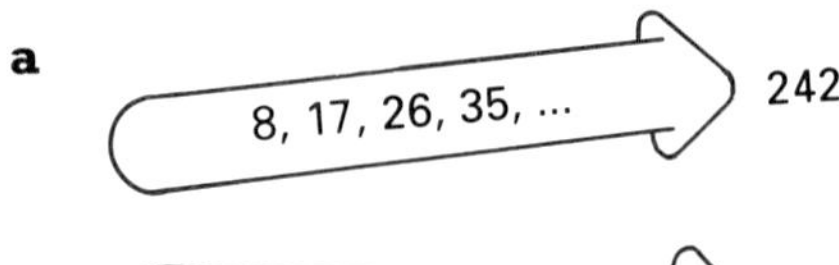

b 10, 17, 24, 31, ... → 290

8 Use the given information about constant differences to find the next three terms of each sequence.
a 2, 8, . . . first differences
b 1, 7, 15, . . . second differences
c 2, 3, 8, 20, . . . third differences
d 1, 9, . . . second differences are 3

9 How many matchsticks are in the eighth design?

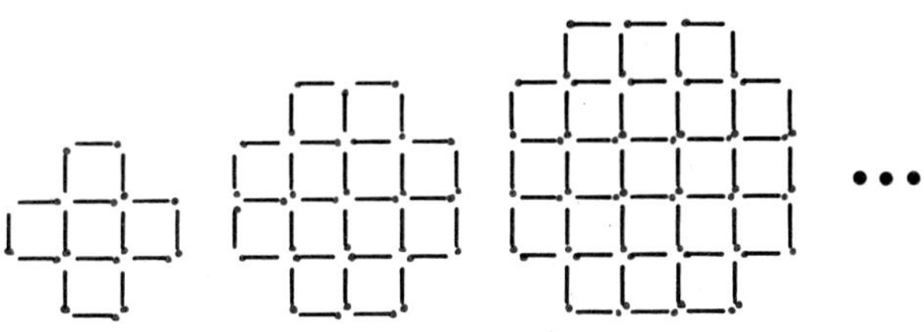

10 For the cuboid, $V = abc$, $A = 2(ab + bc + ca)$, $d = \sqrt{(a^2 + b^2 + c^2)}$.
Calculate V, A and d when $a = 9$, $b = 12$ and $c = 8$ (units are metres).

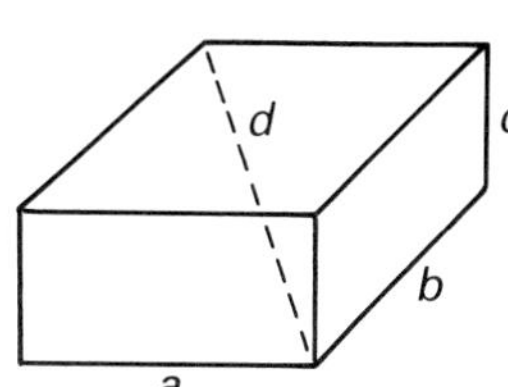

REVISION EXERCISE ON CHAPTER 17: PROBABILITY

1 A ball is thrown, and must finish in one of the holes. You lose on L, or win 10p or £1 as shown.

L	L	L	10p
L	L	£1	L
L	10p	10p	L
10p	L	£1	L
10p	L	10p	L

Copy this 0–1 line, and put arrows to show:
a P(L) **b** P(£1) **c** P(10p) **d** P(a win)

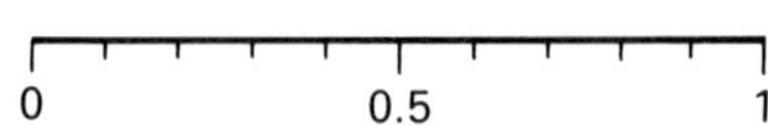

2 A set of 28 dominoes is placed face down, and one is chosen. Calculate:
a P(double blank) **b** P(a double) **c** P(8 dots)

3 **a** What is the probability that a day chosen at random will:
(i) end with Y (ii) start with Z
(iii) have eight letters in it?
b If the probability of catching a cold in winter is $\frac{3}{5}$, what is the probability of avoiding a cold?

4

A tin of sweets contains 20 with mint centres, 50 with chewy centres, 70 with soft centres and 60 with nutty centres.
a Calculate the probability (as a decimal) that a sweet chosen at random will be:
(i) a mint (ii) chewy (iii) soft-centred
(iv) nutty.
b Mark the probabilities on a scale like the one in question **1**.

5 What method would you use to estimate the probability that:
a a box of eggs in the supermarket contains a broken one
b a sweet chosen from a tube of mixed colours is red?

6

Clear View Glazing Company leaflet the town. They estimate that the probability of a response is 3%. From 6500 leaflets, how many enquiries should they expect?

7 From past data, the probability of rain on a day in April at Greenhaven is 40%. How many days in April would you expect to be:
a wet **b** dry?

8 The table shows a week's sales in the Stepping Out shoe shop.

Shoe size	5	6	7	8	9	10	11
Number sold	4	15	25	47	35	21	3

a What is the probability that the next customer wants size 11?
b The shop orders 1000 pairs. How many should be size 6?

9 Three friends carry out a survey of households with cars.

	Hannah	Sophie	Gita
Houses with a car	8	10	120
Number asked	10	20	200

a What would each girl say was the probability that a house had a car?
b Whose estimate is likely to be most accurate?
c Using Gita's result, how many cars would you expect in a city with 3 000 000 homes?

ANSWERS

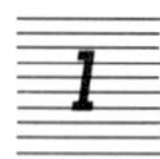

NUMBERS IN ACTION

Page 1 Looking Back

1a (i) 920 km/h (ii) 2000 m (iii) 100° (iv) 110
b 2513 litres **c** 1440 km **d** £138 **2a** 10 **b** 40 **c** 35
d 25 **3a** 20 **b** 48 **c** 50 **d** 2
4a 3530, 3490 **b** 3500, 3500 **5a** £395 **b** £337 **c** £357
6a £16 **b** £1.80 **c** £24.50 **d** £4.58
7a £3.62 **b** £1.04 **c** £11.13 **8** 13

Page 2 Exercise 1A

1a £331 **b** £228 **c** £369 **d** £288 **2a** £319, May
b £288, July **3** £730 **4** £1342 **5** £1722 **6a** 90 000 **b** £20
7a 1750 km **b** (i) 1300 hours (ii) 1030 am, 1 pm

Page 2 Exercise 1B

1 £1400 **2** 45 800 feet **3** (£24 000), £23 760
4a (i) 450 (ii) $7\frac{1}{2}$ **b** 3, £5.50
5 £13 250, £13 900, £14 550, £15 200, £15 850
6a (70 000), 76 800 **b** 15 **7** £187 500 **8** £982.57

Page 3 Exercise 1C

1a 6651 yards **b** 369.5 yards
2 £15 × 170 + £15 × 260 + £15 × 330 + £15 × 290 + £15 × 450, or £15 × (170 + 260 + 330 + 290 + 450). £22 500 **3** 230
4a (i) 60°C (ii) 20°C **b** °C → ×9 → ÷5 → +32 → °F
5a 30, 76, 127 **b** 4, 33, 113
6a (75), 88; (750), 880; (2000), 2112;
b (60), 57; (600), 710; (600 000), 568 181
7a (i) £16 375 million (ii) £675 million **b** £300

Page 4 Exercise 2A

1 1, 3, 4 cm **2a** 7 s **b** 1 s **c** 9 s **d** 4 s **e** 3 s **f** 10 s
3a 2 ml **b** 3 ml **c** 3 ml **d** 4 ml **e** 8 ml **f** 10 ml
4a 8 cm **b** 12 km **c** 37 g **d** 20 m **e** 2 kg **f** 81 s
5a 27.6 cm **b** 21.6 cm
7a 30 cm **b** 40 cm **c** 80 cm **d** 80 cm **e** 40 cm **f** 100 cm
8a 50 km **b** 40 km **c** 70 km **d** 60 km
9a 130 g **b** 250 g **c** 310 g **d** 470 g **e** 820 g **f** 100 g
10a 40 litres **b** 50 kg **c** 340 m **d** 110 m **e** 10 g
f 990 cm **11a** 100 cm **b** 600 cm **c** 400 cm **d** 900 cm
e 200 cm **f** 700 cm
12a 600 g **b** 100 h **c** 300 ml **d** 900 m **e** 300 g **f** 500 s
13a 600 km **b** 500 km **c** 300 km **d** 300 km

Page 5 Exercise 2B/C

1a 120, 180 miles **b** 100, 200 miles
2a £670, £700 **b** £120, £100 **c** £100, £100 **d** £920, £900
e £2350, £2300 **3a** 130 km, 160°C **b** 100 km, 200°C
4a 974, 1606, 427; 3007
b 970, 1000; 1610, 1600; 430, 400; 3010, 3000
5 4410, 4400, 4000 feet; 3560, 3600, 4000 feet; 3210, 3200, 3000 feet
6a People move away, or die, or are born constantly
b (i) 4 597 860 (ii) 4 597 900 (iii) 4 598 000 (iv) 4 600 000
(v) 4 600 000 (vi) 5 000 000 **c** 4.6 million

Page 6 Exercise 3A

1a 136.8 km **b** 20.2 km **c** 451.8 km **d** 2047.1 km
2a 3.2 **b** 5.4 **c** 6.1 **d** 4.8 **e** 8.2 **f** 5.1 **g** 7.6 **h** 11.4
i 9.0 **j** 1.3 **k** 1.1 **l** 0.9
3a 2.5 cm **b** 8.3 g **c** 2.4 km **d** 16.3°
4a (i) 30.4 (ii) 30 **b** (i) 30.5 (ii) 31
5a (9), 9.3 **b** (6), 6.3 **c** (11), 11.1 **d** (3), 2.8
6a £1.75 **b** £8.33 **c** £5.05 **d** £6.12
7a 8.13 **b** 1.51 **c** 2.97 **d** 0.29 **e** 10.13 **8** £5.92
9 0.17, 0.33, 0.67 **10a** 9.03p per km **b** 9.07p per km
c the second journey, by 0.04p per km **11** £5.34 **12** 58.8 s

Page 7 Exercise 3B

1a −273.2°C, −459.7°F **b** −273°C, −460°F **2** 8046.71 m
3 The first is 24.22 days longer. That is why we have a leap year every four years
4a 11.2 s **b** 23.8 s **c** 56.4 s **d** 9.9 s
5a 10.15 s **b** 56.61 s **c** 28.02 s **d** 0.12 s
6a (2), 2.10 **b** (30), 30.79 **c** (8), 8.18 **d** (2), 2.16
e (3), 2.95 **7** £16.23
8a B, E, D, A, C **b** B and E equal; A, C and D equal

Page 7 Exercise 3C

1a 5.5, 6.5 cm **b** 9.5, 10.5 cm **c** 7.5, 8.5 m
d 12.5, 13.5 mm **e** 122.5, 123.5 km **2a** 8.5, 9.5 s
b 31.5, 32.5 g **c** 4.5, 5.5 ml **d** 47.5, 48.5 h **e** 16.5, 17.5 cm²
3a (i) 3.5, 4.5 cm (ii) 35, 45 mm
b (i) 21.5, 22.5 m (ii) 21 m 50 cm, 22 m 50 cm
4 36 cm² is 6 cm² more and 6 cm² less than each of the others
5 106.25 cm², 86.25 cm²
6c 5.3 cm **7a** (i) 4.5 cm (ii) 6.3 cm (iii) 8.9 cm
b (i) 8.1 cm (ii) 11.2 cm

Page 8 Exercise 4A

1a 20 **b** 40 **c** 80 **d** 90 **2a** 230 **b** 130 **c** 510 **d** 320
3 30 000 **4a** 5000 **b** 20 000 **c** 9000 **d** 800 **e** 50 000
5a 1000, 1400 cc **b** 2000, 1800 cc **c** 2000, 2500 cc
d 1000, 1100 cc **e** 7000, 6800 cc **f** 1000, 990 cc
6a 2 **b** 1 **c** 3 **d** 4 **e** 1 **7** 3 **8a** (490), 480
b (640), 660 **c** (270), 260 **d** (400), 440 **e** (5), 5.4
f (10), 9.7 **g** (4), 4.7 **h** (20), 24
9a (240 cm²), 230 cm² **b** (540 mm²), 560 mm²
c (700 m²), 880 m² **d** (3000 cm²), 3500 cm²

Page 9 Exercise 4B/C

1a 2 **b** 2 **c** 3 **d** 4 **e** 1 **f** 4 **g** 3 **h** 2 **i** 4 **j** 3
2a 21 **b** 5.1 **c** 990 **d** 0.72 **e** 0.012
3a 0.91 **b** 1016 **c** 1609.3 **d** 0.4536 **e** 4.55 **f** 25.400
4 To show that the second decimal place is significant (correct to $\frac{1}{100}$ mm)
5a 10 m² **b** 10.7 m². The answer cannot be more accurate than the measurements **c** (33 m), 32.4 m
6a 37 000 mm³ **b** 520 cm³ **c** 205 m³
7a 45, 55 km/h **b** 650, 750 km **c** 8.5, 9.5 s
d 55 000, 65 000 **e** 0.85, 0.95 litre **f** 1500, 2500

Page 10 Exercise 5

1a 11 **b** 7 **c** 24 **d** 3 **e** 35 **f** 2 **g** 60 **h** 0 **i** 9 **j** 7
k 7 **l** 0 **3a** 5 **b** 4 **c** 32 **d** 0 **e** 11 **f** 12 **g** 1 **h** 2
5a 27 **b** 17 **c** 3 **d** 9 **e** 2 **f** 8 **g** 26 **h** 3
6a 9 **b** 25 **c** 14 **d** 2 **e** 2 **f** 5 **g** 40 **h** 10
7a (2 + 3) × 4 = 20 **b** (5 − 3) × 6 = 12 **c** 7 × (5 − 3) = 14
d 12 ÷ (3 + 1) = 3 **8** 63 **9** 45 cm **10a** 8 **b** 375 **c** 4 **d** 15

Page 11 Exercise 6

1a 12 **b** 21 **c** 27 **d** 7 **e** 7 **f** 3
2a 24 **b** 70 **c** 80 **d** 48 **e** 125 **f** 180
3a 59 **b** 80 **c** 58 **d** 24 **e** 28 **f** 25

4a 80 **b** 800 **c** 3500 **d** 9000 **e** 2300 **f** 0
5a 50 **b** 120 **c** 100 **d** 250 **e** 372 **f** 154
6a 12 **b** 2 **c** 4 **d** 15 **e** 5 **f** 200
7a 4 **b** 8 **c** 5 **d** 12 **e** 20 **f** 15
8a 390 **b** 140 **c** 430 **d** 384 **e** 1096 **f** 28
9a 190 **b** 600 **c** 1700 **d** 495 **e** 7992
10a 25 **b** 16 **c** 190 **d** 250
11a 1175 **b** 2456 **c** 525 **d** 528 **e** 47 **f** 57
12a 2820 **b** 4738 **c** 19 845 **d** 25 **e** 8 **f** 7

Page 12 Check-up on Numbers in Action

1 £9142.25 **2a** £7.75 **b** £4.50 **c** £9.80 **d** £30.50
3a 130, 100 km **b** 840, 800 km **c** 570, 600 km
d 610, 600 km **e** 980, 1000 km
4a 2.2 m **b** 8.8 cm **c** 10.1 s **d** 1.0 kg
5a 5.16 km **b** 1.02 litres **c** 0.31 m **d** 10.86 kg
6a 250 **b** 500 **c** 220 **d** 1100 **e** 3600
7 81 cm², 78.32 cm²; 36 cm, 35.4 cm
8a 26.4 **b** 30 **9a** 680 g **b** 45 ml
10a 2 **b** 23 **c** 32 **d** 16
11a $(7-4)\times 9$ **b** $6\times(4+5)$ **c** $20\div(4+1)$ **d** $(55\div 5)\times 5$
12 514 cm

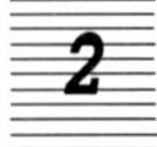

2 ALL ABOUT ANGLES

(Degree symbols are not shown in answer diagrams.)

Page 13 Looking Back

1a Acute **b** right **c** straight **d** reflex **e** obtuse
2 **b**, **d**, **e** are fixed; **a** and **c** can change **3a** Obtuse
b acute **c** acute **d** reflex **e** obtuse **f** right **g** straight
4a 360° **b** (i) $\frac{1}{2}$ (ii) $\frac{1}{4}$ (iii) $\frac{1}{8}$ (iv) $\frac{1}{6}$ (v) $\frac{1}{3}$
5a ∠AOB = 60° **b** ∠RST = 145°
c ∠AEB = 128°, ∠BEC = 52°, ∠DEC = 128°
6 150° **7a** 35° **b** 120° **8a** BC, FG **b** AH, DE
c BC, FG; AH, DE; AB, FE; CD, HG
d BC, DE; DE, FG; FG, AH; AH, BC

Page 14 Exercise 1A

1 ∠DBC = 30°, ∠EFH = 40°, ∠KJM = 45°, ∠PSQ = 15°, ∠TWU = 68°, ∠AYB = 27°, ∠MTR = 55°, ∠PON = 31°
2a 10° **b** 35° **c** 1° **d** 73° **e** 51° **f** 85°
3a 90° **b** (i) 80° (ii) 10° (iii) 70° (iv) 0° **4** 64° **5** 52°
6 78°, 31°; 59°, 68°; 22°, 55°; 35°, 71°; 19°, 44°; 46°, 12°
7a ∠EBF **b** ∠DBC
8 ∠s PQT, RQT; QRT, TRS; RST, TSP; SPT, QPT

Page 15 Exercise 1B/C

1a ∠EOC = 70° **b** ∠EOD = 50° **c** ∠FOB = 65°
2a 80° **b** 65° **c** 42° **d** 13° **e** 0° **3a** ∠MON **b** ∠ROP
4a ∠ABE = 90°, ∠CBD = 180° **b** 47°
5a 39° **b** 63° **c** (i) 160° (ii) 170° (iii) 163°
6a 70°, 50°, 30°, 15°; 80°, 10°; 60°, 30°; 45°, 45°; 30°, 60°
b (i) $(90-x)°$ (ii) $x°$

Page 16 Exercise 2A

1 ∠DBC = 140°, ∠HFG = 65°, ∠PMN = 90°, ∠JMN = 135°, ∠QRT = 70°, ∠URS = 118°,
2a 30° **b** 75° **c** 100° **d** 85° **e** 179° **f** 5°
3a 160°, 20° **b** 55°, 125°
4

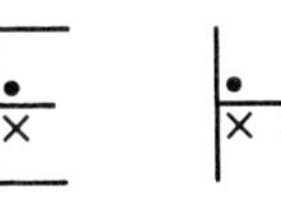
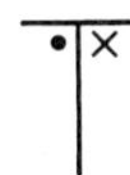

5a ∠PQS = 37° **b** ∠PQT = 115° **c** ∠SQR = 143°
6a 58° **b** 32°
7a Rows: 0°, 180°; 10°, 170°; 20°, 160°; 30°, 150°; 40°, 140°; 50°, 130°; 60°, 120° **b** As angle $a°$ increases by 10°, its supplement decreases by 10° **8** 45°, 105°; 75°, 118°; 62°, 90°; 90°, 29°; 151°, 85°; 95°, 155°

Page 18 Exercise 2B/C

1a ∠VYZ = 80° **b** ∠UYZ = 110° **c** ∠WYX = 150°
d ∠VYX = 100°
2a (i) 28° (ii) 31° **b** (ii) **3** $x = 69$, $y = 45$, $z = 40$
4 $a = 40$, $b = 70$, $c = 70$; 180°
5a 150°, 120°, 90°, 60°; 160°, 20°; 140°, 40°; 120°, 60°; 100°, 80°
b (i) $(180-x)°$ (ii) $x°$ **6a** Ian: 98, 114 at B or 66, 73 at A. Waheed: 8, 8 at B. Shona: 11, 8 at B or 8, 98 at B.
b Ian plays 98, 114 at B

Page 19 Exercise 3A

1

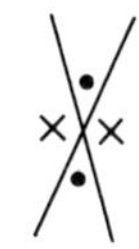
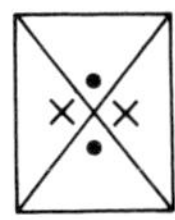
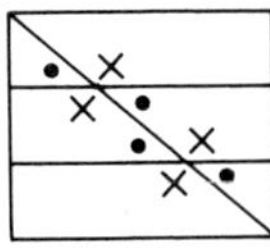

2a ∠s AOC, BOD; AOD, BOC **b** ∠s PQT, RQS; PQS, RQT
3,4b They should be the same size

5a (i) (ii)

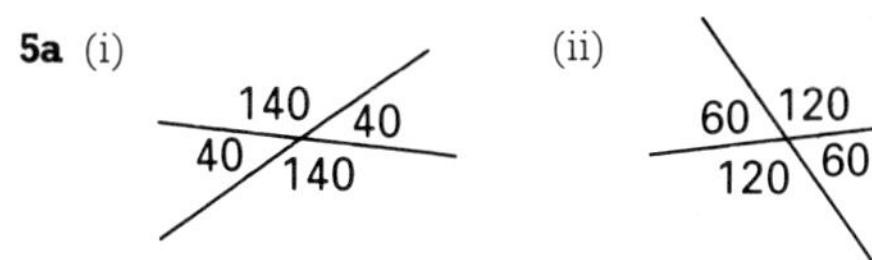

6a **b**

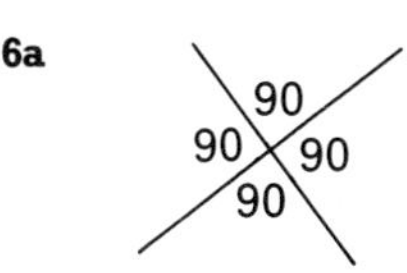

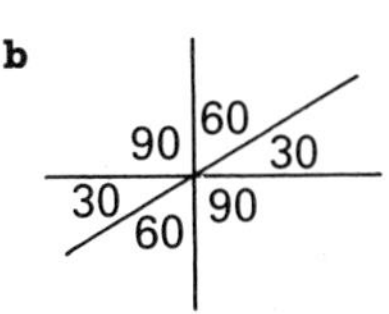

7 ∠CBF = 22°, ∠ABD = 22°, ∠ABG = 68°
8 The angles are 80°, 45°, 55°

Page 20 Exercise 3B/C

1a (i) 155°, 25° (ii) 112°, 112° (iii) 130°, 50°
b The angles in one pair increase to 180°, the others decrease to 0°
2

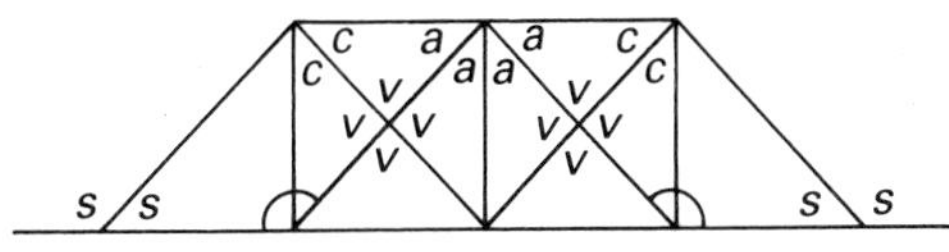

3 Rows: 140°, 40°, 140°; 155°, 25°, 155°; $(180-x)°$, $x°$, $(180-x)°$; decrease by $y°$, increase by $y°$, decrease by $y°$
4a 180° (supplementary), 180° (supplementary), ∠BOC

Page 21 Exercise 4A

1a 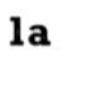 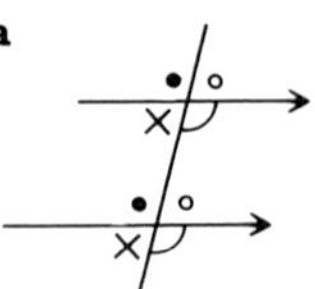 **b** 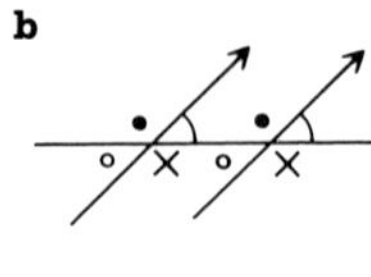

2a ∠s ABE, ACD **b** ∠s JHK, JGF **c** ∠s MNQ, MOP **d** ∠s RST, RUV; RTS, RVU **3a** 3 **b** 4 **c** 7 **d** 8

4a 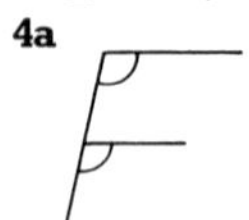 **b** 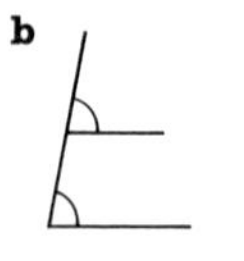 **c** 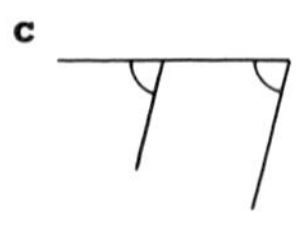

d 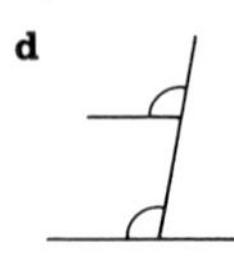 **e** 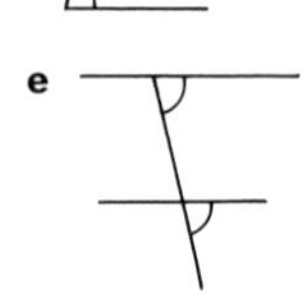 **f** 

5a Each is 40° **b** each is 140°
6 (ii) Yes **8** The angles are equal
9 $a = 100$, $b = 80$, $c = 100$, $d = 70$, $e = 70$, $f = 112$, $g = 68$
10

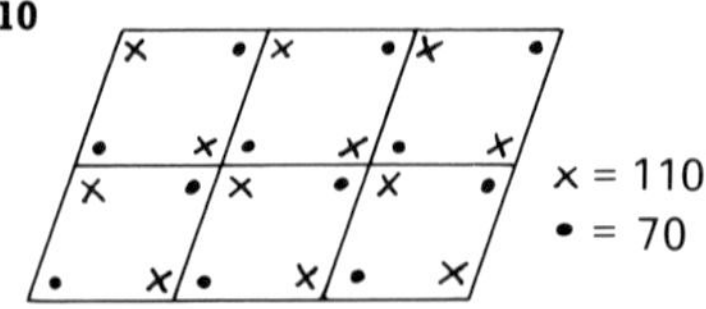

11a

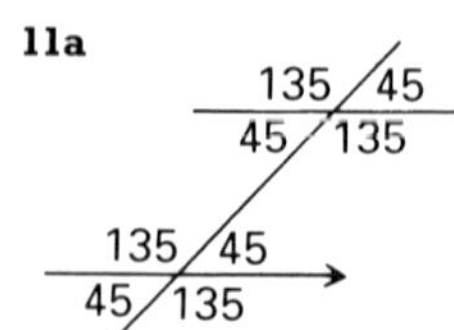

b

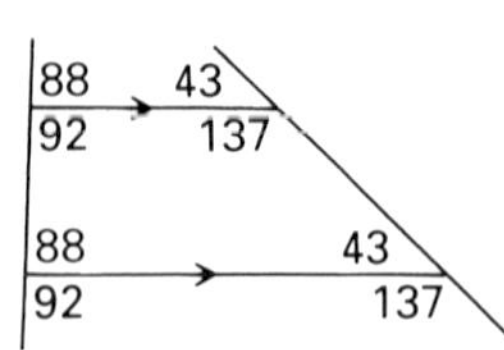

Page 23 Exercise 4B/C

1 $a = 53$, $b = 53$, $c = 40$, $d = 70$, $e = 140$, $f = 110$, $g = 40$, $h = 70$, $i = 78$, $j = 78$, $k = 102$
2 (i) $a = 100$ (corr), $b = 100$ (vert opp), $c = 80$ (supp)
(ii) $a = 66$ (vert opp), $b = 66$ (corr), $c = 114$ (supp)
3 $a = 136$ (corr), $b = 44$ (supp); $c = 82$ (vert opp), $d = 82$ (corr), $e = 98$ (supp), $f = 98$ (corr)
4a $p = 30$ (corr), $q = 70$ (corr), $r = 80$ (supp to $70 + 30$), $s = 80$ (vert opp) **b** 180°
5 $u = 96$, $v = 60$, $w = 120$, $x = 60$, $y = 70$

Page 24 Exercise 5A

1a  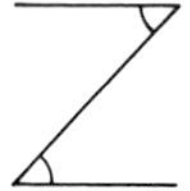 **b** 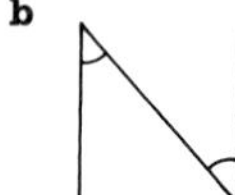 **c** 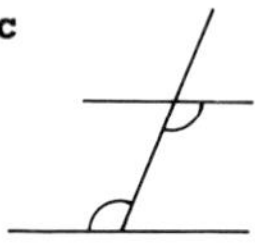

2a Each is 30° **b** each is 150°
3 They are the same **4** They are the same size
5a

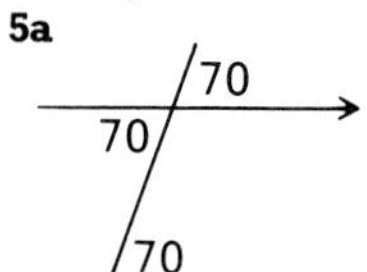

b yes 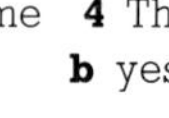

6 (i) **a** **b** yes (ii) **a** **b** yes

(iii) **a** **b** yes

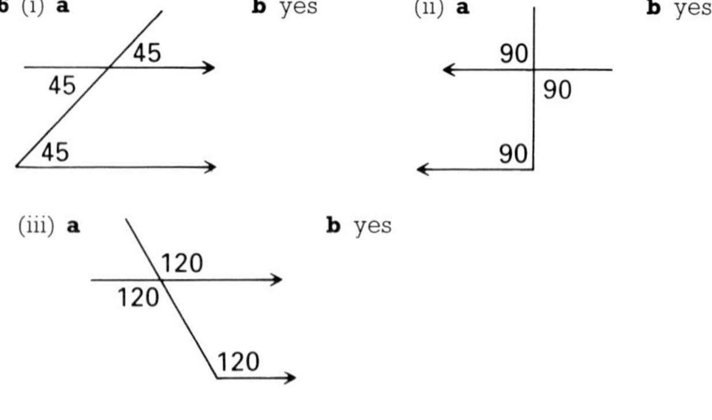

7 $a = 40$, $b = 22$, $c = 35$, $d = 52$, $e = 65$, $f = 65$, $g = 115$, $h = 75$, $i = 105$, $j = 59$, $k = 121$, $m = 117$, $n = 117$
8 23°. Alternate angles are equal for parallel lines
9

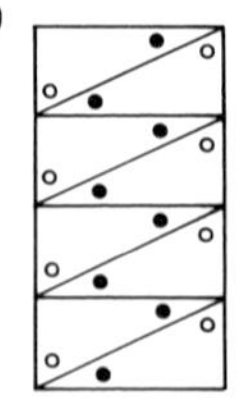

10

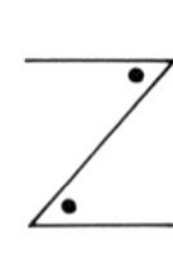 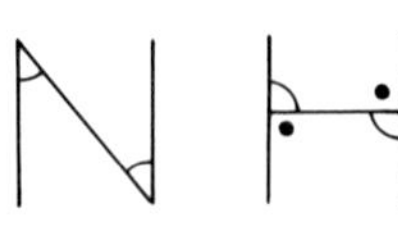 

11 Angles are 115° and 65° **12a** ∠s BAC, ACD; DAC, ACB
b ∠s PQS, QSR; QPR, PRS; PSQ, RQS; SPR, PRQ

Page 26 Exercise 5B/C

1a **b** **c** **d**

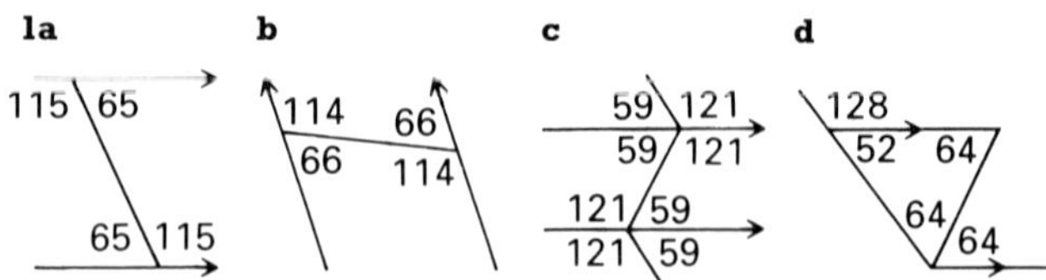

2 $a = 50$, $b = 45$, $c = 50$, $d = 45$ **3** $a = 60$ (vert opp), $b = 60$ (alt or corr), $c = 75$ (alt), $d = 75$ (corr, or vert opp), $e = 105$ (supp), $f = 35$ (alt), $g = 55$ (comp), $h = 55$ (alt, or comp)
4 $x = 100$ **5** $x = 80$
6a ∠ABE = 70° (alt to ∠BED), ∠EBD = 75° (35° + 70° + 75° = 180°), ∠BCD = 70° (corr to ∠ABE) **b** ∠PQR = 45° (corr to ∠PUT), ∠USQ = 60° (alt to ∠SUT), ∠PRQ = 60° (corr to ∠USQ), ∠PTU = 60° (corr to ∠PRQ, or alt to ∠SUT)

Page 27 Exercise 6

1

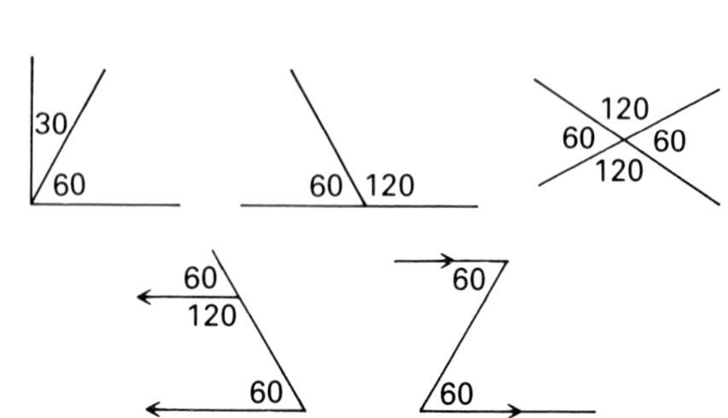

2a Supp 4, 5; vert opp 1, 2; corr 1, 3 and 3, 4; alt 2, 3
b Supp 7, 8; vert opp 9, 10; corr 6, 7 and 8, 10; alt 8, 9
3a ∠AHR **b** ∠CPF **c** ∠SHP **d** ∠BCD or ∠SCE
4 $a = 54$, $b = 126$, $c = 27$, $d = 63$, $e = 115$, $f = 65$, $g = 63$, $h = 77$, $i = 40$

5a

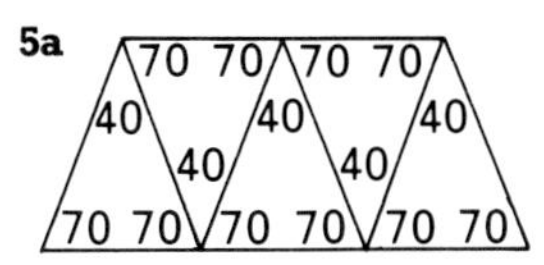

b

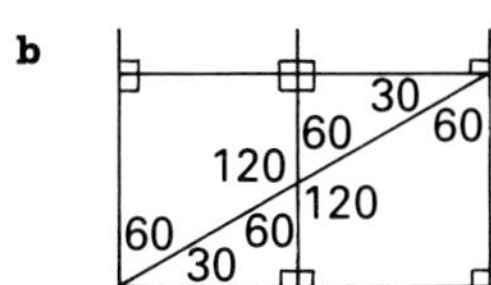

6a

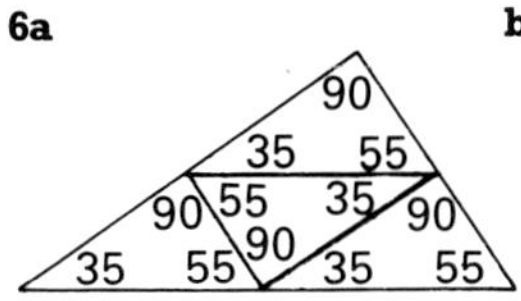

b

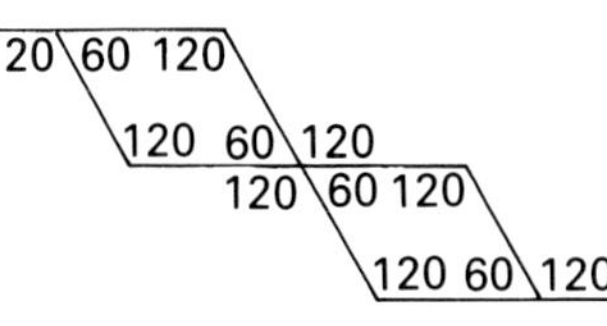

7 ∠ABC = $x°$ (alt), ∠ACB = $z°$ (alt). $x° + y° + z° = 180°$, so ∠ABC + ∠ACB + ∠BAC = 180°

Page 28 Check-up on All About Angles

1a 145° **b** 55° **2a** 2 **b** 4
3a ∠s ABE, DBC and ABD, EBC
b ∠EBC = 115°, ∠ABE = 65°, ∠DBC = 65°
4a 77 **b** 99 **c** 55 **d** 30 **e** 60 **f** 55
5a $a = 42$, $b = 42$, $c = 138$ **b** 48°, 42°; 42°, 138°
6a

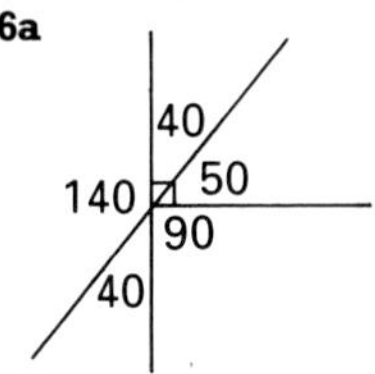

b

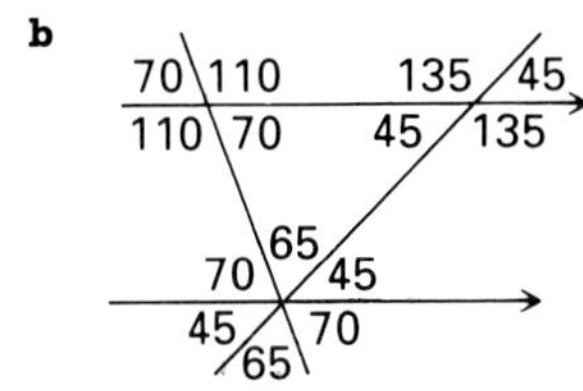

7a ∠s JPS, JQR; KSP, KRQ; MQR, MPS; NRQ, NSP
b

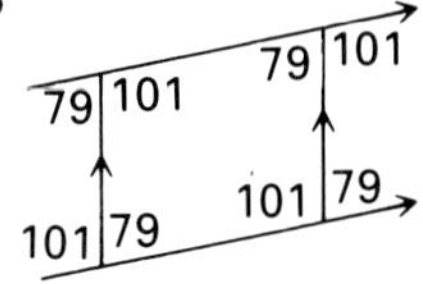

8a 6 **b** Eight right angles, four angles of 125°, four of 55° **9** 140°

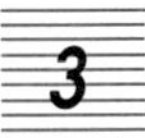

3 LETTERS AND NUMBERS

Page 29 Looking Back

1a 14 **b** 20 **c** 0 **d** 31 **e** 2
2a 8 **b** 2 **c** 15 **d** 30 **e** 19 **3a** 24 cm **b** 27 cm²
4a 24 cm **b** 25 cm **c** $2x$ cm **d** $u+v+w$ cm
5a 0, 35, 56, 7 **b** 7, 13, 19
6a 40 **b** 11 **c** 43 **d** $5x$ **e** $x+3$ **f** $5x+3$
7a 27 **b** 17 **c** 16 **d** 6 **e** 0 **8** 24 cm, 36 cm²
9a 5 **b** 4 **c** 5 **d** 6 **e** 9
10a $2x$ **b** $3x$ **c** $2y$ **d** y **e** $2t+2$ **f** $3+2y$
g $6x+1$ **h** 4 **i** $2x$ **j** $2m+2$ **k** 0 **l** $2+5p$

Page 30 Exercise 1A

1a 3×2 **b** 4×3 **c** 2×7 **d** 5×8 **e** 2×4 **f** 3×9
g 4×11 **h** 6×1
2a $2d$ **b** $3n$ **c** $4k$ **d** $5x$ **e** $3c$ **f** $4d$ **g** $2v$ **h** $6t$
3a $2m^2$ **b** $3x^2$ **c** $4a^2$ **d** $5t^2$
4a $2u$, 30 **b** $4u$, 60 **c** $6u$, 90 **5a** 6^2 **b** 3^4 **c** 4^3 **d** 7^2
e m^2 **f** y^2 **g** x^2 **h** a^2 **i** n^3 **j** t^3 **k** p^2 **l** v^5 **m** t^2
n d^3 **o** c^4 **p** a^6
6a 2×1, 2×2, 2×3, 2×4, $2x$, $2y$, $2a$, $2b$, $2v$
b 3×1, 4×2, 5×3, 2×4, $3a$, $4b$, $2c$, $5d$, $3e$
c 1^2, 2^2, 3^2, 4^2, x^2, y^2, z^2, b^2, c^2 **d** $3a$, $2x$, $3y$, x^2, y^3, d^4, n^2, m^3
e 1^3, 2^4, 3^3, 4^4, a^3, b^4, c^2, d^5, e^3 **f** $2m$, 0, m^2, $3n$, n^3, $4t$, t^4, 1
7a $5+5$, 5^2 **b** $2u$, u^2 **c** $2+2+2$, 2^3 **d** $3t$, t^3 **e** $2n$, n^2
f $3k$, k^3 **g** $4z$, z^4 **h** 5×4, 4^5

Page 31 Exercise 1B

1a $2x$ **b** $3x$ **c** x **d** $5x$ **e** $4a$ **f** $5b$ **g** $10c$ **h** $3a$ **i** $5c$
j 0 **k** a **l** $3y$ **m** $2x^2$ **n** $4y^2$ **o** $6z^2$ **p** t^2 **2a** $5x+4$
b $y+1$ **c** $5x+2$ **d** $5+3m$ **e** $2a+b$ **f** n **g** $3x+y$
h $x+y$ **i** $3a+2$ **j** $4c+3$ **k** $2p$ **l** $s+2t$
3a $3u+3v$, 36 **b** $u+v$, 12 **c** $5v$, 15
4a $2c^2$ **b** $3a^2$ **c** $6y$ **d** $8y$ **e** $15z$ **f** $4s^2$ **g** $4t^2$ **h** $9m^2$
i $10c^2$ **j** $4r^2$ **k** $5y$ **l** $4e^2$ **m** $6y^3$ **n** $10x^3$ **o** $8m^3$ **p** $6a^4$
5 a, $2a$, $3a$, $4a$, $5a$, $6a$, $7a$, a^2, $2a^2$, $3a^2$, $4a^2$, $5a^2$, $6a^2$, a^3, $2a^3$, a^4

Page 31 Exercise 1C

1a $2y+3$ **b** $4k-1$ **c** n^2+7 **d** m^2-2 **e** $8+2d$ **f** $5-3c$
g $1+t^2$ **h** $3-p^2$ **i** $3+2a$ **j** $5b-1$ **k** $4-2n$ **l** $3z+3$
m $3x+3$ **n** $2y+1$ **o** $2p+5$ **p** $s+1$ **q** $2y^2+4$ **r** $3z^2-2$
s $5u$ **t** v **u** 0 **v** $5a+5$ **w** $17a$ **x** $13a$ **y** $30a^2$
2a $\frac{1}{2}a$, $\frac{1}{2} \times a$ **b** $\frac{1}{3} \times y$, $\frac{y}{3}$ **c** $\frac{3}{4}t$, $\frac{3t}{4}$ **d** $\frac{1}{6} \times x$, $\frac{x}{6}$ **e** $\frac{4}{5}m$, $\frac{4}{5} \times m$
f $\frac{1}{12}d$, $\frac{1}{12} \times d$ **g** $\frac{5}{7}b$, $\frac{5b}{7}$
3a Yes, to $2x+1$
b $5 \to 20 \to 22 \to 12 \to 11$; $5 \to 25 \to 29 \to 14 \to 11$

Page 32 Exercise 2A

1a $2t$ cm **b** $4x$ cm **c** $3a$ cm **d** $6n$ cm **e** $5m$ cm **f** $3y$ cm
g $5t$ cm **h** $x+y$ cm **i** $2a+b$ cm **j** $3c+2d$ cm **k** $3u+3v$ cm
2a y^2 cm² **b** m^2 cm² **c** t^2 cm² **d** $2a^2$ cm² **e** $3x^2$ cm²
3a 6 **b** 3 **c** 4 **d** 2 **e** 3 **f** 10 **g** 1 **h** 13 **i** 5 **j** 20
4a 20 **b** 8 **c** 10 **d** 40 **e** 11 **f** 7 **g** 16 **h** 4 **i** 29 **j** 12
5a 3, 6, 9, $3k$ **b** 7, 9, 11, $m+5$ **c** $2x$, $4x$, $6x$ **d** $5y$, $10y$, y^2

Page 33 Exercise 2B/C

1a 20 **b** 12 **c** 56 **d** 6 **e** 30 **f** 100 **g** 16 **h** 144 **i** 0
j 1 **2a** $(2x)^2$ cm², $4x^2$ cm² **b** $(3y)^2$ cm², $9y^2$ cm²
c $(4z)^2$ cm², $16z^2$ cm²
3a 9 **b** 18 **c** 36 **d** 36 **e** 4 **f** 8 **g** 16 **h** 16 **i** 2 **j** 4
k 36 **l** 36 **m** 6 **n** 36 **o** 36
4a $2x$, $4x$, $6x$, $2x^2$ **b** 5, 8, 11, $3n+2$
c $y+1$, $2y+1$, $3y+1$, y^2+1 **d** 5, 6, 7, $n+4$
5a 36 **b** 30 **c** 9 **d** 30 **e** 36 **f** 12 **g** 12 **h** 9
6a and **e**, **b** and **d**, **c** and **h**, **f** and **g**
7a 1 **b** 3 **c** 2 **d** 1 **e** 2 **f** 4 **g** 3 **h** 1 **i** 1 **j** 6 **k** 1
l 3 **m** 2 **n** 1 **o** 7

Page 35 Exercise 3A

1a $x-2$ m **b** $4+y$ m **2a** $2x$ m **b** $x+y$ m
3a $7x$ m **b** $2x-2$ m **4a** 24 m **b** 60 mm **c** 26 cm
5a $P = 2m+2n$ **b** $P = 4x$ **c** $P = 10a$ **d** $P = 18y$
e $P = 8u$ **f** $P = 12x$ **6a** 27 m² **b** 225 mm² **c** 22 cm²
7a $A = mn$ **b** $A = x^2$ **c** $A = 4a^2$ **d** $A = 20y^2$ **e** $A = 3u^2$
f $A = 9x^2$

Page 36 Exercise 3B

1a (i) 9 cm (ii) $4+x$ cm (iii) $9+x$ cm **b** (i) 2 cm
(ii) $3-x$ cm (iii) $5-x$ cm **c** (i) $2+x$ cm (ii) 10 cm
(iii) $8-x$ cm **2a** $P = 8y$ **b** $P = 12x$ **c** $P = 18x$
3a $A = 4y^2$ **b** $A = 5x^2$ **c** $A = 8x^2$

4a 216 m^3 **b** 288 cm^3 **c** 60 cm^3
5a (i) $V = 27a^3$ (ii) $V = 36a^3$ **b** (i) 1728 m^3 (ii) 2304 mm^3
6a $P = 4a+4b+4c$, $V = abc$, $A = 2ab+2bc+2ca$

Page 37 Exercise 3C

1a (i) 3 cm (ii) $y-2$ cm (iii) $y-5$ cm **b** (i) x cm (ii) $2x+7$ cm (iii) $x+7$ cm **c** (i) $3x$ cm (ii) $x+8$ cm (iii) $8-2x$ cm **2a** $P = 26k$ **b** $P = 18z$ **c** $P = 22a$
3a $A = 16k^2$ **b** $A = 8z^2$ **c** $A = 22a^2$
4a $V = 9c^3$ **b** $V = 4x^3$ **c** $V = 20x^3$

Page 38 Exercise 4A

1 (ii) **a** $20+8 = 28 = 4\times(5+2)$ **b** $6+6 = 12 = 3\times(2+2)$
c $80+40 = 120 = 10\times(8+4)$ **d** $35+20 = 55 = 5\times(7+4)$
2a $5\times(4+3) = 35 = 20+15$ **b** $4\times(3+2) = 20 = 12+8$
c $6\times(2+4) = 36 = 12+24$ **d** $3\times(4+2) = 18 = 12+6$
e $10\times(5+5) = 100 = 50+50$ **f** $7\times(9+2) = 77 = 63+14$

3a

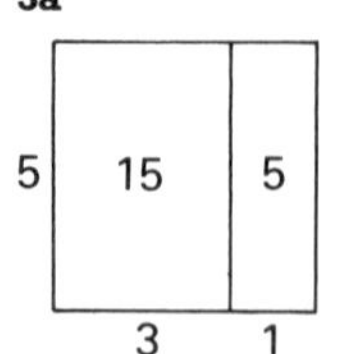

b

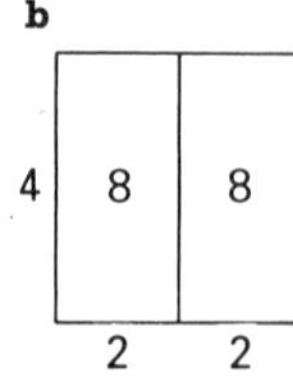

c

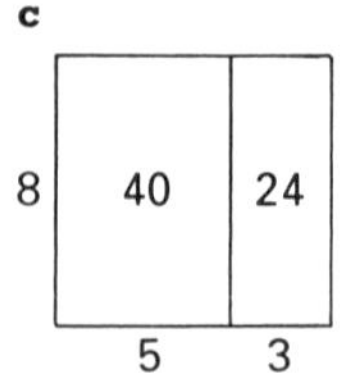

Page 39 Exercise 4B/C

1a $4(x+3) = 4x+12$ m^2 **b** $6(x+2) = 6x+12$ m^2
c $2(x+5) = 2x+10$ m^2

2a

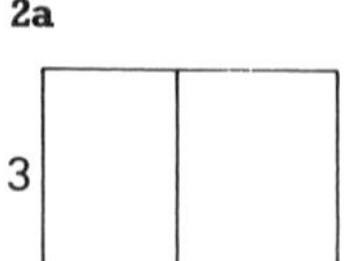

b

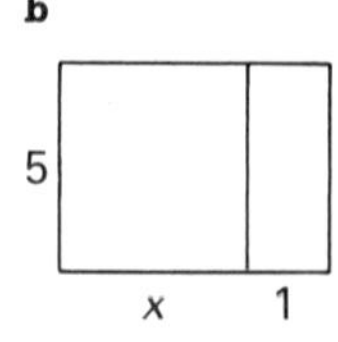

c

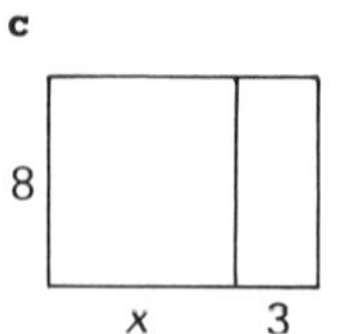

3a $3(x+y) = 3x+3y$ **b** $8(p+q) = 8p+8q$
c $4(d+e) = 4d+4e$ **d** $5(m+n) = 5m+5n$
e $3(2a+4) = 6a+12$ **f** $6(5+3b) = 30+18b$
g $3(8c+2) = 24c+6$ **h** $7(1+5d) = 7+35d$
i $a(b+2) = ab+2a$ **j** $x(y+5) = xy+5x$
k $a(b+c) = ab+ac$ **l** $p(q+r) = pq+pr$

Page 40 Exercise 5A

1a $2x+6$ **b** $3x+12$ **c** $4x+8$ **d** $5x+5$ **e** $8x+16$
f $6y+30$ **g** $3y+6$ **h** $7y+7$ **i** $2y+8$ **j** $4y+24$ **k** $2a+18$
l $3b+21$ **m** $8c+40$ **n** $5k+10$ **o** $4m+36$
2a $3x-3$ **b** $2x-6$ **c** $4y-8$ **d** $6m-6$ **e** $8n-32$
f $7k-21$ **g** $5m-30$ **h** $9n-63$ **i** $10x-10$ **j** $4y-44$
3a $10+2x$ **b** $12+3y$ **c** $5+5t$ **d** $28+4u$ **e** $48+6v$
f $3-3x$ **g** $10-2y$ **h** $8-4m$ **i** $7-7n$ **j** $32-8p$
4a $2x+10$, $3x+6$, $4y+4$, $5y+15$, $6b+12$
b $3x-6$, $2x-10$, $4y-4$, $6m-12$, $8n-8$
5a $6x+3$ **b** $10y+5$ **c** $12a+4$ **d** $24b+6$ **e** $35c+7$
f $6x+8$ **g** $10y+15$ **h** $20y+8$ **i** $12m+15$ **j** $12t+18$
k $5-10x$ **l** $15-6y$ **m** $12-16k$ **n** $12-18u$ **o** $54-45v$
6a $2x+2y$ **b** $7c+7d$ **c** $10p-10q$ **d** $5s-5t$ **e** $7m+7n$
f $6a+3b$ **g** $8c+4d$ **h** $5x+10y$ **i** $2u+6v$

Page 40 Exercise 5B

1a $ax+4a$ **b** $by+2b$ **c** $ck-c$ **d** $dm-5d$ **e** $en+2e$
f $2n+nx$ **g** $3m+my$ **h** $t-tx$ **i** $4u-ux$ **j** $v-2vy$
2a x^2+x **b** y^2-y **c** t^2+2t **d** u^2-3u **e** v^2-8v **f** $t-t^2$
g $3s-s^2$ **h** $k+k^2$ **i** $8m-m^2$ **j** n^2-7n

3a x^2+xy **b** a^2-ab **c** m^2+mn **d** u^2-uv **e** $st+t^2$
f $ab-a^2$ **g** y^2-xy **h** $x-3x^2$ **i** $2y^2+5y$ **j** $4z-7z^2$
4a (i) $4(x+2)$ cm^2 (ii) $4x+8$ cm^2 **b** (i) $5(2x-1)$ cm^2 (ii) $10x-5$ cm^2 **c** (i) $x(x+7)$ cm^2 (ii) x^2+7x cm^2
d (i) $x(2x+3)$ cm^2 (ii) $2x^2+6x$ cm^2 **e** (i) $3(7-y)$ cm^2 (ii) $21-3y$ cm^2 **f** (i) $6(2y+7)$ cm^2 (ii) $12y+42$ cm^2
g (i) $x(3x+2y)$ cm^2 (ii) $3x^2+2xy$ cm^2 **h** (i) $y(y-2x)$ cm^2 (ii) y^2-2xy cm^2

Page 41 Exercise 5C

1a $2x+8$ **b** $3x+9$ **c** $4x+5$ **d** $2y+13$ **e** $3p+7$
f $3y-10$ **g** x^2+3x **h** y^2+2y **i** x^2+5x **j** x^2+x **k** a^2
l y^2 **2a** $2x+2y+6$ **b** $5m+5n-10$ **c** $6x-3y-9$
d $8a+12b-16c$ **e** x^3-x **f** y^3-3y **g** a^3+ab **h** $cd-c^3$
3a $5x+11$ **b** $5y+1$ **c** $7y$ **d** $7a+20$ **e** $12c+1$ **f** $16x+3$
4a Areas $5\times8-3\times6$, $2(6\times1)+2(5\times1)$, $2(8\times1)+2(3\times1)$, $2(6\times1)+2(3\times1)+4(1\times1)$, each 22 m^2 **b** 24 m^2 **c** $2x+10$ m^2

Page 42 Check-up on Letters and Numbers

1a $2m$ **b** $3y$ **c** $4t$ **d** n^2 **e** k^3
2a $3x$ cm **b** $2m$ cm **c** $2y+t$ cm **d** $2a+b$ cm
3a 10 **b** 7 **c** 11 **d** 13 **e** 5
4a $P = 4x$, $A = x^2$ **b** $P = 12a$, $A = 5a^2$ **c** $P = 10t$, $A = 6t^2$
d $P = 16t$, $A = 12t^2$ **5a** $5x+20$ **b** $2y-6$ **c** $3t+3$
d $32-4x$ **e** $10-5m$ **6**

7a $4t+2$ **b** $15n-3$ **c** $4+8k$ **d** $5a+5b$ **e** $12c-18d$
8a $L = 24x$ **b** $A = 22x^2$ **c** $V = 6x^3$ **9a** $4x+5$ **b** $4b$ **c** $8t$
d $6mn$ **e** 0 **f** $4+2t^2$ **g** $\frac{1}{2}p^2$ **h** $\frac{1}{3}a^3$ **i** $9d$ **j** $24d^3$
10a 4 **b** 8 **c** 16 **d** 100 **e** 144 **f** 32 **g** 14 **h** 25 **i** 3
j 1 **11a** 6 **b** $3y+5$ **c** n^2+7n **d** k^2 **e** a^2 **f** t^3
g a^2-b^2 **h** x^2+x

4 MAKING SENSE OF STATISTICS 1

Page 43 Looking Back

1a Fir **b** 42 **c** 12 **3a** January **b** 20 cm **c** 8 **4a** 25
b 45 **c** (i) 8.30 am (ii) 9 am **5a** $\frac{1}{4}$ **b** (i) 4 (ii) 5
6 180°, 120°, 60°

Page 44 Exercise 1A

1a 4 m **b** (i) 3 m (ii) $1\frac{1}{2}$ m **c** $5\frac{1}{2}$, 7 s
2a 3, 5, 2, 1, 4, 2, 1 **c** Tuesday
3a (i) $1\frac{1}{2}$ cm (ii) $6\frac{1}{2}$ cm **b** 3 kg
4a (i) 27 (ii) 31 **b** $\frac{1}{9}$ **c** 31–40 **d** 14 **e** 5

Page 45 Exercise 1B

1a (i) 13°C (ii) 13°C (iii) 13°C **b** (i) 13°C, 11°C (ii) 11°C, 12°C (iii) 11°C, 14°C
c (i) Falling fairly steadily from 17°C to 6°C
(ii) rising from 8°C at noon to 13°C at 3 pm, then falling to 6°C at 7 pm (iii) rising fairly steadily from 9°C to 16°C
2a (i) 1–2.30 pm and after 5 pm (ii) 2.30–4 pm (iii) 4–5 pm
b (i) 0 (ii) 3.5 (iii) 0.5 (iv) 2.2 **c** (i) possibly about 3 (ii) no **3a** (i) 8th (ii) 22nd **b** 14 **c** 15th **d** (i) $\frac{1}{2}$, waxing (ii) $\frac{1}{2}$, waxing

Page 46 Exercise 1C

1a 3, 5, 10, 8, 4 **2a** 2, 6, 8, 10, 4
b they are a little better, with more marks in the 31–40 band
3c Yes. More batteries last longer **d** yes **4b** 10

Page 47 Exercise 2

1 Area of small square is $\frac{1}{4}$ of area of large one **2** No. It is too vague—you need to know how many people were asked **3** The volume of the large cube is 8 times the volume of the smaller one **4** No. The year scale has been changed at 1990 **5** The vertical scales are different **6** No scale for prices, so graph is meaningless **7** The growth is £30 000 in 3 years in each case, so same rate of growth

Page 48 Exercise 3A

1a 4 **b** 50 **c** 5 **d** 5 kg **e** 5.7 cm **f** 24 p **g** 45%
2a 20 p **b** 2 p **c** 29 p **3** 5 cm **4a** 14 **b** 11
5a £77 **b** 7, 4

Page 49 Exercise 3B

1a 13, 39 **b** 33 cm, 4 cm **c** £4.15, £1.25 **d** 4, 5
2a 15.9 hours **b** 16 hours **3a** 5.1 kg, 0.9 kg **b** 4.5 kg
4a 16°C, 10° **b** that the mean and range are actually 16°C and 10°, so they are misleading the public

Page 49 Exercise 3C

1a English 50%, 80%; French 40%, 90% **b** English—average; French—well above average
2 Sample 1, 47; speed up. Sample 2, 53; slow down. Sample 3, 51; no action **3** Old, 110 hours; new 113 hours. Yes

Page 51 Exercise 4A

1a 7, 6 **b** 40 p, 45 p **c** 0, 0.5 goal **2** £52, £62, £51
3 170 mm, 171 mm, 170 mm **4** 3, 3
5a 150, 30, 28, 45 **b** 150, 30, 28, 11

Page 52 Exercise 4B/C

1a 5.6, 6, 6 **b** 5.4, 5.5, 6 **c** 3.5, 3.5, __ **d** __, C, C
2a 1.9, 2, 2 legs **b** No; the mode is the most sensible average here **3a** £188.46, £160, £150
b mean $\frac{2}{13}$, median $\frac{5}{13}$, mode $\frac{8}{13}$
4a 225 g, 150 g, 150 g **b** mode

Page 53 Check-up on Making Sense of Statistics 1

1a 2 g **b** (i) 1 g (ii) 5.5 g **c** 3 cm **d** (i) 0.5 cm (ii) 10 cm
2a 1, 6, 8, 11, 4 **3** 1.13 m, yes
4 8 kg, 13 kg **5a** 26, 11, 11, 136; median or mode
b 18, 11, 9, 51; median **c** 13, 12, 9, 11; mean
6a 1–20, etc: 2, 7, 9, 8, 4; 1–10, etc: 1, 1, 4, 3, 4, 5, 4, 4, 2, 2

5 FRACTIONS, DECIMALS AND PERCENTAGES

Page 54 Looking Back

1a 2 hundreds **b** 2 tens **c** 2 tenths **d** 2 hundredths
2a $\frac{9}{100}$ **b** $\frac{5}{100}$, $\frac{1}{20}$ **3a** £3 **b** £1 **c** £15 **4** £30 **5** £120
6a 100% **b** $\frac{1}{100}$ **7** 99.2% **8** $\frac{2}{5}$
9a 6% **b** 6.5% **c** 7.2% **d** 7.9% **e** 8.5% **f** 9%
10 4, 28 **11a** 0.10 **b** 9.1 **c** $5\frac{1}{4}$
12a $\frac{1}{2}$ **b** $\frac{1}{4}$ **c** $\frac{2}{3}$ **d** $\frac{2}{5}$ **e** $\frac{3}{4}$ **f** $\frac{5}{6}$

Page 55 Exercise 1

1a 16.2 **b** 105.8 **c** 0.14
2a 10, 10.5, 11.2, 11.9, 12.6 **b** 0.01, 0.02, 0.05, 0.08
3a 15.6, 16, 16.1, 16.5, 17.0 **b** 9.68, 9.96, 10.81, 11.12
4a 6, 1 **b** 8.01, 3.33 **5a** 3.45 **b** 0.93 **c** 110.8
6 157.23, 154.08; 3.15 **7** 194.4 km **8** 84.2 cm. Yes
9a 34, 340 **b** 125, 1250 **c** 10.9, 109 **d** 2345, 23 450 **e** 1, 10
10a 40, 39 **b** 27, 28.8 **c** 48, 49.6 **d** 150, 151.8 **e** 6, 5.67
11a 0.34, 0.034 **b** 1.25, 0.125 **c** 0.109, 0.0109
d 23.45, 2.345 **e** 0.01, 0.001
12a 5, 4.9 **b** 5, 5.3 **c** 5, 5.5 **d** 3, 2.6 **e** 0.5, 0.8
13a £5.63 **b** £1.25 **c** £0.09 **d** £7.13 **e** £3.20
14 £2.06 **15** £47.53 **16** 9 pence per km

Page 56 Exercise 2

1a $\frac{1}{2}$ **b** $\frac{3}{4}$ **c** $\frac{2}{5}$ **d** $\frac{1}{2}$ **e** $\frac{1}{2}$ **f** $\frac{5}{8}$
2a month **b** second **c** hour
3a $\frac{3}{4}$ **b** $\frac{1}{2}$ **c** $\frac{2}{3}$ **d** $\frac{3}{4}$ **e** $\frac{3}{10}$ **f** $\frac{2}{5}$ **g** $\frac{8}{9}$ **4a** $\frac{2}{15}$ **b** $\frac{1}{6}$ **c** $\frac{4}{15}$ **d** $\frac{11}{15}$
5a 10 p **b** 60 p **c** 80 p **d** £2.50 **e** £56 **f** £8
6a 45 **b** (i) $\frac{2}{3}$ (ii) $\frac{1}{3}$ **7a** $\frac{1}{1}$ or 1, 100% **b** $\frac{2}{5}$, 40% **c** $\frac{3}{4}$, 75%
d $\frac{1}{2}$, 50% **e** $\frac{1}{4}$, 25% **f** $\frac{1}{10}$, 10%
8a £10 **b** £12 **c** £15 **d** £15 **e** £10 **f** £15
9 540 **10a** £24 **b** £136 **11** Blank, 1; 100%, $\frac{1}{4}$; 25%, $\frac{1}{5}$; 20%, $\frac{3}{4}$; 75%, $\frac{1}{10}$; 10%, $\frac{1}{2}$; 50%, blank **12a** 25% **b** 50%

Page 57 Exercise 3A

1a 45 p **b** 15 p **c** (i) £2.50 (ii) 65 p
2 £30, £1.60, 64 p, 10 p **3a** (i) £8.25 (ii) £46.75
b (i) £14.50 (ii) £275.50 **c** (i) £21.60 (ii) £158.40
4a 80% **b** (i) £36 (ii) £204 **5** The old model, by £5.50
6a £5 **b** £7 **c** £9 **d** £12 **e** £2.50
7a £8 **b** £16 **c** £40 **d** £80 **e** £4
8a £6 **b** £7.50 **c** £9 **d** £3 **e** £11.25 **9a** £10 **b** £260
10a £40.50 **b** £490.50 **11a** £33.75 **b** £483.75
12 National Savings Bank gives £25 more

Page 58 Exercise 3B

1a 6 p, 20% **b** £2, 40% **c** 30 p, 75% **d** £16, 25%
e 60 p, $12\frac{1}{2}$% **f** 20 p, 25% **2** 25% **3** 50 p, 50% **4** £1, 1%
5 33.3% **6a** 10% **b** 25% **7a** Chewy 50%, chocbars $33\frac{1}{3}$%, apples 20%, Fruito 26% **b** 33%

Page 59 Exercise 3C

1 4% **2** 10% **3** 1.25%
4a Alison Roberts **b** Alison 10% increase, Mel 9% decrease
5 0.1% **6a** 100 cm², 121 cm² **b** 21%
7a 1.57% **b** 6.77% **c** 10.13% **d** 12.36%
e 1.01%: 4.5 litres for 1 gallon; 2 pints for 1 litre

Page 60 Exercise 4

1a $\frac{7}{20}$ **b** $\frac{3}{20}$ **c** $\frac{7}{10}$ **d** $\frac{3}{5}$ **e** $\frac{9}{100}$
2a 0.35 **b** 0.15 **c** 0.7 **d** 0.6 **e** 0.09 **f** 0.05
3a £15 **b** 15 cm **c** 3.6 litres **d** 900 pupils **e** 12.5 g
f AB 60 m, BC 48 m, AC 72 m
4a 50% **b** 75% **c** 10% **d** 40% **e** 90% **f** 15%
5a 47% **b** 22% **c** 93% **d** 80% **e** 8% **f** 123%
6a 20% **b** 2% **c** 70% **d** 35% **e** 50% **f** 1%
7a 0.75 **b** 0.4 **c** 0.8 **d** 0.3 **e** 0.5 **f** 0.25 **g** 0.7 **h** 0.2
8a 0.29 **b** 0.63 **c** 0.33 **d** 0.83 **e** 0.09 **f** 0.89
9a $\frac{1}{2}$, 0.5, 50% **b** $\frac{3}{4}$, 0.75, 75% **c** $\frac{3}{8}$, 0.375, 37.5%

10a $\frac{1}{2}$, 50%; $\frac{1}{4}$, 25%; $\frac{1}{3}$, $33\frac{1}{3}$%; $\frac{1}{10}$, 10%; $\frac{3}{10}$, 30%; $\frac{3}{4}$, 75%
b 0.2, 20%; 0.25, 25%; 0.5, 50%; 0.01, 1%; 0.1, 10%; 0.05, 5%
11 $\frac{1}{2}, \frac{3}{4}, \frac{1}{4}, \frac{2}{5}, \frac{3}{10}, \frac{7}{100}$; 0.5, 0.75, 0.25, 0.4, 0.3, 0.07; 50%, 75%, 25%, 40%, 30%, 7%

Page 61 Exercise 5A

1 30% **2a** $\frac{4}{25}$, 16% **b** 84% **3** $\frac{1}{2}$, 0.5, 50%
4a 0.7, $\frac{7}{10}$, 70% **b** 0.3, $\frac{3}{10}$, 30% **5a** 20%, 35%, 20% **b** roof; by insulating it **c** (i) £50 (ii) £100 (iii) £175
6a £40 **b** £440 **7a** (i) $\frac{3}{50}$ (ii) 0.06 **b** 9
8a 59 cm, 199.5 cm^2 **b** 39.8 m, 57.4 m^2
9a 8 **b** 24, 32 **10a** Not easily **b** Test 3, 80%; test 2, 75%; test 1, 72%; test 4, 70%

Page 62 Exercise 5B/C

1a $\frac{4}{5}$, 0.8, 80% **b** $\frac{1}{5}$, 0.2, 20% **2a** More, by 12.7% **b** 40.2%
3a $\frac{1}{10}$, 0.1, 10% **b** $\frac{1}{5}$, 0.2, 20% **c** $\frac{3}{10}$, 0.3, 30% **d** $\frac{2}{5}$, 0.4, 40%
4 108°, 72° **5a** Angles are 216° and 144° **b** 480, 320
6 60% **7** Angles are 108°, 36°, 162°, 36°, 18°
8a 1 **b** 4 **c** 144 **d** 1.8

Page 63 Check-up on Fractions, Decimals and Percentages

1a 260 g **b** £3.50 **c** 9 p **2** 0.39, $\frac{2}{5}$, 41%
3a £4.50 **b** £40.50 **4** £17.50 **5** £150 at 11%
6a £591 **b** £10 441 **7a** $\frac{4}{5}$ **b** 0.8 **8** £31.40
9 63%, 73%, 72%, 74% **10** Oxfam £1000, RSPCA £500, NSPCC £500, RSPB £250, RNLI £250
11 $\frac{4}{5}, \frac{7}{10}, \frac{7}{100}, \frac{3}{8}, \frac{1}{25}, \frac{16}{25}$; 0.8, 0.7, 0.07, 0.375, 0.04, 0.64; 80%, 70%, 7%, 37.5%, 4%, 64% **12a** 14 p **b** 25%
13a 3000 m^2 **b** 2880 m^2 **c** 4% less **14** £40 less
15 Maths, Technology, English, Geography, Science

6 DISTANCES AND DIRECTIONS

Page 64 Looking Back

1a PQ = 8 cm, PR = 5.3 cm, QR = 4.7 or 4.8 cm
b ∠PQR = 40°, ∠QPR = 35°, ∠PRQ = 105°
2b ∠BCA = 34°, ∠CAB = 56°, AC = 10.8 cm
3a 90° **b** 180° **c** 360° **4** West, south, east
5a Castle **b** airport and golf course **c** airport and church
d church
6a

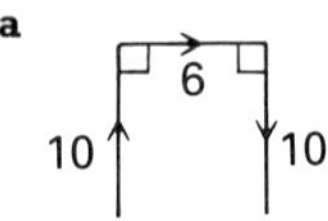

b RIGHT 90, FORWARD 6 **c** (i) FORWARD 5, RIGHT 90, FORWARD 5, RIGHT 90, FORWARD 5, RIGHT 90, FORWARD 5
(ii) FORWARD 8, RIGHT 120, FORWARD 8, RIGHT 120, FORWARD 8

Page 65 Exercise 1A

1a 3 cm, 3 m **b** 4 cm, 4 m **c** 1.5 cm, 1.5 m **d** 5 cm, 5 m
e 2.5 cm, 2.5 m **2a** 12 m **b** 5 m **c** 8 m
3 Lengths **a** (i) 2 cm (ii) 6 cm **b** (i) 2 cm (ii) 5 cm
c (i) 1 cm (ii) 3.5 cm **4a** 2 km N, 3 km E **b** 3 km W, 1 km N
c 4 km E, 4 km S **d** 2 km S, 6 km W

5 (ii) **a** 3.6 cm, 3.6 km **b** 3.2 cm, 3.2 km **c** 2.8 cm, 5.6 km
d 3.2 cm, 6.4 km **6b** MY = 10 cm; 10 km
7b MY = 9.9 cm; 9.9 km

Page 66 Exercise 1B

1b QR = 4.2 cm; 12.6 m
2a 52 m **b** 49 m **3a** 28° **b** 64 m **4** 298 m **5b** 17.4 m

Page 67 Exercise 1C

1 4.7 km **2** 13.9 km **3** 85 km **4** 410 m

Page 68 Exercise 2A

1a North, north-east, east, south-east, south, south-west, west, north-west **2a** 45° **b** 90° **c** 135° **d** 90° **3a** E
b S **c** NW **4a** (i) Kilmaluag (ii) Sligachan
(iii) Dunvegan (iv) Carbost (v) Broadford
b Elgol and Ardvasar
5

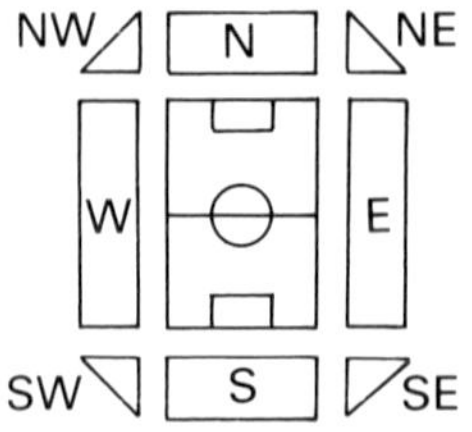

Page 68 Exercise 2B/C

1a S **b** E, SW, NW, W, NE **2c** 13 km
3 Rows: 2 km SE, 3 km E, 1.5 km NE
4a 35 km **b** 40 km **5a** 98 km **b** 18 km

Page 70 Exercise 3A

1 Anstruther 070°, Cupar 025°, Dollar 290°, Edinburgh 180°, Falkirk 255°, Linlithgow 240°, North Berwick 100°, Perth 330°, St Andrews 045° **2** N 000°, NE 045°, E 090°, SE 135°, S 180°, SW 225°, W 270°, NW 315°
3a 040° **b** 120° **c** 270° **d** 045° **e** 330° **f** 240°
4a

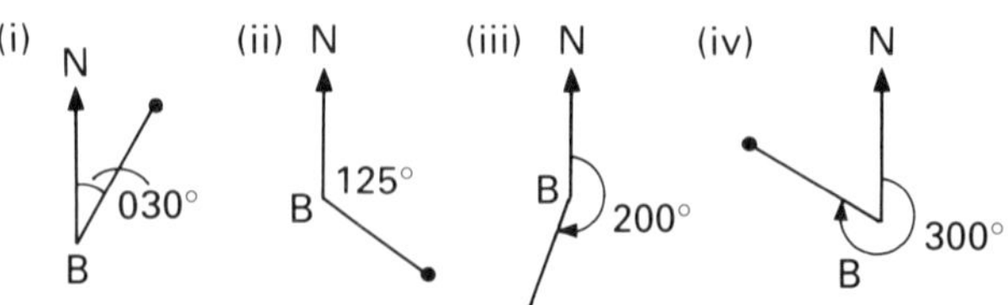

5 A → B → E → F → C → D **7** Ilfracombe 000°, Minehead 025°, Exeter 049°, Torquay 080°, Dartmouth 115°, Land's End 254°, Lundy Island 337°

Page 71 Exercise 3B/C

1 A 065°, B 115°, C 220°, D 290°, E 315°
2 210°, 230°, 240°, 015°, 075°, 015° **3** 35°
4 30°, 65°, 50° (all c'wise); 110°, 130°, 75° (all anti-c'wise)
5a (i) 080° (ii) 260° **b** (i) 100° (ii) 280°. Add 180°
6a 250° **b** 300° **7a** 205° **b** 074° **c** 157° **8a** 340° **b** 150°

Page 72 Exercise 4A

2a 5 km, 035° **b** 4 km, 110° **c** 7.3 km, 247° **3** 136 km, 068°
4b 39 km, 030°

Page 73 Exercise 4B/C

1b 10 km, 120° **2** 57 km, 058° **3** 3.2 km **4** 1100 m, 275°
5b (i) 179° (ii) 11.7 km

Page 74 Check-up on Distances and Directions

1b 7.2 m, 56°, 34° **2b** 58 km
3 A, NW; B, SW; C, W; D, E; E, SE; F, SW; G, S
4a 090° **b** 130° **c** 312° **5a** Cowes 000°, Ryde 073°, Sandown 113°, Shanklin 133°, Chale 193°, Brookgreen 250°, Yarmouth 275° **b** (i) 070° (ii) 250° **c** Cowes 8 km, Ryde 12 km, Sandown 12 km, Shanklin 11.2 km, Chale 11.2 km, Brookgreen 12.8 km, Yarmouth 14 km **6** 305°
7 31.2 km, 281°

7 POSITIVE AND NEGATIVE NUMBERS

Page 75 Looking Back

1 The numbers are −3, −2, −1, 0, 1, 2, 3
2a (i) 2 (ii) −4 **b** (i) 1, 2 (ii) −1, −2, −4
3a 10 **b** 0 **c** 10 **4a** 9, 3 **b** 3, 0 **c** 1, −1
5a A(3, 2), B(−1, 3), C(−3, −3), D(1, −4)
b (i) Right (ii) above **c** (i) (3, −2) (ii) (−3, 2)
6a −3, −1, 2 **b** −2°C **7a** 15 **b** 2 **c** 1 **d** 11
8a 2 **b** 5 **c** 2 **d** 10 **e** 10 **f** 1
9a 10 above **b** 1 below **c** 6 below **d** 2 above

Page 76 Exercise 1A

1a (i) 3, 4, 5 (ii) −1, 0, 1 **b** (i) −2, −3, −4 (ii) 0, 1, 2
2a 9 **b** 0 **c** −1 **d** −5 **e** 9 **f** −1 **g** −2 **h** −5
3a 10 **b** 1 **c** 1 **d** 0 **e** 0 **f** −1 **g** 4 **h** −3
4a (i) 2 (ii) −1 (iii) 4 (iv) −2 **b** (i) −5 (ii) −1 (iii) 0 (iv) −4 **5a** 1°C, −4°C **b** −2°C, −4°C
6 A, −5°C; B, −3°C; C, −1°C; D, 1°C; E, 4°C **7a** (i) 0°C (ii) −2°C (iii) −4°C **b** (i) −2°C (ii) 2°C (iii) 3°C
8 B(1, 3), C(−2, 2), D(−2, −4), E(3, −2), F(2, 0)
9a N **b** V **c** L **d** W
10a −15 m, −30 m **b** −25 m, −10 m

Page 77 Exercise 1B/C

1 Balance: £20.00, 10.00, 5.00, −5.00, +10.00, 0.00
2a (i) 6° (ii) 9° (iii) 14° **b** (i) 4° (ii) 9° (iii) 3° **c** (i) −3° (ii) +3° (iii) −11°
3a (i) Hawaii, Aberdeen (ii) 25° **b** −8°, +10°, −6°, +21°
4a A star shape **b** yes **c** the y-axis **d** yes **e** yes **f** 4
5a (i) 40 m (ii) 20 m (iii) 10 m (iv) −10 m **b** (i) 20 m (ii) 0 (iii) −10 m (iv) −30 m

Page 78 Exercise 2

1a 2+4 = 6 **b** −2+2 = 0 **c** 0+5 = 5 **d** −4+3 = −1
e −1+6 = 5 **f** −5+7 = 2
2a −6+1 = −5 **b** −1+3 = 2 **c** −3+8 = 5 **d** −2+3 = 1
3a −3+4 = 1 **b** −5+7 = 2 **4a** −1 **b** 5 **c** 0 **d** 7
e −3 **f** −8 **g** 0 **h** 10 **i** 2 **j** 1 **k** 3 **l** 0 **m** −4 **n** 6
o 1 **p** 5 **q** 7 **r** 8 **s** 9 **t** 10 **6a** 2°C, −6°C
b −1°C, 0°C, −5°C, 1°C, 2°C; −1°C, 1°C, 0°C, 3°C, 4°C
7a 0 **b** 7 **c** −2 **d** 3 **8a** (20, −20) **b** (−30, 80)
9a 3 **b** 0 **c** −2 **d** 5 **e** 0 **f** 13 **g** −1 **h** 8 **i** 0 **j** −2
k −1 **l** 7
10a 8°C **b** 4°C **c** 2°C **d** −2°C **e** 0°C **f** 4°C

Page 80 Exercise 3

1a 3−4 = −1, 3+(−4) = −1 **b** 1−3 = −2, 1+(−3) = −2
c 0−5 = −5, 0+(−5) = −5
d −1−1 = −2, −1+(−1) = −2
e −4−4 = −8, −4+(−4) = −8
f 5−6 = −1, 5+(−6) = −1
2a −1−2 = −3, −1+(−2) = −3
b −3−4 = −7, −3+(−4) = −7
c 2−7 = −5, 2+(−7) = −5
3a 1−5 = −4 **b** −3+(−6) = −3−6 = −9 **c** −2+4 = 2
4a −2 **b** −2 **c** 1 **d** −5 **e** −5 **f** −9 **g** 5 **h** 0 **i** −1
j −6 **k** −7 **l** −4 **m** −10 **n** −8 **o** −9 **p** −11
q −10 **r** −6 **s** 0 **t** −6
6 10°C, 0°C, −2°C, −5°C, −10°C, −6°C **7a** 10 **b** −6
8 +£10, +£15, −£5, −£10, £0, −£5, £25
9a 1, 2, 3, 4, 5, 6 **b** 4, 5, 6, 7, 8, 9 **c** 4, 5, 6, 7, 8, 9
10a 7 **b** 7 **c** 16 **d** 6 **e** 6 **f** 1 **g** 4 **h** −5 **i** 2 **j** −1
k 2 **l** 15 **m** 1 **n** 10 **o** 0 **p** 0 **q** 12 **r** 8 **s** 1 **t** −8
u 8 **12** 119.5° **13a** +4° **b** −4° **c** −7° **d** +7°

Page 82 Exercise 4A/B

1a 11 **b** 3 **c** 3 **d** −3 **e** −3 **f** 0
2a 2 **b** −5 **c** −2 **d** −10 **e** −7 **f** 0
3a 3 **b** −3 **c** −5 **d** −8 **e** −10 **f** −2
4a 6 **b** 5 **c** 5 **d** 0 **e** 4 **f** 2
5a 1 **b** 6 **c** −4 **d** −11 **e** −4 **f** −3
6a 1 **b** −1 **c** −4 **d** 3 **e** 2 **f** −5 **g** −2 **h** −3
7a West **b** north east **c** east **d** south east
8 Rows: **a** −2, 1, −1; 3, 4, 7; 1, 5, 6 **b** −5, 2, −3; 4, 1, 5; −1, 3, 2 **c** −6, 1, −5; 2, 1, 3; −4, 2, −2 **d** 5, −5, 0; −3, 2, −1; 2, −3, −1 **9a** 1 **b** −1 **c** 3 **d** 1 **e** 5 **f** 7
10a $x+2 = -1$, −3 **b** $x-1 = 0$, 1 **c** $-2+(-1) = x$, −3
d $-3-(-5) = x$, 2 **11a** 3 **b** −3 **c** 5 **d** −5 **e** −7 **f** 2
g 5 **h** −5 **i** −6 **j** −3 **k** −4 **l** 11 **m** −12 **n** 0 **o** 1
p 10 **q** 8 **r** −10 **s** −8 **t** −8 **u** 0
12a $5x$ **b** x **c** $-x$ **d** $-5x$ **e** $5x$ **f** $-5x$ **g** $3x$ **h** 0
13a F **b** F **c** T **d** T **e** F **f** T **14** Rows: **a** 2, 3, −1; 4, 1, 3; −2, 2, −4; **b** −3, 4, −7; 8, 3, 5; −11, 1, −12
c 6, 8, −2; 7, 2, 5; −1, 6, −7 **d** −2, −3, 1; −7, 6, −13; 5, −9, 14

Page 83 Exercise 4C

1a −20 **b** 8 **c** −8 **d** 20 **e** 11 **f** −11 **g** −2 **h** −20
i −2 **2a** −1 **b** 5 **c** 7 **d** −3 **e** −4 **f** 0 **g** 6 **h** 2
i 3 **j** −3 **k** 8 **l** 4 **m** −3 **n** 8
3a (−10, −30), (20, −30), (20, −60)
b (−60, 20), (−30, 20), (−30, −10)
4a 0 **b** 9 **c** −3 **d** 1 **e** −6 **f** 3
5a −5, −9, −8 **b** −3, −6, −4
6a $-2t$ **b** $-6u$ **c** $-v$ **d** $-5a$ **e** $-3b$ **f** $-4c$
7 Rows: **a** −8, −3, −4; −1, −5, −9; −6, −7, −2
b −6, 6, −1, 3; 7, −5, 2, −2; 1, −3, 8, −4; 0, 4, −7, 5
c 8, 0, −8, 4, −4; −11, 6, −7, 10, 2; −5, 12, −1, −9, 3; 1, −12, 5, −3, 9; 7, −6, 11, −2, −10

Page 84 Check-up on Positive and Negative Numbers

1a −7, −6, −5, −4, −3, −2, −1, 0
b −15, −10, −5, 0, 5, 10, 15 **2** B, by 1 square
3a −1 **b** −1 **c** 1 **d** −7 **4a** −3, −2, −1, 1, 3, 5
b −7, −4, −3, −1, 5, 6 **5a** −2+6 = 4 **b** −5+2 = −3
c −1−5 = −6, −1+(−5) = −6
d 2−4 = −2, 2+(−4) = −2
6 4−(−2), 4−(−3), 4−(−4), 4−(−5), 4−(−6); 1, 2, 3, 4, 5, 6, 7, 8, 9, 10

7a 5 **b** −1 **c** −10 **d** 4 **e** −2 **f** −1 **g** −4 **h** 8
i −5 **j** −8 **k** 2 **l** 12 **m** 2 **n** 0 **o** 0
8a 4, 5 **b** −7, −2
9a −3 **b** 8 **c** 2 **d** −2 **e** −4 **f** −2
10a (1, 13) **b** S(1, −15), T(14, −15), U(14, 0), V(1, 0)
11a −4 **b** 6 **c** −12 **d** −2 **e** 2 **f** 12

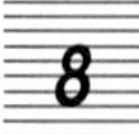

ROUND IN CIRCLES

Page 85 Looking Back

1a Diameter **b** radius **c** circumference
3a 10 mm **b** 20 mm **c** 30 mm
4 *Example:*

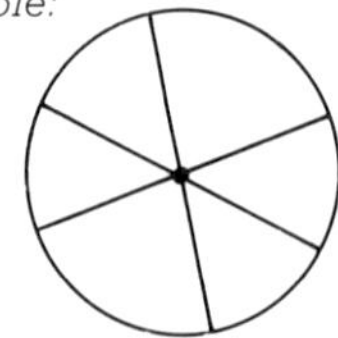

5a 90 mm **b** 70 mm **c** 80 – 100 mm
6b (i) 4 cm (ii) 6 cm (iii) 5 cm
7 About: **a** 1 sq **b** 65 sq **c** 16 sq **d** 4 sq
8a (i), (iii), (vi), (viii) **b** (ii) cone, (iv) sphere

Page 86 Exercise 1

2a 2 **b** (i) 2 cm (ii) 12 inches (iii) 80 cm (iv) 8000 miles
3a 4 cm **b** 140 mm **c** 12 mm **d** 0.3 m, 0.7 m
5 Rows: 16 mm, 3 cm, 1.5 m, 18 cm, 9 cm, 2.8 m;
32 mm, 6 cm, 3 m, 36 cm, 18 cm, 5.6 m
6 *Example:*

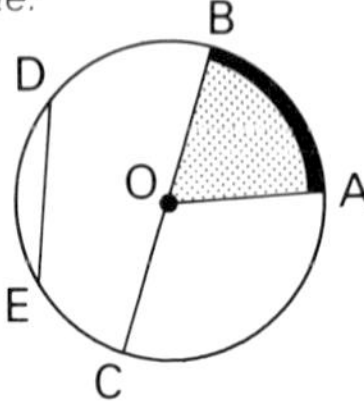

Page 88 Exercise 2A

1 3 × 12 cm = 36 cm **2** 30 cm **3** 60 m **4a** 156 cm
b 15 feet **c** 435 cm **d** 90 mm **e** 81 inches **5a** 45 cm
b 135 cm **6a** 120 cm **b** (i) 360 cm (ii) 3.6 m **7a** 1.5 m **b** 30 m
8a 10 mm, 30 mm **b** 20 mm, 60 mm **c** 30 mm, 90 mm
9a 84 mm, 60 mm, 42 mm, 36 mm, 30 mm
b 14 mm, 10 mm, 7 mm, 6 mm, 5 mm

Page 89 Exercise 2B

1a 47 cm **b** 79 m **c** 60 mm **d** 310 cm **2a** 31 cm
b 94 cm **c** 290 cm **d** 57 inches
3a 25, 22, 19, 16, 13, 10 mm **b** 0 or 2
4a 37.7 cm **b** 50.2 cm **c** 28.3 cm **d** 94.2 mm
5a 20.4 cm **b** 1410 mm **c** 880 mm **d** 509 mm
e 345 mm and 251 mm **f** 56.5, 84.8, 126 cm

Page 90 Exercise 2C

1 20 000, 38 900, 39 900, 21 000, 440 000, 361 000, 160 000, 200 000, 20 000 km **2a** 10 900 km, 4 370 000 km
b the sun is much farther away than the moon
3a 40 090 km **b** 39 930 km **c** it is not quite spherical
4a 7200 km **b** 45 000 km **5** 212 cm, 215 cm
6 40 000 km, 41 000 km

Page 92 Exercise 3A

1 25 cm **2** 40 cm **3** 18 m
4 Rows: 30 cm, 24 m, 6 m, 15 cm, 30 m, 72 cm;
10 cm, 8 m, 2 m, 5 cm, 10 m, 24 cm; 5 cm, 4 m, 1 m, 2.5 cm,
5 m, 12 cm **5a** 48 cm **b** 24 cm **6a** 33.3 cm **b** 100

Page 93 Exercise 3B/C

1a 7.6 cm **b** 3.8 cm **2** Radii are: **a** 4.3 cm **b** 2.9 cm
c 1.9 cm **3a** 17 cm **b** 18 cm **c** 19 cm
4 4.6, 4.8, 4.9, 5.1, 5.3 inches **5** 14 cm **6** 72 cm

Page 94 Exercise 4A

1 300 cm^2 **2** 3 × 1 × 1 cm^2 = 3 cm^2 **3** 108 cm^2
4a 768 cm^2 **b** 192 cm^2 **c** 75 cm^2 **d** 432 mm^2
5a 675 cm^2 **b** square, 1800 cm^2 **6a** 243 cm^2
b square, 648 cm^2 **7** 12 m^2, 4800 cm^2, 3267 mm^2
8 Areas: 108 cm^2, 243 m^2, 1200 mm^2, 3 km^2, 18.75 cm^2
9a 5 mm, 75 mm^2 **b** 10 mm, 300 mm^2 **c** 15 mm, 675 mm^2

Page 95 Exercise 4B

1a 452 cm^2 **b** 962 cm^2 **c** 50.3 cm^2 **d** 227 cm^2
2a (i) 0.75 m (ii) 0.5 m **b** (i) 1.77 m^2 (ii) 0.785 m^2
(iii) 0.985 m^2 **3a** 78.5 cm^2 **b** 628 cm^2
4a 20 100 cm^2 **b** 10 100 cm^2
5a 0.5 m **b** (i) 2 m^2 (ii) 0.393 m^2 (iii) 2.39 m^2
6a 113 m^2 **b** 463 m^2 **7a** A circle **b** 36.3 cm^2

Page 96 Exercise 4C

1a 21.6, 9.62 sq inches **b** 0.785 sq inch
c 20.8, 8.83 sq inches **d** 6.7, 3.41 sq inches
2a 5.31 m^2 **b** 2.4 m^2 **c** 2.91 m^2 **3** 1590 cm^2 **4** 46.3 cm^2
5 466 m^2 **6** 468 (gold), 1400, 2340, 3270, 4210 cm^2

Page 98 Check-up on Round in Circles

1a (i) 28 mm (ii) 10 mm, 12 mm **b** (i) 14 mm
(ii) 5 mm, 6 mm **2a** 1.5 m **b** 24 inches
3 $C \doteqdot 3D$, or $C = \pi D$ **4a** 120 cm (126 cm) **b** 21 cm (22 cm)
5a 78 mm **b** 26 mm (24.8 mm) **6** $A \doteqdot 3r^2$, or $A = \pi r^2$
7a 192 cm^2 (201 cm^2) **b** 432 cm^2 (452 cm^2)
8 84 mm (88 mm), 588 mm^2 (616 mm^2) **9** 36.75 cm^2 (38.5 cm^2)

TYPES OF TRIANGLE

(Degree symbols and units of length are not shown in answer diagrams.)

Page 99 Looking Back

1a ✓ — ✓ — — **b** — ✓ ✓ — — **c** ✓ — — — ✓
d — — — ✓ — **e** — — — — ✓

2 Draw one or two diagonal bars **3a** A(3, 4) **b** congruent
4a 24° **b** 60° **5a** 3 **b** (i) EFG (ii) EFH
6a (i) 5 (ii) 6 (iii) 2, 4 or 6 **b**

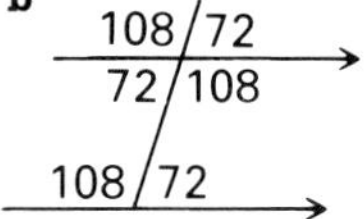

7a 48 cm^2 **b** 37.5 cm^2

Page 100 Exercise 1A

1a ∠BAC = 50°, ∠ABC = 40° **b** 90° **c** 180° **4a** 80° **b** 100° **c** 60° **d** 65° **5a** 87° **b** 100° **c** 43° **d** 75° **e** 20° **6a** $a° = 15°$, $b° = 85°$ **b** $c° = 65°$, $d° = 30°$, $e° = 60°$ **c** $f° = 34°$, $g° = 56°$, $h° = 34°$ **7a** 90°, yes **b** 88°, no **c** 90°, yes **d** 90°, yes **e** 91°, no **f** 90°, yes **g** 90°, yes **h** 100°, no **i** 90°, yes; 76°, no

Page 101 Exercise 1B/C

1a 70° **b** 50°, 60° **c** 180° **2a** 180°
b (i), (ii) alternate angles for parallel lines **c** 180°
3 20°, 70°, 90°; 45°, 90°, 45°; 70°, 60°, 50°; 133°, 37°, 10°; 55°, 60°, 65°; 36°, 69°, 75° **4a** (i) $x°+2x°+3x° = 180°$, $x = 30$
(ii) $2x°+2x°+5x° = 180°$, $x = 20$
(iii) $3x°+4x°+5x° = 180°$, $x = 15$
(iv) $2x°+3x°+5x° = 180°$, $x = 18$ **b** (i), (iv)
5a $x = 30$; 30°, 60°, 90°
b $x = 18$; 36°, 54°, 90°. $y = 18$; 18°, 72°, 90°
c $x = 45$; 45°, 45°, 90° **6** ∠RSU = 28° (alt. to ∠SUT, RS ∥ UT), ∠RUS = 90° (sum of ∠s of △ is 180°)
7a $x+a = 90$, $x = 57$ **b** $x+b+60 = 180$, $x = 66$
c $x+2c = 180$, $x = 46$ **d** $2x+d = 180$, $x = 71$

Page 103 Exercise 2A

1a 30 cm² **b** 32 cm² **c** 12 m² **d** 27.5 mm²
2a 4 sq **b** 6 sq **c** 6 sq **d** 5 sq **3a** 30 cm² **b** 54 cm² **c** 80 cm² **d** 25 cm² **e** 64 cm² **f** 49 cm² **g** 27 cm² **h** 230 cm² **4a** 2 m² **b** 9000 m²
5 40 m², 24 cm², 90 cm², 18 m², 1050 mm², 187.5 cm²

Page 104 Exercise 2B/C

1a 144 cm² **b** 228 cm² **c** 120 cm² **d** 110.5 cm² **2a** 40 m² **b** 1200 m² **c** 200 m² **3b** 21 cm² **4** 72 cm²
5 Missing data: 25 m; 8 m, 2.4 m; 2400 m², 2.7 cm² **6** 14 sq

Page 104 Exercise 3A

1a, b

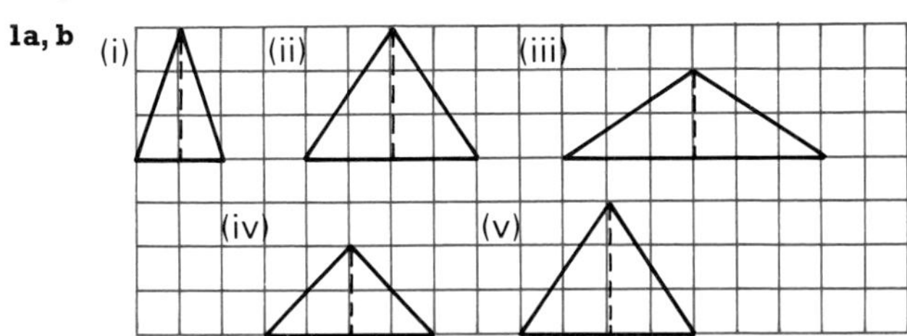

(i)
(ii)
(iii)
(iv)
(v)

3a

5 37 37 5
53 53
3 3

b

5 74 5
53 53
6

4a

9 40 40 9
50 50
6 6

b

8 50 50 8
40 40
6 6

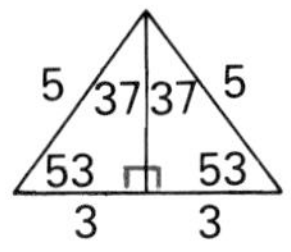

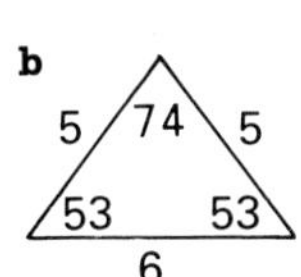

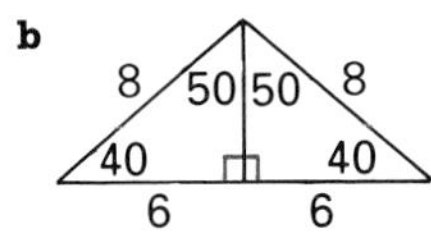

c

5 5
60 60
10 30 30 10

d

10 8 80
3
3
10 8 80

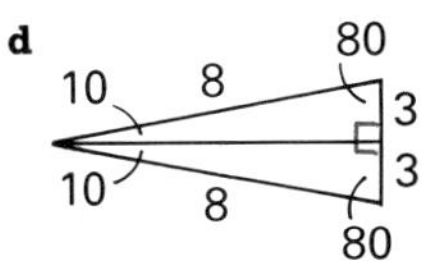

5a

20 20
12 12
70 70

b

8 50 50 8
40 40

c

35 35
6 110 6

d

15 30 15
75 75

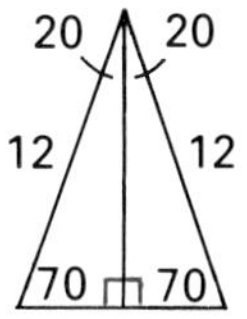

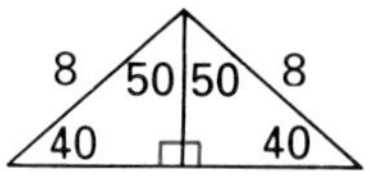

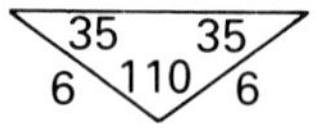

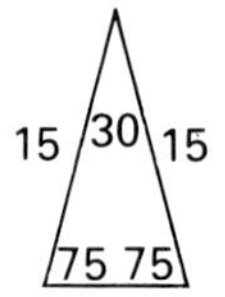

6a

20
250 250
80 80
171

b

11.5 120 11.5
30 30
20

c

1.5 41 2
49
49
1.5 41 2

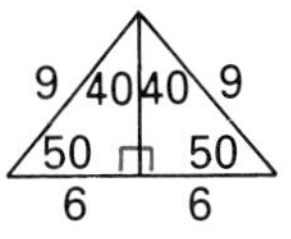

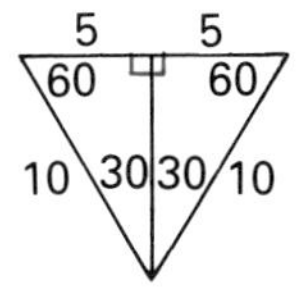

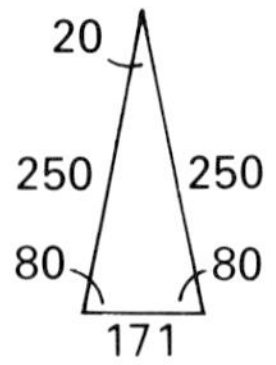

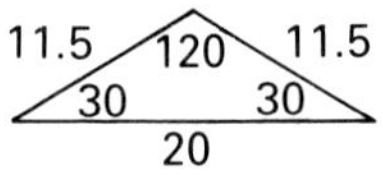

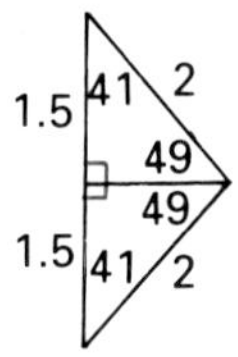

d

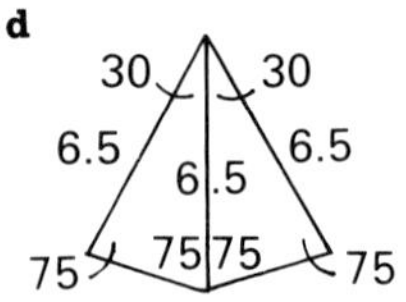

30 30
6.5 6.5
6.5
75 75 75 75

e

100
44 44
70 92 70
70 92 70
44 44
100

f

60
60 60
60 60
60

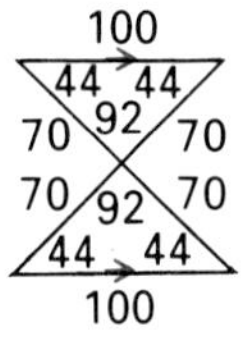

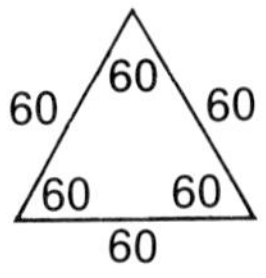

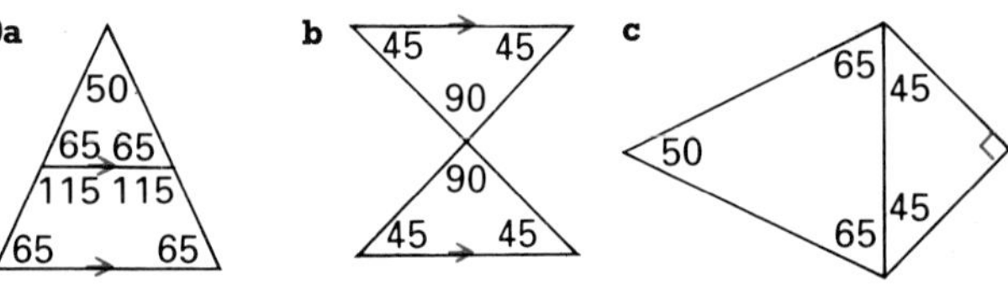

7a Isosceles **b** (5, 0) **8a** Isosceles **b** (0, −3)

9a

50
65 65
115 115
65 65

b

45 45
90
90
45 45

c

65 45
50
65 45

10a 162 cm², 81 cm² **b** 192 cm², 96 cm² **c** 108 cm², 54 cm² **d** 16 m², 8 m²

Page 106 Exercise 3B/C

1 90°, 45°, 45° **2** $x = 73$, $y = 34$ **3a** (i) The y-axis (ii) ∠s ABC, ACB **b** $x = 0$, y is any number
4a (8, 1) **b** (2, 9) **5a** (i) Isosceles (ii) 12 sq units
b (i) 2 sq units (ii) 14 sq units
6a 12 cm² **b** 336 m² **c** 33 cm² **d** 66 mm²
7 ∠BDE = 52° (corr to ∠FBG, AG ∥ DE), ∠DEC = 76° (supp to ∠CEH), ∠DCE = 52° (sum of ∠s of △ is 180°). So △CDE is isosceles. ∠ABC = 52° (vert opp ∠FBG), ∠ACB = 52° (vert opp ∠DCE). So △ABC is isosceles

Page 107 Exercise 4

1a 3 **b** they are equal **c** 60°
2c 3 (another 3 if you turn it over)
3

60
10 10
60 60
10

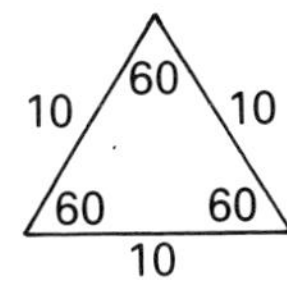

4 Angles are 90°, 45°, 45°, 60°, 60°, 60° (the obtuse angles in the kite are 105° and 105°) **5a** (ii), (iii), (iv), (vii) **b** (i), (viii) **c** (v), (vi)
6a 6 × 60° = 360° **b** 3 × 60° = 180° **c** (i) 9 (ii) 3 (iii) 1
7a 3 **b** (i) turn symmetry (ii) 3 **8a** Its sides are equal, because 5 balls fit each one, or it fits three ways by turning it round **b** 60° **c** 6

Page 109 Exercise 5

1a 70°, acute-∠d **b** 22°, obtuse-∠d **c** 90°, right-∠d
d 28°, obtuse-∠d isosceles **2a** 60°, equilateral
b 35°, 45°, obtuse-∠d **c** 40°, 60°, acute-∠d **d** 90°, right-∠d
3a Obtuse-∠d **b** right-∠d **c** acute-∠d isosceles
d acute-∠d
4a

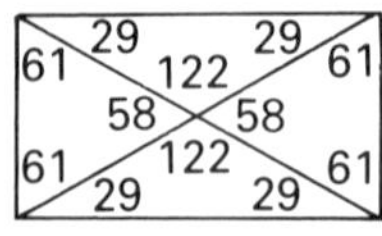

b (i) △s PTS, QTR (ii) △s PTQ, STR
(iii) △s PQS, PQR, RSP, RSQ
5a Obtuse-∠d isosceles **b** equilateral
c right-∠d isosceles **d** equilateral
6a False **b** true **c** false **d** true **e** false

Page 110 Exercise 6

1a 90° **b** 50° **2a** 3.2 cm **b** 5 cm **3a** 41°, 56°, 83°
b 37°, 53°, 90° **4a** 60°, 60°, 60° **b** ∠K = 100°
5a ∠A = 79°, ∠B = 44°, ∠C = 57°
b ∠F = 80°, DF = 4.6 cm, EF = 2.9 cm
c ∠R = ∠Q = 30°, QR = 8.7 cm

Page 111 Exercise 7

1a A(1, 6), B(1, 1), C(7, 6) **b** TRIANGLE(7, 1)
2a A(7, 1), B(7, 4), C(1, 1) **b** TRIANGLE(1, 4) **3b** Isosceles
4a MOVE(3, 6) MOVE(1, 4) TRIANGLE (5, 4) **b** square
5a

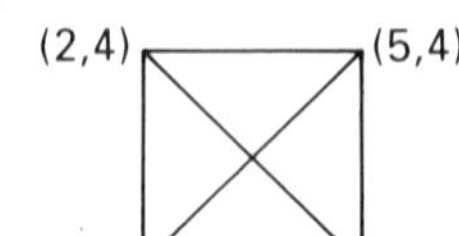

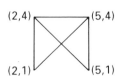

b MOVE(2, 4) MOVE(2, 1) TRIANGLE(5, 4) TRIANGLE(5, 1)

Page 112 Check-up on Types of Triangle

1a, c **2a** 55° **b** 110° **c** 60°, 70° **d** 55°, 55°
3a 60 mm² **b** 30 cm² **c** 42 m² **d** 24 cm²
4a Acute-∠d isosceles **b** right-∠d **c** right-∠d isosceles
d equilateral **e** obtuse-∠d **f** acute-∠d isosceles
5a **c** **d**

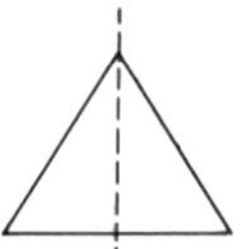

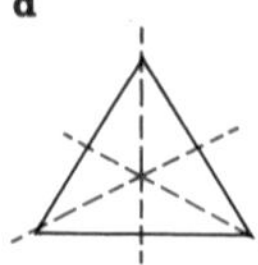

6a 78 **b** 50, 50 **c** 60, 60, 60 **d** 45, 45 **e** 72, 54
7a

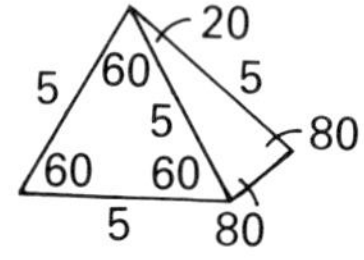

b 360° **8a** 10 sq units **b** 6 sq units
9

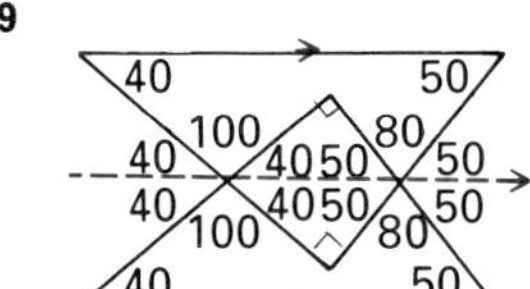

10 METRIC MEASURE

Page 113 Looking Back

1a cm or mm **b** m **c** mm **d** km **e** m
3 8.7, 8.9, 9.1, 9.6, 10.5 cm
4a They are equal **b** (i) 8 sq (ii) 8 sq (iii) 8 sq
5a 18 m² **b** £396 **6a** 24 litres **b** 10 litres
7a 800 cm³ **b** 0.8 litre **8a** 150 ml **b** 125 ml **9a** 25
b 131 **c** 143 **d** 1230 **e** 5.7 **f** 0.85 **g** 3.65 **h** 4.781
10 A car 100 m long, a building 3 km high, a jump of 1 mm

Page 114 Exercise 1A

1a 50 mm **b** 120 mm **c** 48 mm **d** 4 mm
2a 40 cm **b** 6 cm **c** 7.2 cm **d** 0.9 cm
3a 400 cm **b** 250 cm **c** 37 cm **4a** 8 m **b** 4.8 m **c** 0.75 m
5a 5000 m **b** 2600 m **c** 650 m
6a 6 km **b** 8.4 km **c** 0.12 km **7a** 13 cm **b** 75 mm
8a 27 mm, 38 mm **b** (i) 130 mm (ii) 13 cm
9 AB 25 mm, AC 30 mm, AD 25 mm, AE 15 mm, AF 15 mm
10 382 m; no **11** 53 cm **12** 5 mm **13a** 30 **b** 8

Page 115 Exercise 1B/C

1a 1.2 cm **b** 300 cm **c** 70 mm **d** 6500 m **e** 1.89 m
f 9.6 cm **g** 1.2 km **h** 150 cm **2a** 4200 m **b** 600 m
c 70 mm **d** 50 mm **e** 240 cm **f** 380 cm
3a 91 mm **b** 25.4 cm **4** 48 km/h, 112 km/h **5** 16
6 22.5 km **7a** 378 mm **b** 896 mm **8a** 10 **b** 25
9a 1.42 m **b** 32 cm **c** Panda, Metro
10a (i) 1 230 000 m (ii) 1230 km **b** 499 seconds

Page 117 Exercise 2A

1a m² **b** cm² **c** cm² **d** mm² **e** hectares **f** km²
3a 20 sq **b** 36 sq **c** about 47 sq **4** Length × breadth; no
5a 48 cm² **b** 247 mm² **c** 20.25 cm² **d** 9.72 m² **e** 196 km²
6a 40 mm, 20 mm **b** (i) 8 cm² (ii) 800 mm² **c** 100
7a (i) 240 cm, 90 cm (ii) 2.4 m, 0.9 m **b** (i) 21 600 cm²
(ii) 2.16 m² **8a** 16 **b** 6 **c** 96
9 100 cm² for the door, 10 hectares of grass, 9 mm² of floor
10b The area of each is 16 cm²

Page 119 Exercise 2B

1a 12 m² **b** 400 cm, 300 cm **c** 120 000 cm². 1 cm² is smaller than 1 m², so more are needed
2a (i) 2000 m (ii) 40 000 m² **b** 4 hectares
3a (i) 15 (ii) 15, 1; 5, 3 units **b** (i) 14 (ii) 14, 1; 7, 2 units
c (i) 16 (ii) 16, 1; 8, 2; 4, 4 units **4a** 32 **5** 4.5 hectares
6 11 m² **7** 19 cm² **8** 26.8 m²

Page 119 Exercise 2C

1a 4 units **b** 5 units **2a** (i) 10 (ii) 100 (iii) 500 **b** (i) 100
(ii) 10 000 (iii) 80 000 **3** 100 **4** 60 000 m², 6 hectares
5 Area is 76 cm² **a** 40 + 36 cm² **b** 56 + 20 cm²
c 126 − 50 cm² **6a** 2.56 cm² **b** 343 mm²

Page 121 Exercise 3A

1a Litre or ml **b** mm³, cm³ or ml **c** m³ **d** ml **e** cm³
f litre **2a** 300 ml **b** 850–900 ml
3a 5000 ml **b** 25 000 ml **c** 200 ml
4a 7 litres **b** 12 litres **c** 4.5 litres
5a 50 000 ml **b** 800 ml **c** 10 000 ml
6a 864 cm³ **b** 2970 cm³ **7a** 36 cm³ **b** 3 cm³

8a (i) 34 800 cm³ (ii) 34 800 ml (iii) 34.8 litres **b** £16.70
9a 250 **b** 500 **c** 750 **d** 420 **e** 700
10 750, 650, 500, 400, 350 ml

Page 122 Exercise 3B

1 16 days, plus two doses **2** 1500 **3** Shed A
4a 3750 **b** 7500 **c** 4 **d** yes **5a** 1000 **b** 12.5 cm

Page 122 Exercise 3C

1a 216 cm³ **b** 6 cm by 6 cm by 6 cm **2** 46.3 cm
3a 160 mm³ **b** 336 m³ **4a** (i) 4000 cm³ for each
(ii) the second one **b** the first one, less material needed
5 20 × 1 × 1, 10 × 2 × 1, 5 × 4 × 1, 5 × 2 × 2: 5 × 2 × 2, as it has least surface area, so needs least material

Page 123 Exercise 4A

1a kg **b** g **c** tonnes **d** mg **e** g **f** kg **g** mg
h tonnes **2** About: **a** 500 g ($\frac{1}{2}$kg) **b** 50 kg **c** 50–100 g
d 10 g **e** 5 mg **f** 3 kg
3 2 kg pencil, 5 tonnes weight, 5 mg box of chocolates
4a 5.7 g, 64 g, 10.3 g **b** 58.3 g
5a 11.1 kg, 17.8 kg **b** 28.9 kg
6a (i) 8200 kg (ii) 7800 kg **b** (i) 2100 kg (ii) 3500 kg

Page 124 Exercise 4B/C

1a 7 g **b** 15 g **c** 0.8 g **2a** 1000 g **b** 6000 g **c** 2300 g
3a 3 kg **b** 0.3 kg **c** 10 kg
4a 1000 kg **b** 5000 kg **c** 6400 kg **5** 1275 g **6** 4570 kg
7a 1550 g **b** 1.55 kg **8** 63.8 kg
9a 40 m **b** 96 kg **c** £35.96 **10a** 2 kg **b** 2 kg **c** 4500 kg
11a 1 g **b** its density is greater than that of water **c** float
12 6500 **13** 12.06 kg **14a** £31.66 **b** £42.44

Page 125 Check-up on Metric Measure

1a × 10 **b** × 1000 **c** × 1000 **d** ÷ 100 **e** ÷ 1000 **f** × 1
2a 476 cm **b** 4.76 m **3a** (i) 30 000 m² (ii) 3 hectares
b (i) 46 800 m² (ii) 4.68 hectares **4** 1000 **5a** 55 g **b** 13 g
6 86.32 kg
7a 3.75 **b** 2.304 **c** 100 × 100 = 10 000 **d** 1 000 000
8a 990 ml **b** 11 cm to 12 cm **9** 12, 1; 6, 2; 4, 3 **10** 480

11 EQUATIONS AND INEQUATIONS

Page 126 Looking Back

1a $5n$ **b** $4k$ **c** 0 **d** $4t$ **e** $5d+4$ **f** $n+1$
2a $2x+2$ **b** $2y-6$ **c** $15+3x$ **d** $4x-2$ **e** $12w+16$
f $5-10t$ **3a** $2x+12$ cm **b** $6x+2$ cm **c** $6y+8$ cm
4a $y=1$ **b** $u=3$ **c** $t=18$ **d** $x=5$ **e** $m=5$ **f** $w=5$
5a (i) $x+3$ (ii) 3 **b** (i) $2t$ (ii) t **c** (i) $3y+2$ (ii) 2
d (i) $x-2$ (ii) x **6a** $x=2$ **b** $x=3$
7a $x=5$ **b** $y=4$ **c** $w=5$ **d** $m=3$ **e** $x=7$
8a $n=24$ **b** $n=15$

Page 127 Exercise 1A

1a $x=4$ **b** $t=3$ **c** $e=6$ **2a** $y=6$ **b** $p=6$ **c** $n=11$
3a $k=1$ **b** $m=7$ **c** $t=6$ **4a** $n=7$ **b** $u=7$ **c** $v=4$
5a $x=7$ **b** $x=5$ **c** $y=2$ **6a** $w=4$ **b** $y=5$ **c** $t=1$
7 $n=4$ **8a** $4n-3=17, n=5$ **b** $6n+7=19, n=2$
9a $7(2x+6=20)$ **b** $5(3x-1=14)$
10a $x=5$ **b** $a=3$ **c** $x=1$ **11a** $x=4$ **b** $y=2$ **c** $x=1$
12a $w=2$ **b** $t=3$ **c** $x=4$ **13a** $t=1$ **b** $u=4$ **c** $v=1$
14a $x=3$ **b** $y=3$ **c** $t=2$ **15a** $x=4$ **b** $y=1$ **c** $k=2$
16a $5w=25, w=5$ **b** $8+x=23, x=15$
17a (i) $4y=2y+8, y=4$ (ii) 4 m **b** (i) $6x=2x+48, x=12$
(ii) 12 m **18** $4n-6=2n+8, n=7$

Page 129 Exercise 1B

1a $x=2$ **b** $x=-1$ **c** $x=-3$
2a $x=3$ **b** $y=3$ **c** $t=1$
3a $u=-3$ **b** $v=-1$ **c** $w=-6$
4a $t=7$ **b** $m=5$ **c** $k=5$ **5a** $n=4$ **b** $p=6$ **c** $t=3$
6a $y=5$ **b** $t=2$ **c** $g=4$ **7a** $x=2$ **b** $t=3$
8a $x=3$ **b** $x=7$ **9a** $y=2$ **b** $t=3$
10a $x=3$, 11 cm **b** $x=6$, 24 cm **c** $y=3$, 13 cm
d $n=2$, 14 cm **11a** $t=3$, 20 wts **b** $t=3$, 15 wts
c $m=9$, 19 wts **d** $x=3$, 5 wts **12a** (i) $3n=n+100, n=50$
(ii) 150 **b** (i) $2n-1=n+19, n=20$ (ii) 39
c (i) $5n=n+52, n=13$ (ii) 65
d (i) $4n+10=2n+22, n=6$ (ii) 34

Page 130 Exercise 1C

1 $x=9$; 10, 6, 4: $x=1$; 2, 6, 4 **2** $x=12$; 7, 11, 4: $x=4$; 7, 3, 4
3 $x=4$; 4, 8, 4 **4** $x=1$; 2, 1, 3 **5** $x=2$; 8, 6, 2
6 $x=1$; 2, 6, 8: $x=2$; 4, 12, 8 **7** $x=2$; 3, 8, 5
8 $x=2$; 3, 9, 6: $x=8$; 15, 9, 24
9 $x=6$; 12, 2, 14: $x=2$; 4, 2, 2
10 $x=5$; 2, 8, 10: $x=1$; 2, 4, 2 **11** $x=3$; 9, 2, 7: $x=2$; 6, 1, 7
12 $x=6$; 14, 2, 12: $x=2$; 2, 2, 4
13 $x=3$; 1, 11, 10: $x=1$; 1, 5, 6
14 $x=1$; 5, 3, 2: $x=5$; 9, 15, 6

Page 131 Exercise 2A

1a $x=3$ **b** $x=0$ **c** $x=1$ **2a** $x=6$ **b** $x=3$ **c** $x=5$
3a $x=2$ **b** $x=2$ **c** $x=5$ **4a** $y=7$ **b** $y=0$ **c** $y=0$
5a $x=2$ **b** $y=3$ **6a** $t=4$ **b** $m=6$
7a $x=2$; $4\times5=20$ **b** $5(x-1)=40, x=9$; $5\times8=40$
c $3(x+7)=30, x=3$; $3\times10=30$
d $6(x-2)=48, x=10$; $6\times8=48$ **8a** $n=4$
b $3(n-1)=12, n=5$ **c** $5(n+4)=20, n=0$
d $7(n-3)=35, n=8$ **9** $6(n+8)=120, n=12$
10 $5(n-10)=120, n=34$

Page 132 Exercise 2B

1a $x=2$ **b** $x=2$ **c** 0 **2a** $x=-3$ **b** $x=3$ **c** $x=12$
3a $x=2$ **b** $x=16$ **c** $x=11$ **4a** $x=10$ **b** $x=2$
c $x=5$ **5a** $x=1$, 26 cm **b** $x=7$, 58 cm **c** $x=4$, 32 m
d $x=9$, 44 m **6a** $3(a+1)=24, a=7$ **b** $5(a-1)=40, a=9$
c $6(a-2)=3a, a=4$ **d** $2(3a-2)=1(2a+4), a=2$

Page 132 Exercise 2C

1a $x+1$ **b** $6(x+1)$ km **c** $6(x+1)=144$; $x=23$
2a $x+3$ **b** £$6(x+3)$ **c** $6(x+3)=210$; $x=32$
3a (i) $P=4x+22$ (ii) $A=8(2x+3)$ **b** (i) $x=2, A=56$
(ii) $x=1, P=26$ **4a** (i) $P=6x+16$ (ii) $A=10(3x-2)$
b (i) $x=5, A=130$ (ii) $x=7, P=58$ **5** 12 and 42
6 620 (220 10ps, 200 5ps, 200 2ps)

Page 133 Exercise 3A

1a $4>3$ **b** $1<2$ **c** $3>0$ **d** $8>7$ **e** $-1<1$ **f** $5>-5$
g $9<11$ **h** $0<2$ **i** $0>-2$ **j** $3>-3$ **k** $-2<-1$
l $-3>-6$ **2a** $x<y$ **b** $x>y$ **c** $x=y$ **d** $x=y$
e $x>y$ **f** $x<y$

3a 4 is greater than 2 **b** 3 is less than 5
c -1 is greater than -3 **d** -2 is less than 0 **4a** $p < q$
b $q > r$ **c** $r < s$ **d** $s > t$ **e** $p < r$ **f** $r > t$ **g** $q < s$
h $p > t$ **5a** 5 **b** 1 **c** 9 **d** 0 **6a** F **b** T **c** T **d** T
e F **f** F **g** T **h** T **i** F **j** F **k** F **l** T **m** T **n** F **o** F

Page 134 Exercise 3B

1 $W \leqslant 40$ **2** $S \leqslant 20\,000$ **3** $Y \geqslant 18$ **4** $T \geqslant 50$ **5** $B \geqslant 10$
6 $S > 70$ **7** $P > 11$ **8** $P \leqslant 245$ **9** $D \leqslant 3.8$ **10** $L \leqslant 4$
11a $5 < x$ or $x > 5$ **b** $y > 12$ or $12 < y$ **c** $a > b$ or $b < a$
d $x+1 < 8$ or $8 > x+1$ **12a** (i) 10 (ii) 0, 1, 2, 3
(iii) 7, 8, 9, 10 (iv) 0, 1 **b** (i) $n < 6$, or $n \leqslant 5$
(ii) $n > 7$, or $n \geqslant 8$ (iii) $n < 1$, or $n \leqslant 0$

Page 135 Exercise 3C

1 Rows: 5, 6, 7, 8, 9, 10; no, no, no, yes, yes, yes: $y = 3, 4, 5$
2 Rows: 10, 9, 8, 7, 6, 5; yes, yes, no, no, no, no: $y = 0, 1$
3a 4, 5 **b** 3, 4, 5 **c** 5 **d** 0, 1, 2, 3, 4 **e** 3, 4, 5 **f** 0 **g** 5
h 2, 3, 4, 5 **i** 4, 5

4a (i) 0, 1, 2 (ii)

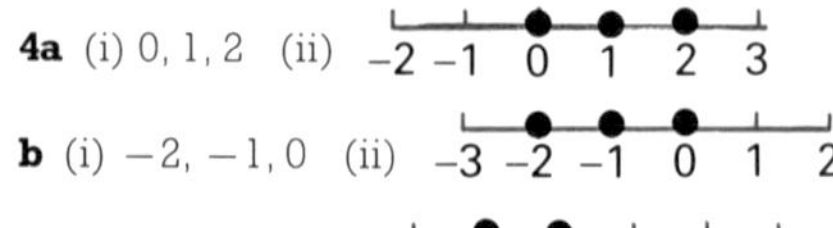

b (i) $-2, -1, 0$ (ii)

c (i) $-2, -1$ (ii) −3 −2 −1 0 1 2

d (i) $-1, 0, 1, 2$ (ii) −2 −1 0 1 2 3

e (i) 2 (ii) −2 −1 0 1 2 3

f (i) 1, 2 (ii) −2 −1 0 1 2 3

5a $L \geqslant W, N \leqslant 16$ **b** $L \geqslant W+2$ **c** 3, 4, 5 m

Page 136 Check-up on Equations and Inequations

1a $x = 5$ **b** $y = 3$ **c** $t = 10$ **d** $x = 6$ **e** $n = 2$ **f** $k = 14$
2a $x = 6$ **b** $y = 4$ **c** $t = 2$
3a $2n+5 = 11, n = 3$ **b** $7n-1 = 20, n = 3$
4a $x = 0$ **b** $x = 9$ **c** $x = 5$
5a $x = 6$; 30 cm **b** $x = 2$; 10 wts
6a $8(3x-1) = 40, x = 2$ **b** $6(2x+1) = 3(x+2), x = 0$
7a (i) $3n-2$ (ii) $n+40$ **b** $3n-2 = n+40, n = 21$: both 61
8a (i) $4(x-6)$ (ii) $4(x-6) = 2x, x = 12$ **b** (i) $2x-4$ cm
(ii) $2x-4 = 50, x = 27$
9a $x < 10$ **b** $y > 8$ **c** $x \geqslant 20$ **d** $y < 3$
10a 3, 4, 5 **b** 0, 1, 2, 3 **c** 2, 3, 4, 5 **d** 5 **e** 0, 1, 2, 3 **f** 3, 4, 5
11 The cat 49.9 m, the elephant 0.1 m
12 $x = 13$

12 RATIO AND PROPORTION

Page 137 Looking Back

1a $\frac{1}{2}$ **b** $\frac{1}{3}$ **c** $\frac{2}{5}$ **d** $\frac{1}{4}$ **2a** $\frac{1}{2}$ **b** $\frac{1}{4}$ **c** $\frac{1}{10}$ **d** $\frac{3}{4}$
3a £5 **b** £3 **c** £2 **d** £10
4a 10 **b** 100 **c** 1000 **d** 1000 **e** 1000
5a 4 cm **b** 2.5 cm **c** 3 cm **d** 0.5 cm **6** £1.50
7a $\frac{1}{2}$ **b** $\frac{1}{3}$ **c** $\frac{3}{4}$ **d** $\frac{1}{5}$ **e** $\frac{3}{4}$ **f** $\frac{2}{3}$ **g** $\frac{1}{2}$ **8a** $\frac{1}{2}$ **b** $\frac{1}{4}$ **c** $\frac{1}{3}$ **d** $\frac{5}{6}$

9a £3 **b** £6 **c** £30 **d** £75
10a True **b** false **c** true **d** false

Page 138 Exercise 1A

1a $\frac{2}{3}$ **b** $\frac{4}{5}$ **c** $\frac{2}{3}$ **d** $\frac{3}{4}$ **e** $\frac{1}{3}$ **f** $\frac{4}{3}$ **g** $\frac{2}{1}$ **2a** $\frac{1}{2}$ **b** $\frac{1}{2}$ **c** $\frac{2}{3}$ **d** $\frac{3}{1}$
3a $\frac{2}{3}$ **b** $\frac{2}{1}$ **c** $\frac{1}{2}$ **d** $\frac{4}{3}$ **e** $\frac{1}{3}$ **f** $\frac{3}{2}$ **4b, c**
5a $\frac{1}{5}$ **b** $\frac{1}{10}$ **c** $\frac{5}{1}$ **d** $\frac{4}{1}$ **e** $\frac{1}{10}$ **f** $\frac{1}{2}$ **g** $\frac{1}{20}$

Page 139 Exercise 1B

1a $\frac{7}{3}$ **b** $\frac{3}{2}$ **c** $\frac{13}{7}$ **d** $\frac{1}{1}$ **2a** $\frac{1}{3}$ **b** $\frac{2}{15}$ **c** $\frac{5}{2}$ **d** $\frac{15}{2}$ **3** $\frac{50}{1}$
4a $\frac{25}{1}$ **b** $\frac{1}{2}$ **c** $\frac{10}{1}$ **d** $\frac{1}{100}$
5a $\frac{1}{10}$ **b** $\frac{1}{1000}$ **c** $\frac{10}{1}$ **d** $\frac{1}{4}$ **e** $\frac{3}{10}$ **f** $\frac{1}{7}$ **g** $\frac{1}{1000}$ **h** $\frac{1}{4}$

Page 139 Exercise 1C

1a 3:1 **b** 1.5:1 **c** 2:1 **d** 3.5:1
2a 1:3 **b** 1:1.5 **c** 1:1.2 **d** 1:8.5 **3** 1.33:1, 1.78:1
4a 1:3 **b** 1.8:1 **c** 0.6:1, or 1:1.67 **d** 1:1
5a UK:USA:China = 1:4.07:17.41 **b** 1:38.75:39.58; the UK

Page 140 Exercise 2A

1 12 shares, £4, £20, £28, £48 **2a** 3 **b** £4 **c** £4, £8
3 £20, £16 **4** £40, £60 **5a** 5 cm, 15 cm **b** 2 litres, 3 litres
c 20 kg, 4 kg **d** 35 min, 25 min **6a** 15 m **b** 6 min **c** 60 kg
7a 8, 4 parts **b** 9, 3 parts **c** 2, 10 parts **d** 6, 6 parts
8 9 parts copper, 3 parts nickel **9** Kim £12, Curtis £8
10 2 litres **11a** 270°, 90° **b** 300°, 60° **c** 280°, 80°
d 252°, 108° **e** 340°, 20°

Page 141 Exercise 2B

1a 6 **b** 3 **c** 9 **d** 10 **e** 56 **f** 12
2a 12 **b** 42 **3a** 8 **b** 10 **c** 20, 10
4a 48 cm, 64 cm **b** 4:3 **5a** (i) 12 (ii) 20 **b** 6, 4 **c** 9, 6, 15

Page 142 Exercise 2C

1 £300 000, £400 000, £200 000, £100 000
2a 90 g, 90 g, 180 g **b** 520 g
3a (i) 1:7 (ii) 1:14 (iii) 1:2.4 **b** (i) 6 m (ii) 9 m (iii) 8 m
4a (i) 150 ml (ii) 350 ml **b** (i) 1500 ml (ii) 1000 ml
c 360 ml, 240 ml **d** $166\frac{2}{3}$ litres

Page 143 Exercise 3

1a 10 cm, 200 cm **b** 4.5 cm, 90 cm **c** 3 cm, 60 cm
d 2 cm, 40 cm **e** 2.5 cm, 50 cm **f** 1.5 cm, 30 cm
2a 6 cm, 4 cm; 6 m, 4 m **b** 1.5 cm, 1 cm; 1.5 m, 1 m
c 2 cm, 1 cm; 2 m, 1 m
3a 160 cm **b** 15 cm **4a** 60 m **b** 13 cm
5a (i) 4 km (ii) 7 km (iii) 12–13 km **b** 14 cm
6a 1:30 **b** 40 cm, 30 cm

Page 145 Exercise 4A

1a 60p, 75p **b** 60p, 80p, 100p
2a 75, 150, 225, 300, 375 **b** 30, 60, 90, 120, 150
c 24, 48, 72, 96, 120 **d** 15, 30, 45, 60, 75
3a 4 km **b** 8 km **c** 20 km **4a** 20 km **b** 40 km **c** 60 km
5a 30 **b** 150 **c** 450 **6a** 3 **b** 300 **c** 2400
7a 4 **b** 32 **c** 60 **8a** 100 **b** 400 **c** 750
9a £2.25 **b** £11.25 **c** £20.25 **10a** 94p **b** £1.88 **c** £6.58
11a (i) £7.20 (ii) £50.40 (iii) £86.40 **b** £28.80

Page 146 Exercise 4B/C

1 £3.60 **2** £66 **3** £18 **4** 500 **5** 21.6 **6** 248 km
7 £563.40 **8** £600
9a,b: to encourage people to buy more items

10a 10 feet **b** 51 feet **c** (i) $7\frac{1}{2}$ feet (ii) $12\frac{3}{4}$ feet
11a (i) 20p (ii) 48p **b** (i) 96 m^3 (ii) 2 m **c** £3.12

Page 147 Exercise 5

1a 60, 80, 100, 120 **c** yes **2a** 30, 40, 50, 60 **c** yes
3a 120, 160 **c** 240 km **4a** 30, 40 **c** (i) £50 (ii) £15
5c (i) 180 (ii) 90 **d** 25

Page 148 Exercise 6A

1a 50, 25, 10, 5, 2 **b** 72, 36, 24, 18, 12 **c** 96, 80, 60, 48, 40
d 18, 9, 6, 4, 3 **e** 2, 3, 5, 6, $7\frac{1}{2}$ **f** 6, 3, 2, $1\frac{1}{2}$, $1\frac{1}{5}$ **2** $\frac{1}{2}$ hour
3 2 hours **4** 8 minutes **5a** 4 **b** 16 **c** 12

Page 149 Exercise 6B/C

1a 80 min **b** 16 min **2a** 60 min **b** 12 min **c** 3 min
3a 360 min **b** 72 min **c** 45 min **4** 3 days **5** $4\frac{1}{2}$ hours
6 96 km/h **7a** 6 **b** 2 days **8a** 60 ohms **b** 8 amps
9a 30 days **b** 24 **10** 210 people per km^2

Page 150 Exercise 7

1a Direct **b** inverse **c** neither **d** inverse **e** direct
f (i) direct (ii) neither **2** £700 **3** 20 min
4a 30 days **b** 6 days **5** £104.25 **6** 96 **7a** 25 **b** 45 days
8 £37.44 **9** 160 **10a** Inversely **b** 1.875 kg

Page 151 Check-up on Ratio and Proportion

1a 1:3 **b** 3:5 **c** 5:1 **2a** 40p, 20p **b** 36p, 24p
c 10p, 50p **3** 6 **4** 9 litres of blue, 3 of yellow **5a** 47 cm
6a 240 cm **b** 20 cm **7a** (iii) $y:x$ is constant (1.5) or doubling x doubles y, etc **b** (i) xy is constant (240) **8** 48
9 25 min **10a** 15, 30, 45, 60, 75, 90p **b** cost is directly proportional to time **c** the cost is zero when the time is zero

13 MAKING SENSE OF STATISTICS 2

Page 152 Looking Back

1a (i) Wednesday (ii) Sunday
2b 8.56–8.58 am **3a** 2, 4, 6, 9, 7, 2 **b** 1950–1999 kg
4a 4, 4, 4, 8 **b** 18, 16, 16, 26 **c** 3.24, —, 2.8, 3.5
d £3.72, —, £2.86, £8.77 **5** Maths 9 above, English 8 below, French 19 below, Geography 12 above, Science 10 above

Page 153 Exercise 1A

1b 3 h **c** 3, 3 h **2a** 6.5 marks **b** 7, 7 marks
3a 74p **b** 9p **c** 74p, 74p **4a** 45.3, 7 matches **b** yes
5b 1.8, 0, 1 hour **c** (i) $\frac{1}{3}$ (ii) $\frac{1}{3}$

Page 154 Exercise 1B/C

1a 1, 4, 5, 9, 8, 5, 4, 1 **b** yes; mean is 250.5 ml
2a July **b** June **3** 7.1

Page 155 Exercise 2A

1 2 **2** 45 **3** 0, 1 **4b** 1°C, 1°C
5a 15, mode **b** (i) 10% (ii) $7\frac{1}{2}$% **c** 187
6a 8, 8, 7.9 **b** 3, 4, 4.2

Page 157 Exercise 2B/C

1c 11–15, 35–40 g **2b** (i) Sunday (ii) Friday
3b (i) Term 3 (ii) Term 1

Page 158 Exercise 3

1b (i) 55–59 (ii) 55–59 **2c** (i) 30–39 (ii) 35–39 **d** $\frac{1}{6}$
3a (i) 12–14 hours (ii) 12–14 hours **b** 8 **4b** 8

Page 159 Exercise 4A

1a As the temperature rises, the ice cream sales rise fairly steadily **b** As the temperature rises, the sale of scarves tends to fall **c** Change in temperature has little or no effect on newspaper sales **d** As rainfall rises, the number of seaside visitors tends to fall

2 *Examples:*

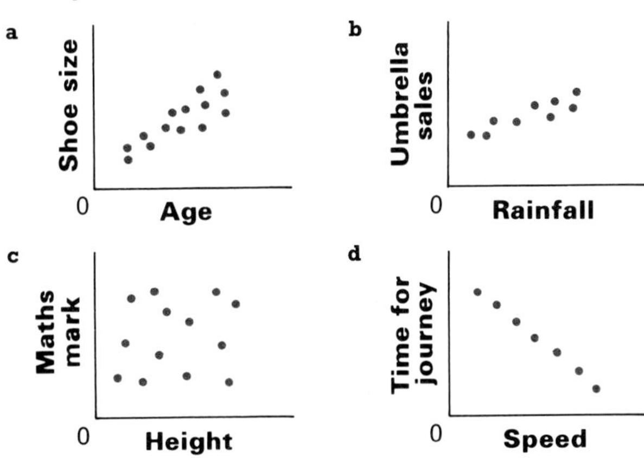

3a The general connections are: (i) height increases with age (ii) weight increases with age (iii) weight increases with height **b** (i) 170–190 cm, 60–70 kg (ii) 140–160 cm, 45–55 kg **4a** As speed increases, length of skid increases, slowly at first, then fairly steadily **b** 45 km/h **c** 100 m

Page 161 Exercise 4B/C

1a (i) 15 (ii) 12 **b** (i) 48 kg (ii) 68 kg
c (i) 180 cm (ii) 150 cm **2a** (i) 6 (ii) 2 **b** (i) 4 (ii) 1
3a (i) 35 mph (ii) 15 mph **b** (i) 4 h (ii) 2 h
4b (i) 55 mm (ii) 165 mm **c** (i) 4 kg (ii) 10 kg

Page 162 Check-up on Making Sense of Statistics 2

1b 37 cm **c** 3.7 cm **d** 3 cm, 3.5 cm **2a** (i) 25% (ii) 27%
3 9.8 minutes **4a** 15, 25, 25.0; 15, 25, 24.7 **b** No, especially as the number of cards was to the nearest 5 **5** 9.5 minutes
6a 8 **b** 3

14 KINDS OF QUADRILATERAL

Page 163 Looking Back

1a Squares, rectangles, triangle, parallelogram **b** rhombus (diamond) **c** parallelograms, trapezium, semi-circle
d kite, triangle

2a ∠ACB = 35°; ∠DEF = 70°, ∠DFE = 40°; ∠GHK = ∠HKG = ∠KGH = 60° **b** EF = 5 cm; GK = HK = 6 cm **c**

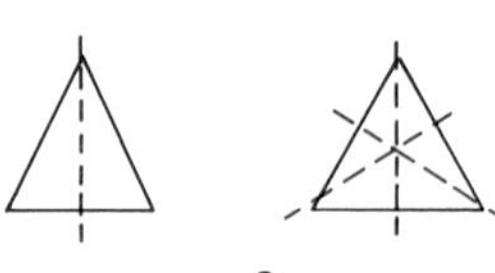

3a

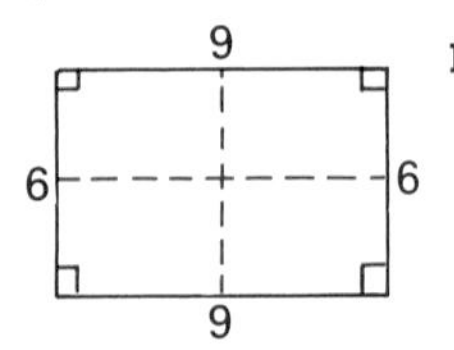

b 36 cm², 54 cm²

4a 21 m² **b** 16 cm² **c** 50 mm²
5a All angles 60° or 120° **b** all angles 140° or 40°
c

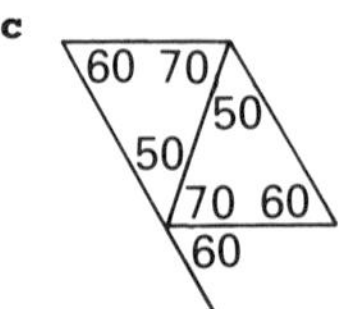

6a (6, 5) and (6, 1), or (−2, 5) and (−2, 1)
b (7, 8) and (1, 8), or (7, −2) and (1, −2)

Page 164 Exercise 1A

1a Two rods at right angles, the longer through the middle of the shorter **b** kite-shaped fabric **c** run into the wind
2a

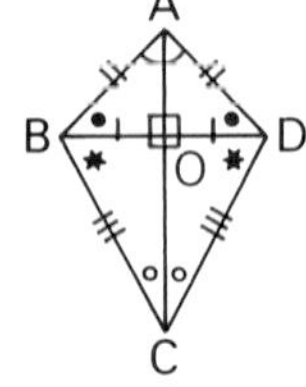

b △s ABD, CBD, **c** △ADC

3a

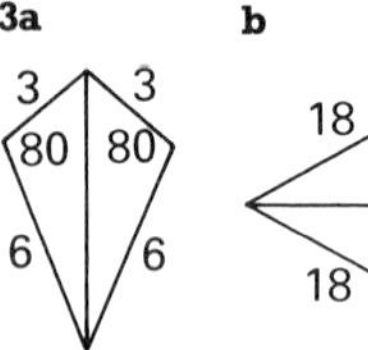

b **c**

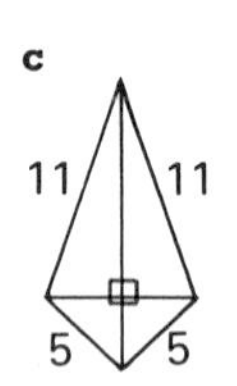

d

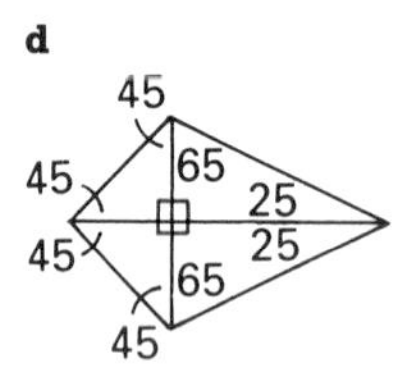

4b 1st and 5th, 2nd and 4th **c** 3rd: all its sides are equal
5a (7, 5) **b** (4, 5)
6a, b

c 360° **d** all quadrilaterals can be split into two triangles, and the sum of the angles of each triangle is 180°
7a **b**

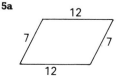

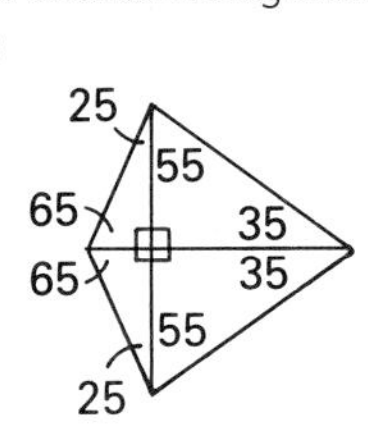

Page 165 Exercise 1B/C

1a **b** **c** **d**

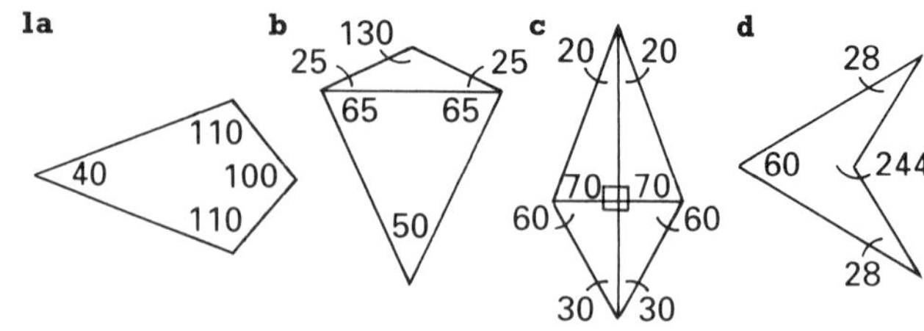

2a AECF, ABEG, AGFD
b angles at A are equal.
So $x = \frac{1}{4}$ of $90 = 22\frac{1}{2}$

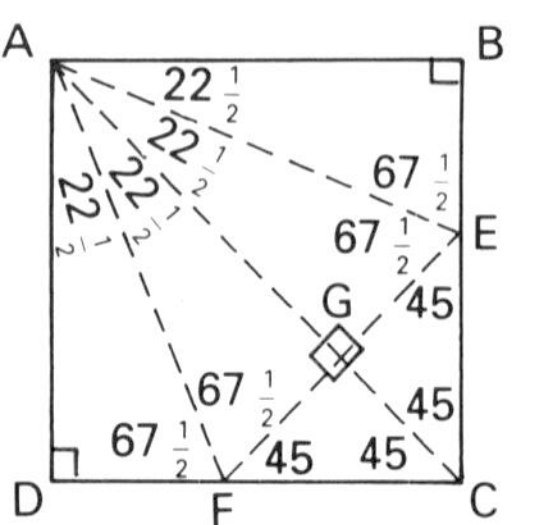

3a 320 cm **b** 133 cm

4a (i) 30° (ii) 72° **b** (i) 25.7°, 60° (ii) $\frac{180°}{n+1}, \frac{360°}{n}$

Page 166 Exercise 2A

1a AD **b** DC **c** ∠ADC **2a** CB **b** CD **c** ∠BCD
3a 4 **b** ABEF; all its angles are right angles; a square
4a (i) OD (ii) 90° **b** (i) OC (ii) ∠BOC = 90°
5 **6a** **b**

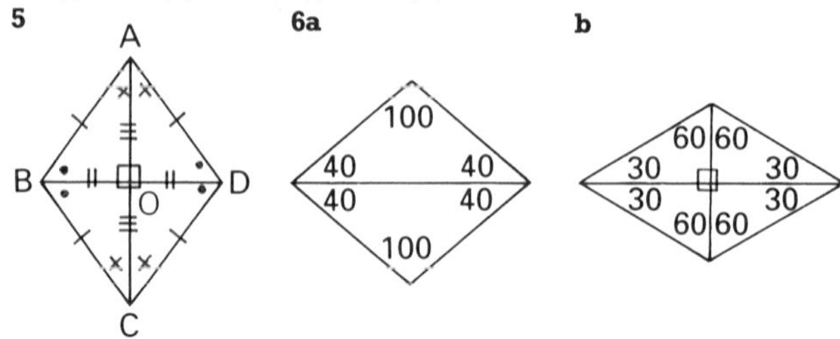

7b 5 cm, two of 74°, two of 106°
8a Half diagonals are 10 cm and 15 cm long **b** each side is 12 cm long **9a** (2, 0) **b** (12, 2)
10a (5, 2) **b** 4, 8 units **c** (5, 4)
11 ∠QPS = ∠QRS = 89°; ∠PQR = ∠PSR = 91° **12c** (i) Yes (ii) no

Page 167 Exercise 2B/C

1a 17 **2** 120 cm **4a** 72° **5a** 200 cm **b** 280 cm
6 A→C, B→D, AB→CD. CD . . . AD is equal and parallel to CB . . . equal and parallel

Page 168 Exercise 3

1b (i) All its sides are equal (ii) diagonals bisect each other
2b BT is an axis of symmetry

Page 169 Exercise 4A

1a (i) C (ii) D (iii) CD **b** (i) C (ii) B (iii) CB
2a 4 **b** ABEF; all its angles are right angles; a rectangle
4a DC **b** BC **c** OC **d** OB
5a **b**

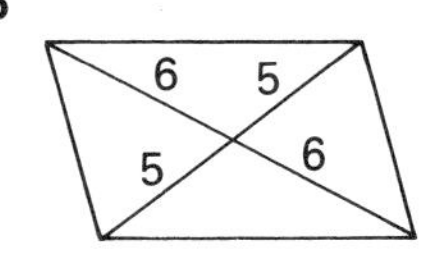

6a ∠CDA, ∠ABC = ∠CDA **b** ∠DCB, ∠BAD = ∠DCB
7b Because all the lines are parallel **c** all the angles are 65° or 115° **8a** ∠RSQ **b** ∠QSP **c** ∠RTQ

9a **b** **c** **d**

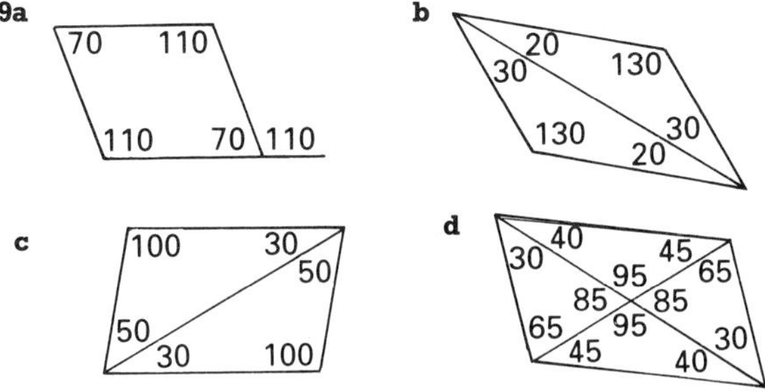

10a 48 cm **b** 4 **c** 77°, 103° **11a** 5 **b** 4
12 The sum of the angles of each triangle is 180°

Page 171 Exercise 4B/C

1 Its angles are right angles, its diagonals are equal and it has two lines of symmetry **2** A trapezium
3a Its opposite sides are parallel (and equal) **b** SUTP, SUQT, URQT, URTP, UVTW, PQRS
4a ABQP, CBQR **b** 3 m **c**

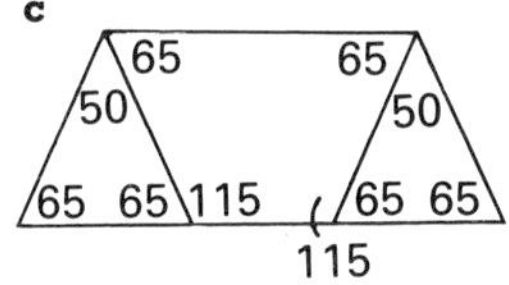

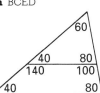

5a 6 **b** 3
6a (i) ∠SUT (ii) ∠UTS **b**
c Isosceles

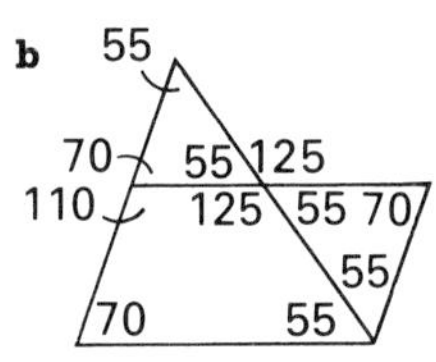

Page 172 Exercise 5

2b It has two sides parallel to the y-axis
4a, b

4 4
108 108
5 5
72 72
6 6

5a (10, 4) **b** (1, −4), (3, −1), (8, −1), (10, −4)
6a BCED **b** PUTS

60
40 80
140 100
40 80

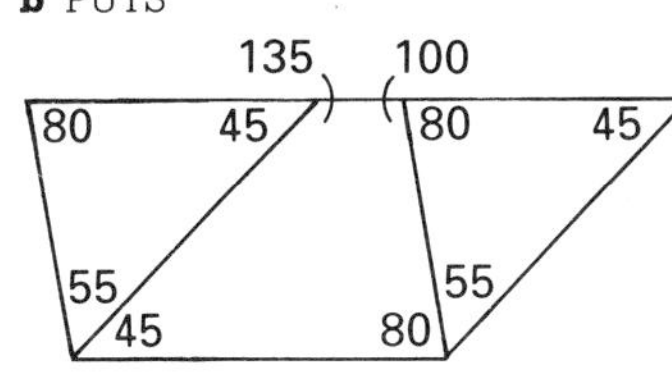

7b All the angles are 88° or 92° **c** 6
d $h_2 = 33.1$ mm, $h_3 = 31.2$ mm

Page 173 Exercise 6

1a Square, rhombus **b** square **c** parallelogram, rectangle, square, rhombus **d** parallelogram **e** kite **f** square, rhombus **g** rhombus, parallelogram, square, rectangle **h** square, rhombus **2a** (i) D(3, 5) (ii) H(13, 4) (iii) L(21, 4) (iv) R(6, 7) (v) V(18, 9) (vi) B(24, 10)

Page 175 Check-up on Kinds of Quadrilateral

1a Kite **b** parallelogram **c** rhombus **d** rectangle **e** trapezium **f** square **2a** Rhombus, rectangle **b** parallelogram, trapezium **c** rhombus, parallelogram, rectangle, square
3

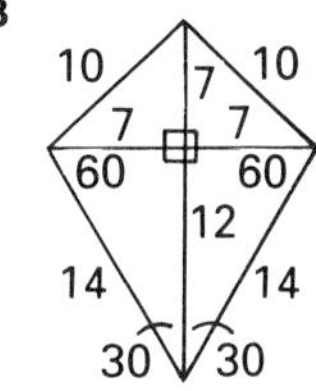

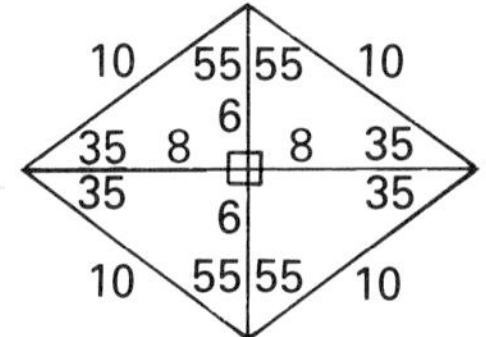

4a (i) FO (ii) FG (iii) ∠GDE (iv) ∠OFG (v) △FOE
b

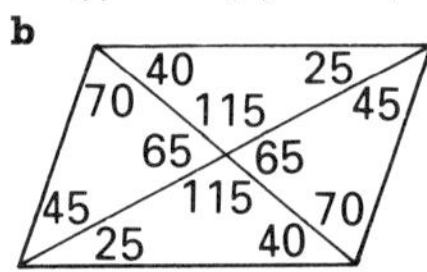

5a Its opposite sides are equal; its diagonals bisect at right angles **b** (6, 4), (10, 4), (10, 2), (6, 2); all its angles are right angles **6a** 67°, 67°, 113°, 113° **b** 71.2 cm
7a (i) BCDE (ii) BCDF **b**

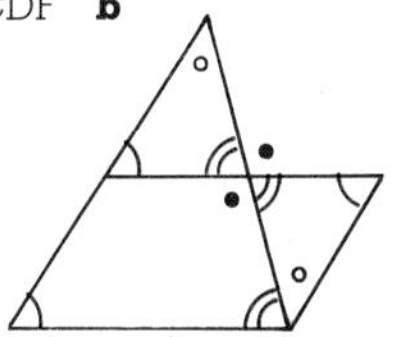

15 SOME SPECIAL NUMBERS

Page 176 Looking Back

1a 8, 10 **b** 12, 15 **c** 20, 25 **d** 36, 45
2a Yes **b** yes **c** no **d** yes
3a 3 × 3 **b** 3 × 3 × 3 **c** 10 × 10
4a 81 **b** 125 **c** 100 **d** 10 **5a** True **b** true **c** false **d** true **e** false **f** false **g** true **h** true
6a 17, 21 **b** 20, 24 **c** 64, 128
7 6 − 2, 3; 51 − 3, 17; 12 − 2, 3, 4, 6; 8 − 2, 4
8a 10 × 1, 5 × 2 **b** 16 × 1, 8 × 2, 4 × 4
9 After 6, 12, 18, . . . minutes

Page 177 Exercise 1A

1a 2^3 **b** 7^2 **c** 3^4 **d** 8^5 **2a** $2^2, 3^2$ **b** $2^3, 3^3$ **c** $7^3, 4^5$
3a 16, 25, 36 **b** the dot pattern is a square **4a, b** 144
6a 100, 1000, 10 000, 100 000, 1 000 000
b one thousand, one million
7 $1 \times 1 \times 1 \times 1 = 1$, $2^2 = 4$, $2 \times 2 \times 2 = 8$, $3^2 = 9$, $4 \times 4 = 16$, $5^2 = 25$, $3 \times 3 \times 3 = 27$ **8** 4, 8, 16, 32, 64, 128, 256
9a £30 **b** £90 **c** £270 **d** £810 **e** £2430
10a 2^2 **b** 3^2 **c** 2^3 **d** 3^3 **e** 2^4

Page 178 Exercise 1B

1a 4^3 **b** 5^4 **c** 6^5 **2a** 625 **b** 512 **c** 100 000 **d** 64 **e** 27
3a 3^2 **b** 3^4 **c** neither; they are equal **d** 5^3
4a 7 **b** 4 **c** 1 **d** 9 **e** 5
5 $1^4, 2^3, 3^2, 2^4, 5^2, 3^3, 2^5, 6^2, 7^2, 4^3, 9^2$
6a 8 **b** (i) 8 (ii) 16 (iii) 32

Page 178 Exercise 1C

1a 36 **b** 1 000 000 **c** 3 **d** 5 **e** 4 **f** 3
2a $2^3 = 8\,\text{cm}^3$, $3^3 = 27\,\text{cm}^3$, $4^3 = 64\,\text{cm}^3$, $5^3 = 125\,\text{cm}^3$
b $2^3\,\text{cm}^3$ gives the volume of a cube **3a** 5 years **b** 7 years
4a 3 **b** 3 **c** 4 **d** 4 **5a** 16 **b** (i) 1024 (ii) 6561
(iii) 15 625 (iv) 1 000 000 (v) 1
6a 10 **b** 15 **c** 25

Page 179 Exercise 2A

1a 25 **b** 49 **2a** 64 **b** 100 **3a** 6 **b** 8 **c** 9
4a 4 **b** 12 **c** 2
5a 64, 121, 225, 400, 1521, 5929 **b** 5, 9, 10, 13, 16, 25
6a (i) 2 by 2 square (ii) 4 by 4 square (iii) 3 by 3 square
b (i) 2, 4 (ii) 4, 16 (iii) 3, 9 **7a** (iv) 324 **8a** (iii) 19
9a 256 **b** 15 625 **c** 11.56 **d** 0.81 **e** 0.01 **f** 22 **g** 9
h 2.8 **i** 0.8 **j** 75 **10a** Each number is squared, then the square root is taken, so the first number is obtained **b** 27
11a (i) 3, 4 (ii) 5, 6 (iii) 2, 3 (iv) 8, 9 **b** (i) 3.2 (ii) 5.5
(iii) 2.8 (iv) 8.4 **12a** $7.84\,\text{m}^2$ **b** 13 **13a** 60 m **b** $3600\,\text{m}^2$

Page 181 Exercise 2B/C

1a 5 by 5 square **b** not possible
2a 1.82 **b** 0.04 **c** 1.41 **d** 31.69 **e** 0.15
3a 1225 **b** 27 **c** 40 **d** 784
4b (i) 3.6 (ii) 12.3 (iii) 4.5 (iv) 7.1 **5** 22, 16
6a 9, 7 **b** 31, 39 **7a** (i) 80 (ii) 180 **b** (i) 3 (ii) 3.5
8a $8.4^2 = 70.56$, $8.3^2 = 68.89$, so $\sqrt{70} = 8.4$ to 1 d.p.
b $8.36^2 = 69.89$, $8.37^2 = 70.06$, so $\sqrt{70} = 8.37$ to 2 d.p.
c $8.366^2 = 69.990$, $8.367^2 = 70.007$, so $\sqrt{70} = 8.367$ to 3 d.p.
9 6.3, 6.32, 6.325; 9.7, 9.75, 9.747; 20.5, 20.49, 20.494
10a 5.477 **b** 9.512

Page 182 Exercise 3A

1a 0, 2, 4, 6, 8, 10, 12, 14, 16, 18, 20 **b** 0, 3, 6, 9, 12, 15
c 0, 4, 8, 12, 16, 20, 24
2a Yes **b** yes **c** yes **d** no **e** no **f** yes **g** yes **h** yes
3a 2, 4, 6 **b** 4, 8 **c** 3, 6, 9, 12 **d** 21, 28 **e** 10, 15, 20
f 48, 54 **g** 1, 2, 3, 4, 5, 6, 7, 8 **h** 24, 32, 40, 48, 56
4a 0, 5, 10, 15, 20 **b** 0, 10, 20, 30, 40
5b (i) 10, 15, 20 (ii) 10, 20 **c** 10, 20
6b (i) 2, 4, 6, 8, 10, 12 (ii) 3, 6, 9, 12 **c** 6, 12
7a 5 **b** 3 **c** 2, 4
8a 1, 2, 3, 5, 6, 10, 15, 30 **b** 1, 2, 3, 4, 6, 9, 12, 18, 36

Page 183 Exercise 3B/C

1a 0, 2, 4, 6, 8, 10, 12; 0, 3, 6, 9, 12; 6
b 0, 2, 4, 6, 8, 10, 12, 14, 16, 18, 20; 0, 5, 10, 15, 20; 10
c 0, 4, 8, 12, 16, 20, 24; 0, 6, 12, 18, 24; 12
d 0, 3, 6, 9, 12, 15, 18, 21, 24, 27, 30; 0, 5, 10, 15, 20, 25, 30; 15
2a 12 **b** 18 **c** 10 **d** 20
3a 2 **b** 6 **c** 20 **d** 36 **4a** 45 **b** 24
5a (i) 3, 6, 9, 12, 15, 18, 21, 24, 27, 30; 4,8, 12, 16, 20, 24, 28
(ii) 12, 24 (iii) 12th **b** (i) 6, 12, 18, 24, 30; 8, 16, 24 (ii) 24th
(iii) 24th **c** 24th **6a** 34 **b** 18 **c** 40 **d** 70 **e** 36

Page 184 Exercise 4A

1a Yes **b** no **c** yes **d** no **e** yes **2a** 1, 11 **b** 1, 3, 7, 21
c 1, 3, 5, 15 **d** 1, 3, 9,27 **e** 1, 5, 7, 35 **f** 1, 2, 4, 5,10,20
g 1, 2, 3, 5, 6, 10, 15, 30 **h** 1, 2, 3, 4, 6, 8, 12, 24
3a 1×26, 2×13 **b** 1×33, 3×11
c 1×40, 2×20, 4×10, 5×8 **d** 1×45, 3×15, 5×9
e 1×60, 2×30, 3×20, 4×15, 5×12, 6×10
4a 8×1, 4×2 **b** 5×1 **c** 9×1, 3×3 **d** 12×1, 6×2, 4×3
e 18×1, 9×2, 6×3 **5a** (i) 42×1, 21×2, 14×3, 7×6
(ii) 32×1, 16×2, 8×4 (iii) 50×1, 25×2, 10×5
(iv) 55×1, 11×5 (v) 25×1, 5×5
(vi) 100×1, 50×2, 25×4, 20×5, 10×10 **b** 5×5 and 10×10

Page 185 Exercise 4B

1a 1, 2, 11, 22 **b** 1, 2, 4, 11, 22, 44 **c** 1, 3, 5, 9, 15, 45
d 1, 2, 3, 4, 5, 6, 10, 12, 15, 20, 30, 60
e 1, 2, 3, 4, 5, 6, 8, 10, 12, 15, 20, 24, 30, 40, 60, 120
2a 1; 1, 2, 4; 1, 3, 9; 1, 2, 4, 8, 16; 1, 5, 25;
1, 2, 3, 4, 6, 9, 12, 18, 36 **b** odd
3 1; 1, 2, 4, 8; 1, 3, 9, 27; 1, 2, 4, 8, 16, 32, 64; 1, 5, 25, 125 **4a** 3
b 2, 4 **c** 2, 5, 10 **d** 2, 3, 6 **e** 2, 3, 4, 6, 12 **f** 2, 5, 10 **g** 7
h 2, 4, 5, 10, 20 **i** 2, 3, 4, 6, 8, 12, 24 **j** 2, 3, 4, 6, 8, 12, 24

Page 185 Exercise 4C

1 6 **2** 16 **3** 36 **4** 17 **5** 12 **6** 5 **7** 4 **8** 7 **9** 18

Page 186 Exercise 5

1a 1; 2; 1, 3; 1, 2, 4; 1, 5; 1, 2, 3, 6; 1, 7; 1, 2, 4, 8; 1, 3, 9;
1, 2, 5, 10; 1, 11; 1, 2, 3, 4, 6, 12; 1, 13; 1, 2, 7, 14; 1, 3, 5, 15;
1, 2, 4, 8, 16; 1, 17; 1, 2, 3, 6, 9, 18; 1, 19; 1, 2, 4. 5, 10, 20
b 2, 3, 5, 7, 11, 13, 17, 19 **2** 2, 3, 5, 7, 11, 13, 17, 19, 23, 29, 31, 37, 41, 43, 47, 53, 59, 61, 67, 71, 73, 79, 83, 89, 97
3a It is divisible by 2 **b** divisible by: (i) 3 (ii) 5
(iii) 3 (and 11) **4** Divisible by: **a** 2 **b** 5 **c** 2 **d** 7 **e** 3
5 11, 23, 43, 97, 101 **6a** Tim: 3, 11, 19, 43, 59, 67, 83;
Kate: 5, 11, 17, 23, 29, 41, 47, 53, 59, 71, 83, 89;
Salim: 11, 19, 29, 31, 41, 59, 61, 71, 79, 89 **b** Kate's **7** Yes

Page 187 Exercise 6A

1a 2×3 **b** 2×7 **c** $2 \times 2 \times 3$ **d** $2 \times 3 \times 3$
2a $2 \times 2 \times 5$ **b** $2 \times 7 \times 7$ **c** $3 \times 2 \times 5$ **d** $3 \times 5 \times 5$
3a $2 \times 2 \times 3 \times 3$ **b** $2 \times 2 \times 3 \times 3$
4a 3×7 **b** 5×7 **c** $3 \times 3 \times 3$ **d** $3 \times 3 \times 5$ **e** $2 \times 5 \times 5$
5a $2 \times 2 \times 5 \times 5$ **b** $2 \times 2 \times 2 \times 3 \times 5$ **c** $5 \times 5 \times 5$
d $2 \times 2 \times 2 \times 2 \times 3 \times 3$ **e** $2 \times 2 \times 3 \times 3 \times 5$

Page 188 Exercise 6B/C

1a 2^5 **b** $2^2 \times 3^2$ **c** $2 \times 3^2 \times 5$ **d** $2^3 \times 11$ **e** $2^3 \times 3$
2a $2^4 \times 3$ **b** $2^2 \times 5^2$ **c** 3^4 **d** $2^6 \times 3$ **e** $2^4 \times 5^2$
3a 24 **b** 140 **c** 90 **d** 96 **4** 300 seconds
5a July 25th **b** September 23rd

Page 189 Check-up on Some Special Numbers

1a 2^4, 5^3, 10^2 **b** 16, 125, 100 **2a** 2^4 **b** 3^3 **c** equal
3a, d, e **4a** 196 **b** 59.29 **c** 30 **d** 17 **e** 2.2 **5** 35 cm
6a $1^2, 2^2, 3^2, 4^2$ **b** n^2 **c** 1000th
7a 0, 2, 4, 6 **b** 0, 6, 12, 18 **c** 0, 9, 18, 27 **d** 0, 10, 20, 30
8a 3 **b** 2 **c** 5 **d** 7 **9a** 1, 7 **b** 1, 3, 5, 15
c 1, 2, 3, 4, 6, 12 **d** 1, 19 **e** 1, 2, 4, 5, 10, 20 **10** 2, 3, 29, 31
11a $2 \times 2 \times 3$ **b** $2 \times 3 \times 3$ **c** $2 \times 2 \times 2 \times 2$ **d** $2 \times 2 \times 2 \times 2 \times 3$
e $3 \times 3 \times 3 \times 5$ **12a** $2^2 \times 3$ **b** 2×3^2 **c** 2^4 **d** $2^4 \times 3$ **e** $3^3 \times 5$
13a 0, 3, 6, 9, 12, 15; 0, 5, 10, 15, 20, 25 **b** 15 **14** 40
15 After 12 seconds; after 60 seconds

FORMULAE AND SEQUENCES

Page 190 Looking Back

1a 9, 11, 13 **b** 20, 24, 28 **c** 12, 10, 8 **d** 10, 5, 0 **e** 16, 32, 64 **f** 12, 17, 23 or 13, 21, 34 **2a** $x = 6$ **b** $y = 13$ **c** $n = 18$ **d** $t = 2$ **e** $u = 5$ **f** $v = 3$ **3** $x = 7, y = 9, b = 12, a = 12, k = 14, m = 18$ **4a** 10 **b** 5 **c** 7 **d** 0 **e** 4 **f** 12 **g** 36 **h** 1 **5a** -2 **b** 0 **c** 2 **d** 4 **e** 6
6a 1, 4, 9, 16 **b** 25, 36, 49 **7a** $P = 4x$ **b** $P = 2m+2n$ **c** $P = p+q+r$ **d** $P = 11a$
8 102 cm^3 **9a** $1+2+3+4+5+4+3+2+1$, etc.
b (i) 1, 4, 9, 16 (ii) 10 000

Page 191 Exercise 1A

1a 4 **b** 7 **c** $x+1$ **2a** £10 **b** £18 **c** £$2y$
3a 8 m **b** 200 cm **c** $4w$ m **4a** 90 mm **b** 45 cm **c** $3d$ m
5a 48 cm^2 **b** 300 cm^2 **c** $3r^2$ cm^2
6a 62°F **b** 80°F **c** $2x+30$°F **7a** 11 **b** $T = p+q+r+s$
8a £7 **b** $A = a+b+c+d+e-f$
9a (i) £1.50 (ii) £1.75 **b** £s; $p = s-c$
10a (i) £45 000 (ii) £55 500 **b** $L = 3A$
11a (i) £9800 (ii) £10 600 (iii) £11 400 **b** $s = 9000+800n$
12a (i) 68 kg (ii) 64 kg **b** $W = \dfrac{x+y+z}{3}$

Page 193 Exercise 1B/C

1a 60 cm^2 **b** 300 cm^2 **c** $\frac{1}{2}pq$ cm^2 **2a** 20 m **b** 125 m **c** $5t^2$ m **3a** 27 m^2 **b** 132 m^2 **c** $3(v+u)(v-u)$ m^2 **d** $9r^2$ m^2
4a (i) £41 (ii) £39 **b** $S = D+np$
5a (i) 140 (ii) 220 (iii) 300 **b** (i) $t = 60+40x$ (ii) $2\frac{1}{2}$
6a $y = 2x$ **b** $y = \frac{1}{2}x$ **c** $y = x-1$ **d** $y = x^2$

Page 194 Exercise 2A

1 $A = l^2$, 144 cm^2 **2** $V = lbh$, 90 m^3 **3** $C = \pi D$, 78.5 mm
4 $P = 2(l+b)$, 300 m **5** $A = \frac{1}{2}bh$, 72 cm^2 **6** $A = \pi r^2$, 452 m^2
7 $A = \pi r^2$, 113 cm^2 **8** $V = l^3$, 3380 mm^3 **9** $A = lb$, 16.3 cm^2
10 $A = \frac{1}{2}bh$, 300 mm^2

Page 195 Exercise 2B/C

1 12 cm^2 **2** 113 cm^2 **3** 2460 cm^2, 11 500 cm^3
4a 204 cm^2 **b** 283 cm^2 **c** 314 cm^3 **5a** 15 cm **b** 8.5 cm
6a 326 cm^2 **b** 44.4 cm

Page 196 Exercise 3A

1 (ii) Add 1 (iii) add 7 (iv) subtract 1 (v) subtract 6
2a Add 5 **b** add $\frac{1}{2}$ **c** (i) add 10 (ii) subtract 10 or 20
d (i) add 3 (ii) multiply by 2 **e** (i) add 15 (ii) divide by 3
3a 1st, 4th, 7th, 10th, 13th **b** 0, 4, 8, 12, 16 **c** A, 3, 5, 7, 9
d 1996, 1992, 1988, 1984, 1980 **e** 2, 9, 16, 23, 30
f 600, 625, 650, 675, 700 **4a** 5, 8, 11, 14 **b** 20, 15, 10, 5
c 2, 4, 8, 16 **d** 32, 16, 8, 4 **e** 1, 3, 7, 15 **f** 10, 10, 10, 10

Page 197 Exercise 3B/C

1a 5, 8 **b** 0, -1, -2 **c** 2, 5, 13 **d** 1, -1, -3
2a A **b** C **c** B **d** F **e** A **f** D **g** E **h** F
3 In the boxes: **a** 3, add 4 **b** 30, subtract 5 **c** 1, add 9
d 1, multiply by 10 **e** 19, subtract 2 **f** 81, divide by 3
4a 1, 2, 4, 8, 16, . . . **b** 1, 2, 5, 26, 677, . . .
c 9, -5, 4, -1, 3, 2, 5, 7, . . . **d** 1, 4, 2, 1, 4, 2, . . .
5a 15, 21, 28, 36, 45 **b** 25, 36, 49, 64, 81 **c** 35, 51, 70, 92, 117

Page 198 Exercise 4A

1a 3, 6, 9, 12 **b** 4, 7, 10, 13 **c** 2, 5, 8, 11 **d** 8, 11, 14, 17
2a 5, 8, 11 **b** 3, 5, 7 **c** 0, 2, 4 **d** 15, 20, 25 **e** 15, 19, 23
f 3, 7, 11 **g** 9, 8, 7 **h** 10, 8, 6 **3** $n+3$, D; $7n$, E; $2n+1$, C; $3n+1$, B; $4n-1$, A; $5n+2$, F; $6n-3$, G

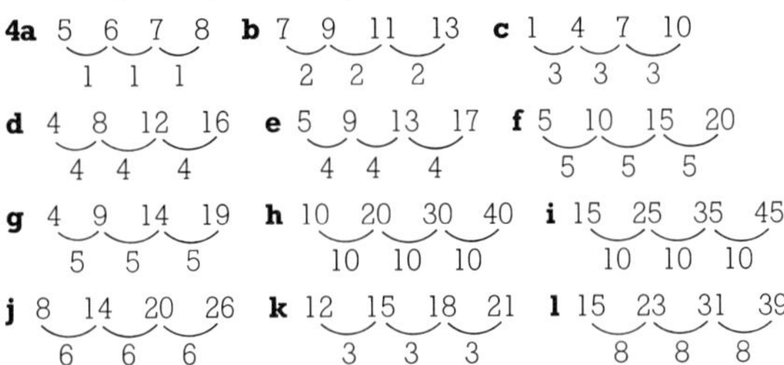

The number with n (coefficient of n) is the same as the difference between terms
5a $\frac{1}{2}$, 1, $1\frac{1}{2}$ **b** 1, 8, 27 **c** 5, 4, 3 **d** 4, 6, 8

Page 199 Exercise 4B/C

1a $2n$ **b** $4n$ **c** $6n$ **d** $n+2$ **e** $2n+7$ **f** $3n+1$ **g** $6n-2$
h $6n+4$ **i** $n-1$
2a $3n+5$, 65 **b** $5n-3$, 57 **c** $4n+1$, 101 **d** $7n-5$, 695
3a $3n+4$ **b** £64 **4a** $n+25$ **b** £125
5a Rows: 4, 6, 8, 10; 5, 8, 11, 14 **b** (i) $2n+2$ (ii) $3n+2$
c 26 H atoms, 38 H and C atoms
6a $2n+25$ **b** 41 tonnes **7** 42 mm

Page 201 Exercise 5A

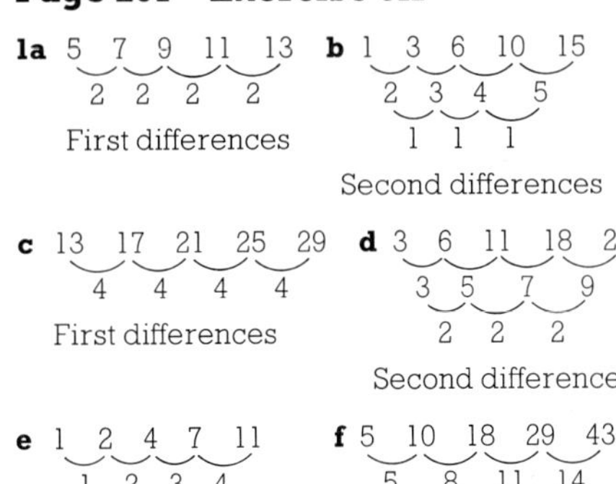

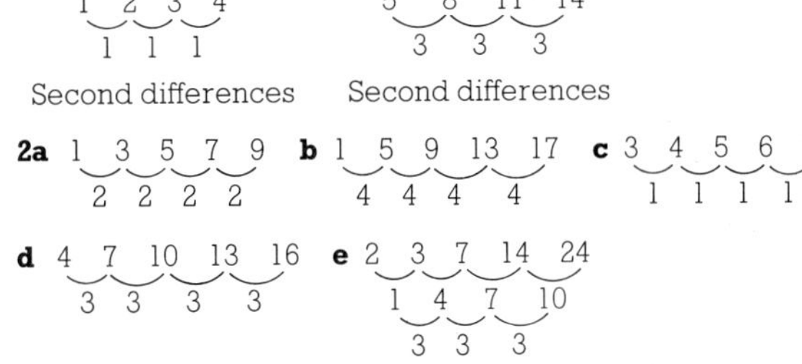

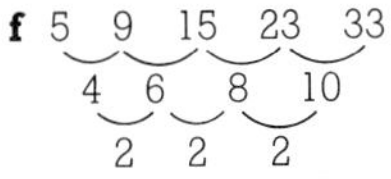

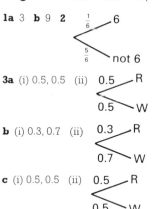

3a 5 18 31 44 57 70 83
13 13 13 13 13 13

b 7 24 41 58 75 92 109
17 17 17 17 17 17

c 3 14 27 42 59 78 99
11 13 15 17 19 21
2 2 2 2 2

d 11 25 42 62 85 111 140
14 17 20 23 26 29
3 3 3 3 3

e 14 41 68 95 122 149 176
27 27 27 27 27 27

f 4 47 91 136 182 229 277
43 44 45 46 47 48
1 1 1 1 1

4a 10 17 24 31 38
7 7 7 7
38 matchsticks

b 3 9 18 30 45
6 9 12 15
3 3 3
45 matchsticks

Page 202 Exercise 5B

1 (Second differences constant) **a** 101 **b** 96 **c** 91 **d** 106 **e** 86 **f** 77 **2a** 266 **b** 117 **c** 156 **d** 220 **e** 476
3a −2, −7, −12 (first differences −5)
b 10, 16, 24 (second differences 2)
c −5, −7, −10 (second differences −1)
d −30, −47, −67 (second differences −3)

Page 202 Exercise 5C

1a (i) 2 (ii) 6 **b** 2, 6, 12, 20 **c** (i) 30 (ii) 56 **2** 99
3b 505 **c** yes **4a** 1, 5, 14, 30, 55 **b** 204

Page204 Check-up on Formulae and Sequences

1a (i) $P = m+2n$ (ii) $P = 4t$ (iii) $P = 6r+\pi r$ **b** (i) 42 cm (ii) 60 cm (iii) 45.7 cm **2** $A = 4r^2+\frac{1}{2}\pi r^2$
3a (i) 24 km (ii) 36 km (iii) $12H$ km **b** $D = HS$
4a $n = p+q-r$ **b** 12 **5a** Add 5 **b** multiply by 3
c subtract 1, then 2, then 3, . . . **6a** 9, 13, 17 **b** 5, 14, 29
7a $5n-4$ **b** $4n+5$ **c** $6n+1$ **8a** 20th **b** 31st
9a 19, 22, 25 **b** 53, 72, 95 **c** 56, 92, 141 **10** 27
11a 1, 4, 10, 20 **b** third differences constant **c** 56

17 PROBABILITY

Page 205 Looking Back

1a, c **2a** Heads **b** both the same **3a** $\frac{5}{8}$ **b** $\frac{3}{8}$
4a $\frac{1}{11}$ **b** $\frac{2}{11}$ **c** $\frac{4}{11}$ **5** $\frac{9}{10}$ **6a** $\frac{5}{12}$ **b** $\frac{1}{3}$ **c** $\frac{1}{4}$
7a $\frac{1}{12}$ **b** $\frac{1}{3}$ **c** $\frac{1}{6}$ **d** $\frac{5}{12}$
8a Experiment **b** survey **c** past data

Page 206 Exercise 1A

1a H, T **b** $\frac{1}{2}, \frac{1}{2}$ **c**

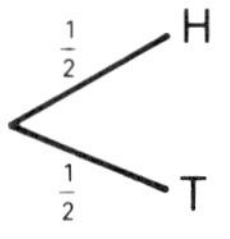

2a 3 **b** $\frac{1}{3}, \frac{2}{3}$ **c**

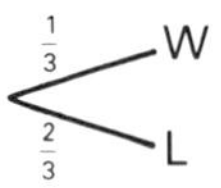

3a 1, 2, 3, 4, 5, 6 **b** (i) $\frac{1}{2}$ (ii) $\frac{1}{2}$ **c**

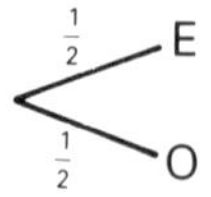

4a (i) $\frac{2}{3}$ (ii) $\frac{1}{3}$ **b**

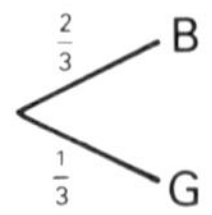

5a (i) $\frac{3}{5}$ (ii) $\frac{2}{5}$ **b**

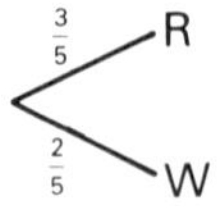

6a (i) $\frac{3}{4}$ (ii) $\frac{1}{4}$ **b**

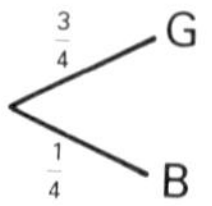

7a (i) $\frac{1}{5}$ (ii) $\frac{4}{5}$ **b**

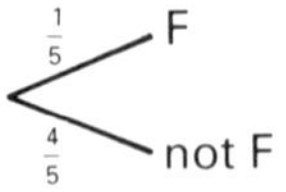

8a (i) $\frac{1}{2}$ (ii) $\frac{1}{2}$ **b**

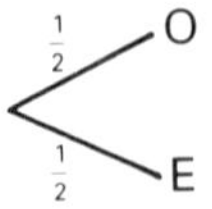

Page 207 Exercise 1B/C

1a 3 **b** 9 **2**

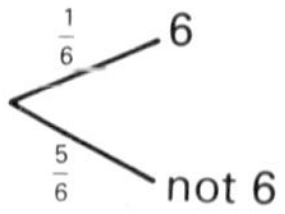

3a (i) 0.5, 0.5 (ii)

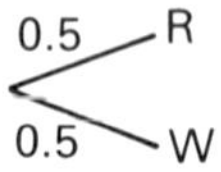

b (i) 0.3, 0.7 (ii)

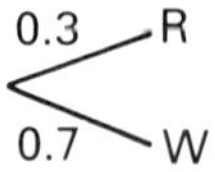

c (i) 0.5, 0.5 (ii)

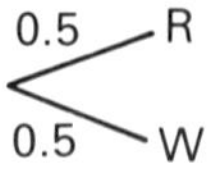

d (i) 0.4, 0.6 (ii)

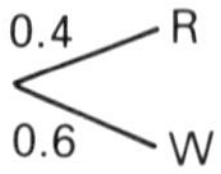

4a (i) 0.2 (ii) 1
b 4 black, 12 red, 16 green, 8 orange
5a 60 **b** (i) 0.3 (ii) 0.7 **c**

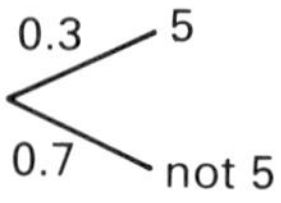

6a $\frac{1}{2}$ **b** $\frac{1}{3}$ **c** $\frac{1}{6}$

Page 208 Exercise 2A

1

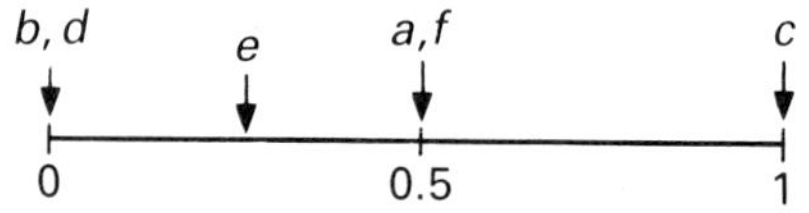

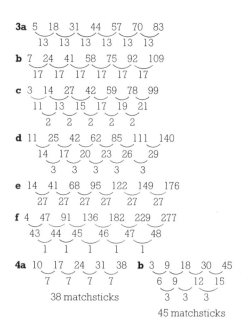

2a (ii) **b** (iii) **c** (iv) **d** (i) **e** (v)
3a 0.1 **b** 0.3 **c** 0.7 **d** 0.2 **e** 0.4 **4** Computer: 0.2; TV: 0.9; Video: 0.55; Microwave: 0.4; Phone: 0.85

Page 208 Exercise 2B/C

1a (i) $\frac{1}{2}$ (ii) $\frac{1}{2}$ **b** 1; total probabilities = 1
2a (i) $\frac{1}{6}$ (ii) $\frac{1}{6}$ **b** $\frac{1}{3}$; P(4), P(3), P(2) andP(1) are also possible
3a $\frac{1}{4}$ **b** $\frac{1}{4}$ **c** $\frac{1}{4}$ **d** $\frac{1}{4}$ **e** 1
4

a c e(i) e(ii) d b
0 1

5 0.1 **6** 0.85 **7a** 28% **b** a **8** 0.05
9a C **b** A **c** B **d** B **e** A and C **f** C **g** B

Page 210 Exercise 3

1 Survey **2** past data **3** survey
4 calculate (or experiment) **5** past data
6 calculate (or experiment) **7** experiment **8** past data
9 past data **10** past data **11** experiment
12 calculate (or experiment) **13** calculate **14** survey
15 past data

Page 211 Exercise 4A

1a $\frac{1}{2}$ **b** 5, 25, 50 **2a** $\frac{1}{6}$ **b** 1, 10, 20 **3a** $\frac{1}{5}$ **b** 5 s: 2, 20, 200; Evens: 4, 40, 400 **4a** $\frac{1}{10}$ **b** 2, 10, 30 **5** 1920 **6** 624, 576
7a 3 **b** the machine's owner; £21 **8a** 6000 **b** 45 000 **9** 54

Page 212 Exercise 4B

1 4000 **2** 2850 **3a** $\frac{9}{10}$ **b** 27, 180, 720 **c** 90, 10
4a 4, 8 **b** 10, 20 **c** 15, 30 **d** 40, 80 **5a** $\frac{1}{5}$ **b** 4
6a 10 **b** 100 **c** 250 **7a** 30 **b** 150 **8a** 10 **b** 17
9 No. It should be about 60 times

Page 213 Exercise 4C

1a 0.85 **b** (i) 425 (ii) 8500 **2a** 4 **b** 12 or 13
3a 400 **b** 15 or 16 **4a** (i) $2\frac{1}{2}$% (ii) $37\frac{1}{2}$% **b** £30
5a 750 **b** 2 **6a** 660 **b** 3080

Page 214 Check-up on Probability

1a (i) $\frac{1}{3}$ (ii) $\frac{2}{3}$ **b**

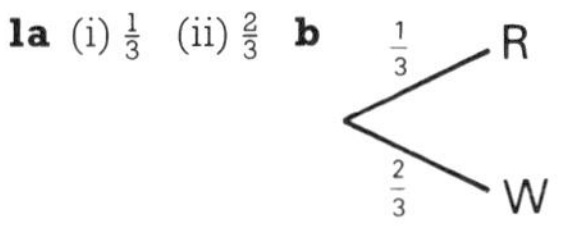

2

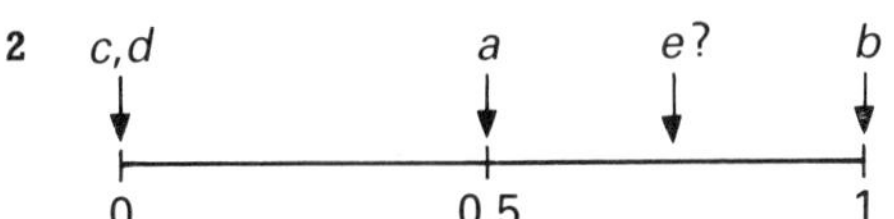

3 95% **4a** Use a survey **b** past data **5** 60, 90 **6** 70%
7a 5% **b** £1 **8a** $\frac{1}{2}$ **b** $\frac{1}{2}$ **c** $\frac{1}{4}$ **d** $\frac{1}{2}$ **e** $\frac{5}{8}$
9a 0% **b** 3% **c** 63%

18 INFORMATION TECHNOLOGY

Page 215 Exercise 1

1a 1 **b** 5 **c** 10 and 11 **d** 10, 11, 13, 14 and 15
e 10, 11, 12, 13, 14, 15 **2a** Black, red **b** 0.50, 550
3a 6 **b** £1870.50 **4a** 11 **b** 14 **c** 9 **5a** 1840, black; 1840, blue; 1841, blue; 1847, lilac; 1847, brown; 1847, green; 1854, blue; 1857, blue; 1858, blue **b** 1857, 1d; 1858, 1d; 1858, $1\frac{1}{2}$d **c** VR, 6d; none, 10d; none, 1/-; small crown, 1d; small crown, 2d
6a (i) Saturn, Uranus, Neptune, Pluto (ii) Mercury, Mars (iii) Mercury (iv) Mercury, Venus, Mars, Jupiter, Saturn, Uranus, Neptune, Pluto **b** They are the same

Page 217 Exercise 2

1a Item (label) **b** sold (label) **c** 12 (number)
d item (p) (label) **e** 12 (number) **f** total (label)
g 480 (formula = B6 × C6) **h** 1632 (formula = sum (D3, D6))
2 Cell C4 should be 26, D4 becomes 312 and E7 becomes 1644
3

D	E
Total	Takings
cost	
494	
390	
372	
494	
	1750

4a (i) B2 × C2 (ii)

E
Floor area
15
24
35
30
45

b (ii)

F
Wall area
48
66
72
66
84

c Cost per m² (carpet); Cost per m² (paper); Cost of room (carpet); Cost of room (walls); Total cost

5a

E
Total
(180)
126
135
108
90
45
81

b C9 = Sum (C3, C8), D9 = Sum (D3, D8)
c (i) F3 = (C3 + D3) × 100 ÷ 180
(ii)

F
Percentage
%
70
75
60
50
25
45

REVISION EXERCISES

Page 220 Revision Exercise on Chapter 1: Numbers in Action

1a £84.50 **b** £44.28 **2a** (i) 100 km (ii) 90 km (iii) 86 km (iv) 86.1 km **b** (i) 200 km (ii) 210 km (iii) 215 km (iv) 214.7 km **c** (i) 1100 km (ii) 1080 km (iii) 1083 km (iv) 1083.2 km **3a** 13, 13.1 **b** 60, 57.2 **4a** (i) 21 cm² (ii) 28 cm² **b** (i) 20.3 cm² (ii) 29.1 cm² **5a** 14 **b** 14.29
6a (i) 50 (ii) 50 (iii) 30 (iv) 2 (v) 0 (vi) 5
7a 1.533 . . . , 1.56, 1.528 . . . p **b** 140 g jar; 2 **8a** 750 **b** 8
9a 0.11 mm **10a** 7.429 **b** 0.088

Page 221 Revision Exercise on Chapter 2: All About Angles

1a 15° **b** 105° **c** 75°
2a ∠ADB = 90°, ∠ABX = 145°, ∠DAC = 35° **b** (i) 3 (ii) 3
3a 27 **b** 117 **c** 30 **d** 45
4a

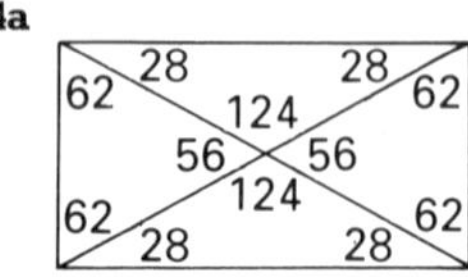

b

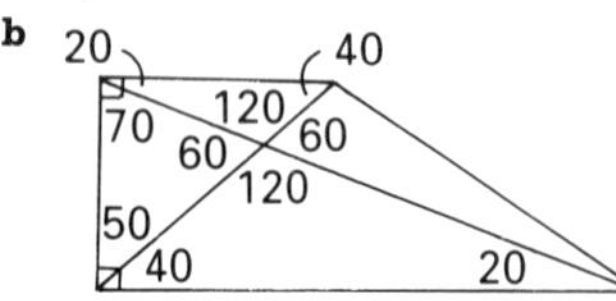

5 $a = 98$ (supp), $b = 98$ (symmetry), $c = 82$ (alternate), $d = 98$ (corresponding), $e = 98$ (corr, or symm)
6a $a = 65, b = 65, c = 75$ **b** $a = 34, b = 34, c = 84, d = 62, e = 118, f = 146, g = 146$ **7** $x = 15$ **8** Use alternate angles to show that $a° + b° + c° = 180°$ at the vertex where angle c is
9 45°

Page 222 Revision Exercise on Chapter 3: Letters and Numbers

1a $5n$ **b** $4t$ **c** $2k$ **d** m^2 **e** 5^4, or 625 **f** x^2
2a $4x+2$ **b** $4y-2$ **c** $8a^3$
3a 6 **b** 6 **c** 9 **d** 4 **e** 1 **f** 8 **g** 5 **h** 36
4a $2+t, 2t, 6t$ **b** $2t, t^2, 5t^2$ **5a** $5x$ **b** $x-4$ **c** $3x-2$ **d** $y+5$
6a $4n, n^2$ **b** $6t, 2t^2$ **c** $12x+4y, 8xy$ **7a** 36 m² **b** $5(x+4)$ m²
8a $5x+10$ **b** $6y+6$ **c** $3z-12$ **d** $2p-2$ **e** $8+8t$ **f** $14+7n$ **g** $12-4v$ **h** $36-9w$ **i** $4a+2$ **j** $12b+6$ **k** $10c-5$ **l** $42-63d$ **9** $300uv$ cm³
10a 16 **b** 32 **c** 64 **d** 0 **e** 4 **f** 16 **g** 8 **h** 36 **11a** $3x$ **b** $2y$ **c** x^2+2x **d** x^2+2xy **e** $x-2x^2$ **f** x^3+x^2+x

Page 223 Revision Exercise on Chapter 4: Making Sense of Statistics 1

1a 60 papers **b** Thursday **c** sales rose from Friday to Sunday each week, and fell from Sunday to Tuesday **d** slightly upwards overall **2a** 14.1°C, 13°C, 13°C, 5° **b** 14.9°C, 15°C, 17°C, 5° **c** 14.5°C, 14.5°C, 13°C, 5°
3b In the first week the temperature was lower in the first part, and reached its highest point on day 5. In the second week the warmest spell was from day 2 till day 5
4a 55, 47 **b** 1.8, 1.9 **c** goal average is a little better
5a 19 millionths of a metre **b** 18.9, 17.4 millionths of a metre

Page 224 Revision Exercise on Chapter 5: Fractions, Decimals and Percentages

1a $\frac{1}{4}$, 0.25, 25% **b** $\frac{5}{8}$, 0.625, 62.5% **c** $\frac{1}{8}$, 0.125, 12.5%
2a £4.50 **b** £260 **c** 3p **3a** 553.8 m² **b** 102.2 m
4a £9 **b** £51 **5a** £21 **b** £70 **c** £5.25
6a (i) £300 (ii) 12% **b** £2850 **7a** $\frac{3}{5}$ **b** $\frac{4}{11}$ **c** $\frac{7}{11}$ **d** $\frac{8}{11}$
8a (i) BBC (ii) ITV (iii) BBC **b** (i) 6 hours (ii) $\frac{3}{4}$ hour (iii) $2\frac{1}{4}$ hours **9** 20% **10a** 15% **b** 69 g
11a (i) £3 (ii) £6 (iii) £5 **b** 60%, 60%, $33\frac{1}{3}$%
c 40%, 44%, $33\frac{1}{3}$%, . . .

Page 225 Revision Exercise on Chapter 6: Distances and Directions

1b 330–340 cm, 63° **2** 28 m **3** 12 m **4a** North **b** west **c** turn 90° anti-clockwise, and walk north
5a (i) K (ii) P (iii) M (iv) B **b** (i) 24 km (ii) 27 km
6a 19 km **b** 194° **7a** 40 m **b** 43 m

Page 226 Revision Exercise on Chapter 7: Positive and Negative Numbers

1a (i) London (ii) Inverness **b** −10°C, −9°C, −8°C, −7°C, −6°C **2a** (i) 0 (ii) −5 **b** (i) −5, −4, −3 (ii) −1, 0, 1, 2
3a −4, −2 **b** −1, 0, 1 **4a** −4 + 5 = 1 **b** −3 − 5 = −8, or −3 + (−5) = −8 **c** −3 − 2 = −5, or −3 + (−2) = −5
d −3 + 4 = 1 **5a** 3, 2, 1, 0, −1; −8 **b** 3, 2, 1, 0, −1; −7
c 1, 2, 3, 4, 5, 6; 12 **6a** −4 **b** −4 **c** 10 **d** −10 **e** −10
f 4 **g** 4 **h** −7 **i** −8 **j** 8 **k** 10 **l** −10
7a −1 **b** 6 **c** −2 **8a** C(0, −4), D(2, 0) **b** 8 and 4 units
9 Rows: **a** 0, −1, −1; −2, 3, 1; −2, 2, 0 **b** 0, −1, 1; −2, 3, −5; 2, −4, 6 **10a** −8 **b** 2 **c** −10 **d** −4

Page 227 Revision Exercise on Chapter 8: Round in Circles

1a 2 m **b** 20 m **c** 200 m **d** 1 m **2a** 11 cm **b** 4 cm
3 33 cm, 24 cm **4a** 180 cm **b** 4.5 cm **c** 108 cm
5a 44.9 cm **b** cup 229 mm, saucer 408 mm, side plate 512 mm, soup plate 691 mm, dinner plate 691 mm
6 27 cm², 300 cm² **7** 5030 mm² **8a** 145 mm² **b** 36 000 mm²

Page 228 Revision Exercise on Chapter 9: Types of Triangle

1 $a = 125$; $b = 38$; $c = 65$, $d = 50$; $e = f = 80$; $g = h = i = 60$
2a 180 m² **b** 3.75 m² **c** 80 m² **d** 6 m² **3** △1 acute-angled, △2 right-angled, △3 right-angled isosceles, △4 obtuse-angled, △5 right-angled

4a **b** **c** **d**

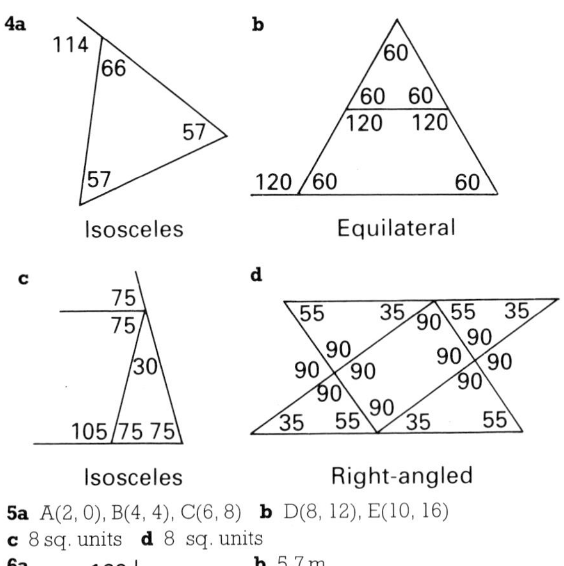

5a A(2, 0), B(4, 4), C(6, 8) **b** D(8, 12), E(10, 16)
c 8 sq. units **d** 8 sq. units
6a 162 18 108 72 **b** 5.7 m

Page 229 Revision Exercise on Chapter 10: Metric Measure

1a 4000 m **b** 1.5 cm **c** 1200 g **d** 5000 ml
2a 75 mm **b** (i) 210 mm (ii) 21 cm
3a 47 cm² **b** 3100 mm² **4a** 49 cm², 48 cm **b** 49 cm², 28 cm
5a 4000 m **b** £189 900 **c** 200 m³
6a 12 000 ml **b** 400 ml **c** 600–650 ml
7a 9.7 g, 5.9 g **b** (i) 3.8 g (ii) 3.8 ml **8** 1.14 kg **9** 6 cm

Page 230 Revision Exercise on Chapter 11: Equations and Inequations

1a $x = 4$ **b** $x = 5$ **c** $x = 3$ **d** $y = 4$ **e** $y = 10$ **f** $y = 2$
2a (i) $4x = x + 12$, $x = 4$ (ii) £16 **b** (i) $4x - 16 = 2x$, $x = 8$
(ii) 16 kg **3a** $x = 3$ **b** $x = 1$ **c** $x = 6$ **4a** $t < 6$ **b** $n < 1$
5a $x = 5$ **b** $y = 6$ **6a** $3(1 + 4x) = 3 \times 17$, $x = 4$
b $3(5x - 10) = 5(2x - 1)$, $x = 5$ **7a** $x = 4$ **b** $x = 4$ **c** $t = 3$
8a (i) T (ii) T (iii) T (iv) F **b** (i) $-1, 1, 2$ (ii) $-1, -2$
9 $20(x - 1) = 16x$, $x = 5$: 20 cm, 4 cm; 8 cm, 5 cm
10 20 cm by 2 cm, 8 cm by 3 cm
11a $-2, -1$ **b** $-2, -1, 0$ **c** $-2, -1$

Page 231 Revision Exercise on Chapter 12: Ratio and Proportion

1a $\frac{3}{4}$ **b** $\frac{2}{3}$ **c** $\frac{1}{2}$ **d** $\frac{2}{5}$ **2a** 50p, 50p **b** 75p, 25p **c** 60p, 40p
3a 200, 50, 100 g **b** (i) 25 g; 50 g (ii) 175 g
4a 180 mm **b** 54 mm **c** 76 mm
5a £5.50 **b** £16.50 **c** £27.50
6a £100, £200, £300, £500, £1000 **b** £5, £2.50, £1, 50p, 25p
7a 70 **b** 50 cm **8a** No. of miles per gallon = 392 ÷ 14 = 28 = 42 ÷ 1.5 **b** (i) 448 (ii) 406 (iii) 8.5 (iv) 17.5

Page 232 Revision Exercise on Chapter 13: Making Sense of Statistics 2

1 4k **2a** 4.5, 4.5, none **b** 4.8, 5, 5; it is biased towards the scores of 4, 5, 7 and 8, so overall the scores are too high
3b £41 700 (nearest £100), £40 000, £40 000 **4b** 25 **c** 21–25
5b 5.3 **6a** 0.58 mm **b** 56 g **c** 0.5 mm

Page 233 Revision Exercise on Chapter 14: Kinds of Quadrilateral

1 **2**

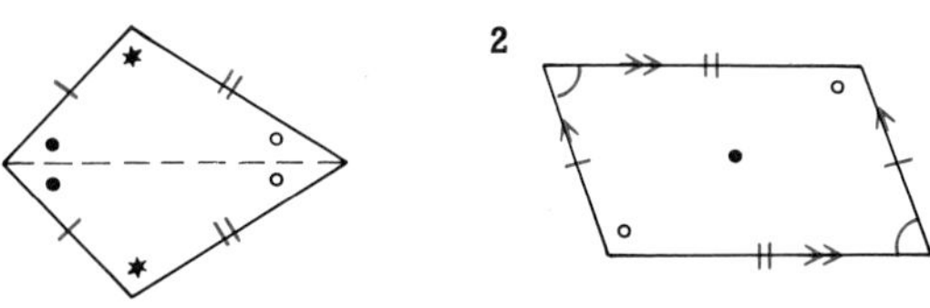

3a Rectangle, rhombus **b** rectangle, square, rhombus, parallelogram, trapezium **c** all quadrilaterals **d** square, rectangle, rhombus, parallelogram
4

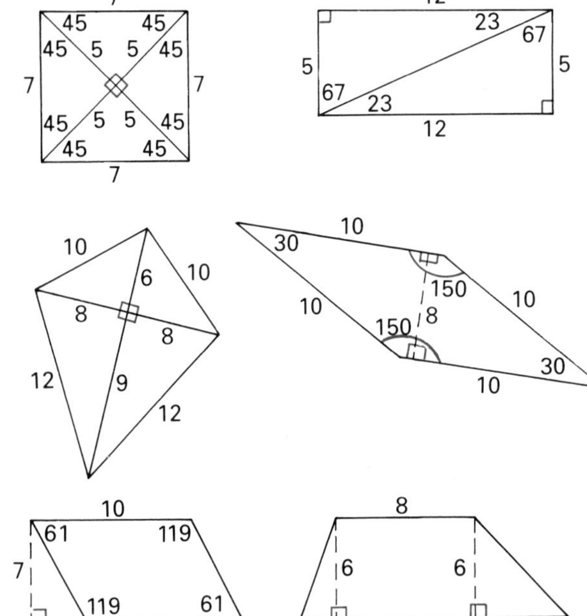

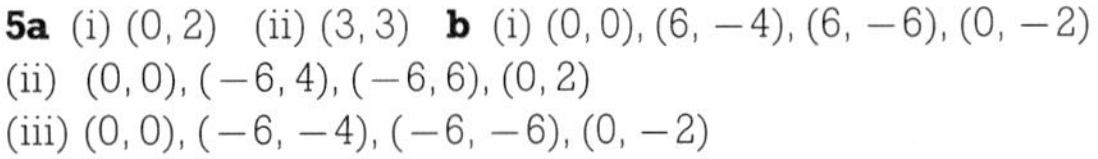

5a (i) (0, 2) (ii) (3, 3) **b** (i) (0, 0), (6, −4), (6, −6), (0, −2)
(ii) (0, 0), (−6, 4), (−6, 6), (0, 2)
(iii) (0, 0), (−6, −4), (−6, −6), (0, −2)
6

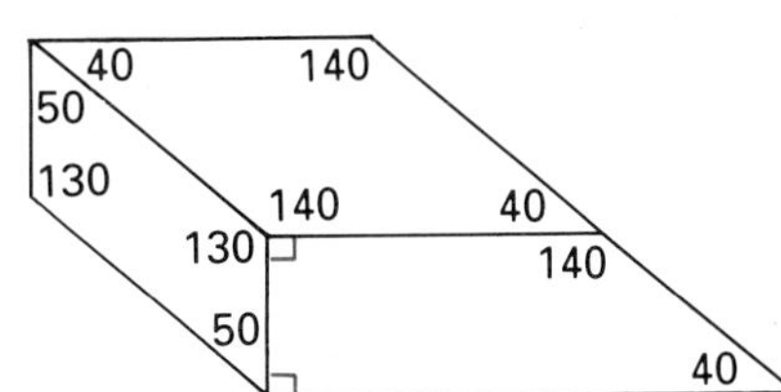

7a 6, 6 **b** all 50° at points, 60° at centre, others 125°
8

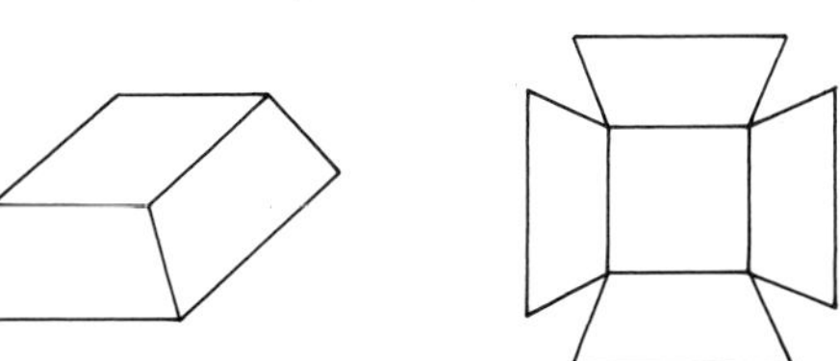

Page 234 Revision Exercise on Chapter 15: Some Special Numbers

1a 25 **b** 81 **c** 32 **d** 1000 000 **2a** 7^3 **b** 2^6 **c** 25^2
3 25, 36, 49, 64, 81 **4a** 10 mm **b** 30 m **c** 15 cm **d** 1 cm
5a 2, 3 **b** 5, 6 **c** 9, 10 **d** 8, 9
6a 2.2 **b** 5.6 **c** 9.8 **d** 8.4 **7a** 3, 6 **b** 8, 16
8a 1, 2, 5 **b** 1, 4, 8 **9a** 9, 16 **b** 9, 16 **c** 14, 21
d 11, 13, 17, 19 **e** 11, 22 **10** 29, 31
11a 5×7 **b** $2\times2\times5$ **c** $2\times2\times3\times5$ **d** $2\times3\times3\times3$
12a 1, 2, 4, 7, 14, 28 **b** 1, 2, 3, 5, 6, 10, 15, 25, 30, 50, 75, 150
13a 88 **b** 18 **c** 84
14a 3×3 **b** $2\times2\times3\times3$ **c** $2\times5\times11$ **d** $2\times2\times2\times2\times5\times5$
15a 3^2 **b** $2^2\times3^2$ **c** $2\times5\times11$ **d** $2^4\times5^2$ **16** 60 seconds

Page 235 Revision Exercise on Chapter 16: Formulae and Sequences

1a $d = 3a+3b$ **b** $d = 600$
2a (i) $P = 2s+2\pi r$, $A = 2rs+\pi r^2$
(ii) $P = 2d+2e$, $A = c(a+b)$ **b** (i) $P = 474$, $A = 15\,900$
(ii) $P = 70$, $A = 300$
3a 8, 16, 24, 32 **b** 8, 0, −8, −16 **c** 5, 4, 2, −2
4a 11, 17 **b** 11, 19 **5a** 6, 8, 10 **b** 25, 32, 39
6a $5n+1$ **b** $9n$ **c** $8n-1$ **7a** 27th **b** 41st
8a 14, 20, 26 **b** 25, 37, 51 **c** 42, 77, 128 **d** 20, 34, 51
9 212 (second differences 4) **10** $V = 864$, $A = 552$, $d = 17$

Page 236 Revision Exercise on Chapter 17: Probability

1 Arrows at: **a** 0.6 **b** 0.1 **c** 0.3 **d** 0.4 **2a** $\frac{1}{28}$ **b** $\frac{7}{28}$ **c** $\frac{3}{28}$
3a (i) 1 (ii) 0 (iii) $\frac{2}{7}$ **b** $\frac{2}{5}$
4a (i) 0.1 (ii) 0.25 (iii) 0.35 (iv) 0.3
5a Survey **b** calculation **6** 195 **7a** 12 **b** 18
8a $\frac{1}{50}$ **b** 100 **9a** 0.8, 0.5, 0.6 **b** Gita's **c** 1 800 000